KT-199-557

MODELING COMPONENTS
OF HYDROLOGIC CYCLE

Edited by
V. P. Singh

WATER RESOURCES PUBLICATIONS

UNIVERSITY LIBRARY
- 5 JIII 1983
LANCASTER

For information and correspondence:
WATER RESOURCES PUBLICATIONS
P.O. Box 2841
Littleton, Colorado 80161, U.S.A.

MODELING COMPONENTS OF HYDROLOGIC CYCLE

Proceedings of the International Symposium on Rainfall
Runoff Modeling held May 18-21, 1981 at Mississippi State
University, Mississippi State, Mississippi, U.S.A.

Edited by
Vijay P. Singh

Associate Professor of Civil Engineering
Department of Civil Engineering
Louisiana State University
Baton Rouge, Louisiana 70803, U.S.A.

ISBN-0-918334-46-2

U.S. Library of Congress Catalog Card Number-81-71291

Copyright © 1982 by Water Resources Publications.
All rights reserved. Printed in the United States of
America. The text of this publication may not be
reproduced, stored in a retrieval system, or transmitted in any form or by any means, without a written permission from the Publisher.

This publication is printed and bound by BookCrafters,
Inc., Chelsea, Michigan, U.S.A.

82 009993

PREFACE

In the last three decades there has been a proliferation of research on rainfall-runoff modeling. As a result there exists an abundance of literature in this area. As we enter into a new decade with new possibilities and challenges, it appears appropriate to pause and determine where we are, where we are going, where we ought to be going, and what the most outstanding problems are that ought to be addressed on a priority basis for rapid progress of hydrology. To address these issues in a scientific forum is what constituted essentially the rationale for organizing the International Symposium on Rainfall-Runoff Modeling which was held May 18-21, 1981 at Mississippi State University, Mississippi State, Mississippi.

The objectives of this Symposium were therefore (1) to assess the state of the art of rainfall-runoff modeling, (2) to demonstrate the applicability of current models, (3) to determine directions for future research, (4) to assemble unreported research, (5) to establish complementary elements of seemingly different approaches, and (6) to augment interdisciplinary interaction.

We received an overwhelming response to our call for papers. It was indeed a difficult task to select among the many excellent papers that were submitted, and we regret that we could not include all of them in the Symposium program. The sole criterion for selection of a paper was its merit in relation to the Symposium objectives. The subject matter of the Symposium was divided into 26 major topics encompassing virtually the entire spectrum of rainfall-runoff cycle. Each topic entailed an invited state-of-the-art paper and a number of contributed papers. These contributions blended naturally to evolve a synthesized body of knowledge on that topic. Extended abstracts of all the invited and contributed papers were assembled in a pre-Symposium proceedings volume. Each registered Symposium participant was given this volume. This helped stimulate discussion and exchange of ideas during the Symposium.

The papers presented at the Symposium were refereed in a manner similar to that employed for publishing a journal article. As a result, nearly 40 percent of the papers did not pass the review and were therefore eliminated from inclusion in the final procedings. The accepted papers were divided in four parts. The papers contained in this book, MODELING COMPONENTS OF HYDROLOGIC CYCLE, represent one part of the Symposium contributions. The other parts are embodied in three separate books, RAINFALL-RUNOFF RELATIONSHIP, APPLIED MODELING IN CATCHMENT HYDROLOGY and STATISTICAL ANALYSIS OF RAINFALL AND RUNOFF, which are being published simultaneously. Arrangement of papers in these books under four different titles was a natural consequence of the diversity of techical material discussed in these papers. These books can be treated almost independently, although some overlap does exist between them.

This book contains seven sections. Each section starts normally with an invited state-of-the-art paper followed by contributed papers. Beginning with acquisition and management of hydrologic data the papers go on to discuss infiltration, evapotranspiration, geomorphology, runoff, water quality and sediment yield, thus encompassing virtually all major components of hydrologic cycle.

The book will be of interest to researchers as well as those engaged in practice of Civil Engineering, Agricultural Engineering, Hydrology, Water Resources, Earth Resources, Forestry, and Environmental Sciences. The graduate students as well as those wishing to conduct research in rainfall-runoff modeling will find this book to be of particular significance.

I wish to take this opportunity to express my sincere appreciation to all the members of the Organizing Committee and the Mississippi State University administration for their generous and timely help in the organization of the Symposium. A lack of space does not allow me to list all of them here, but I would like to single out Dr. Victor L. Zitta who chaired the local arrangements pertaining to the Symposium. Numerous other people contributed to the Symposium in one way or another. The authors, including the invited speakers, contributed to the Symposium technically and made it what it was. The session chairmen and co-chairmen administered the sessions in a positive and professional manner. The referees took time out from their busy schedules and reviewed the papers. I owe my sincere gratitude to all these individuals.

If the success of a Symposium is measured in terms of the quality of participants and presentations then most people would agree that this Symposium was a resounding success. A very large number of inter-nationally well-known people, who have long been recognized for their contributions and have long been at the forefront of hydrologic research, came to participate in the Symposium. More than 25 countries, covering the five continents and most of the countries of the world active in hydrologic research, were represented. It is hoped that many long and productive friendships will develop as a result of this Symposium.

Vijay P. Singh
Symposium Director

TABLE OF CONTENTS

ACKNOWLEDGEMENTS

The International Symposium on Rainfall-Runoff Modeling was sponsored and co-sponsored by a number of organizations. The sponsors supported the Symposium financially without which it might not have come to fruition. Their financial support is gratefully acknowledged. The co-sponsors extended their help in announcing the Symposium through their journals, transactions, newletters or magazines. This publicity helped increase participation in the Symposium, and is sincerely appreciated. The following is the list of Symposium sponsors and co-sponsors.

SYMPOSIUM SPONSORS

National Science Foundation

U.S. Department of Agriculture, Science and Education Administration

U.S. Army Research Office

U.S. Department of the Interior, Office of Water Research and Technology

United Nations Educational Scientific and Cultural Organization

Mississippi State University, Agricultural and Forestry Experiment
    Station, Department of Civil Engineering, Office of Graduate
    Studies and Research, Water Resources Research Institute

Mississippi-Alabama Sea Grant Consortium

SYMPOSIUM CO-SPONSORS

American Geophysical Union

American Society of Civil Engineers, Hydraulics Division and Irrigation
    and Drainage Division

Institute of Hydrology, Great Britain

International Association of Hydrological Sciences

International Union of Forestry Research Organizations

International Water Resources Association

Mississippi State Highway Department

Regional Committee for Water Resources in Central America

Tennessee-Tombigbee Water Development Authority

U.S. Army Corps of Engineers, Vicksburg District

U.S. Department of Agriculture, Forest Service

U.S. Geological Survey, Water Resources Division, Gulf Coast
    Hydroscience Center; and Water Resources Division, Jackson,
    Mississippi

Universidad de San Carlos de Guatemala

Section 1
HYDROLOGIC DATA

ACQUISITION AND MANAGEMENT OF WATER QUALITY AND QUANTITY DATA

L. F. Huggins, Professor
Agricultural Engineering Department,
Purdue University, W. Lafayette, IN 47907

ABSTRACT

A brief review of recent trends in hydrologic research is presented to provide a basis for projecting future needs in relation to data acquisition and management systems. The fundamental recommendation is that water quantity and quality monitoring must be broadened in its approach in order to avoid seriously hampering research in the science of hydrology. Both philosophic considerations and practical guidelines for the development of data acquisition/management systems of increased utility are presented.

INTRODUCTION

The charge for this paper was to "reflect on the current state of thinking with particular regard to where we are, where we are going, where we ought to be going, and what are the most outstanding problems that ought to be investigated on a priority basis for rapid progress of hydrology." The paper is organized with brief reflections on recent historical trends in hydrology that impact field data collection needs followed by some basic philosophy, interspersed with occasional specific guidelines, concerning the design of data collection/management systems.

The single, fundamental recommendation which underlies all sections of this paper is that the scope and nature of field data collection efforts undertaken as an integral part of hydrologic research should be broadened. Collection of hydrometeorological data is an expensive and time-consuming process. Therefore, the utility and availability of information collected under such programs should be maximized. This can be accomplished only if broad hydrometeorologic needs are considered when a data collection/management system is developed.

The first consideration in the development of any data acquisition or management system is its effectiveness at meeting individual project needs. While that is certainly appropriate, a critical examination of strategies currently in use leads to the conclusion that only rarely are factors other than project exigencies given any consideration. Broad hydrometeorologic research goals and large scale acquisition system efficiencies are usually considered "beyond the scope" of narrowly defined, problem oriented research projects. This philosophy has become pervasive because of pressures from funding agencies; however, hydrologic researchers cannot be absolved from blame for their failure to more

3

vigorously resist this narrow, short-sighted use of public monies and their time.

Modern instrumentation technology and the comparatively high cost of technical personnel are such that, for very modest incremental costs, the overall utility of hydrometeorological and water quality data collected could be greatly magnified. However, these benefits can be attained only with proper design of the data collection/management system at a project's inception. The intention of this paper is to suggest concepts which are believed relevant to broadening the scope and utility of hydrologic data collection efforts.

Development of data bases which are of broader utility than individual project necessities requires a high degree of collection and management efficiency to avoid unacceptable cost burdens to a project. Project goals as well as general hydrologic research trends need to be considered in order to design such systems. A brief review of recent historical trends in hydrology can serve as the basis of extrapolation to identify future needs.

HISTORICAL PERSPECTIVE ON CURRENT RESEARCH NEEDS

The general concepts and principles of hydrology have been known for several years. Thus, it is not the development of new principles that has dominated hydrologic research during the past 40 years; rather, it is the adaptation of these generally accepted concepts into methodologies for making quantitative predictions of hydrologic behavior that is difficult and still far from satisfactory resolution.

The rapid evolution and widespread availability of modern computers have increased computational capacity by several orders of magnitude and spawned a virtual explosion of the mathematical modeling approach in hydrology. However, before this symposium of modelers becomes too ostentatious, it should be noted that the current widespread reliance on nomographic techniques and simple equations developed prior to the availability of modern computers raises questions about the efficacy of modern hydrologic modeling research.

Two philosophically different mathematical modeling approaches have evolved, systems identification and component process modeling. Amorocho and Orlob (1961) were among the early proponents of the system identification methodology, a sophisticated statistical technique to elucidate trends in historical hydrologic records and to identify those characteristics with which they correlate. While fundamentally similar to nomographic procedures developed for the pre-computer era, system identification techniques greatly increased the efficiency of the approach and the sophistication of resulting relationships. Its fundamental weaknesses are the requirement of a relatively long historical record and the theoretical implication of a time invariant system.

Crawford and Linsley (1962) developed one of the first detailed component relationship models to be used on a practical scale. Research in hydrologic component modeling has subsequently divided into two fundamentally different directions, lumped and distributed parameter modeling. A lumped model is one which uses coefficients computed as averaged values intended to account for spatial variations that normally exist in all systems being modeled. The lumped approach is founded on the calculus law of the mean. A distributed model attempts to directly

4

analyze the actual distribution of each non-uniform parameter. Such an approach to hydrologic prediction was proposed by Bernard (1936) long before the availability of computers made it feasible. Early applied distributed models were developed by Laurenson (1964) and by Huggins and Monke (1966).

The relative merits of lumped versus distributed component relation approaches to quantitative hydrologic analysis are computational efficiency versus potential increases in model authenticity that should be obtainable when static spatial weighting functions are eliminated. Computational requirements of distributed parameter models are such that simulation of large hydrologic systems is still practical only with lumped models.

During the past decade the emphasis in hydrologic research has focused more on water quality than quantity predictors. The need to develop methodologies for analyzing and devising solutions to control water pollution has required new and more accurate techniques for quantifying water movement. Flowing water not only acts as the transporting agent for pollutants, but also contributes to their generation and chemical interactions. Thus, the ability to quantify the complete hydrologic system is a prerequisite to developing models that can be used to analyze water pollution conditions.

Heightened concern about water quality is one of the primary reasons for the increasing dominance of component process models over system identification models and for the attention being given to distributed parameter concepts within the former category of models. The component process approach provides a building block structure to model evolution which allows research on individual hydrologic processes and associated pollutants to be chronologically added to an existing model without destroying the integrity of earlier work.

The integration of more comprehensive hydrologic component models with chemical fate models will probably continue to be a dominant trend in hydrologic research in the foreseeable future. The increasing utilization of data base management principles will facilitate this overall effort by allowing individual applications to select appropriate submodels from a large pool wherein the output of each submodel is automatically shared so it can serve to drive subservient processes. If these projections are valid, the minuscule amount of available comprehensive and accurate field data concerning hydrologic processes and pollutant fate will be the primary impediment to more rapid progress in the science of hydrology.

FIELD MONITORING SYSTEM DESIGN

The availability of low cost, low power micro-processor technology and the continued rapid evolution of improved electronic components have revolutionized the feasible scope of field monitoring systems. However, if commercially available equipment is a reliable barometer, neither equipment manufacturers nor the researchers who purchase and suggest new product configurations yet appreciate the true dimensions for improvement that this technology offers. In short, with limited but noteworthy exceptions, the primary thrust has been to simply automate procedures that were developed for the strip-chart, single parameter recorder era.

Fundamental Concepts. How does one achieve the "broader perspective for their data base" that was the appeal of the introduction? The

5

answer is primarily by increasing the number of parameters monitored and by changing the mode in which these variables are stored.

The recommendation to increase the number of parameters monitored under each hydrometeorological project is fairly straightforward. Modern, micro-processor based data acquisition systems have made this quite feasible by greatly reducing the per channel cost of collecting information. The primary need is to anticipate that other researchers in hydrology as well as other disciplines, e.g. entomological and crop growth modeling, may desire to use the data base collected for purposes other than outlined in the project which is funding the field stations. The monitoring of a few otherwise non-critical parameters can usually be added at a very modest cost to greatly expand the utility of the result-ing data base. Specific parameters required for this expanded utility include: instantaneous rainfall intensities, solar radiation, wind characteristics, soil temperatures and evaporation conditions. Water quality parameters are also needed, although laboratory costs make comprehensive determinations difficult unless they are a prime project objective. As a minimum, determinations on nutrients and heavy metals, plus any specifically suspected toxics should be included.

Langham (1971) eloquently developed the mathematical basis for a departure from the current mode in which we collect and store hydrometeorological data. The traditional daily discrete observations (occasionally the frequency is increased to hourly intervals) of a parameter's magnitude are not only of limited utility for many needs, they represent an extremely inefficient means of recording information. Hydrometeorological data are characterized by wide ranges in their rates of change; typically, long intervals of minor change followed by brief periods of rapid fluctuation. The classical timed interval recording approach for data requires either the selection of a very short record-ing interval with large volumes of redundant data or the acceptance of a high probability that rapid perturbations will not be recorded. Langham's solution to this dilemma is a recording mode he called the incremental integral concept. This approach requires recording the elapsed time required for a variable to change by a preselected incre-ment. This mode, sometimes referred to as event recording, not only yields a great improvement in storage efficiency, it captures all short duration perturbations of significant magnitude and still permits com-plete reconstruction of the temporal pattern of the variable. Informa-tion on acceleration rates, important to certain disciplines, is also preserved, Barrett, et al (1975).

Additional data storage efficiencies beyond the incremental integral mode can be implemented in data storage formats. As an exam-ple, consider the single channel, all solid state event recorder manufactured by Omnidata International, Inc. Time-of-event information is stored in a 2048 byte programmable-read-only memory (PROM) which is erasable with ultra-violet light. The operating software for monitoring rainfall intensity data is designed to achieve a one minute resolution for the time of each rain gauge bucket tip over a period of several months. Instead of recording the number of minutes elapsed from when the recorder starts, necessitating 3 bytes of storage for each bucket tip, a unique number is "burned" in a single byte each time 4 hours elapses. Every tip of the rain gauge records the number of minutes elapsed since the start of the last 4-hour data mark, a value that can also be stored in a single byte of PROM. This scheme allows the time of up to 1500 "events" to be stored over a 3 month period, a data capacity of 2.2 times that which could be obtained in the same PROM with the more conventional format.

Equipment and Operational Considerations. Selection of a particular configuration of field transducers and data recording equipment to operate in an unattended location involves many considerations. While arranged in order of decreasing priority, the following list of system attributes is suggested as almost essential requirements to be augmented with individual project considerations:
1. Unattended operational reliability over environmental extremes.
2. Compatibility with a wide variety of transducer outputs.
3. Battery powered operational capability.
4. An ability to communicate data and/or operational status of individual transducers to a central location.
5. Low unit cost for transducers and data transmission links.
Each of these items needs further elaboration to be viewed in proper perspective.

Field sensors and acquisition equipment must operate over a wide range of adverse environmental conditions with only infrequent site visits to provide maintenance. Until problems of keeping a sizeable network of field instruments operational have been experienced firsthand it is difficult to appreciate the importance of the first selection criteria. Central computer monitoring can detect only gross failures; it cannot detect deterioration of calibration accuracy nor effect the repair of a nonfunctional sensor. Unfortunately, it is difficult to evaluate the operational reliability of prospective system components. Some of the factors that merit consideration in making such an evaluation are: (1) quality of material and workmanship in the product; (2) design simplicity, especially for components utilizing electromechanical elements; (3) for sensors, the stability of the transducing element, i.e., the principle by which the variable being measured is converted into electrical form; (4) if available, field experience of other users of the equipment. Finally, it can hardly be overemphasized that one of the most effective means of assuring operational continuity and data integrity is to include as much redundancy in the system as possible.

The ability to accept "data" from transducers in more than one format can be a significant factor in the economic viability of a data acquisition system. This flexibility is important in three ways: incorporation of existing transducers into a new recording/communication system, ability to monitor a variety of parameters with low cost transducers, and delay of system obsolescence. While almost all modern systems require electronic signal inputs, this can take the form of a voltage, current or frequency proportional to the monitored variable or of a switch closure which occurs after a preselected increment of change. The ability to accept a variety of these signal forms can significantly affect the feasibility of modifying an existing network of instruments to incorporate an automatic acquisition capability. Similarly, this flexibility lengthens the useful life of the acquisition/recording unit, normally the item of dominant expense, because the probability is increased of its being compatible with new transducers that might be added in the future or with monitoring a wide range of parameters.

Numerous examples could be given of locations which require environmental monitoring equipment to be battery powered because line power is not economically available. However, even when line power is available, battery operation is a very desirable, even essential, feature. Battery power can greatly increase the likelihood of an uninterrupted record of data in two ways. First, much hydrologic data is storm associated; it must be collected during periods when the reliabil-

7

ity of of line power is poorest. Second, the probability of lightning damage to field transducers and recording equipment is greatly increased if it is electrically connected to line power, especially in remote regions. It is recommended that, if the use of line power is unavoidable for certain transducers, they be powered only intermittently for the shortest possible interval. This can be accomplished using battery powered relays which concurrently break the connection to all power lines to the transducer (hot, neutral and safety ground).

Costs for technically trained field personnel continue to increase rapidly relative to costs for computer-to-computer communications. Providing a mechanism for communication between a central computer and the remote field station(s) can simultaneously reduce the number of routine site visits required, improve the continuity of the data record by quickly detecting equipment malfunction and provide an enhanced operational capability via selective activation of field equipment upon command from the central computer. The latter can be particularly effective in terms of a closed-loop control situation. Multiple parameter condition tests of incoming data can readily be programmed into the central computer so that modified operational commands can be transmitted to the field monitoring equipment. A communications capability also provides an opportunity for a redundant means of recording data.

The ability of modern, micro-processor based data acquisition systems to accomodate a large number of transducers at minimal incremental cost provides the opportunity to economically broaden the scope of data collected from many field systems. Since gage houses, acquisition/recording equipment and stream control sections together with wages for personnel to service a station usually dominate costs when transducer prices are reasonable, an increase in the number of parameters monitored can greatly increase the utility of the resulting data base and the economic justification for the entire station.

DATA MANAGEMENT SYSTEM DESIGN

There are three major functional requirements for a hydrologic data management system: (1) verification and editing of field data, (2) archival storage and retrieval, and (3) analysis and interpretation of data. The separation of functions (2) and (3) is especially noteworthy. Attempting to accomodate the needs of an analysis function into the structure of the storage scheme can seriously compromise this vital element in a successful data management system.

Verification/Editing. All data bases derived from field monitoring stations are subject to erroneous entries, missing information and periods when individual transducers are operating out of calibration. The job of eliminating these problems is substantial. Normally a combination of manual editing and automated range checking is desirable. If the data acquisition network is connected to a central computer via a communications link as recommended above, the existence of software to automatically examine all incoming data for out-of-range conditions can substantially reduce problems with the data base by early identification of field difficulties that require on-site attention. Several specific recommendations in these areas have been presented by Wong, et al (1976).

Despite the best intentions, gaps will ultimately occur for some or all parameters. If a decision is made to fill these gaps by estimating values, e.g. data from the closest alternate station, an efficient means must be incorporated for flagging such information as an estimate rather than observed values.

Archival Storage and Retrieval. One of the most important and frequently overlooked requirements for a good data management system is a space efficient scheme for storing the information. The most frequent mistake made is to use a scheme that is convenient for accessing the data in order to analyze it. The analysis function is usually of a very short term nature in comparison with archiving requirements. Therefore, storage efficiency must be paramount. This recommendation is in direct conflict with current, admittedly immature, data base management structures. While general data base management systems have much to offer for data analysis requirements and for interfacing separate simulation models, most are unsatisfactory for archival/retrieval applications because of their copious storage demands.

To the extent practical, archived data should be stored in a "raw" form. That is, only clearly erroneous data should be modified before permanent storage. The use of alternate calibration coefficients, zero shift corrections, etc. should be accommodated, but stored in separate linked files or clearly identified in the original file.

Finally, it is crucial that documentation be developed concerning field equipment and laboratory analysis techniques employed. This should be summarized and reported in the form of an uncertainty analysis for each monitored parameter. The importance of a quantitative estimate of inaccuracies or uncertainties associated with physical measurements cannot be over-emphasized, Kline and McClintock (1953). An explicit and comprehensive analysis of uncertainty is the only rational means by which potential users can determine the suitability of monitored data for a desired application. Similarly, provision should be made for machine-readable storage of all field notes.

Analysis and Interpretation. Data analysis should involve a distinct and separate process of transferring data from archival storage into a form convenient for the desired type of processing. This avoids compromising storage efficiency needs in order to accommodate short term or partial record analyses. The varied needs of this field make any detailed discussion beyond the scope of this paper. Only two recommendations are made: modern data base management systems can be very useful and considerable attention should be directed toward graphical data presentations.

INTERPRETING WATER QUALITY DATA

The analysis and interpretation of hydrologic data is too broad for the scope of this paper. Therefore, remarks are restricted to the specific field of non-point source (NPS) pollution data. This discussion concerning limitations of comprehensive water quality monitoring programs is included because, in my opinion, misconceptions in this area are so pervasive. While specific illustrations concentrate on NPS pollution, several of the concepts are equally applicable to other areas of hydrology.

Partly as a result of demonstrated deficiencies of water quality models and partly because of naivete, the overwhelming majority of both the scientific and informed lay communities believe that the only reli-

9

able way to evaluate NPS pollution and the effectiveness of control measures is by field monitoring programs. The perception, though often not explicity stated, is that bottles of water must be collected and subjected to sophisticated laboratory analyses in order to establish "truth". Modeling results are inferred to represent "theoretical or hypothetical" numbers which are manipulated at the whim of the modeler to suit individual biases.

The primary problems of monitoring NPS pollution are not associated with the laboratory analysis of a collected sample, although there are still significant difficulties with certain constituents. Rather, they are with determining the source of pollutants present in a sample, assessing the true significance of individual component levels, separating out influences of individual storm characteristics and assessing the contributions of installed or proposed treatments on pollutant yields, i.e. determining cause-effect relationships.

There is a valid need for a quantitative determination of cause-effect impacts between alternative NPS control measures and the resulting pollution. Unfortunately, no economically feasible monitoring program can be devised which is capable of establishing cause-effect relationships between NPS pollution and control measures on a watershed scale, even for areas as small as a few square kilometers in size, especially on a short-term basis. This situation prevails because of the storm-induced nature of NPS pollution, seasonal variations in weather patterns and the uncontrolled nature of the many factors which profoundly influence levels of such pollution.

The broadly held misconception about the role of NPS monitoring appears to be based upon an inadequate differentiation between statistical correlation and conditions of cause-effect. Two of more variables are statistically correlated when the numerical value or level of one, the independent variable(s), infers a likely value for the other, the dependent variable. Two variables are related by cause-effect when there are governing physical laws which require that a change in level of the independent variable directly causes a prescribed change in the dependent variable. Variables which are related by cause-effect considerations will certainly show a high degree of statistical correlation. However, statistical correlation does not, of itself, infer that a cause-effect relationship exists between the correlated variables.

Statistical correlation will occur when variables are either related by physical laws or are simultaneously influenced by common unspecified variables. To illustrate the latter situation, consider the statistical correlation of Figure 1. Assume a need exits to predict the level of consumption of alcohol and the most readily available data pertains to average professorial salaries. Figure 1 shows the resulting "model" which is quite valid and useful for the stated purpose.

Now consider the situation if this model is deemed to depict a cause-effect relation and a national decision were made to reduce alcohol consumption. It would then follow that the appropriate course of action would be to reduce professorial salaries to attain this goal. Such a recommendation would, I hope, be considered ludicrous. Unfortunately, there seems to be little reticence to use only slightly less obvious correlation models for cause-effect projections in the field of water pollution abatement.

The point to be made from this example is that it is much easier to compute statistical correlations for variables which are influenced by

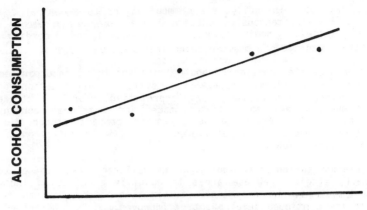

Figure 1. Hypothetical statistical correlation.

many unknown factors than to establish true cause-effect relationships. A real hazard exists that these correlation models will be interpreted as demonstrating cause-effect, especially when there is justification for assuming at least part of the correlation is due to physical interaction between variables.

This discussion is not meant to infer that field monitoring efforts have no role in evaluating water pollution. They can, especially when directed toward biological community determinations and habitat evaluation, determine overall water quality conditions on a watershed scale.. Furthermore, when restricted to field size areas with a single land use, monitoring can quantify the benefits of individual control measures, i.e. establish cause-effect for individual control measures in a specific location. Such information is vital to the development of accurate simulation models and is in woefully short supply.

SUMMARY AND CONCLUSIONS

Recent historical trends in hydrologic research and their impact on data acquisition and management needs are presented. Both philosophical and practical considerations are given for the design of such systems. The general recommendations are summarized below.

The lack of better quality hydrometeorologic and water quality data from various geographic regions is a primary impediment to more rapid progress in the science of hydrology. The scope and utility of hydrologic monitoring programs need to be broadened to meet the requirements for not only a range of hydrologic and water quality modelers, but of other disciplines as well. This can be accomplished at modest cost by increasing the number of parameters monitored and departing from the outdated mode of recording parameter levels at fixed time intervals.

Micro-processor based data acquisition systems offer the opportunity to obtain expanded data bases at reasonable costs; however, this will only come to fruition by properly designing monitoring programs to take advantage of this technology. The selection of data acquisition equipment should be based on its flexibility and the opportunity to

11

increase data base integrity, via communications to a central computer system, while simultaneously reducing field labor requirements. The range of parameters monitored must be increased and recording modes which take advantage of computer technology must be adopted.

Three separate functional requirements should be incorporated into any hydrologic data management system: verification/editing, archival/retrieval and analysis/interpretation. In the trade-offs always involved between storage space and retrieval convenience, increasing emphasis needs to be given to compact storage formats. Analysis requirements should be viewed as totally separate from archival storage considerations.

Field monitoring of non-point source pollution needs to be expanded for small, single land use areas to quantify benefits of individual treatment systems. In this manner, information can be obtained which is vital to the continued development of improved models. Watershed scale monitoring is useful for establishing general levels of pollution and for testing the overall accuracy of comprehensive models, but not for establishing cause-effect relationships or component relationships for models.

ACKNOWLEDGEMENTS

Many of the concepts and techniques discussed were developed and field tested under research concerned with non-point source pollution control from agricultural lands, sponsored by the U.S. Environmental Protection Agency and the Purdue Agricultural Experiment Station and coordinated by the Allen County Soil and Water Conservation District and the Indiana Heartland Coordinating Commission. AES Journal Paper No. 8559.

REFERENCES

Amorocho, J. and G.T. Orlob. 1961. Nonlinear Analysis of Hydrologic Systems. Univ. of California. Water Resources Ctr. Contr. No. 40.

Barrett, J.R., L.F. Huggins and W.L. Stirm. 1975. Environmental Data Acquisition. Envir. Entomology. 4:855-860.

Bernard, M. 1936. Giving Areal Significance to Hydrologic Research on Small Areas. Headwaters Control and Use. Proc. Upstream Engineering Conf. U.S. Gov. Printing Office.

Crawford, N.H. and R.K. Linsley. 1962. The Synthesis of Continuous Streamflow Hydrographs on a Digital Computer. Stanford Univ. Tech. Rept. No. 12.

Huggins, L.F. and E.J. Monke. 1966. The Mathematical Simulation of the Hydrology of Small Watersheds. Purdue Univ. Water Resources Res. Ctr. Tech. Rept. No. 1.

Kline, S.J. and F.A. McClintock. 1953. Describing Uncertainties in Single-sample Experiments. Mech. Eng. 75:3-8.

Langham, E.J. 1971. New Approach to Hydrologic Data Acquisition. Proc. ASCE, J. Hydro. Div. HY12:1965-78.

Laurenson, E.M. 1964. A Catchment Storage Model for Runoff and Routing. J. of Hydrology. 2:141-163.

Wong. G.A, S.J. Mahler, J.R. Barrett and L.F. Huggins. 1976. A Systematic Approach to Data Reduction using GASP IV. Proc. Winter Simulation Conf. pp 403-410.

RAINGAUGE NETWORK DESIGN - A   REVIEW

P. E. O'Connell
Principal Scientific Officer
Institute of Hydrology
Wallingford
Oxon, UK

ABSTRACT

Research carried out over the past several years in the field of raingauge network design is reviewed, mainly from a statistical view-point, and the practical relevance of the research is assessed.

Central to any raingauge network design procedure is the quantification of the estimation errors associated with point inter-polations and areal averages; to allow these estimation errors to be calculated, a statistical description of the rainfall process in time and space is required. In this context, the various assumptions which can be made about the stationarity of the rainfall process are discussed. The covariance (or correlation) function and the structure function (or variogram) represent the two main alternatives. The former assumes stationarity of the rainfall process, while the latter involves the less restrictive assumption of stationarity of its increments. Various forms for the correlation function which have been employed in the literature are discussed, and applications involving the fitting of these functions to observed correlations are described.

Procedures for calculating the mean square error of point inter-polations and areal estimates consisting of linear weightings of a set of rainfall measurements at points in space are considered when either the spatial correlation function or the variogram are used to charac-terize the spatial dependence in rainfall. Expressions for the minimum (or optimal) mean square error are given, together with expressions for the optimal weights. The case where averaging is carried out over both space and time is also considered.

A number of different approaches to network design involving the use of optimal estimation procedures are discussed: some of these involve consideration of accuracy only in the design procedure, while other, more formal analytical approaches view the network design problem as minimization of accuracy for a given cost, or vice versa. A real world case study carried out in the UK, in which optimal estimation procedures were employed, is discussed, and the merit of such procedures over more empirical approaches is illustrated.

Finally, some possible directions for future research are contem-plated. On the statistical side, there is scope for the development of techniques for the estimation and fitting of spatial models; however, the application of existing methodology in some real world case studies is seen as of paramount necessity if the practical relevance of recent research is to be judged, and priorities for future research identified.

INTRODUCTION

The general purpose of a raingauge network is to provide data which can be used as a basis for a description of the rainfall process in time and space. While, from the research viewpoint, the collection of rainfall data is desirable in order to gain a better understanding of the spatial and temporal distribution of rainfall, the sampling, collection and processing of rainfall data must be justified primarily in terms of the economic benefits which can derive from their more immediate use. Thus, the design and operation of raingauge networks must be closely linked with the various phases in the planning and operation of water resources systems; in this context, it has been suggested that three levels of information may be considered in relation to data network design (Rodda et al., 1969; Rodriguez - Iturbe and Mejia, 1974):

Level 1 provides a base level of information for wide regional and national planning, and can be used to provide background information for the design of more intensive and specific networks;

Level 2 corresponds to the provision of general data for water resources planning while

Level 3 is restricted to specific planning and management activities.

While the above three level classification of networks is conceptually useful, it is usually found in practice that such a classification cannot be made for existing networks. In many countries raingauge networks have evolved historically in an ad-hoc manner, with gauges tending to be concentrated close to centres of human activity and with relatively poor coverage in upland and mountainous areas where gauges may be needed most. This serves to emphasize that the design of raingauge networks is not a process which can be tackled totally within the framework of statistical and economic analysis; due recognition must be given to the important role which the 'human' element plays in the operation of a network.

Level 1 and 2 networks as classified above can be regarded as providing data for regional use. Such networks must make provision for the emergence of unanticipated needs in the future, as well as fulfilling their major role of providing data for the initial assessment of the hydrological characteristics of large areas. As a result it is difficult to assess what benefits might be derived from such levels of information; there are then three possible approaches to network design:

(a) design a network to minimize the estimation error of rainfall quantities without explicit consideration of cost, but with some notional constraint on the number of gauges;

(b) for a fixed cost, design a network to minimize estimation error;

(c) for a minimum acceptable criterion of accuracy, design a network at minimum cost

Most of the research on raingauge network design which has been carried out over the past ten years falls under (a), (b) or (c). Since nearly all approaches rely on a statistical description of the rainfall process in time and space, relevant work in this area is reviewed first. Procedures for deriving the accuracies of point and areal rainfall estimates (spatial averaging) are then considered in the following sections and optimal estimation techniques are discussed; averaging in space and time is also treated. Some general approaches to network design are then covered, and some applications, including a case study carried out in the UK, are described. Finally, the current state of the art as reflected in the review is discussed and some directions for future research are contemplated.

THE STATISTICAL STRUCTURE OF RAINFALL
IN TIME AND SPACE

General Considerations

In the context of rainfall network design, there are a number of problems of concern to hydrologists:

(i)   estimation of the total rainfall over a particular interval of time at a given point: often this is a point at which no rain gauge exists (interpolation in space);

(ii) the long-term average rainfall total over a time interval of given length for a given point (time averaging)

(iii) the total rainfall over a particular interval of time, averaged over a given region (space averaging);

(iv) the long-term mean total rainfall over a time interval of given length, averaged over a given region (space-time averaging)

In order to calculate the accuracies with which the above quantities can be estimated for various network configurations and densities, it is necessary to quantify the statistical structure of rainfall in time and space. This can be approached in a number of ways (Ord and Rees, 1975):

Covariance specification

If $X(\underline{x},t)$ represents the rainfall at $(\underline{x},t)$ where $\underline{x} = (x_1, x_2)$ denotes the coordinates of the point in two dimensional space, then

$$E\{X(\underline{x},t)\} = \mu(\underline{x},t) \qquad (1)$$

$$\text{cov}\{X(\underline{x},t), X(\underline{x+d}, t+\tau)\} = c(\underline{x}, \underline{x+d}, t, t+\tau).$$

If X is a Gaussian process, then the mean and covariance functions serve to specify the process completely. Equation (1) constitutes a wide sense specification of the process; a strict sense version requires knowledge of the joint density functions. The specification given by (1) covers all points within the area of interest, not just those for which measurements are available.

Conditional expectations or regression functions

A spatial autoregressive model for the rainfall at n sites with coordinates $\underline{x}_1, \underline{x}_2, \ldots \underline{x}_n$ is written as

$$\nabla_t\{X(\underline{x}_j,t) - \mu(\underline{x}_j,t)\} = \sum_{i \neq j} \beta(\underline{x}_i)[X(\underline{x}_i,t) - \mu(\underline{x}_i,t)]$$

$$+ \varepsilon(\underline{x}_j,t) \quad j = 1,2, \ldots N \qquad (2)$$

where ∇_t represents a difference operator with respect to time and $\varepsilon(\underline{x},t)$ is an independently distributed noise term satisfying

$$E\{\varepsilon(\underline{x},t)\} = 0, \tag{3}$$

$$E\{\varepsilon(\underline{x},t), \varepsilon(\underline{x}'t')\} = \begin{cases} \sigma_\varepsilon^2 \text{ if } \underline{x} = \underline{x}' \text{ and } t = t', \\ 0, \text{ otherwise} \end{cases}$$

Such an autoregressive scheme only accounts for temporal dependence in $X(\underline{x}_j,t)$ through the difference operator; the scheme could be generalized to include temporal dependence at the other points and higher order spatial and temporal differences. This specification is based directly on the measurements of rainfall at the set of points $\underline{x}_1$, $\underline{x}_2$, ... $\underline{x}_n$ and does not characterize the structure of rainfall at all points within the area of interest.

Spectral representation

To illustrate this, it is generally more convenient to operate in continuous space and time. For example, a first order spatio-temporal Markov process can be written in the form

$$\{D_t + \alpha^2 - D_1^2 - D_2^2\} X(\underline{x},t) = \varepsilon(\underline{x},t) \tag{4}$$

where $D_t = \frac{\partial}{\partial t}$ and $D_i = \frac{\partial}{\partial \underline{x}_i}$, $i = 1, 2$, for which the spectral density function is

$$f(\omega_t, \underline{\omega}) = \sigma^2 H(\omega_t,\underline{\omega}) H^*(\omega_t, \underline{\omega}) \tag{5}$$

where

$$H(\omega_t,\underline{\omega}) = (\alpha^2 + i \omega_t + \omega_1^2 + \omega_2^2)^{-1} \tag{6}$$

and H^* is the complex conjugate of H (Bartlett, 1975). If $f(\omega_t, \underline{\omega})$ is integrated over t, the purely spatial spectrum is obtained as

$$f(\underline{\omega}) = (\alpha^2 + \omega_1^2 + \omega_2^2)^{-1}. \tag{7}$$

Most of the work which has been carried out to date by hydrologists and meteorologists on characterizing the statistical structure of rainfall in time and space in relation to network design has centred on representation (1) although Mejia and Rodriguez-Iturbe (1973) have used spectral representations as a basis for the sampling of spatial processes. However, before the specification given by (1) can be rendered into a usable form, it is necessary to consider what properties of the rainfall process the covariance or correlation structure should reflect. In this context, stationarity is an important issue. The process $X(\underline{x},t)$ is said to be stationary in the wide sense if

$$E\{X(\underline{x},t)\} = \mu$$

$$\text{Var}\{X(\underline{x},t)\} \text{ is finite for all } \underline{x} \text{ and } t \tag{8}$$

and $\text{cov}\{X(\underline{x},t), X(\underline{x} + \underline{d}, t + \tau)\} = c(\underline{d}, \tau)$.

If it is further assumed that the spatial covariance structure is invariant for any given time interval (and vice versa), then the covariance may be written as

$$c(\underline{d}, \tau) = c_1(\underline{d}) c_2(\tau). \tag{9}$$

The extent to which these assumptions are satisfied by the rainfall process will depend very much on the extent of the area in question, the type of rainfall affecting the area, topographic variation etc. For example, orographic effects will tend to violate the assumptions in (8), since rainfall in mountainous areas is higher and more variable than on lowland terrain, and the rate of decay of correlation with distance may also be found to vary with altitude. Rainfall exhibiting such variation over space is said to be non-homogeneous, while 'non-stationarity' may imply variation either spatially or temporally, or both. Further properties which the rainfall process might be expected to exhibit are anisotropy i.e. the rate of decay of correlation with distance may be a function of direction, while measurements of rainfall might not be expected to correlate perfectly at zero distance due to measurement errors and microclimatic irregularities. The observed correlation structure of rainfall would also be a function of the sampling interval (hour, day, month, year).

A less restrictive assumption than that given by (8) is to assume that the spatial increments $X(\underline{x},t) - X(\underline{x} + \underline{d}, t)$ are stationary (the 'intrinsic hypothesis' Matheron, 1973); the process $X(\underline{x},t)$ is then characterized by

$$E\{X(\underline{x} + \underline{d},t) - X(\underline{x},t)\} = \mu(\underline{d})$$

$$\text{Var}\{X(\underline{x+d},t) - X(\underline{x},t)\} = 2\gamma(\underline{d})$$

(10)

where $\mu(\underline{d})$ is referred to as the linear drift and the function $\gamma(\underline{d})$ is called the variogram. If $X(\underline{x},t)$ is second order stationary, then

$$\gamma(\underline{d}) = c_1(0) - c_1(d) \tag{11}$$

The assumptions inherent in (10) may not be sufficient when $X(\underline{x},t)$ exhibits spatial trends; in such cases, the process $X(\underline{x},t)$ is considered to be the sum of a drift component $\mu(\underline{x})$ and another process $z(\underline{x},t)$ such that

$$E\{X(\underline{x},t)\} = \mu(\underline{x}) \tag{12}$$

with

$$\mu(\underline{x}) = \sum_{\ell=0}^{k} a_\ell f_\ell(\underline{x}) \tag{13}$$

$$\text{Cov}\{X(\underline{x+d},t), X(\underline{x},t)\} = \text{Cov}\{z(\underline{x+d},t), z(\underline{x},t)\}$$

$$= c_1(\underline{d}) \tag{14}$$

or, alternatively,

$$\text{Var}\{X(\underline{x+d},t) - X(\underline{x},t)\} = E\{z(\underline{x+d},t) - z(\underline{x},t)\}^2 = 2\gamma(\underline{d}) \tag{15}$$

The $f(\underline{x})$ are $(k+1)$ independent functions and the a_ℓ are unknown coefficients to be estimated.

Equations (8), (10) and (12) - (15) provide three alternative hypotheses which the rainfall process in time and space might be taken to satisfy. However, most of the work which has been carried out to date has tended to assume that a particular hypothesis is tenable, since the data or the techniques of statistical inference required to distinguish between the various hypotheses do not generally exist. Considerable effort has been devoted to the study of the spatial covariance function $c_1(\underline{d})$ of the rainfall process $X(\underline{x},t)$ through

17

empirical data analysis; as a result a variety of functional forms have been proposed, some of which are discussed in the following section.

Spatial Correlation Analysis

Spatial correlation functions

The most important functions in the isotropic class which have been suggested and used by various authors have been classified by Stol (1981) as follows. If $\rho(d)$ denotes a spatial correlation function where d now refers to the scalar measure of distance between two points in space, irrespective of direction, then the various forms are:

(a) polynomial functions of the form

$$\rho(d) = \alpha + \beta d + \gamma d^2 + \delta d^3 \quad 0 \leqslant d \leqslant d_{max}$$

$$\tag{16}$$

$$= 0 \qquad\qquad d > d_{max}$$

suggested by Hutchinson (1970) and Chang (1977); with $\delta = 0$ by Boyd (1939), Stenhouse and Cornish (1958), Cornish et al (1961), Hutchinson (1969) and Sneva and Calvin (1978); and with $\gamma = \delta = 0$ by Holland (1967), Sneyers (1968), Hendrick and Comer (1970), De Bruin (1975), Buishand (1977) and Sharon (1978);

(b) power functions of the form

$$\rho(d) = (k + ad)^{-b} \tag{17}$$

used, for instance, with $k = 0$ by Fisher and MacKenzie (1922); and with $k = 1$ by Yevjevich and Karplus (1973) and Richardson (1977);

(c) exponential functions of the form

$$\rho(d) = a \exp(- bd^k) + c \tag{18}$$

with $c = 0$ and $k = 1$ have been used by Caffey (1965), Kagan (1966, 1972), Guscina et al. (1967), Stol (1972), Rodriguez-Iturbe and Mejia (1974) and Longley (1974); with $k = 2$ by Stol (1972), and with $k = 1$ and $c > 0$ by Yevjevich and Karplus (1973);

(d) Bessel functions of the form

$$\rho(d) = \exp(- a^b) \, J_0(cd) \text{ and } \rho(d) = adK_1(ad) \tag{19}$$

where J_0 is a Bessel function of the first kind of order zero suggested Chemerenko (1975) and K_1 is a modified Bessel function of the second kind as used by Rodriguez-Iturbe and Mejia (1974) and Gottschalk (1978).

All of the above forms assume that the rainfall process is stationary and isotropic within the area in question.

Anisotropic correlation functions may be expressed as a function of distance d and orientation θ; an example given by Caffey (1965) is

$$\rho(d,\theta) = a_1 \exp(- a_2 d + a_3 d \cos 2\theta + a_4 d \sin 2\theta) \tag{20}$$

An alternative form, used by O'Connell et al (1977, 1978) may be defined as follows. For two points with coordinates $\underline{x} = (x_1, x_2)$ and $\underline{y} = (y_1, y_2)$ and d_1, d_2 given by

$$d_1 = x_1 - x_2, \quad d_2 = y_1 - y_2,$$

the correlation function is given by

$$\rho(\underline{x},\underline{y}) = \rho(d_1,d_2) = \rho(\underline{d})$$

$$= a + (1-a) \exp[- b\{(d_1 + c_1 d_2)^2 + c_2 d_2^2\}^{0.5}] \quad (21)$$

for $a \leqslant 1$; $b \geqslant 0$; $c_2 \geqslant 0$.

Applications

The identification and fitting of spatial correlation functions of the form given above might be expected to involve the following steps:

(i)   preliminary analysis of the basic data: plots of the decay of estimated correlations with distance around central stations can help to reveal some salient features such as anisotropy or non stationarity;

(ii) identification and fitting of an appropriate functional form;

(iii) calculation of standard errors of estimated parameters and some measures of goodness of fit.

However, most applications have tended to concentrate on (i), since, in general, there is a lack of formal methodology for carrying out steps (ii) and (iii), and fitting has usually been a rather ad hoc procedure. A brief review of some past applications will serve to illustrate the effects of factors such as storm type, season, topography etc on rainfall as well as the range of procedures used for fitting correlation functions.

The work of Kagan (1966) and Guscina et al (1967) exemplifies some of the earliest applications of spatial correlation analysis to rainfa l; they presented plots of correlation as a function of distance from a central station for 12 hour, 24 hour, 10 day, monthly and seasonal rainfall for locations in the Valdai area of Russia; the rate of decay was observed to increase with decreasing duration and was assumed to conform to

$$\rho(d) = \rho(o) \exp(- bd) \quad (22)$$

where $b^{-1} = d_0$ is defined as the correlation radius or the distance at which the correlation decays by a factor of e, and $\rho(0)$ is the value of the correlation function when extrapolated to zero distance. The parameter $\rho(0)$ reflects rainfall measurement errors and micro-climatic irregularities. No details were given on how the function (23) was fitted, although Guscina et al (1967) comment that (22) decayed more rapidly than observed correlations at large distances.

Hershfield (1965) analysed rainfall data for 15 storms for each of 15 watersheds with a total of 400 raingauges and found that plots of correlation around key gauges showed evidence of anisotropy. The

19

rates of decay of correlation with distance around different key
stations in the same watershed were found to be different, suggesting
non-stationarity in the rainfall process. No functional forms were
suggested.

Caffey (1965) analysed the spatial correlation structure of
annual rainfall from 1141 stations from the Western U.S. and Southwestern Canada with an average length of record of 54 years. The
regional variation in inter-station correlation was found to be
explained reasonably well by (20) which was fitted directly by least
squares to the correlations between a central station and surrounding
stations in a region. Approximately 60 per cent of the variation in
inter-station correlation coefficients was explained by (20), and the
effects of topography, general wind circulation and frontal activity
upon the orientation of the axis of maximal correlation were noted.

Huff and Shipp (1969) carried out an extensive spatial analysis
of rainfall from three dense raingauge networks in Illinois; data
ranging from one minute rates to total storm, monthly and seasonal
amounts were analysed. The effects of rain type, synoptic storm type
and other factors on spatial correlation were studied. Correlation
decay with distance was greatest for thunderstorms, rainshowers and
air mass storms, and least for steady rain and the passage of low
pressure centres. Summer decay rates were also much greater than those
in winter. Anisotropy in correlation contours was also observed
and the direction of least decay was observed to coincide with
preferred storm paths. No functional representations of spatial
correlation were suggested.

Hutchinson (1969) analysed monthly and annual rainfall data from
two areas in New Zealand, one relatively flat and the other with
variable topography. Plots of correlation around key gauges showed
distinct anisotropy for both areas as well as dependence of the rate
of decay on topography. Correlation functions of the form (16) were
fitted using regression in which additional terms were included to
account for measures of topography (differences in elevation, exposure
and aspect); however, the improvement in explained variance due to the
inclusion of topographic variables was significant only for a number
of calendar months. Further work by Hutchinson (1970) for the same
areas showed a distinct relationship between relief and the magnitude
of spatial correlation for a given distance for monthly rainfall.

Hendrick and Comer (1970) analysed the spatial structure of daily
rainfall data from the Sleepers River Watershed in Northern Vermont.
Data for days on which the rainfall at one or more gauges was ≥ 0.10,
≥ 0.5, ≥ 1.00 inches were selected for a winter season and a summer
season. The correlation field around a central gauge showed strong
evidence of anisotropy; dependence on daily rainfall amount and season
was also evident. No dependence on elevation difference or slope was
found; correlation decay with distance was approximated by a linear
segment

$$\rho(d) = a + (b - c \sin (220 - 20))d \quad 0.5 \text{ miles} < d < 8 \text{ miles} \quad (23)$$

where ϕ is interstation azimuth angle, and a, b and c were defined as
functions of daily rainfall minimum. Fitting was carried out using
simple graphical techniques.

Stol (1972) analysed daily rainfall data from 3 groups of
stations aligned along different directions for an area in the

Netherlands with no relief. Data for days with rainfall > 0.5 mm at all
sites were analysed on a month by month basis; no evidence of anisotropy was found. The function

$$\rho(d) \ = \ \rho_0 \ \exp(- \ \beta d) \tag{24}$$

was fitted for each month, apparently by least squares regression.
The parameter β exhibited strong seasonality while ρ_0, reflecting
measurement error, was relatively constant at about 0.95. A quadratic
function (with d in (24) replaced by d^2), which has a much flatter
shape near the origin that (24), was also fitted; no improvement in the
explained variances for the regressions were obtained, but ρ_0 exhibited
greater seasonality, dropping below 0.8 in some months which would
seem unreasonably low. The results for the two models suggest some
dependence on the shape of the correlation function near the origin;
Stol expressed a preference for (24) which is similar in this respect
to the model (20) finally chosen by Caffey (1965).

Sharon (1974) discussed some of the limitations of correlation analysis
as a basis for network design, particularly in relation to localized
storm rainfall; he suggested, however, that it could play an important
role in studying the spottiness of rainfall and its general spatial
organization. In analysing daily data he suggested that a criterion
based on the exceedence of a threshold amount of at least one gauge,
rather than all gauges, would allow days with 'spotty' rainfall to be
included, as correlation is systematically lower for localized rainfall;
he argued further that rainfall should be analysed on the basis of
distinct rainfall types. Sharon also questioned whether correlation
functions should decrease monotonically, on the basis of the work of
Austin and Houze (1972), the correlation function might be expected
to go negative before approaching zero. Examples of estimated non-
monotonic correlation functions were given; bumps in the correlation
field such as observed in Hershfield (1968) may reflect the sizes of
single cells and small meso-scale systems, and the existence of
preferred distances between storm cells.

As a basis for some theoretical work on network design (to be
discussed later) Rodriguez-Iturbe and Mejia (1974) employed an
exponential function

$$\rho(d) \ = \ \exp(- \ hd) \tag{25}$$

and a modified Bessel function of the second kind

$$\rho(d) \ = \ b \, d \ K_1(bd) \tag{26}$$

to represent the correlation structure of rainfall in space. They
noted that the correlation function (26) derives from a spatial
process

$$\{D_1^2 + D_2^2 - \kappa^2\} \ X(\underline{x},t) \ = \ \varepsilon(\underline{x},t) \tag{27}$$

while (2.25) derives from

$$\{D_1^2 + D_2^2 - \alpha^2\}^{\frac{3}{4}} \ X(\underline{x},t) = \ \varepsilon(\underline{x},t), \tag{28}$$

Whittle (1954) pointed out that it is difficult to visualize a
physical process satisfying (28), and expressed a preference for the
function (26). The function (26) was fitted to some annual rainfall

totals for a catchment in Venezuela using the equation

$$\sum_{i=1}^{N}\sum_{j=1}^{N}\sum_{k=k_i(i,j)}^{k_f(i,j)} z_{i,k}\, z_{j,k} = \sum_{i=1}^{N}\sum_{j=1}^{N} [k_f(i,j) - k_i(i,j) + 1]$$

$$d_{i,j}\; bK_1(d_{i,j}\, b) \qquad (29)$$

where $k_i(i,j)$, $k_f(i,j)$ represents the first and final years, respectively, for which the records of both stations i and j exist, $z_{i,k}$ is a standardized rainfall value during year k at station i, $d_{i,j}$ is the distance between stations i and j, and N is the number of stations. The estimate of b derived by solving (29) will reflect the weighting of individual stations according to their length of record. The exponential function (25) was then fitted by calculating the value of $\rho(d)$ at the 'characteristic correlation distance' (defined as the average distance between two randomly chosen points in the region of interest) using the fitted function (26) and then solving for h in (25) at this distance. Thus, the fitting of the exponential function depended on that of the Bessel function. Values of b and h were also derived by fitting the respective functions on the basis of values quoted in the literature for the 'correlation radius' of a rainfall event, defined as the distance corresponding to $\rho(d) = 0.5$. However, the distance thus derived will be dependent on the particular rainfall type, while the characteristic correlation distance defined above is a function only of the geometry of the region. The two characteristic measures of distance were approximately equal for an example from Eagleson (1967) but this need not necessarily be true in other cases.

O'Connell et al (1977) analysed an extensive volume of daily, monthly and annual rainfall data for two regions, one in the east and one in the north of England. Data for a number of categories of daily rainfall were analysed; days on which daily rainfall at a selected number of gauges within each region exceeded 2 mm, 5 mm and 10 mm represented three categories, while a fourth category consisted of data for every 20th day. Plots of the correlation fields around a number of key stations showed evidence of anisotropy, and an increase in the rate of decay with decreasing threshold; the data for every 20th day showed the most rapid rate of decrease. The parameters a, b, c_1, c_2 in (21) were estimated by minimizing the function $g(a,b,c_1^2, c_2^2)$ defined as follows. Let the jth sample correlation, r_j be calculated between the ith station at coordinates (x_{1i}, x_{2i}) and the central station at coordinates (x_1, x_2), based on n_i pairs of values. Then, if N is the number of stations, the function $g(.)$ is defined as

$$g(a, b, c_1, c_2) = \sum_{i=1}^{N} (n_i - k)(z_i - f_i)^2 \qquad (30)$$

where

$$z_i = \tfrac{1}{2} \log_e \left(\frac{1 + r_i}{1 - r_i}\right) \quad i = 1, 2, \ldots p$$

$$f_i = \tfrac{1}{2} \log_e \left(\frac{1 + \rho_i}{1 - \rho_i}\right) \quad i = 1, 2, \ldots p$$

$$\rho_i = \rho(x_{1i} - x_1, x_{2i} - x_2) \quad i = 1, 2, \ldots p$$

The constant k was taken as 3 for daily and yearly and as 36 for monthly data. Equation (30) represents one method of giving higher weight to correlations based on more observed values.

The above approach was adapted further by O'Connell et al. (1978) to describe the correlation structure of an area of 10,000 km² in the south-west of England as a basis for network design. An extra parameter was included in the correlation function to account for measurement errors i.e.

$$\rho(\underline{d}) = a + (1 - a - \varepsilon) \exp[- b\{(d_1 + c_1 d_2)^2 + (c_2 d_2)^2\}^{0.5}] \quad (31)$$

and this function was then fitted on a grid square basis over the region by minimizing the functional (30) with the parameter ε included. The r_j were calculated for all pairs of gauges within a square of side 35 km, with the analysis being repeated for all overlapping squares covering the region. For an analysis relevant to any one point the parameters referring to the square with the nearest centre point were taken. This procedure tries to ensure that the correlation structure reflects any non-stationary behaviour in rainfall over the region, and that the fitted parameters vary smoothly in space.

Space-time Correlation Analysis

Zawadzki (1973a) analysed the space-time correlation structure of a widespread convective storm. He used an optical device to compute the space-time correlation structure of radar data stored on film. Autocorrelation functions of storm rainfall for an observer moving with the storm (the Lagrangian time frame) and an observer at a fixed location on the ground (the Eulerian time frame), were calculated to check the validity of the Taylor hypothesis that

$$\rho_E(\tau) = \rho_L(U\tau) \quad (32)$$

where $\rho_E(\tau)$ and $\rho_L(\tau)$ denote the Eulerian and Lagrangian autocorrelation functions, and $U = (U_x^2 + U_y^2)^{0.5}$ where U_x and U_y are the velocity components of the storm in the x and y directions; the hypothesis was found to be approximately valid for time periods shorter than 40 minutes. The spatial correlation function (zero time lag) was found to be isotropic for scales of the order of 10 km but showed an elliptic pattern for larger scales. No functional forms were fitted to the space and time correlation functions.

Rodriguez-Iturbe and Mejia (1974) assumed that the covariance function of the rainfall process in time and space could be expressed in the form (9) i.e. that the spatial and temporal covariances were separable. The spatial component $c_1(d)$ was described using a modified Bessel function of the second kind and and an exponential function as described above; the temporal component was assumed to follow a first order Markov process with parameter ρ which Mejia and Rodriguez-Iturbe suggested could be estimated from the following relationship

$$\sum_{i=1}^{N} \sum_{j=1}^{N} \sum_{k=k_i'(i,j)}^{k=k_f'(i,j)} z_{i,k} z_{i,k+1} = \rho \sum_{i=1}^{N} \sum_{j=1}^{N} [k_f'(i,j) - k_i'(i,j) + 1]$$

$$v_{i,j} b K_1(v_{i,j} b) \quad (33)$$

where $k_i^!(i,j)$ and $k_f^!(i,j)$ denote the initial and final time point for which both the record of station i and the record of station j at the following time point exist. For the annual data from a Venezuelan region, the value of ρ obtained was zero.

Brady (1975) proposed that the space-time correlation structure of 1 minute storm rainfall on the Goose Creek network in Illinois could be described using the correlation function

$$\rho(\underline{d},\tau) = a_1\left(\frac{d_1}{d}\right) + a_2\left(\frac{d_2}{d}\right) + (a_3 - a_1\left(\frac{d_1}{d}\right) - a_2\left(\frac{d_2}{d}\right)\}$$
$$\exp\left[-\left\{\left(\frac{d_1}{a_4}\right)^2 + \left(\frac{d_2}{a_5}\right)^2 + \left(\frac{\tau}{a_6}\right)^2\right\}\right] \tag{34}$$

where a_1, a_2 allow the function to become negative, as this had been found to be both reasonable and necessary; $d = d_1 + d_2$; a_3 is the point at which the function cuts the origin at zero lag in space and time, and a_4, a_5 and a_6 are the scale sizes in x, y, t, respectively and are a measure (in terms of correlation) of the precipitation extent as reflected in the observations. The function (34) was fitted by minimizing

$$Q = \sum\{\rho(\underline{d},\tau) - r(\underline{d},\tau)\}^2 \quad n(\underline{d},\tau) \tag{35}$$

where $n(\underline{d},\tau)$ denotes the number of observation pairs used in calculating the sample correlation $r(\underline{d},\tau)$. Having used 'objective analysis' (Eddy, 1967, 1973) to reconstruct the original data from the fitted function (34) and found that the fitted correlation function for the reconstructed data was similar to that of the original data, Brady suggested that (34) provided an adequate representation for the 71 observations of 1 minute rainfall from the chosen rain storm.

ESTIMATION OF ACCURACY OF POINT AND AREAL RAINFALL ESTIMATES

General

The estimation of rainfall at an ungauged point or over an area has occupied the attention of meteorologists and hydrologists for many years, and many schemes have been devised for deriving such estimates. Most of these schemes are empirical in that the weights assigned, for example, to individual gauges in deriving the spatial average of a rainfall event are derived on a non-theoretical basis; nonetheless, they do seek to represent such factors as the variation in rainfall with altitude or the known long term behaviour of rainfall over an area in deriving the estimate. However, most schemes of this type cannot provide an estimate of the accuracy of the derived estimate, since the methods do not involve establishing the underlying spatial statistical structure of the rainfall.

Gandin (1965) was apparently the first to provide a theoretical basis for calculating the accuracy (in terms of mean square error) of a spatial average over a meteorological field, and to advocate the concept of 'optimum interpolation' i.e. the determination of weights which minimize the mean square error of the estimated quantity. However, as Gandin points out, the problem of optimum interpolation was first formulated in detail by Kolmogorov (1941) and further developed and popularized by Wiener (1949) who was apparently the first to use the term 'optimum interpolation'. Similar problems were addressed by Matheron (1965) and his co-workers who were concerned with determining the best linear unbiased estimators of areal averages of geological variables treated as realizations of stochastic processes. Since then, further theoretical work on the properties of spatial and spatio-temporal averages has been carried out, and the main results which have been derived to date are discussed in the following sub-sections.

Spatial Averaging

Point interpolation

Theory: A number of results for optimal point interpolation have been presented by O'Connell et al (1977). If measurements of rainfall at each of p gauges are denoted by X_1, X_2, ... X_p, and an estimate of Y, the rainfall at an ungauged point over the same time interval is required, then the linear estimates of Y which can be obtained are of the form

$$\hat{Y} = a + b_1 X_1 + b_2 X_2 \ldots + b_p X_p \tag{36}$$

where a, b_1, ..., b_p have known values. The accuracy of such an estimator may be quantified by its mean square error (mse) defined as

$$mse(\hat{Y}) = E[(\hat{Y} - Y)^2] \tag{37}$$

while, if the estimator is to be unbiased, then it must satisfy

$$E\{Y - \hat{Y}\} = 0 \tag{38}$$

Suppose that the long-term average rainfalls (for the time interval being considered) are known to be μ_1, μ_2, ... μ_p at each of the gauged sites and μ_Y at the ungauged site: then estimators $\hat{Y}$ of the form

$$\hat{Y} = \mu_Y + b_1(X_1 - \mu_1) + \ldots + b_p(X_p - \mu_p) \tag{39}$$

are unbiased for any constants b_1, ... b_p, and it remains to choose these constants.

The mean square error of the estimator (39) is

$$mse(Y) = \sigma_{YY} - 2\underline{b}^T \underline{\sigma}_{XY} + \underline{b}^T \underline{\sigma}_{XX} \underline{b} \tag{40}$$

where

$$\sigma_{YY} = \text{Var } Y$$

$$\underline{\sigma}_{XY} = (\sigma_{X_1 Y}, \sigma_{X_2 Y}, \ldots \sigma_{X_p Y})$$

$$\underline{\sigma}_{XX} = (\sigma_{X_i X_j})$$

$$\underline{b} = (b_1, b_2, \ldots b_p)^T \tag{41}$$

and where $\sigma_{X_i Y} = \text{cov}(X_i, Y)$, $\sigma_{X_i, X_j} = \text{cov}(X_i, X_j)$.

For any particular choice of constants, (40) gives the mse of the

25

estimator (39); however, if the vector $\underline{b}$ is chosen to minimize the mse, then the resulting coefficients of the optimal linear estimator

$$\hat{Y}* \;=\; \mu_Y + b_1^*(X_1 - \mu_1) + \ldots + b_p^*(X_p - \mu_p)$$

are defined by

$$\underline{b}* \;=\; \underline{\sigma}_{XX}^{-1}\, \underline{\sigma}_{XY}, \qquad\qquad (42)$$

and the corresponding minimum mse is

$$\text{mse}(\hat{Y}*) \;=\; \sigma_{YY} - \underline{\sigma}_{XY}^T\, \underline{\sigma}_{XX}^{-1}\, \underline{\sigma}_{XY} \qquad\qquad (43)$$

It is clear that to calculate $\text{mse}(\hat{Y}*)$ knowledge of the covariances between rainfall at gauged points and the ungauged point are required; these can be derived from fitted correlation functions of the type discussed in the previous section.

The assumption that $\mu_Y, \mu_1 \ldots \mu_p$ are known exactly does not hold in practice. One possible assumption then is that the long-term average rainfall at all places is constant and equal to μ; this may be reasonable if rainfall is fairly homogeneous over the region. Equation (40) then becomes

$$\hat{Y} \;=\; \sum b_i\, X_i + \mu(1 - \sum b_i) \qquad\qquad (44)$$

for which the mse of the optimal estimator is given by (43)

If there is no information about μ available, then the only unbiassed linear estimators which can be employed are of the type

$$\hat{Y} \;=\; \sum b_i X_i \qquad\qquad (45)$$

and for which

$$\sum b_i \;=\; 1. \qquad\qquad (46)$$

The optimal estimator in this class can be found by minimizing the mse (40) subject to the constraint (46). This gives, for example, by the method of Lagrange multipliers,

$$\underline{b}** \;=\; \underline{\sigma}_{XX}^{-1}\, (\underline{\sigma}_{XY} + \theta\underline{1}) \qquad\qquad (47)$$

where $\underline{1}$ denotes a vector of ones and

$$\theta \;=\; (1 - \underline{1}^T\, \underline{\sigma}_{XX}^{-1}\, \underline{\sigma}_{XY})/(\underline{1}^T\underline{\sigma}_{XX}^{-1}\, \underline{1}) \qquad\qquad (48)$$

and the corresponding mse of the estimator

$$\hat{Y}** \;=\; \underline{b}**^T\underline{X}$$

$$\text{mse}(Y**) \;=\; \sigma_{YY} - \underline{\sigma}_{XY}^T\, \underline{\sigma}_{XX}^{-1}\, \underline{\sigma}_{XY} + (1 - \underline{1}^T\underline{\sigma}_{XY}^{-1}\, \underline{1})^2/(\underline{1}^T\underline{\sigma}_{XX}^{-1}\, \underline{1}) \quad (49)$$

and clearly this is always larger than (43).

Estimators of the form (45) have the attractive property that if the rainfalls at each of the gauges are equal, then this same value is produced as the interpolated value. However, Schaake (1979) suggests that a point estimator should have the property that the weights decrease as the point moves away from the set of gauges, but presumably this is only reasonable if the mean μ is known and included in the estimator.

So far, it has been assumed that the measurements X_i are error free; however this will not generally be the case. One possible hypothesis is that the true rainfalls, denoted by Y_i, are related to the measured rainfalls through

$$X_i = Y_i + \varepsilon_i \qquad (50)$$

where $\varepsilon_i (i = 1,2, \dots p)$ are uncorrelated amongst themselves and uncorrelated with the true rainfalls Y, Y_j ($j = 1,2, \dots p$). The covariances (41) then become

$$\sigma_{X_i Y} = \sigma_{Y_i Y}$$

$$\sigma_{X_i X_j} = \sigma_{Y_i Y_j} \qquad (i \neq j) \qquad (51)$$

$$\sigma_{X_i X_i} = \sigma_{Y_i Y_i} + \sigma_{\varepsilon_i \varepsilon_i}$$

If the variances of rainfall and measurement errors are constants $(\sigma_{YY}, \sigma_{\varepsilon\varepsilon})$ and if

$$\underline{R}_{YY} = \rho(Y_i, Y_j) \qquad (52)$$

$$\underline{r}_Y = \{\rho(Y_1, Y), \rho(Y_2, Y), \dots, \rho(Y_p, Y)\}^T$$

are arrays of correlations of the true rainfall at pairs of points, then the mean square error of the linear estimator (39) is, with $\eta = \sigma_{\varepsilon\varepsilon}/\sigma_{YY}$ and $\underline{I}$ defined as the identity matrix,

$$\text{mse}(\hat{Y}) = \sigma_{YY} \{1 - 2\underline{b}^T \underline{r}_Y + \underline{b}^T (\underline{R}_{YY} + \eta\underline{I})\underline{b}\} \qquad (53)$$

while the mean square error of the optimal linear estimator $\hat{Y}*$ is

$$\text{mse}(\hat{Y}*) = \sigma_{YY} \{1 - \underline{r}_Y^T (\underline{R}_{YY} + \eta\underline{I})^{-1} \underline{r}_Y\} \qquad (54)$$

with coefficients

$$\underline{b}* = (\underline{R}_{YY} + \eta\underline{I})^{-1} \underline{r}_Y \qquad (55)$$

The corresponding quantities in the absence of measurement error (i.e. (43) and (42) written in terms of correlations rather than covariances) are

$$\text{mse}(\hat{Y}*) = \sigma^2 \{1 - \underline{r}_{XY}^T R_{XX}^{-1} \underline{r}_{XY}\} \qquad (56)$$

and

27

$$\underline{b}^* = \underline{R}_{XX}^{-1}\, \underline{r}_{XY} \tag{57}$$

Similar results can be derived for the optimal estimator (45) employed when the means are assumed constant but unknown.

Three points are worth noting in relation to the effect of measurement error:

(i)   mse $(\hat{Y}^*)$ is increased over the no measurement error case;

(ii) mse $(\hat{Y}^*) \geqslant \dfrac{1}{p}\, \sigma_{\varepsilon\varepsilon}$; this means that if a fixed number of gauges are to be used there is a limit to the accuracy with which rainfall can be interpolated no matter how close the gauges are - essentially because, even at a gauged site, the rainfall is only known to within the measurement error.

(iii) a consequence of (ii) is that it is possible to derive a better estimate of the rainfall at a gauged point than the rainfall measurement at that site.

The necessary covariances (or correlations) between the rainfall at gauged and ungauged points in the above expressions would be derived from a fitted correlation function; in the presence of measurement error it would be necessary to allow for correlation less than unity at zero distance to ensure consistency with the expressions (53) - (55).

It is assumed above that the rainfall process is stationary in the wide sense (as defined previously) over the region of interest; if, however, this is not the case, the assumptions (12) - (15) may be used as the starting point for the derivation of optimal point estimators; the resulting optimal estimation technique is known as Universal Kriging, and has been developed by Matheron (1969).

Let $Y(\underline{x}_0)$ denote the unknown rainfall at the point of interpolation $\underline{x}_0$, and let $X(\underline{x}_i)$, $i = 1, 2, \ldots p$ denotes the measured rainfalls (assumed error free) at the gauged points; then the estimate

$$\hat{Y}(\underline{x}_0) = \sum_{i=1}^{p} \lambda_i\, X(\underline{x}_i) \tag{58}$$

is desired which minimizes the mean square error $E\{\hat{Y}(\underline{x}_0) - Y(\underline{x}_0)\}^2$ or

$$E\{\hat{Y}^*(\underline{x}_0) - Y(\underline{x}_0)\}^2 = \mathrm{Var}\,\{\hat{Y}^*(\underline{x}_0) - \hat{Y}(\underline{x}_0)\} + [E(Y^*(\underline{x}_0) - Y(x_0))]2 \tag{59}$$

The first term on the rhs of (59) is

$$\mathrm{Var}\{\hat{Y}^*(\underline{x}_0) - Y(\underline{x}_0)\} = \sum_{i=1}^{p}\sum_{j=1}^{p} \lambda_i \lambda_j\, \gamma(x_i - \underline{x}_j) + 2\sum_{i=1}^{p} \lambda_i \gamma(\underline{x}_i - \underline{x}_0) \tag{60}$$

under the assumption that $\Sigma\lambda_i = 1$, while the mean error is

$$E\{\hat{Y}^*(\underline{x}_0) - Y(\underline{x}_0)\} = \sum_{i=1}^{p} \lambda_i \mu(\underline{x}_i) - \mu(\underline{x}_0) \tag{61}$$

where $\mu(\underline{x}_i)$ denotes the value of the mean at site i with coordinates $\underline{x}_i = (x_{1i}, x_{2i})$. By assuming that the mean can be represented using (13) then

$$E\{\hat{Y}^*(\underline{x}_0) - Y(\underline{x}_0)\} = \sum_{\ell=0}^{k} a_\ell [\sum_i \lambda_i f^\ell(\underline{x}_i) - f^\ell(\underline{x}_0)] \tag{62}$$

and that

$$\sum_i \lambda_i f^\ell(x_i) = f^\ell(\underline{x}_0)$$

then the mean error is exactly zero whatever the values of the coefficients a_i, thus assuring that the estimator (57) will be unbiased. By minimizing the mean square error under these conditions, the result is

$$E\{\hat{Y}^* - Y\}^u = \sum_{j=1}^{n} \lambda_j \gamma(\underline{x}_j - \underline{x}_0) + \sum_\ell u_\ell t_0^\ell(\underline{x}_0) \tag{63}$$

where the u_ℓ are Lagrangian multipliers.

Areal estimation

The mean square error of the spatial average of a meteorogical field was first derived by Gandin (1965); his results have formed the basis of further work by Zawadzki (1973b),Bras and Rodriguez Iturbe (1976), Lenton and Rodriguez-Iturbe (1977), O'Connell et al (1977, 1978) and Jones et al (1979).

The ensuing formulation is based on that presented by Jones et al (1979). The average rainfall Y_A over an area A is given by

$$Y_A = \frac{1}{A} \int_A Y(u_1, u_2) \, du_1 \, du_2 = \frac{1}{A} \int_A Y(\underline{u}) \, d\underline{u} \tag{64}$$

where $Y(u_1, u_2) = Y(\underline{u})$ denotes the true total rainfall over a given duration at each point $(u_1, u_2) = \underline{u}$ in a region

Estimators $\hat{Y}_A$ of Y_A of the following form are considered:

$$\hat{Y}_A = b_1 X_1 + \ldots + b_p X_p$$

with $\sum b_i = 1$. The true rainfall $Y(\underline{u})$ is treated as a random quantity, and the expression for Y_A is considered as a stochastic integral. The variance of Y_A, the true average total rainfall is given by

$$\sigma_{YY} = \text{Var}(Y_A) = \text{Var} \left(\frac{1}{A} \int Y(\underline{u}) d\underline{u} \right)$$

$$= \frac{1}{A} \int_A \int_A \text{cov} \{Y(\underline{u}), Y(\underline{v})\} \, d\underline{u} \, d\underline{v}$$

29

$$= \frac{1}{A^2} \int_A \int_A \sigma_{YY}(\underline{u},\underline{v}) \ d\underline{u} \ d\underline{v} \qquad (65)$$

The vector of covariances of X_i with Y_A is $\underline{\sigma}_{XY} = \{cov(X_i, Y_A)\}$ where

$$cov\{X_i, Y_A\} = cov[Y(\underline{u}_i) + \eta_i, \frac{1}{A} \int Y(\underline{u})d\underline{u}]$$

$$= \frac{1}{A} \int_A cov\{Y(\underline{u}_i), Y(\underline{u})\} \ d\underline{u}$$

$$= \frac{1}{A} \int_A \sigma_{YY} (\underline{u}_i, \underline{u}) \ d\underline{u} \qquad (66)$$

assuming that the true rainfalls X_i at the points $\underline{u}_i$ are subject to measurement errors as defined by (50). With the quantities σ_{XY}, σ_{YY} defined by (66) and (65) and σ_{XX} as defined previously in (41), equation (49) then provides the mse of the optimal estimate.

To evaluate such expressions, the four dimensional integral in (65) and the p 2-dimensional integrals in (66) must be calculated. The most straightforward case is when the area under consideration is rectangular and the correlation function is isotropic; Bras and Rodriguez-Iturbe (1976a) and Lenton and Rodriguez Iturbe (1977) have used simple numerical techniques to evaluate the necessary integrals. Jones et al (1979) show how, for the case of the anisotropic function (31), the necessary integrals can be simplified without any restriction on the shape of the area A.

In the case of Universal Kriging, the estimator of the average rainfall over the area A is (assuming error free observations):

$$\hat{Y}_A = \sum_{i=1}^{p} \lambda_i X(\underline{x}_i) \qquad (67)$$

Proceeding as in the case of Kriging for point values, it may be shown that the λ_i are solutions of the system of equations

$$\sum_j \lambda_j \gamma(x_i - x_j) + \sum_\ell u_\ell f^\ell(\underline{x}_i) = \frac{1}{A} \int_A \gamma(x_i - \underline{x})d\underline{x} \quad (i = 1,2, \ldots p)$$

$$\qquad (68)$$

$$\sum_j \lambda_j f^\ell(\underline{x}_j) = \frac{1}{A} \int_A f^\ell(\underline{x})d\underline{x} \qquad (\ell = 0, 1, \ldots k)$$

As a constant function identically equal to unity is always chosen for $f^0(\underline{x})$, the first condition on the functions f^ℓ (for $\ell = 0$) is simply

$$\sum_j \lambda_j = 1.$$

The optimal estimation variance is given by

$$Var(\hat{Y}_A - Y_A) = \frac{1}{A} \sum_i \lambda_i \int_A \gamma(\underline{x}_i - \underline{x})d\underline{x} + \frac{1}{A} \sum_\ell u_\ell \int_A f^\ell(\underline{x}) \ d\underline{x}$$

$$- \frac{1}{A^2} \int_A \int_A \gamma(\underline{x} - \underline{x}')d\underline{x} \ d\underline{x}' \qquad (69)$$

Space-time Averaging

In a significant contribution to network design methodology, Rodriguez Iturbe and Mejia (1974) derived the variances of space-time averages in the context of trading-off spatial against temporal information. The problem posed is the estimation of the long-term areal average defined as

$$\mu_A = \lim_{T \to \infty} \frac{1}{AT'} \sum_{t=1}^{T'} \int_A z(\underline{x},t)d\underline{x} \tag{70}$$

by means of

$$\bar{P} = \frac{1}{NT} \sum_{i=1}^{N} \sum_{t=1}^{T} z(\underline{x}_i,t) \tag{71}$$

where $z(\underline{x}_i,t)$ is the difference between rainfall depth at the point x_i during year, month or season t and the mean of the process, N is the number of stations in the network and T is the number of years months or seasons that the network is in operation. Rodriguez-Iturbe and Mejia show that the variance of the regional mean is given by

$$\text{Var}(\bar{P}) = \sigma^2 \{F_1(T)\} \{F_2(N)\} \tag{72}$$

where σ^2 is the variance of the point process defined as

$$\sigma^2 = E\{z^2(\underline{x},t)\} \tag{73}$$

and $F_1(T)$ and $F_2(N)$ are two reduction factors, one of which is due to sampling in space, and the other $F_1(T)$ due to sampling in time. Since the spatial and temporal correlation functions were assumed to be separable as given by equation (9) $F_1(T)$ was calculated on the basis of a Markov correlation function with lag one correlation ρ_1, and $F_2(N)$ for both the modified Bessel and exponential correlation functions. $F_1(T)$ depends only on ρ and T, while $F_2(N)$ depends on the correlation structure in space, the geometry of the network and the number of stations. Two types of sampling scheme, random sampling and stratified random sampling were considered, and graphs were presented showing (i) $F_2(N)$ as a function of N, the number of gauges, a quantity characterizing the area and the parameter of the correlation function for each sampling scheme, and (ii) $F_1(T)$ as a function of the parameter of the Markov correlation function. Equation (71) thus provides a very useful tool for comparing the performance of different spatial and temporal sampling schemes for estimating the long-term mean μ_A.

Rodriguez-Iturbe and Mejia (1974) advocate that, in calculating $\bar{P}$ in practice, an optimal weighting scheme should be used which minimizes $\text{Var}(\bar{P})$ i.e.

$$\bar{P} = \sum_{i=1}^{N} \alpha_i \frac{1}{T_i} \sum_{t=T_{i,i}}^{T_{f,i}} z(\underline{x}_i,t) \tag{74}$$

where α_i are the weights, and $T_{i,i}$ and $T_{f,i}$ denote the initial and

31

final time points for the record at gauge i. The weights α_i are derived using a Lagrange multiplier technique; however, Rodriguez Iturbe and Mejia used simple arithmetic averaging in space in deriving the expression for $F_2(N)$. If (74) is to be used in practice, then it would appear that the weights should be derived first and then that $F_2(N)$ should be derived with these weights incorporated.

Bras and Colon (1978) have also analysed the space-time averaging problem. They pointed out that the work of Mejia and Rodriguez-Iturbe (1974), which is based on zero-mean rainfall quantities, implies knowledge of the mean a priori. The approach of Bras and Colon, which employed state estimation procedures, is more general and explicitly acknowledges that the mean is unknown, while also allowing for a non-random distribution of rainfall stations and measurement error.

APPROACHES TO NETWORK DESIGN

General

The procedures described in the previous section for deriving the accuracies of point and areal rainfall estimates provide the basic elements around which any general procedure for designing a network can be structured. A number of such procedures have been described in the literature; these can be divided into two categories:

(i) approaches where the desired accuracy in estimating point and areal rainfall is sought without explicitly considering the costs associated with achieving that accuracy;

(ii) approaches where costs are explicitly considered.

As noted in the introduction, both the costs and benefits associated with achieving different levels of accuracy should be considered within any comprehensive approach to network design; however, lacking information on the economic worth of rainfall data, hydrologists have tended to reduce the problem to one where the costs of achieving the desired accuracy of estimation are minimized. Some examples of both approaches are given in the following sub-sections.

Rather than focussing directly on the problem of designing a network to provide some stated accuracy of estimation in point or areal rainfall, the problem can be tackled indirectly by attempting to quantify the accuracy with which streamflow can be estimated as a function of network density and configuration; this approach is also considered here.

Approaches Based on Accuracy Criteria

Most of the approaches to network design reported in the literature have used some accuracy criterion as a basis for designing a network. Early contributions were based on the notion that the spatial correlation between gauges should be not less than some aribtrarily chosen level; for example, Hershfield (1965) suggested that the level should be 0.9, and derived average gauge spacings on this basis. Hendrick and Comer (1970) followed a similar procedure, but attempted to take account of anisotropy by centring gauges within ellipses corresponding to the 0.9 correlation contour.

Cislerova and Hutchinson (1972) used optimal point interpolation error for pairs of gauges (Gandin, 1965) as a basis for the redesign

of the raingauge network of Zambia aimed at bringing the density up
to the WMO recommended standard of one gauge per 900 km^2. Maps of
interpolation error for annual rainfall, expressed in absolute terms
and as a percentage of mean rainfall, were used as a basis for
identifying areas of deficient accuracy through reference to predetermined accuracy criteria, taken as 10 and 15 per cent of mean
annual rainfall. It was suggested that a criterion of 10% be adopted
for design purposes in areas where human, agricultural and water
resources activity was likely to be concentrated, ranging to 15% in
areas of limited activity. Actual locations of new gauges were not
specified since implementation of the proposed redesign would depend
on a large number of voluntary observers offering their services.

Delhomme and Delfiner (1973) used Universal Kriging to inter-
polate rainfall on a regular grid for a large storm over an arid
region in Chad. In order to identify potential sites for new gauges
to reinforce the network, they calculated the gain in accuracy in
the estimation of mean rainfall during a storm resulting from siting
a new fictitious gauge at point M within the basin; this gain
in accuracy was defined as

$$G_M = \frac{\sigma_0^2 - \sigma_M^2}{\sigma_0^2} \tag{75}$$

where σ_M^2 and σ_0^2 are the estimation variances with and without the
new gauge included calculated using (63). The quantity G_M was
contoured over the basin and the maximum gain in accuracy from siting
a new gauge was found to be 13% as opposed to 3% given by an empirical
analysis; no a priori specification of required accuracy was given.

Morin et al (1979) advocated the use of principal component
analysis in conjunction with optimal interpolation as an approach to
raingauge network design. The former technique was applied to ten-
day rainfall totals from 30 stations sited within the Eaton River
basin in Quebec (area 642 km^2) and it was found that the stations
could be divided into three groups, the composition and geographic
distribution of which varied from season to season. However, since
insufficient data were available to adequately define a structure
function (or variogram) for each group, optimal interpolation was
applied to all 30 stations and to a reduced network of 5 stations,
and it was shown that the accuracy of point interpolation varied
more from season to season for the same network than from the
30 station to the 5 station network for the same season. Errors
of interpolation deriving from microclimatic irregularities and
observational errors were found to be greater than those resulting
from the reduction of the network from 30 to 5 stations.

Crawford (1979) describes an experimental design model which
has been developed to evaluate trade-offs involved in the optimal
sampling of rainfall. The model reveals that the deployment of
statistically adequate 'sensor arrays' depends critically on the
storm physics, the sensor engineering aspects and the procedure
chosen for analysis of information from the deployed network.
Deployment along and across a preferred storm track is related to
convective system anisotropy, raingauge density, temporal sampling
intervals, availability of radar data, and interrelationships among
the multivariate prediction data sets. Any proposed sensor array
is shown to represent a balance between the demands of economy and
climatology.

O'Connell et al (1978, 1979) employed optimal estimation procedures in the redesign of a raingauge network for an area of about 10,000 km^2 in the South of England. Their overall approach involved the following steps:

(i)   evaluation of the accuracy requirements of users;

(ii)   evaluation of the adequacy of individual gauges to provide reliable measurements;

(iii) calculation of the accuracies of point and areal rainfall estimates for existing and proposed networks, and comparison with the requirements of users.

The accuracy requirements of users were established by interview and were expressed in terms of mse for point and areal rainfall estimates (specific areas and their extent were designated for the latter) over various durations. By fitting the correlation function on a 5 km x 5 km grid square basis, the optimal point interpolation error was mapped on a 1 km grid within each square for various daily and monthly data; the resulting maps could then be compared with the accuracy requirements of users.

A procedure was evolved for building up a design network through a series of steps which took account of the quality of the data provided by individual gauges, the percentage of area over which the accuracy criterion for point interpolation was met, and any constraints associated with the maintenance or deletion of existing gauges in any new network. Three alternative design networks of 75,220 and 297 gauges were produced to illustrate the extent to which networks of varying density could meet the accuracy requirements of users. It was found that the design network of 220 gauges could provide the required point interpolation accuracy over approximately the same percentage of the area (roughly 90%) as the existing network of 333 gauges; Figure 1 shows a map of the optimal point interpolation error for the former network. Figure 2 is a similar map for a network of 133 gauges with an approximately uniform spacing of 10 km, illustrating that the interpolation error is greater towards the west of the region where the topography varies most, and that the notion of a uniform density in rainfall network design is not generally tenable.

To illustrate the merit of the design procedure, the percentage of area over which the desired accuracy was achieved was plotted as a function of the numbers of gauges in the design networks, and also as a function of the numbers of gauges in networks derived by deleting gauges arbitrarily from the existing network. Figure 3 shows that the design networks give significantly better results; the design network of 220 gauges has since been implemented, thus achieving a significant reduction in the cost of network operation and data processing.

Approaches Involving Both Cost and Accuracy

Bras and Rodriguez Iturbe (1976b)and Lenton and Rodriguez Iturbe (1977) present two different approaches to network design for the estimation of the areal mean of rainfall events (spatial averaging) in which both cost and accuracy are explicitly considered. Both approaches use the same objective function as a starting point which may be written as

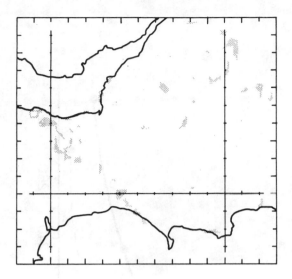

Figure 1:   Regions (shaded) for which the root mean square error
            of point interpolation is greater than 1.5 mm for
            days with widespread rainfall of over 1 mm, for
            design network of 220 gauges.

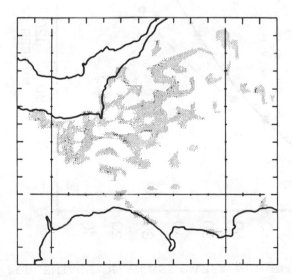

Figure 2:   Regions (shaded) for which the root mean square error
            of point interpolation is greater than 1.5 mm for days
            with widespread rainfall of over 1 mm, for a network
            of 133 gauges with an approximately uniform spacing of
            10 km.

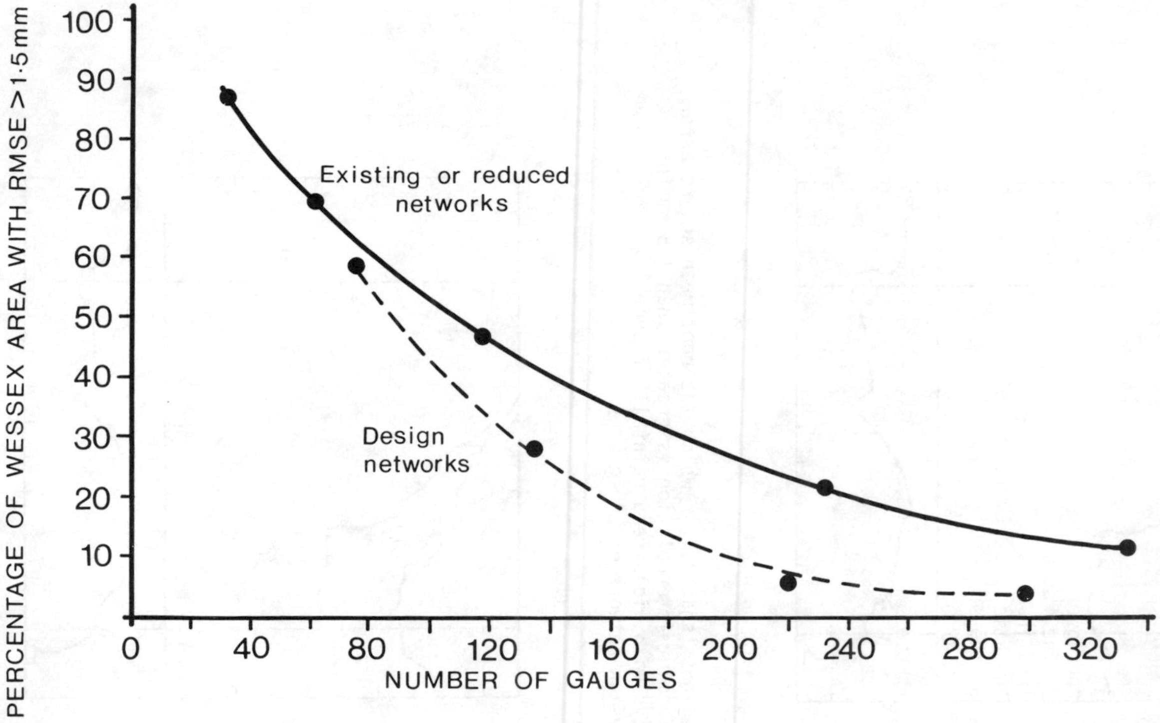

Figure 3: Percentage of area have a root mean square error of point interpolation greater than 1.5 mm for days with widespread rainfall of over 1 mm, for (a) existing or arbitrarily reduced networks and (b) networks redesigned using optimal estimation procedures.

$$U(\underline{x},\underline{b}, n) = E(\underline{x}, \underline{b}, n) + fC(\underline{x},n) \qquad (76)$$

where $\underline{x}$ denotes the locations of the gauges, $\underline{b}$ denotes the set of weights, n the number of gauges, $E(\underline{x}, \underline{b}, n)$ denotes the mean square error of the areal average, f is a factor equivalencing cost and accuracy, and $C(\underline{x},n)$ the annual operating and data processing costs of the set of gauges into which capital costs have been absorbed by distributing them among amortization items.

Bras and Rodriguez (1976b) define the term $E(\underline{x}, \underline{b}, n)$ as a sum of three terms. The first term represents model error and represents the error involved in approximating the continuous integral in (64) by a discrete summation of a set of weights applied to rainfall at points in space; the second term derives from estimating the discrete summation with noisy measurements, and a third term accounts for dependence between these two types of error. Bras and Rodriguez Iturbe then represent the data collection network as a measurement equation

$$\underline{Z} = \underline{H}\underline{X} + \underline{V} \qquad (77)$$

where

$\underline{Z}$ is an (m x 1) vector of noisy observations at each discrete point in space (m $\leqslant$ n);

$\underline{X}$ is an (n x 1) vector of true values of rainfall $X(\underline{x}_i)$ at the n discrete grid points;

$\underline{V}$ is an (m x 1) vector of white noise representing measurement error;

$\underline{H}$ is an (m x n) matrix defining the data collection network,

with non zero elements h_{ij} if $\underline{x}_j$ is the location of an observation, and zero otherwise. The mean square error of estimation of $\underline{X}$ is then a well known result in estimation theory and is independent of the observations $\underline{Z}$; its component terms (defined above) are evaluated by Bras and Rodriguez Iturbe (1976b). This then defines the term $E(\underline{x},\underline{b},n)$ as a function of an a priori selected grid size and the number of gauges; optimization of (76) was carried out using a search procedure suitable for the problem for values of f defined a priori. A hypothetical case study was carried out to illustrate the technique for an area with gauges sited on a rectangular grid. The approach does not provide an optimal solution but merely illustrates how cost and accuracy can be traded off against each other, and what the minimum cost solution (with the corresponding number and locations of gauges) is for a given value of f.

Lenton and Rodriguez-Iturbe (1977) formulated the minimization of (76) as a mathematical programming problem with the decision variables being the total number of raingauges, the coordinates of each of them and the estimator weights.

Alternative formulations of the mathematical programming problem considered by Lenton and Rodriguez-Iturbe are

$$\min E(\underline{x}, \underline{b}, n) \qquad (78)$$

subject to

$$C(\underline{x},n) = B \qquad\qquad (79)$$

where B is the annual budget established for network operation, or

$$\min C(\underline{x}, n) \qquad\qquad (80)$$

subject to

$$E(\underline{x}, \underline{b}, n) \leqslant E_{max} \qquad\qquad (81)$$

where E_{max} is the maximum acceptable mse and the $\underline{b}$ vector must be fixed a priori. This is considered to be the data collection stage of network design; the second or data analysis stage is taken to be the derivation of the optimal set of weights $\underline{b}^*$ as described in a previous section. Lenton and Rodriguez-Iturbe describe a case study in which they determine the least cost network whose accuracy was as good as the existing network.

Approaches Based on Rainfall-Runoff Modelling

The network design problem posed here is to determine the density and configuration of a network of gauges such that the required accuracy in estimating streamflow through a rainfall-runoff model is obtained. The approaches which have been described in the literature fall into two categories:

(i) experiments in which rainfall in space and time is generated and fed through an analytical rainfall-runoff model; the rainfall field is then sampled at discrete points in space and the effect of gauge location and density is studied;

(ii) experiments in which observed rainfall and runoff data are used in conjunction with a rainfall-runoff model.

The advantage of (i) is that the true rainfall input in time and space is known, and the effect of any number of sampling points can, in principle, be studied; with (ii) the true rainfall input is not known, only its sampled values at a limited number of points in space, and the effects of introducing new gauges cannot be studied.

Bras (1978) has reviewed a number of selected papers which fall mainly in category (i). Early work by Eagleson (1967) in this area involved describing the rainfall field with a radially symmetric characteristic storm structure described in the case of a convective storm as

$$\frac{X_T(d)}{X_T(0)} = 1 - 0.72 \left| \frac{d}{d_0} \right| \qquad\qquad (82)$$

where $X_T(0)$ denotes the depth of the storm of duration T at the origin, $X_T(d)$ is the depth at distance d, and d_0 is defined as the distance corresponding to a correlation radius of 0.5. The temporal properties of rainfall storms of the form (82) were described using the autocorrelation function; Eagleson then used deterministic linear systems theory in the frequency domain to derive the frequency spectrum of streamflow response. The effect of sampling rainfall in space at different finite densities was then determined by truncating the spectrum at different wave numbers and calculating the error in peak discharge prediction. Further applications of this approach are described by Eagleson and Goodspeed (1973).

Grayman and Eagleson (1971) used a stochastic model based on the meso-scale and synoptic levels of storm structure to generate rainfall in time and space; the rainfall was then sampled using simulated radar and raingauge combinations and routed through a model based on a spatially distributed solution of the kinematic wave equations. Sampling requirements in space and time were studied in terms of the number of gauges required for calibrating the radar, averaging area for the radar signal and sampling interval for accurate peak flow prediction; confidence limits for such predictions were also quoted for particular storm types.

Bras and Rodriguez-Iturbe (1976c) used a non-stationary multi-dimensional stochastic model of rainfall to generate the input to a deterministic rainfall-runoff model; noisy 'measurements' of the rainfall were then sampled at a number of points, and the Kalman filter was used to derive minimum mean square error estimates of rainfall intensity. This uncertainty was then propogated within a state space model of the rainfall-runoff process to derive the mean square error of estimated discharge.

The multi-dimensional rainfall model was written in the form

$$\underline{i}(t) = \underline{A}(t-1)\ \underline{i}(t-1) + \underline{B}(t-1)\ \underline{w}(t-1) \qquad (83)$$

where

$\underline{i}(t)$ = (Nx1) vector of zero mean rainfall intensities

$\underline{A}(t)$ - (NxN) matrix of coefficients

$\underline{B}(t)$ = (NxN) matrix of coefficients

$\underline{w}(t)$ = (Nx1) vector of white noise, N being the number of discrete points describing the area of interest.

The underlying correlation structure of (83) was based on the Taylor hypothesis (discussed previously) which allows the covariance function of rainfall to be written in the form

$$\sigma(\underline{x}_i, t';\ \underline{x}_j, t'') = \sigma(\underline{x}_i, t')\ \sigma(\underline{x}_j, t'') \qquad (84)$$

$$r[||\underline{x}_j + \underline{U}t'' - (\underline{x}_i + \underline{U}t')||]$$

where $\underline{x} = (x,y)$, $||\underline{x}|| = (x^2 + y^2)^{0.5}$, $\underline{x}_i$ and $\underline{x}_j$ are vectors of coordinates for points i and j, $\underline{U}$ is the process velocity vector, $\sigma(\underline{x}_i, t)$ is the point variance at coordinate $\underline{x}_i$ and time t and r(.) is the correlation function. For a storm with velocity U in the direction of the x axis, (84) becomes

$$\sigma(\underline{x}_i, t';\ \underline{x}_j, t'') = \sigma(\underline{x}_i, t')\ \sigma(\underline{x}_j, t'')\ r[\{(y_j - y_i)^2 + [(x_j + U\ t'')$$

$$- (x_i + Ut')]^2\}^{\frac{1}{2}}] \qquad (85)$$

Once the form of r(.) is specified, and the $\sigma(\underline{x}_i, t')$ time evaluated, the matrices $\underline{A}(t)$ and $\underline{B}(t)\ \underline{B}(t)^T$ can be solved for at each time point from the space time covariance function (85), since Bras and Rodriguez Iturbe point out that, for the network design problem, it is not necessary to solve for $\underline{B}(t)$. The rainfall is then sampled through an observation equation of the form of (77), and the Kalman filter applied to derive the minimum mean square error estimate $\hat{\underline{i}}(t|t)$ together with its error covariance matrix $\sum(t|t)$.

The rainfall-runoff model used was the kinematic wave equation applied to a schematized basin consisting of 3 overland flow and 2 channel flow segments; the equation was solved on a finite difference grid and written in state-space form as

$$\underline{x}(t) = \underline{\zeta}[\underline{x}(t-1)] + \underline{\beta}[\underline{i}(t-1)] \qquad (86)$$

where $\underline{x}(t)$ is a state vector of depths and cross sectional areas of flow and $\underline{\beta}$ is a matrix with the time step of the solution in its non zero elements. By linearizing (86) through a Taylor expansion about the mean solution, defined as the response to a mean rainfall event, a linearized state equation

$$\underline{D}(t) = \underline{\zeta}'[\underline{x}_\mu(t-1)] \, \underline{D}(t-1) + \underline{\beta}[\underline{i}(t-1)] \qquad (87)$$

results in which

$$\underline{D}(t) = \underline{x}(t) - \underline{x}_\mu(t)$$

$\underline{x}_\mu(t)$ = mean of state vector $\underline{x}(t)$

$\underline{\zeta}'[\underline{x}_\mu(t-1)]$ = first derivative of functional matrix ζ evaluated at the mean solution

$\underline{i}(t)$ = zero-mean true rainfall intensity

The mean square error matrix associated with the estimation of $\underline{D}(t)$ is then derived as

$$mse\{\underline{D}(t)\} = \underline{\zeta}'[\underline{x}_\mu(t-1)] \, [mse \, \underline{D}(t-1)] \, \underline{\zeta}'^T[\underline{x}_\mu(t-1)]$$
$$+ \underline{\beta} \sum(t-1|t-1)\underline{\beta}^T$$

where $\sum(t-1|t-1)$ is derived for the noisy rainfall observations as outlined above. By linearizing the non-linear relationship between discharge and depth, or cross-section, the mean square error of estimated discharge is obtained.

Bras and Rodriguez-Iturbe (1976c) applied the approach to a hypothetical basin (Area 82 km^2), and concluded that raingauge location was important in determining the accuracy with which discharge could be simulated. It was observed that sampling in the upper catchment areas resulted in the deterioration of the simulation of the hydrograph rising limb compared to sampling the lower areas. Simulation accuracy was also found to be sensitive to the number of stations.

O'Connell et al (1977, 1978) carried out some experiments in category (ii) to assess the effect of network density and configuration on discharge simulation and forecasting. Two sets of experiments were conducted: the first involved calibrating the CLS model with thresholds (Todini and Wallis, 1977) on a daily basis for a number of catchments in the UK ranging in area from 75 km^2 to about 500 km^2. The CLS model was first calibrated using all the available gauges to define the average daily rainfall input for each catchment; then, by fixing the model parameters, and successively reducing the number of gauges used to define the average rainfall input, the effect on the accuracy of discharge simulation was studied. It was found that, on some catchments, this accuracy was limited by network density, but not on others; a sufficiently larger number of catchments was not analysed to allow more general conclusions to be drawn.

In a second set of experiments, the effect of network density and configuration on real-time flow forecasting for a sub-catchment on the River Dee in North Wales, the Hirnant (area 33.9 km^2) was analysed. The model employed assumed that streamflow could be written as the sum of a deterministic component q_t and a stochastic component n_t such that

$$q_t = q_t + n_t$$

where the deterministic component (or process model) is the linear transfer function

$$q_t = \delta_1 q_{t-1} + \delta_2 q_{t-1} + \cdots + \delta_r q_{t-r} + \omega_0 p_{t-b} + \omega_1 p_{t-b-1}$$

$$+ \cdots \omega_{s-1} p_{t-b-s-1}$$

where h is a pure time delay, r is the number of autoregressive terms and s is the number of moving average terms. The above model can be readily extended to accommodate multiple inputs (O'Connell et al., 1977); once values for r, s and b have been identified using cross-correlation analysis, the parameters of such models can be estimated using the instrumental variable (IV) algorithm (Young et al, 1971) which takes account of autocorrelation in the n_t series and provides consistent estimates of the model parameters.

Half-hourly rainfall data from six telemetring gauges in and around the Hirnant catchment were employed in the study together with half-hourly discharge values. Three separate categories of model were considered:

(i)   where the rainfall input was defined by each of the six gauges separately (single gauge models);

(ii)   where the rainfall input was defined by averaging a subset of gauges (lumped model);

(iii) where the rainfall input was defined by treating the rainfall at each gauge as a separate input (spatially distributed models).

By comparing explained variances for the single gauge models, it was found that the most important gauge was sited in the lower areas of the catchment contributing to the rising limb and peak of the hydrograph which agrees with the conclusions of Bras and Rodriguez-Iturbe (1976c). Among the lumped models, an input based on two gauges gave the best results, which suggests that for this input, a stronger causal link may exist with runoff then an input defined using all gauges, some of which may be unrepresentative. Comparison of the results from the lumped models with those from the distributed models suggested that little was gained through having spatially distributed inputs.

Jettmar et al (1979) carried out some experiments with a rainfall-runoff model (also falling in category (ii)) to determine the value to

41

river flow forecasting of possible changes in existing rainfall and streamflow networks operated by the National Weather Service. A surrogate measure of benefits, called the 'mean forecast lead time' (MFLT), was used as a basis for assessing the effects of network changes since the value of a river flow forecast depends on the lead time available for the flood plain dweller to respond to the forecast.

Historical rainfall records were used to produce mean areal rainfall data which were assumed error free; these data were then fed through a rainfall-runoff model to produce base hydrographs that were assumed error free at various points in a river basin. Noise was then added to the mean areal rainfall data (reflecting the number of gauges operational at any time) and replicated sequences of these rainfall data were fed through the rainfall-runoff model. Each runoff forecast was then compared with the base hydrographs to analyse the MFLT's and their variances at the points of interest. The benefits to river flow forecasting from increasing the density of gauges were then measured in terms of the reduction in variance of MFLT. Some results were derived for rainfall and flow data generated by two hurricanes on the Susquehana River Basin.

The approach could be extended to study the impact of quantitative precipitation forecast (QPF) on MFLT as well as the effect of model error (Jettmar et al., 1979).

DISCUSSION AND FUTURE PERSPECTIVES

Over the past ten years, a considerable amount of effort has been devoted to the development of methodology for raingauge network design; here the progress which has been made is discussed, some gaps in existing methodology are identified, and the potential of the methodology for real world application is assessed.

Firstly, it was noted that, as a prerequisite to network design, a statistical description of the rainfall process in time and space is required. Spatial correlation functions and variograms (or structure functions) are the two main alternatives for describing the spatial structure of rainfall, with the use of correlation functions predominating in the literature. In choosing a form for the spatial correlation function, it is possible to justify the choice either on the basis of the differential equation describing the underlying physical process or through data analysis. Rodriguez-Iturbe and Mejia (1974) argue in favour of a modified Bessel function over an exponential function on the basis that the governing differential equation for the latter is not physically justifiable. While time does not explicitly enter into the choice of a spatial correlation function for rainfall, it does implicitly, since the spatial correlation functions of rainfall for different durations exhibit different decay rates at least, and might not be expected to obey the same functional form. If it is assumed that the Bessel function describes the spatial structure of instantaneous rainfall, then because of aggregation. the same functional form would not be expected to apply at the annual level. Furthermore,aggregation could conceivably result in a spatial process with a governing differential equation of the form of (28), since a number of different rainfall types would have been aggregated in deriving an annual total. The conditions under which the Bessel function might apply need to be clarified, since it has been used in much of the recent theoretical work on network design.

In the absence of physical grounds, the choice of a correlation function can in principle proceed through data analysis: as in the case of fitting any statistical model, a process of model identification, parameter estimation and calculation of measures of goodness of fit is indicated. However, with a few exceptions most of the studies discussed above have adopted a very empirical approach in which the form of correlation function is assumed, perhaps with the aid of some graphical data analysis, and the function then fitted without employing any formal parameter estimation procedure. One approach which has been used is to fit the correlation function at some arbitrarily chosen 'characteristic distance'; however, as noted in a previous section, two different definitions of this measure can be found in the literature, one of which is defined as the average distance between two randomly chosen points in the region of interest, and the other which is defined as the distance corresponding to some specified value of spatial correlation, usually 0.5. The arbitrariness of these criteria, and the fact that the function is fitted at one point only may mean that the function as a whole does not fit particularly well. Little has also been done on checking the goodness of fit of spatial correlation functions, for example, by checking the randomness of deviations from the fitted correlation function.

While the theory of probabilistic models for the design of raingauge networks has been developed to some extent, virtually nothing has been done to develop the necessary techniques of statistical inference for model fitting, and work in this area is definitely required. David (1978), in reviewing progress in spatial sampling and estimation in the mineral industry, presented a very useful review of work on the estimation and parameterization of the variogram. He notes that, in general, there is a lack of work on parameter estimation for spatial models and that, for example, no tests are available whereby the goodness of fit of two alternative models may be compared. While the quantities to be estimated may be robust to errors deriving from model choice and parameter estimation, little has been done to show that this might be the case. One overall way in which this might be done is to perform point interpolations at points for which measurements are available, and then compare the theoretical estimate of mean square error with the actual estimate derived from the available measurements. This approach was tested on a limited scale by O'Connell et al (1978) who found that the distribution of observed interpolation errors agreed reasonably well with that predicted by the theory. This approach has also been suggested by Switzer (1979); however, it cannot be applied in the case of areal averaging.

Although there is scope for further research to improve and extend the methodology for raingauge network design (indeed many topics could be enumerated because of their intrinsic research interest), considerable advances in this research area have been made in recent years and it is pertinent to observe if any of this methodology is finding practical application. Langbein (1979), in his overview of the Chapman Conference on Hydrologic Data Networks, summarized the situation then admirably when he noted that "the products of research and critical inquiry appear to find few applications" and that "unless greater use is found, network design may have little purpose other than to furnish learned journals with articles by researchers to be read by their colleagues". Not much progress on this situation appears to have been made in the interim, and there is clearly a need for some effort to bridge the gap between

43

research and practice. Langbein (1979) suggested that, as a way forward, existing networks be audited to see how they fulfil their objectives as stated at their inception, and their objectives as now perceived. This approach has much to recommend it but its implementation will not always be clear cut. Raingauge networks have frequently not been installed to meet formal objectives; in the UK, at least, they have tended to evolve historically in an ad-hoc manner in response to local needs, according to whether or not suitable observers could be found at the locations in question. The resulting historical records have then found multiple uses as and when required in the fields of agriculture, water resources, rainfall forecasting, climatological research etc. (O'Connel et al., 1977) to name but a few; any attempt at redesigning such networks is faced with the problem of how to quantify the economic benefits deriving from the use of rainfall data, and of how the costs of acquiring and processing rainfall data can be distributed equitably among users. Lacking information on benefits, one way forward, adopted in a recent UK case study, (O'Connell et al., 1978, 1979) is to try to extract from users statements of accuracy (in terms of mean square error) for the estimates of rainfall data which they require. But this is also fraught with difficulty since many users cannot provide reliable statements of the accuracies they require. One way around this problem is to use the relevant point and areal estimation procedures to establish the accuracies currently being obtained from an existing network which can then serve as a reference level for any redesigned network.

While it has been found possible to use optimal estimation procedures in a recent case study in the UK where an existing network was redesigned to eliminate redundancy and to try to meet the requirements of users in a more cost effective manner (O'Connell et al., 1978, 1979), the overall design procedure which was employed involved a number of empirical steps. These were dictated largely by considerations of the quality of the data provided by existing gauges (gauges providing good quality data to be retained as far as possible) the roles of various organizations in operating the network and the necessity of finding observers for new raingauge sites. These considerations may be particular to the UK, but it is likely that attempts at redesigning raingauge networks elsewhere will encounter similar problems.

A further case study in the Northwest of England is currently being carried out jointly by the Meteorological Office and the Institute of Hydrology, where the techniques used previously (O'Connell et al. 1978, 1979) are being refined and extended. Research has been carried out on how the variance and correlation structure of rainfall can be modelled in an area of highly variable topography, and on how the overall network design procedure can be made more objective (Nicholass et al., 1982). Future work will investigate how rainfall data from a conventional network and radar data can be used jointly in the design of raingauge networks.

Finally, it is perhaps salutary to remind those who are involved in research in the field of network design that, in those parts of the world where little or no data exist, raingauge networks have to be designed on the basis of the recommended gauge spacings in the WMO Guide to Hydrometeorological Practices (1974) which represent largely the best judgements of experienced hydrometeorologists and hydrologists. It is a well recognised paradox that quantitative network design procedures cannot be employed without an existing data

base: nonetheless, there must now be a number of sufficiently dense
networks in existence in different climatic regimes of the world
which could form the basis of worthwhile case studies; such case
studies would allow promising techniques such as trading off
information in space and time to be investigated at the practical
level. Out of such case studies would hopefully come some messages
about the relevance of recent research to the practical design
of raingauge networks, and recommendations about the design of
networks in various climatic regimes which would represent an
advance on the present state of the art as reflected in the WMO
recommendations.

REFERENCES

Austin, P.M. and Houze, R.A. 1972. Analysis of the structure of
    precipitation patterns in New England. Journal of Applied
    Meteorology, Vol. 11, pp 926-935

Bartlett, M.S. 1975. The Statistical Analysis of Two Dimensional
    Point processes. Biometrika, 51, pp 299-311

Boyd, D.A. 1939. Correlations between monthly rainfall at eleven
    stations in the British Isles. Mem. Royal. Meteorological
    Society [Nos. 31-40 (1931-1939)], Vol. 4, pp 143-160.

Brady, P.J. 1975. Matching raingauge placement to precipitation
    patterns. Proceedings of the National Symposium on Precipita-
    tion Analysis for Hydrologic Modeling, Davis, California,
    American Geophysical Union, pp 111-122.

Bras, R.L. 1979. Sampling of interrelated random fields. The
    rainfall runoff case. Water Resources Research, Vol. 15,
    No.6, pp 1767-1780.

Bras, R.L. and Colon, R. 1978. Time-averaged areal mean of precipi-
    tation: estimation and network design. Water Resources Research,
    Vol. 14, No.5, pp 878-888.

Bras, R.L., and Rodriguez-Iturbe 1976a. Evaluation of mean square
    error involved in approximating the areal average of a rainfall
    event by a discrete summation, Water Resources Research, Vol. 12,
    No.2, pp 181-184.

Bras, R.L. and Rodriguez-Iturbe, I. 1976b. Network design for the
    estimation of areal mean of rainfall events. Water Resources
    Research, Vol. 12, No. 6, pp 1185-1196.

Bras, R.L. and Rodriguez-Iturbe, I. 1976c. Rainfall network design
    for runoff prediction. Water Resources Research, Vol. 12,
    No.6, pp 1197-1208.

Buishand, T.A. 1977. De variantie van de gebiedsneerslag als functie
    van puntneerslagen en hun onderlinge samenhang. Meded.
    Landbouwhogesch. Wageningen, 77-10, 12 pp.

Caffey, J.E. 1965. Inter-station correlations in annual precipita-
    tion and in annual effective precipitation. Hydrology Papers,
    No.6, 47 pp Colorado State University Fort Collins, Colorado.

Chang, M. 1977. An evaluation of precipitation gauge density in a mountainous terrain, Water Resources Bulletin, Vol. 13, No.1, pp 39-46.

Chemerenko, Ye. P. 1975. Spatial averaging of data on the water equivalent of snow and problems in the rationalisation of the observation network. Soviet Hydrology, Selected Papers, No.2, pp 55-59.

Cislerova, M. and Hutchinson, P. 1974. The redesign of the raingauge network of Zambia. Hydrological Sciences Bulletin, Vol. XIX, No.4, pp 423-434.

Cornish, E.A., Hill, G.W. and Evans, M.J. 1961. Interstation correlations of rainfall in Southern Australia. CSIRO, Canberra, A.C.T., Division of Mathematics, Statistics and Technology Papers, No. 10, 16 pp.

Crawford, K.C. 1979. Considerations for the design of a hydrologic data network using multi variate sensors. Water Resources Research, Vol. 15, No.6, pp 1752-1762.

David, M. 1978. Sampling and estimation problems for three dimensional spatial stationary and non stationary stochastic processes as encountered in the mineral industry, Journal of Statistical Planning and Inference, Vol. 2, pp 211-244.

De Bruin, H.A.R. 1975. Over het interpoleren van de neerslaghoogte. K. Ned. Meteorol. Inst., De Bilt, Wetensch. Rap. WR 75-2, 34 pp.

Delhomme, J.P. and Delfiner, P. 1973. Application du krigeage à l'optimisation d'une campagne pluviométrique en zone aride. In: Proceedings of the Symposium on the Design of Water Resources Projects with Inadequate Data, Vol. 2, UNESCO, Madrid, pp 191-210.

Eagleson, P.S. 1967. Optimum density of rainfall networks. . Water Resource Research, Vol. 3, pp 1021-1033.

Eagleson, P.S. and Goodspeed, M.J. 1973. Linear systems techniques applied to hydrologic data analysis and instrument evaluation. Technical Report 34, Commonwealth Science and Industry Research Organisation - Division of Land Use Research. Melbourne, Australia.

Eddy, Amos 1967. The statistical objective analysis of scalar data fields, Journal of Applied Meteorology, Vol. 6, No.4, pp 567-609.

Eddy, Amos 1973. The objective analysis of atmospheric structure. Journal of the Meteorological Society of Japan, Vol. 51, No.6, pp 450-457.

Fisher, R.A. and MacKenzie, W.A. 1922. The correlation of weekly rainfall. Quarterly Journal of the Royal Meteorological Society, Vol. 48, pp 234-245.

Gandin, L.S. 1965. Objective Analysis of Meteorological Fields, Israel Program for Scientific Translations, Jerusalem, pp 242.

Gottschalk, L. 1978. Spatial correlation of hydrologic and physiographic elements. Nordic Hydrology, Vol. 9, No.5, pp 267-276.

Grayman, W.M. and Eagleson, P.S. 1971. Evaluation of radar and raingauge systems for flood forecasting. M.I.T. Dept of Civil Engineering, Report 138, 427 pp.

Guscina, M.V., Kagan, R.L. and Polishchuk, A.I. 1967. Accuracy in determining the mean precipitation depth over an area. Soviet Hydrology selected Paper No. 6, pp 585-596.

Hendrick, R.I. and Comer, G.H. 1970. Space variations of precipitation and implications for raingauge network design. Journal of Hydrology, Vol.10, pp 151-163.

Hershfield, D.M. 1965. On the spacing of raingauges. Symposium on Design of Hydrological Networks, I, Quebec, W.M.O. - I.A.S.H. Publication No.67, pp 72-81.

Hershfield, D.M. 1968. Rainfall input for hydrological models. Symposium on Geochemistry, Precipitation, Evaporation, Soil Moisture, Hydrometry. Bern, 1967. International Association de Hydrologie Scientifique Pub. No. 78, pp 177-188.

Holland, D.J. 1967. The Cardington rainfall experiment. Meteorological Magazine, Vol 96, pp 193-202.

Huff, F.A. and Shipp, W.L. 1969. Spatial correlations of storm, monthly and seasonal precipitation. Journal of Applied Meteorology. Vol. 8, pp 542-550.

Hutchinson, P. 1969. Estimation of rainfall in sparsely gauged areas. Bulletin International Association of Scientific Hydrology, Vol. 14, No. 1, pp 101-119.

Hutchinson, P. 1970. A contribution to the problem of spacing raingauges in rugged terrain. Journal of Hydrology, Vol. 12, pp 1-14.

Jettmar, R.V. and Young, G.K. 1979. Design of operational precipitation and streamflow networks for river forecasting. Water Resources Research, Vol. 15, No. 6, pp 1823-1832.

Jones, D.A., Gurney, R.J. and O'Connell, P.E. 1979. Network design Using optimal estimation procedures. Water Resources Research, Vol. 15, No. 6, pp 1801-1812.

Kagan, R.L. 1966. On the evaluation of the representiveness of raingauge data Gl. Geofiz. Obs. 191, pp 22-34.

Kagan, R.L. 1972. Precipitation: statistical principles in: Casebook on Hydrological Network Design Practice, Ch. 1, W.M.O. Geneva No. 324, pp 1-11.

Kolmogovov, A.N. 1941. Interpolated and Extrapolated Stationary Random Sequences, Izvestya ANSSSR, Seriya Matematicheskaya, Vol. 5, No.1.

Langbein, W.B. 1979. Overview of conference on hydrologic data networks, Water Resources Research, Vol. 15, No.6, pp 1867-1871.

Lenton, R.L. and Rodriguez-Iturbe, I. 1977. Rainfall network systems analysis: the optimal estimation of total areal storm depth.

Water Resources Research, Vol. 13, No.5, pp 825-836.

Longley, R.W. 1974. Spatial variation of precipitation over the Canadian prairies. Monthly Weather Review, Vol. 102, pp 307-312.

Matheron, G. 1965. Les Variables Regionalisees et Leur Estimation, Masson, Paris, 306 pp.

Matheron, G. 1969. Le Krigeage Universal, les Cahiers du Centre de Morphologie Mathematique de Fontainebleau, No.1.

Matheron, G. 1973. The intrinsic random functions and their applications. Advances in Applied Probability, 5, No. 439-468.

Mejia, J.M. and Rodriguez-Iturbe, I. 1973. Multidimensional characterisation of the rainfall process Part I: Synthetic generation of hydrologic spatial processes, Part II: On the transformation of point rainfall to areal rainfall M.I.T., Dept of Civil Engineering, Report No. 177, 53 pp.

Morin, G., Fortin, J.P., Sochanska, W. and Lardeau, J.P. 1979. Use of principal component analysis to identify homogeneous precipitation stations for optimal interpolation. Water Resources Research, Vol. 15, No.6, pp 1841-1850.

Nicholass, C.A., Stewart, N.J., Jones, D.A., Hosking, J.M., Haylock, S., Walsh, P.D., O'Connell, P.E., Hall, B.A., Miller, J.B. 1982. Rationalization of the Northwest Water Authority Raingauge Network, in preparation.

O'Connell, P.E., Beran, M.A., Gurney, R.J., Jones, D.A. and Moore, R.J. 1977. Methods for evaluating the U.K. raingauge network. pp 262. Institute of Hydrology, Report No. 40, Wallingford, Oxfordshire.

O'Connell, P.E., Gurney, R.J., Jones, D.A., Miller, J.B. Nicholass, C.A. and Senior, M.R. 1978. Rationalisation of the Wessex Water Authority raingauge network. Institute of Hydrology Report No. 51, pp 179, Wallingford, Oxfordshire.

O'Connell, P.E., Gurney, R.J., Jones, D.A., Miller, J.B., Nicholass, C.A. and Senior, M.R. 1979. A case study of rationalization of a raingauge network in Southwest England, Water Resources Research, Vol. 15, No.6, pp 1813-1822.

Ord, K. and Rees, M. 1979. Spatial processes: Recent developments with applications to hydrology in: Lloyd, E.H., O'Donnell, T., Wilkinson, J.C. (editors), The Mathematics of Hydrology and Water resources. Academic Press, London, pp 95-118.

Richardson, C.W. 1977. A model of stochastic structure of daily precipitation over an area. Hydrology Paper No.91, 46 pp, Colorado State University, Fort Collins, Colorado.

Rodda, J.C., Langbein, W.B., Kovzel, A.G., Dawdy, D.R. and Szesztay, K. 1969. Hydrological network design-needs, problems and approaches. World Meteorological Organisation, Report No. 12, pp 57.

Rodriguez-Iturbe, I. and Mejia, J.M. 1974. The design of rainfall networks in time and space. Water Resources Research, Vol. 10, No. 4, pp 713-735.

Schaake, J.C., Jr. 1979. Accuracy of point and mean areal precipitation estimates from point precipitation data. National Oceanographic and Atmospheric Administration Technical Memorandum NWS-HYDRO. Washington D.C.

Sharon, D. 1974. On the modelling of correlation functions for rainfall studies. Journal of Hydrology, Vol. 22, pp 219-224.

Sharon, D. 1978. Rainfall yields in Israel and Jordan and the effect of cloud seeding on them. Journal of Applied Meteorology, Vol. 17, No.1, pp 40-48.

Sneva, F.A. and Calvin, L.D. 1978. An improved Thiessen grid for Eastern Oregon: an interstation correlation study determining the effect of distance, bearing and elevation between stations upon the precipitation correlation coefficient. Agricultural Meteorology, Vol. 19, No.6, pp 471-483.

Sneyers, R. 1968. De l'utilisation des séries brèves dans la description du climat d'une région - Une application. les conditions de l'enneigement du sol en Belgique. International Association of Scientific Hydrology. Publication No. 78 pp 274-281.

Stenhouse, N.S. and Cornish, E.A. 1958. Inter-station correlations of monthly rainfall in South Australia. CSIRO, Canberra, A.C.T. Division of Mathematics, Statistics and Technology Papers, No 5, pp 22.

Stol, Ph. Th. 1972. The relative efficiency of the density of rain-gauge networks. Journal of Hydrology, Vol. 15, pp 193-208.

Stol, Ph. Th. 1981. Rainfall interstation correlation functions, I. An analytic approach. Journal of Hydrology, Vol. 50, pp 45-71.

Switzer, P. 1979. Statistical considerations in network design, Water Resources Research, Vol. 15, No.6, pp 1712-1716.

Todini, E. and Wallis, J.R. 1977. Using CLS for daily and longer period rainfall-runoff modelling in: Ciriani, T.A., Maione, U. and Wallis, J.R. (editors). Mathematical Models in Surface Water Hydrology, Wiley, London, pp 423.

Whittle, P. 1954. On stationary processes in the plane. Biometrika, Vol. 41, pp 434-449.

Wiener, N. 1949. Extrapolation, Interpolation and Smoothing of Stationary Time Series, New York.

World Meteorological Organization 1974. Guide to Hydrological Practices, W.M.O. - No. 168, 3rd Edition, Geneva.

Yevjevich, V. and Karplus, A.K. 1973. Area - time structure of the monthly precipitation process. Hydrology Paper No. 64, 45 pp, Colorado State University, Fort Collins, Colorado.

Young, P.C., Shellswell, S.H. and Neethling, C.G. 1971. A recursive approach to time series analysis, CUED/B - Control/TR16. Dept of Engineering, Univ. of Cambridge, pp iv + 69.

Zawadzki, I.I. 1973a. Statistical properties of precipitation patterns. Journal of Applied Meteorology, Vol. 12, pp 459-472.

Zawadzki, I.I. 1973b. Errors and fluctuations of raingauge estimates of areal rainfall. Journal of Hydrology, Vol. 18, pp 243-255.

HYDROLOGIC DATA NETWORK DESIGN
BY
MODIFIED LANGBEIN METHOD

Vulli L. Gupta
Professor, Department of Civil Engineering
University of Texas at El Paso
El Paso, Texas, 79968, USA

INTRODUCTION

The phrase "Hydrologic data network design," as implied in this paper, deals with efforts toward obtaining rational answers to questions such as: In a specific basin to be gaged, how many gages are needed to describe the areal variability of hydrologic and/or water quality parameters?; How many gages should be operated as base or primary stations?; How many secondary stations are needed?; How long should each site be gaged to provide accurate and adequate information? and, What sampling frequency should be employed so as to minimize the repetition or redundacy of information (sampling too frequently) or the loss of information (sampling not often enough)? Several agencies are continually involved in monitoring activities for a variety of hydrologic and water resources decision-making endeavors, and systematic answers for some of the foregoing questions will be of assistance.

The intent of this paper is to present the conceptual framework of "modified Langbein method," and to illustrate its applicability to regional data network case studies, one dealing with precipitation gaging, and the other dealing with selected water quality parameters.

LITERATURE REVIEW

Some of the earliest efforts in the area of data network design were reported by Langbein (1954, 1960, 1972) essentially consisting of guidelines to estimate the optimum number of base stations and operational time frames

for secondary stations relative to precipitation data
acquisition in a region. The criteria was to maximize the
number of gaged stream-years in a hydrologic basin. Langbein
futher detailed (1960) a number of correlation techniques
and their utilization for extrapolation of streamflow data.
Matalas (1969) suggested the utilization of the information
content for network design efforts, and Hershfield (1965)
and Hutchinson (1969) dealt with the problem of raingage
spacing for best estimates of regional parameters. Hendrick
and Comer (1970) illustrated the utilization of correlation
fields in deciding whether to add or delete precipitation
gages in a region. Significant advances in the subject
area are also reported by Quimpo (1970), Salas-La Cruz
(1972), Rodriguez-Iturbe (1972, 1976), Solomon (1972).

LANGBEIN METHOD

Consider two stations, as illustrated in Figure 1,
of unequal data lengths 'n' and 'k' years register a
simple parametric correlation coefficient "r". Let the
station with the longer record be known as the "primary"
station and the one with the shorter data record be considered as "secondary" station. If data at secondary station
is augmented by simple correlation techniques, the resulting
effective data at secondary station can be expressed using
the concepts of Langbein (1954) and Fiering (1963):

$$n_e = \frac{n}{1 + \left(\frac{n-k}{k-2}\right)\left(1-r^2\right)} \quad \cdots \quad \cdots \quad (1)$$

where n_e = effective data length at secondary station;
n = primary station data length; k = historic data length
of secondary station; and r = simple parametric correlation
coefficient.

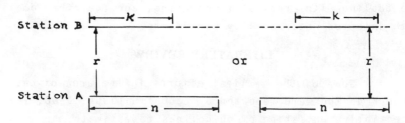

FIGURE 1. General Schematic.

It follows from Eq. 1, that $n_e < n$ when $r < 1$, which is usually the case in correlation exercises. In the ideal situation where $r = 1$ (i.e., perfect correlation), Eq. 1 reduces to $n_e = n$ implying a perfect and unbiased data augmentation. In the other extreme situation when $r = 0$ (i.e., no correlation exists), n_e would equal $\left\{\frac{n(k-2)}{n-2}\right\}$ which leads to the result, $n_e < n$. In either case, the gain of information can be expressed by

$$G = n_e - k \quad . \quad . \quad . \quad . \quad . \quad (2)$$

where G = gain of information relative to the secondary station; n_e = effective data length at the secondary station; and k = historic data length of secondary station. Combining Eqs. (1) and (2) yields:

$$G = \frac{n}{1 + \left(\frac{n-k}{k-2}\right)\left(1-r^2\right)} - k \quad . \quad . \quad . \quad (3)$$

At this stage, Langbein (1954) expresses the unexplained variance $(1-r^2)$, in Eq. 3, in terms of drainage basin area and the number of primary stations, by the following framework:

$$(1-r^2) = \frac{(0.4)\ C^2 A}{B} \quad . \quad . \quad . \quad . \quad (4)$$

where $(1-r^2)$ = unexplained variability; r = simple parametric correlation coefficient; A = basin area; B = number of primary or base stations; and C = increase in standard error of estimate of correlation per unit interstation distance.

Langbein (1954) utilizes a value of 0.1 for "C" in Eq. 4 and employs a trial and error solution for maximizing the information gain in Eq. 3 in order to obtain an optimum number of primary or base stations. In spite of the apparent simplicity of approach, Langbein's method is characterized by inherent empiricism. Very little effort was demonstrated toward the incorporation of applicable physiographic, hydrologic and/or land use characteristics.

These drawbacks are the principal motivation for modification while retaining the overall simplicity of the method.

MODIFIED LANGBEIN METHOD

The term $(1-r^2)$ in Eqs. (3) and (4) represents the
unexplained variance attributable to spatial differences
of the gaging or monitoring sites. If water quality
parameters such as TDS, chlorides, silica, etc., the
unexplained variance can be postulated as:

$$1-r^2 = \phi(X_1, X_2, X_3, X_4, X_5) = \phi(X_i) \quad . \quad . \quad (5)$$

where X_1 = Total number of primary and secondary stations;

X_2 = Average data length of station(s);

X_3 = Urbanization factor;

X_4 = Average data length at secondary stations;

X_5 = Average data length at primary stations;

r = Average interstation correlation coefficient;

ϕ = Functional symbol.

Combining Eqs. (3) and (5), the information gain
can be expressed as:

$$G = \frac{n}{1 + \left[\frac{n-k}{k-2}\right] \phi\left(X_i\right)} - k \quad . \quad . \quad . \quad . \quad (6)$$

If information gain, G, is to be maximized, Eq. 6
becomes the working equation. Thus, if the optimum data
length of the secondary station is desired, Eq. 6 needs to
be differentiated with respect to "k" and set to zero,
as indicated below:

$$\frac{\partial G}{\partial K} = \frac{\partial G}{\partial X_4} = 0 \quad . \quad . \quad . \quad . \quad . \quad (7)$$

$$\text{or } \frac{\partial}{\partial k}\left[\frac{n(k-2)}{(k-2)+(n-k)\phi(X_i)} - k\right] = 0 \quad . \quad . \quad . \quad (8)$$

Rearranging the terms in Eq. 8 and simplification will
result in a quadratic in "k" such as:

$$ak^2 + bk + c = 0 \quad . \quad . \quad . \quad . \quad . \quad (9)$$

where $a = \left[\phi(X_4)-1\right]^2 - n\,\frac{\partial\phi}{\partial k}$

$b = -4+2n\ \phi(X_4)+(n^2+2n)\frac{\partial\phi}{\partial k} - 2n\ \phi^2(X_4)+4\phi(x_4)$

and $c = 4+n^2\ \phi^2(X_4) - n^2\ \phi(X_4) - 2n^2\ \frac{\partial\phi}{\partial k} - 2n\ \phi(X_4)$

Solution of Eq. 9 is possible when the values of the variables n, $\phi(x_5)$, and $\frac{\partial\phi}{\partial k}$ are known. The result maximizes the information gain as well as it determines the optimum data length at the secondary gages.

Similarly, if the optimum length of data at the primary gages is desired, Eq. 6 needs to be differentiated with respect to "n" and set to zero. Therefore,

$$\frac{\partial u}{\partial n} = \frac{\partial u}{\partial X_5} = 0 \quad . \quad . \quad . \quad . \quad . \quad (10)$$

or $\frac{\partial}{\partial n}\left[\frac{n(k-2)}{(k-2)+(n-k)\ \phi(X_5)} - k\right] = 0$

which results in a quadratic expression in "n" such as:

$$n^2\ \frac{\partial\phi}{\partial n} - n(k\ \frac{\partial\phi}{\partial n})+\left\{k\phi(X_5) - k+2\right\} \quad . \quad . \quad . \quad (11)$$

Eq. 11 can be evaluated for "n" knowing the remaining variables k, $\frac{\partial\phi}{\partial n}$, and $\phi(X_5)$.

The foregoing methodology is structured to (a) eliminate the empiricism in the Langbein method, (b) to incorporate physiographic hydrologic and land use descriptors in network design exercises, and (c) to eliminate trial and error approaches.

Eq. 5 illustrates a possible formulation of variables applicable to water quality data networks. If precipitation data networks are being planned, the expression for the unexplained variance can be structured as:

$$(1 -r^2) = \phi(X_1, X_2, X_3, \ldots\ldots\ldots X_9, X_{10}) \quad . \quad . \quad (12)$$

where X_1 = Differences of distances to moisture sources;
X_2 = Differences of station elevations;

X_3 = Interstation distance differences;

X_4 = Station 'exposure' difference;

X_5 = Area;

X_6 = Total number of stations;

X_7 = Number of primary stations;

X_8 = Number of secondary stations;

X_9 = Data length of primary stations;

X_{10}= Data length of secondary stations

r = Interstation correlation coefficient;

ϕ = Functional symbol.

The procedure for estimating the optimum number stations and optimum sampling periods, essentially parallels the indicated procedure for water quality data networks in terms of maximizing the information gain.

PRECIPITATION DATA NETWORKS

The first case study, reported herein, deals with a dense network of precipitation gages situated in the Tuy River drainage basin, Venezuela, with an area of 2550 mi^2 (6,608 Km^2). A total of 73 rain gages were included in the study and based on topography and coastal proximity considerations, the area was divided into two zones containing 42 gages in one and 31 in the other. Figure 2 represents the general layout of the study area with a random illustration of some of the gages. Elevation of the gages range from 33 feet (10 m) to 6,462 feet (1970 m) and the data length ranges from 13 years to 83 years. Since area of coverage is one of the parameters, the zones were subdivided into sub-areas with the help of concentric circles. Zone I was divided into 15 sub-areas and Zone II into 19 sub-areas. A typical arrangement is illustrated in Fig. 2.

In each of the two zones, the following ten parameters were evaluated:

x_1 = coastal distance difference (Km)

x_2 = elevation difference (m)

x_3 = interstation distance (Km)

x_4 = exposure difference (degrees)

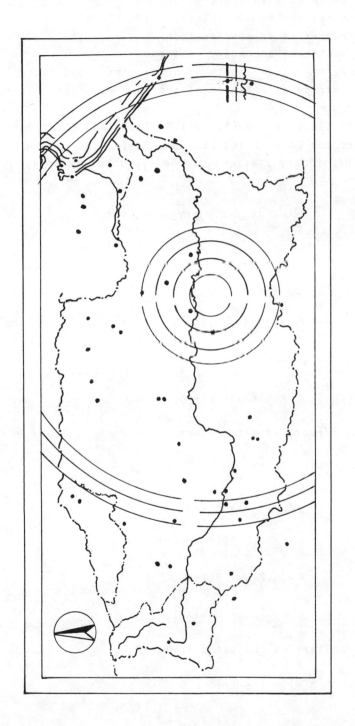

FIGURE 2. Case Study: Precipitation Data Network, Tuy River Basin,
Venezuela.

x_5 = sub-area size (Km^2)
x_6 = total number of stations
x_7 = number of primary stations
x_8 = number of secondary stations
x_9 = average data length of primary stations (years)
x_{10} = average data length of secondary stations (years)

Unexplained variance ($1 - \bar{r}^2$) Tables 1 and 2 represent the parameters for Zones I and II. Similarly, the optimum data length of primary stations can be evaluated by applying Eq. 11, where

$$\frac{\partial \phi}{\partial n} = \frac{\partial \phi}{\partial x_9} = -0.0172$$

$$\phi(x) = 0.74$$
$$K = 19.16$$

and

$$n^2 - 19.593n - 173.26 = 0$$

or solving for 'n',

$$n = 26.17 \text{ years}$$

Eq. 12 can be structured with reference to Zone I, as:

$$1 - (\bar{r})^2 = 0.67 + 0.01x_1 - 0.002x_2$$

$$+ 0.013x_3 - 5x10^{-5}x_4$$

$$- 25x10^{-6}x_5 + 8.2x_6$$

$$- 8.2x_7 - 8.21x_8$$

$$-0.02x_9 + 0.01x_{10}$$

Using the parameters as indicated in Table 1,

$$\phi(x) = 1 - (\bar{r})^2 = 0.74$$

and
$$n = 29.92$$

TABLE 1.

MODIFIED LANGBEIN METHOD PARAMETERS
ZONE I, TUY RIVER BASIN, VENEZUELA

Serial Number	Coastal Distance Diff. (Km)	Elevation Diff. (m)	Inter-Station Distance (Km)	Exposure Diff. (Degrees)	Area (Km²)	Number of Stations	Primary Stations	Secondary Stations	Avg. Data Length, Primary (Years)	Avg. Data Length, Secondary (Years)	$1-(\bar{r})^2$	Circle No.
1	8.0	14.7	22.03	16.07	572.30	3	3	0	30.3	0.00	.39	3
2	6.38	130.3	11.70	21.70	1,017.87	4	3	1	30.3	25.00	.36	4
3	6.26	112.4	13.97	26.50	1,552.88	5	4	1	31.0	25.00	.40	5
4	10.82	103.86	23.97	52.86	2,951.62	8	5	3	30.0	20.00	.52	7
5	17.75	115.93	29.81	125.22	3,767.75	12	6	6	30.6	19.33	.53	8
6	20.33	101.61	33.50	112.37	4,232.99	16	7	9	30.0	19.67	.66	9
7	23.88	104.53	36.64	106.02	4,697.47	20	9	11	29.44	19.73	.70	10
8	25.57	112.54	41.75	101.93	5,499.19	25	12	13	29.92	20.46	.73	11
9	26.87	111.55	43.67	97.76	5,944.88	27	12	15	29.92	19.93	.74	12
10	29.27	120.36	46.80	111.56	7,363.82	29	12	17	29.92	19.29	.75	13
11	30.24	125.73	47.82	97.19	7,381.53	30	12	18	29.92	19.00	.75	14
12	30.12	125.71	48.72	97.02	7,413.36	31	12	19	29.92	19.16	.74	15

TABLE 2.   MODIFIED LANGBEIN METHOD PARAMETERS
ZONE II, TUY RIVER BASIN, VENEZUELA

Serial Number	Coastal Distance Diff. (Km)	Elevation Diff. (m)	Inter-Station Distance (Km)	Exposure Diff. (Degrees)	Area (Km²)	Number of Stations	Primary Stations	Secondary Stations	Avg. Data Length, Primary (Years)	Avg. Data Length, Secondary (Years)	$1-(\bar{r})^2$	Circle No.
1	18.67	724.67	20.10	64.00	1,964.0	3	0	3	0.0	15.67	.64	5
2	19.67	617.57	17.97	157.50	2,711.9	4	0	4	0.0	17.50	.57	6
3	29.66	536.67	30.32	147.00	3,648.4	6	0	6	0.0	16.50	.60	7
4	32.38	456.86	32.55	122.14	4,694.5	8	1	7	84.0	15.71	.72	8
5	29.83	424.05	35.47	134.00	5,924.0	10	1	9	84.0	16.44	.71	9
6	29.74	416.40	37.23	126.00	7,182.5	11	1	10	84.0	17.30	.74	10
7	28.34	319.70	39.71	133.64	8,820.8	12	1	11	84.0	17.90	.79	11
8	23.96	365.32	40.62	146.87	10,354.0	14	3	11	51.0	17.90	.77	12
9	23.16	353.01	42.31	136.72	13,852.0	15	3	12	51.0	18.50	.78	13
10	21.50	411.44	44.60	139.97	15,775.5	17	4	13	49.25	19.07	.81	15
11	20.84	398.33	44.83	145.30	17,071.5	18	5	13	46.40	19.07	.81	16
12	21.50	384.40	47.28	144.74	18,571.5	19	6	13	43.17	19.07	.82	17
13	21.36	377.65	48.89	143.29	21,860.3	20	7	13	41.86	19.07	.84	19

$$\frac{\partial \phi}{\partial K} = \frac{\partial \phi}{\partial x_{10}} = 0.01$$

As an illustration of the applicability of Modified Langbein method with reference to the optimum data length at secondary stations, substituting for $\phi(x)$, n, and $\frac{\partial \phi}{\partial K}$ in Eq. 9,

$$K^2 \left(-0.2316\right) + K \left(20.024\right) - 230.41 = 0$$

and solving the Quadratic,

$$K = 13.67 \text{ years.}$$

WATER QUALITY DATA NETWORKS

The second case study, reported herein, deals with a network of 30 sampling points in the Truckee River Basin, California-Nevada where monthly parameters of inorganic constituents such as chlorides, TDS, Silica were monitored on a monthly basis. Based on the relative degree of urbanization, the study area was divided into three zones, namely, upper, middle, and lower reaches of the river. Eighteen sampling sites were included in the upper zone and 6 sites each were considered in the middle and lower zones.

Urbanization factors were assigned somewhat qualitatively by dividing the urbanized area above a particular site by the total drainage area above the site. Table 3 identifies some of the features of the sampling network. A numerical value of 1.00 means no urbanization influence on the water quality regime whereas a factor of 2.00 was assigned for total urbanization of the area. Consequently, the urbanization factors listed in Table 3 for the thirty stations in the study area were estimated based on the degree of urbanization prevalent in the region upstream of each of the sampling sites.

In view of the relative differences of urbanization, the study area was divided into three zones as indicated in Table 3; and to accomplish the regression analysis each zone was further subdivided into a number of sub-areas. Zone I comprises ten sub-areas whereas six sub-areas each make up the Zones II and III. A typical subdivision is illustrated in Figure 3 with reference to Zone II including Stations 19 through 24.

TABLE 3.   WATER QUALITY DATA NETWORK

Reach	Sta. Number	Name	Data Length (Years)	Urbanization Factor	Station Category
Upper	1	Below Tahoe Dam	8	1.08	Primary
	2	Wright Spring No. 1	8	1.70	Primary
	3	Wright Spring No. 2	8	1.75	Primary
	4	Above Squaw Creek	4	1.15	Secondary
	5	Above Donner Creek	8	1.18	Primary
	6	Donner	8	1.05	Primary
	7	Above Martis Creek	4	1.20	Primary
	8	Martis Creek	3	1.08	Secondary
	9	Below reservoir	3	1.05	Secondary
	10	Below Martis Creek	4	1.22	Secondary
	11	Prosser Creek (Above)	2	1.05	Secondary
	12	Prosser Creek (Below)	7	1.05	Primary
	13	Little Truckee (Above)	3	1.02	Secondary
	14	Sagchen Creek	2	1.02	Secondary
	15	At reservoir	2	1.02	Secondary
	16	Little Truckee (Below reservoir)	9	1.02	Primary
	17	Juniper Creek	8	1.28	Primary
Middle	18	Farad	8	1.28	Primary
	19	Below Verdi	8	1.28	Primary
	20	Idlewild	8	1.30	Primary
	21	Reno	5	1.32	Secondary
	22	Boynton Lane	3	1.38	Secondary
	23	Steamboat Creek	5	1.10	Secondary
	24	North Truckee	7	1.60	Primary
Lower	25	Below STP	8	1.90	Primary
	26	Vista	7	1.70	Primary
	27	Tracy	3	1.71	Secondary
	28	Derby	8	1.71	Primary
	29	Wadsworth	4	1.73	Secondary
	30	Nixon	8	1.75	Primary

Once the sub-areas were delineated, an average value
for each 'r' (correlation coefficient between two sites for
a given water quality constituent) within a given sub-area
is determined. This exercise can be demonstrated with a
four site, say, configuration as illustrated in Figure 4
where the letters 's' and 'p' represent the secondary and
primary stations, respectively.

Average data lengths of primary stations were 7.56,
7.67, and 7.75 years in Zones I, II, and III, respectively.
Similarly, secondary stations records ranged in years as
3.00, 4.33, and 3.50 in the Zones I, II, and III, respectively.
Just as in the precipitation data network design

case study, the unexplained variance, $(1-r^2)$, was postulated as:

$$(1-r^2) = \phi(X_1, X_2, X_3, X_4, X_5) = \phi(X_i) \quad \cdot \quad \cdot \quad \cdot \quad (13)$$

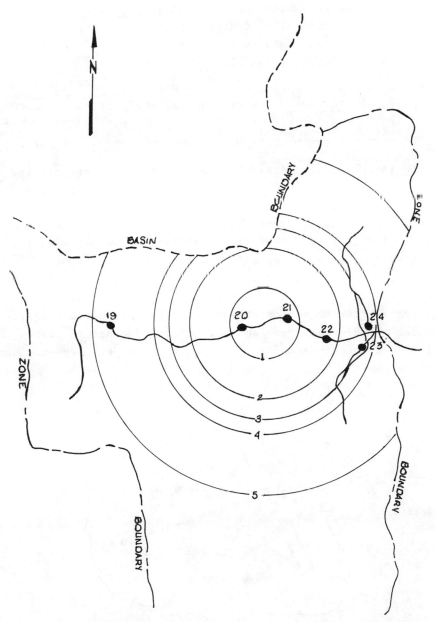

FIGURE 3.   Subarea Division, Zone II, Water Quality Data
            Networks (Truckee River Basin, Nevada - Case
            Study).

where X_1 = Total number of stations (primary and secondary),
       X_2 = Average data length of station (primary and
              secondary),
       X_3 = Urbanization factor,
       X_4 = Average length of data at secondary stations,
       X_5 = Average data length at primary stations,
       r = Average interstation correlation coefficient,
       ϕ = Functional symbol.

The four station grouping yield six possible combinations of pairs of stations. The average value of 'r', thus, becomes:

$$r = \frac{r_{s_1 s_2} + r_{s_1 p_1} + r_{s_1 s_3} + r_{s_2 p_1} + r_{s_2 s_3} + r_{s_3 p_1}}{6} \quad \dots \quad (14)$$

In Equation 14, a typical term such as $r_{s_1 s_3}$ represents the interstation correlation coefficient relative to Stations S_1 and S_3 as illustrated in Figure 4. Tables 4 and 5 list the formulations of simple regression results with reference to total dissolved solids and chlorides as the water quality parameters. As an illustration of the application of modified Langbein method for this case study, the following refers to Zone II with reference to the middle reach of the river relative to concentrations of total dissolved solids. As can be inferred from Table 4, the unexplained variance, $(1-r^2)$, with reference to the data length of primary stations follows as:

$$(1-r^2) = 5.774 - 0.654 \, X_5 \quad \dots \quad (15)$$

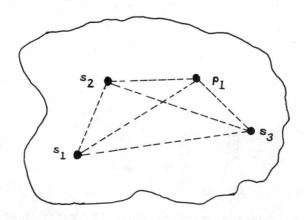

FIGURE 4. Sample Subarea of Four Stations.

TABLE 4: REGRESSION COEFFICIENTS

Equation of Form : $\hat{y} = \alpha + \beta x$

α = constant β = coefficient

$\hat{y} = (1 - \bar{r}^2)$ $x = \phi(x_i)$

Total Dissolved Solids

Reach	x	α	β
Upper	$\phi(x_1)$	0.275	0.051
	$\phi(x_2)$	-0.926	0.346
	$\phi(x_3)$	2.343	-1.468
	$\phi(x_4)$	3.904	-1.005
	$\phi(x_5)$	-0.985	0.265
Middle	$\phi(x_1)$	0.086	0.141
	$\phi(x_2)$	2.687	-0.289
	$\phi(x_3)$	0.076	0.436
	$\phi(x_4)$	2.549	-0.387
	$\phi(x_5)$	5.774	-0.654
Lowe	$\phi(x_1)$	0.187	0.139
	$\phi(x_2)$	4.374	-0.515
	$\phi(x_3)$	3.838	-1.789
	$\phi(x_4)$	6.287	-1.540
	$\phi(x_5)$	5.757	-0.644

Therefore

$$\frac{\partial \phi}{\partial X_5} = \frac{\partial \phi}{\partial K} = -0.654 \quad . \quad . \quad . \quad . \quad . \quad . \quad (16)$$

Substituting the essence of Equation 16 and K, and $\phi(X)$ in the working equation of Modified Langbein Method such as Eq. 11,

$$-0.654 \, n^2 + 2.832 \, n + 0.965 = 0$$

which leads to:

$$n = 4.648 \quad . \quad . \quad . \quad . \quad . \quad . \quad . \quad (17)$$

implying that the optimum data length of primary stations

TABLE 5 : REGRESSION COEFFICIENTS

Equation of Form : $\hat{y} = \alpha + \beta x$

α = constant β = coefficient

$y = (1 - \bar{r}^2)$ $x = \phi(x_i)$

Cloride

Reach	x	α	β
Upper	$\phi(x_1)$	0.242	0.056
	$\phi(x_2)$	-0.824	0.326
	$\phi(x_3)$	2.223	-1.355
	$\phi(x_4)$	3.993	-1.008
	$\phi(x_5)$	-0.806	0.238
Middle	$\phi(x_1)$	-0.076	0.194
	$\phi(x_2)$	3.349	-0.376
	$\phi(x_3)$	-2.008	2.005
	$\phi(x_4)$	3.226	-0.515
	$\phi(x_5)$	7.393	-0.854
Lower	$\phi(x_1)$	0.193	0.149
	$\phi(x_2)$	3.294	-0.354
	$\phi(x_3)$	0.565	0.129
	$\phi(x_4)$	6.009	-1.450
	$\phi(x_5)$	5.239	-0.572

in order to maximize the Information Gain is 4.648 years.

Similarly, the procedure can be illustrated with reference to the estimation of the optimum data length for secondary stations in Zone II, as follows:

$$\frac{\partial \phi}{\partial X_4} = \frac{\partial \phi}{\partial K} = -0.387 \quad \ldots \ldots \quad (18)$$

and   n = 7.67 years as $\phi(X) = 0.761$.

Substituting for $\frac{\partial \phi}{\partial K}$, n, and $\phi(X)$ into Eq. 9,

$$3.025 \ K^2 - 26.869 \ K + 27.16 \quad \ldots \ldots \quad (19)$$

which results in K = 1.163 implying that the optimum data length at secondary stations in Zone II is 1.163 years.

$$k^2 \left[(\phi(x_4)-1)^2 - n\frac{\partial\phi}{\partial k} \right] + k \left[-4 + 2n\phi(x_4) + (n^2+2n)\frac{\partial\phi}{\partial k} - 2n\phi^2(x_4) + 4\phi(x_4) \right]$$
$$+ \left[4 + n^2\phi^2(x_4) - n^2\phi(x_4) - 2n^2\frac{\partial\phi}{\partial k} - 2n\phi(x_4) \right] = 0$$
$$n^2 \left[\frac{\partial\phi}{\partial n} \right] - n \left[k\frac{\partial\phi}{\partial n} \right] + \left[k\phi(x_5) - k + 2 \right] = 0$$

Constituent: TDS

Reach	k (mon.)	$\phi(x)$	$\partial\phi/\partial n$	n (mon.)
Upper	3.00	0.929	0.265	ind.
Middle	4.33	0.761	-0.654	4.648
Lower	3.50	0.861	-0.644	4.077

Reach	n (mon.)	$\phi(x)$	$\partial\phi/\partial n$	k (mon.)
Upper	7.56	0.920	-1.005	1.717
Middle	7.67	0.761	-0.387	1.163
Lower	7.75	0.861	-1.540	1.801

Ind. = indeterminant

$$k^2 \left[(\phi(x_4)-1)^2 - n\frac{\partial\phi}{\partial k} \right] + k \left[-4 + 2n\phi(x_4) + (n^2+2n)\frac{\partial\phi}{\partial k} - 2n\phi^2(x_4) + 4\phi(x_4) \right]$$
$$+ \left[4 + n^2\phi^2(x_4) - n^2\phi(x_4) - 2n^2\frac{\partial\phi}{\partial k} - 2n\phi(x_4) \right] = 0$$
$$n^2 \left[\frac{\partial\phi}{\partial n} \right] - n \left[k\frac{\partial\phi}{\partial n} \right] + \left[k\phi(x_5) - k + 2 \right] = 0$$

Constituent: Chloride

Reach	k (mon.)	$\phi(x)$	$\partial\phi/\partial n$	n (mon.)
Upper	3.00	0.921	0.238	ind.
Middle	4.33	0.911	-0.854	4.730
Lower	3.50	0.955	-0.572	4.257

Reach	n (mon.)	$\phi(x)$	$\partial\phi/\partial n$	k (mon.)
Upper	7.56	0.921	-1.005	1.714
Middle	7.67	0.911	-0.515	1.453
Lower	7.75	0.955	-1.450	1.813

ind = indeterminant

Summary of results concerning the optimum values of 'n' and 'K' (n = data length of primary stations and K = data length of secondary stations) are presented in Tables 6 and 7.

SUMMARY

A framework of concepts labelled as "Modified Langbein Method" is presented herein for the purpose of optimizing data networks in a region. Two case studies, one related to precipitation data networks and the other with reference to water quality data networks, are presented to illustrate the potential features and limitations of the methodology.

REFERENCES

1. Amisial, R., "Correlacion y Regresión con aplicaciones en la Hidrologia", unpublished notes, CIDIAT, Merida, Venezuela, July 1976, 30 p.

2. Benson, M.A., and Carter, R.W., "Area Extension of Stream-flow Information by Multiple Regression Techniques", Proceedings of the Symposium on the water balance of North America, American Water Resources Association, Urbana, Illinois, July 1970.

3. Caffey, J.E., "Interstation Correlations in Annual Precipitation and in Annual Effective Precipitation", Hydrology Paper N° 6, Colorado State University, June 1965.

4. Federici, J.M., "Statistical Design of Hydrologic Data Networks", M. S. Thesis, University of Nevada, Reno, NV, Dec. 1977, 77p.

5. Gupta, V.L., "Information Content of Time - Invariant Data", Journal of the Hydraulics Division, American Society of Civil Engineers, Vol. 99, N° HY3, March 1973, pp. 383-394.

6. Gupta, V.L., and James M. Federici, "Precipitation Data Network Design by Maximizing the Information Gain", paper presented at the Fall Annual Meeting, Hydrology Session, American Geophysical Union, San Francisco, California, Dec. 1976, EOS, p. 11.

7. Gupta, V.L., and Domingo Fossi, "Design of Hydrometeorological Data Networks in Tuy River Watershed, Venezuela (in Spanish), Vols. I and II, 1976.

8. Hendrick, R.L., and George H. Comer, "Space variations of Precipitation and Implications of Raingage Network Design", Journal of Hydrology, Vol. 10, 1970, pp. 151-163.

9. Hershfield, D.M., "On the Spacing of Raingages", International Association of Scientific Hydrology, Publ. N° 67, Symposium on Design of Hydrologic Networks, 1965, pp. 72-79.

10. Hutchinson, P., "Estimation of Rainfall in Sparsely Gaged Areas", Bulletin, International Association of Scientific Hydrology, Vol. 14, March 1969, pp. 101-119.

11. Langbein, W.B., "Hydrologic Data Networks and Methods of Extrapolating or Extending Available Hydrologic Data", Hydrologic Networks and Methods, United Nations - Flood Control series, N° 15, WMO, Bangkok, 1960, pp. 13-41

12. Linsley, R.K., "The Relation Between Rainfall and Runoff", Journal of Hydrology, Vol. 5, 1967, pp. 297-311.

13. Matalas, N.C., and M.A. Benson, "Effect of Interstation Correlation on Regression Analysis", Journal of Geophysical Research, Vol. 66, N° 10, Oct. 1961, pp. 3285-3293.

14. Matalas, N.C., and E.J. Gilroy, "Some Comments on Regionalization in Hydrologic Studies", Water Resources Research, Vol. 4, N° 6, pp. 1361-1369, 1968.

15. Matalas, N.C., "Optimum Gaging Station Location", The Progress of Hydrology, Proceedings of the First International Seminar for Hydrology Professors, pp. 473-489, Urbana, Illinois, 1969.

16. Matalas, N.C., "Statistical Design of Data Collection Systems", Paper presented at the ASCE - 17th Annual Specialty Conference, Hydraulics Division, Logan, Utah, August 1969.

17. Quimpo, R.G., and J.Y. Yang, "Sampling Considerations in Stream Discharge and Temperature Measurements", Water Resources Research, Vol. 6, Dec. 1970, pp. 1771-1774.

18. Rodríguez - Iturbe, I., General Report on Data Network Design in Proceedings of the International Symposium

on Uncertainities in Hydrologic and Water Resources
Systems, Tucson, Arizona, 1972, pp. 1457-1475.

19. Rodríguez - Iturbe, I., J.M. Mejía, "The Design of
    Rainfall Networks in Time and Space", Water Resources
    Research, Vol. 10, N° 4, pp. 713-728.

20. Salas - La Cruz, J.D., "Information Content of the
    Regional Mean", International Symposium on Uncertain-
    ities in Hydrologic and Water Resources Systems,
    Tucson, Arizona, pp. 646-660.

21. ——————"Case Book on Hydrologic Network Design Practice——————
    WMO Publication N° 324, Geneva, Switzerland, 1972.

22. Solomon, S.I., "Multi-regionalization and Network Strategy",
    WMO Publication N° 324, Geneva, Switzerland, 1972, 11p.

URBAN STORMWATER DATA MANAGEMENT SYSTEM

W. Harry Doyle, Jr., Hydrologist, U.S. Geological Survey,
Gulf Coast Hydroscience Center, NSTL Station, Mississippi 39529

Joy A. Lorens, Computer Specialist, U.S. Geological Survey,
Gulf Coast Hydroscience Center, NSTL Station, Mississippi 39529

ABSTRACT

The U.S. Geological Survey and the U.S Environmental Protection Agency are jointly conducting urban stormwater studies in 16 cities in the United States. The general objective of these studies is to provide local managers with urban-hydrology data and methods of analysis suitable to support management decisions.

A large amount of information has been collected in the urban hydrology studies. It was recognized that a computerized data storage system and a data management system was needed to do the following:
1. Store and update information in the data base;
2. Retrieve information from the data base for publication, analyses, etc.; and
3. Interface files from the data base with user computer programs, statistical procedures, and rainfall/runoff/quality models.

The data management system, created for the joint cooperative study, is used to retrieve and combine data from U.S. Geological Survey data files for use in rainfall/runoff/quality models and for data computations such as storm loads. This system is based on the data management aspect of the Statistical Analysis System and was used to create all the data files in the data base. The statistical Analysis System is used for storage and retrieval of basin physiographic, land-use, and environmental-practices inventory data. Also, storm event water-quality characteristics are stored in the data base.

INTRODUCTION

The U.S. Geological Survey (USGS) and the U.S. Environmental Protection Agency (USEPA) are jointly conducting urban stormwater studies in 16 cities in the United States. The general objective of these studies is to provide local managers with urban-hydrology data and methods of analysis suitable to support management decisions. More specifically, objectives are:
1. To establish a consistent and accessible data base for typical urban watersheds in each study area. Information will consist of rainfall, runoff, water quality and other environmental factors.
2. To determine the magnitude and frequency of storm runoff loads of water-quality constituents from typical urban watersheds.

71

3. To develop methods for estimating storm and annual loads of water-quality constituents for unsampled watersheds in each urban-study area and to identify pollution sources.
4. To test the effectiveness of stormwater management alternatives such as street sweeping and detention storage.
5. To determine data base needs and to evaluate methods of transferring the information to ungaged watersheds in other regions.

A large amount of information has been collected in the urban hydrology studies. It was recognized that a computerized data storage system and a data management system was needed to do the following:
1. Store and update information in the data base;
2. Retrieve information from the data base for publication, etc; and
3. Interface files from the data base with user computer programs, statistical procedures, and rainfall-runoff quality models.

This paper describes the Urban Stormwater Data Management (USDM) System and the following: (1) General system requirements, (2) General system design, (3) Specific components of the data management system using that design, and (4) Several example application programs.

SYSTEM REQUIREMENTS

An important component of any hydrologic study is the data base that contains the information collected during the study. There are several important concepts that are considered when a data base system is being designed.
1. The computer system where the data base resides.
2. The type of storage device for the data base--disk, tape, drum, etc. The chosen storage device is influenced by the frequency of use and cost of storing the information.
3. Users of the data base--who will put the information in and who will retrieve it?
4. The information to be included in the data base. The data base users need certain information which must be included, if possible. Usually programs set forth guidelines as to what information should be included, however, it is not always feasible or practical to include everything requested.
5. Ease of entry and retrieval is very important. If the system is too complicated, users will be confused and discouraged and proportionally more errors will creep into the system. Programs should be easy to execute and efficient and economical when executed.
6. Last, but by far the most important, is a system and data base documentation. Without a good documentation the system is useless.

If these concepts are considered and implemented then the best utilization of information in the data base will be obtained.

SYSTEM DESIGN

The above general requirements are associated with the design of any computerized data management system. However, a specific requirement imposed upon the design of the USDM system was that it be able to interact with the USGS data management system, WATSTORE (Showen, 1978). The following discussion describes individual parts of the USDM system and its interrelationship to WATSTORE. Figure 1 shows a flow chart that depicts data as it moves into and through WATSTORE to the USDM system.

The individual operations in figure 1, designated by letters A through J, show how data enter and flow through the system.

Operation A.   An urban hydrology monitoring system (UHMS) records storm rainfall and runoff quantity and quality information on 16-channel punched paper tape. The UHMS also controls an automated water sampling device that collects samples at preselected sampling time intervals which may vary in length from 30 seconds to 1 hour. A punched data record on the 16-channel tape includes the following:

1. Time in hours, minutes, and seconds.
2. Julian day.
3. Stage or discharge parameter(s).
4. Accumulated rainfall (one or more sites).
5. Sequential sample number if water-quality sample was taken.

As many as eight data channels are available for recording items 3 and 4 above. Between storms the UHMS operates on a standby mode recording only at a pre-selected interval such as twice a day or when rainfall occurs. The UHMS switches to a continuous mode when a threshold value of stage corresponding to a preselected discharge is reached.

Operations B,C.   The punched tape is converted to a magnetic tape by using software developed by USGS.

Operations D,E.   Data conversion programs are then used to put the data into WATSTORE. The three WATSTORE files that provide for the general storage of meteorological, discharge, and water-quality data are:

1. Daily-Values File.--Accommodates one value per day for a given parameter such as mean discharge, total rainfall, daily evaporation, etc.
2. Unit-Values File.--Accommodates information collected at uniform time intervals for as many as 2880 observations per day per parameter (30-second data), such as rainfall and runoff collected during a storm event.
3. Water-Quality File.--Accommodates analyses of water-quality parameters. Each water-quality value is referenced as to time, date, site identification, and parameter code.

Standard USGS data input programs are used to input these data (1-3) into WATSTORE.

Operations F,G.   Next, USGS data programs retrieve first-time data or updated data for input into the USDM system User File (fig. 1). The data are from three WATSTORE files, the Daily-Values File, the Unit-Values File, and the Water-Quality File. Three corresponding files comprise the User File. The User File contains information for application programs and has the capacity to store:

1. About ten long-term daily time series per site in a daily-values file component. Example, five years of daily rainfall data, five years of daily evaporation data, etc. Information for all sites in a state program will be within one file.
2. As many as eight unit-values time series (30 seconds to hourly time interval) per

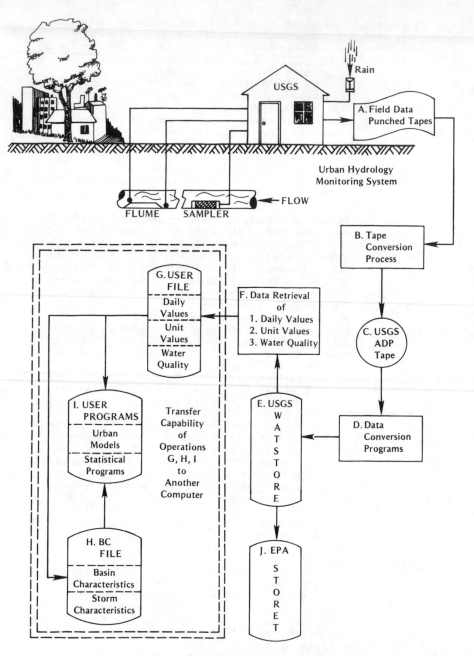

Figure 1.--Flow Chart of Data Management System for USGS/USEPA
Urban Hydrology Studies Program

site in a unit-values file component.
Storage accommodates information for each
unit value in time series for as many as
100 events for all sites in a state program.

3. As many water-quality or other parameters per water sample as needed. The total maximum capacity of the water-quality file component for water sample information is

n x m x o x p.

where the variables are as follows:

n = number of sites;
m = number of storms per site;
o = number of water samples per storm; and
p = number of parameters for the water samples.

The User File accommodates multiple data collection sites; each site includes as many storm events as necessary of as long as several days duration at given time intervals. The User File is operational on the USGS computer system and it with other associated components (Operations H and I) is capable of transfer to other compatible systems.

Operation H.   The Basin Characteristics File (BC File) contains specific-site information about basin characteristics such as generalized physiographic, land-use, and climatic characteristics of storms. Environmental practices information that are used to establish cause and effect relationships between management techniques and water-quality processes are stored in the BC File. Also, computed storm and dry weather characteristics are stored.

Operation I.   The User File is interfaced with user programs, urban storm water models, and statistical packages.

Operation J.   Finally, all data are transferred from WATSTORE to USEPA's storage system, STORET. Information in WATSTORE and STORET are available for retrieval by USGS and USEPA system users.

Although this section mainly described a system that was designed for a particular program and for certain users, this does not preclude or limit its usefulness for other areas of application.

STATISTICAL ANALYSIS SYSTEM

For the User File and other associated files, the concept that allows efficient transfer and storage of information for the respective operations is based on the Statistical Analysis System (SAS) (Barr et al., 1979). This system was developed at North Carolina State University and is available on the USGS computer system as well as on many other systems. The files that were created for the USDM system, such as the User File and the BC File were created by SAS programs. SAS also has a very comprehensive statistical package that can be used for statistical analyses. Therefore, SAS is not only used as a data manager in this system, but also as a data analyzer.

A few of the advantages of SAS are:
1. It has available documentation;
2. It is simple and easy to use; and
3. It is cost-effective.

The USDM system created for the USGS/USEPA Urban Hydrology Studies Program has many interactive users. Adopting a standardized package like SAS made it easier not only to develop the system, but also to implement and use it.

Since SAS is such an integral part of the USDM system, it will be most informative if some information about SAS be given. Only a few rudiments of the system will be presented and readers are encouraged to consult the references at the end of this report.

First of all, what is SAS? The letters stand for Statistical Analysis System, but SAS is much more than that. In broad terms it is a computer system for managing data, analyzing the data statistically, and writing reports. A system is a set or arrangement of processes that are interconnected as to form a whole. Because SAS is a system, you don't have to prepare one computer job to plot data values, another job to perform a regression, and another to print the report. SAS does it all within one job. Most common statistical procedures are available within SAS. Moreover, SAS is one of the most advanced proprietary packages in the area of linear statistical models (Showen, 1978). SAS can provide data management functions such as sorting, merging, copying, modifying, and condensing sets of data. SAS also functions as a word processor, allowing users to print their results in custom-made reports.

Information that comes from punched cards, disk, or tape is analyzed by SAS. SAS checks for errors, provides excellent error diagnostics and syntax checking, performs statistical tests, and prints results. Users can learn SAS in a tutorial approach, by first attempting simple SAS jobs and then following instructions issued by the SAS programs, should errors occur. Basic terminology pertaining to information that are analyzed by SAS are as follows:

Data Values--single measurements, such as rainfall amounts;
Observations--a set of data values for the same event, like
rainfall amount, discharge, total evaporation, etc.

SAS is used to establish the data files for the USDM system and for manipulating the data. An examination of the steps in designing a system will help in understanding the reasons behind the selection of SAS. These steps include:

1. Design the files for the data base. Decide whether file organization will be sequential, indexed sequential, direct access, etc., and then compute the size of the records and blocksizes that contain the records.
2. Write computer programs to create the data files. All the necessary file information would be included in the computer programs.
3. Create the necessary Job Control Language (JCL) to establish the files.
4. Create computer programs and associated JCL to update the files.
5. Create the data analysis programs necessary to manage and monitor the system.
6. Create a backup system to safeguard the data files.

With SAS steps 1 and 2 are available for creating any file that is needed. Most, if not all, of steps 4-6 are available, too. SAS uses simple commands which can be entered in any position on the program cards. The commands are very similar to the commands used by the computer language PL/I. Whereas FORTRAN is very scientifically orientated and COBAL highly applicable to business-related activities, PL/I uses the strengths of both. It can compute and process as FORTRAN does and it can handle tables and store data the way COBAL functions.

The most important feature however, is the ability that SAS has to modify files. In what would be referred to as a "static system", one created by other computer programs and in a fixed-file format, a major effort is required to change or reorganize data files and analysis programs. SAS addresses data values within observations by names and all positioning within the records to locate the data values is done by SAS. The relative ease with which one can create and manage data sets with SAS resulted in its selection for the USDM System.

APPLICATION PROGRAMS

A most important part of any data management system are the computer programs that establish and update the data files, and those that retrieve and analyze information from the files. Equally important are the associated program documentations that inform users about program and system requirements.

The following section describes the naming convention adopted for identifying data sets, JCL and SAS control commands, and several application programs that input, retrieve, and analyze information that are in the USDM system. The discussion is oriented toward USGS users and the USGS computer system.

A standardized naming convention was adopted for data sets that are a part of the USDM system: Every data set name begins with these three common nodes,

AG4yyyz.STATEXX.URBAN.etc.,

where: AG4yyyz   is an assigned TSO/ONLINE USGS User ID;
      STATEXX   is a state identification with XX being a two-digit state code; and
      URBAN   refers to the particular program.

An example daily-values file component in the User File for Pennsylvania might be:

AG4yyyZ.STATE42.URBAN.DAILY.VALUES

with XX=42 being the state code. This procedure eliminates any confusion about what the data are and who the data belong to.

The JCL and SAS commands for creating and updating files are relatively simple. Fig. 2 illustrates JCL for establishing a daily-values file for Pennsylvania. Card 1 is the Job card particular to the computer installation. Cards 2-5 request retrieval of daily-value information from the USGS WATSTORE system. Cards 6-11 transform the retrieved information into a SAS data set that is input for SAS commands following card 11. The "Other" SAS commands may include a command

PROC PRINT;

which lists the information as seen in fig. 3. This command calls out a SAS routine that formats and lists the associated data set.

Application programs were created to input watershed and storm characteristics information into the Basin Characteristics File (BC File) shown in Operation H (fig. 1). Previous studies by USGS in Portland, Oregon and Miami, Florida have indicated that storm loading of selected water-quality constituents can be related statistically (by multiple regression) to storm and watershed characteristics describing the generalized physiography and land use. For the purpose of regional and national transferability a consistent set of basin characteristics are compiled for each basin sampled in the USGS/USEPA Urban Hydrology Studies Program. The physiographic and land use characteristics that can be used in statistical analyses are listed in table 1. This detailed explanation of the BC File is presented so that the reader has a better understanding of the concepts that were used in developing one of the USDM system components.

Urban study areas may have environmental practices which impact water quantity and quality. These practices should be identified and documented to support both modeling and statistical techniques of analysis to establish cause and effect relationships between management techniques and water-quality processes. A list of environmental-practice information for each urban study is listed in table 2. This information is also stored in the BC File in the same data set as table 1 information.

Climatological or hydrologic factors which affect stormwater

```
        COL.1      COL.12
          •          •

Card 1../ /        JOB

     2../*PROCLIB WRD.PROCLIB

     3..// EXEC DVRETR,AGENCY=USGS

     4..//HDR.SYSIN DD *

        Retrieval cards for DVRETR

     5../*

     6..//EXEC WRDSAS,DSN='&&BKREC',MACRO=DV2,DSN1=NULLFILE,DSN2=NULLFILE

     7..//DVFILE DD DSN=AG4yyyz.STATE42.URBAN.DAILY.VALUES,

     8..// DISP=(NEW,CATLG),DCB=DSORG=DA,

     9..//SPACE=(TRK,(N1,N2)),UNIT=3330,VOL=SER=CCD810

    10..//SYSIN DD *

    11.. DVINPUT

Other SAS commands

    12../*

    13..//

    14..$$$
```

Figure 2.--JCL for Establishing Daily-Values File for Pennsylvania

runoff quality include rainfall depth, rainfall intensity, runoff
and antecedent conditions. The actual constituent accumulation and
washoff processes can be explained by combinations of these hydrologic
variables and physiographic characteristics, existing drainage
patterns, and the environmental practices. Table 3 contains a list
of storm characteristics, many of which have been found to be
significant in previous statistical analyses or model applications.

     The programs that enter information into the BC File have
several editing features, for example, total land use and total
street drainage are computed for the basin (fig. 4). The programs
also list the other physiographic and land-use characteristics as
seen in fig. 4.
     In any data management system there is a need to store routinely
used computer programs. These programs process information from the
data base for use in stormwater models, statistical analyses, etc.
In essence, the file contains all programs that the user might
need. This is the concept of the User Programs File (UP File).
     The UP File (Operation I, fig. 1) contains application programs
such as: user programs, urban stormwater models, utility programs
to perform chemical load calculations, report listings, etc. The
programs are loaded as executable load modules either in the USGS
program library or individual user libraries. Some of the programs
are written as SAS instructions while others are written in a
scientific program language such as FORTRAN.

Table 1.--Physiographic, land use and water-quality characteristics
in the Basin Characteristics File

Variable name	Numeric values	Variable identification
SITE	1.	Site identification number which is an 8 or 15 digit station ID.
TCAREA	2.	Total drainage area, in square miles (exclude noncontributing areas).
IAREA	3.	Impervious area in percentage of drainage area.
EAREA	4.	Effective impervious area in percentage of drainage area. Include only impervious surfaces connected directly to a sewer pipe or principal conveyance.
BSLOPE	5.	Average basin slope, in feet per mile, determined from an average of terrain slopes at 50 or more equispaced points using best available topographic map.
CSLOPE	6.	Main conveyance slope, in feet per mile, measured at points 10 and 85 percent of the distance from the gaging station to the divide.
PAHOR	7.	Permeability of the A horizon of the soil profile, in inches per hour.
AWAABC	8.	Available water capacity as an average of the A, B, and C soil horizons, in inches of water per inch of soil.
SWPHA	9.	Soil-water pH of the "A" horizon (in H_2O).
HYSGR	10.	Hydrologic soil group (A, B, C, or D) according to SCS methodology. Use numeric codes, A=1, B=2, etc.
POPDEN	11.	Population density in person/mi^2.
STDEN	12.	Street density, in lane miles per square mile (approximately 12 ft. lanes).
	13.	Land use of the basins as a percentage of drainage area including:
LURUPA		a. Rural and pasture.
LUAGRI		b. Agricultural.
LULOWD		c. Low density residential (1/2 to 2 acres/dwelling).
LUMEDD		d. Medium density residential (3 to 8 dwellings/acre).
LUHIGD		e. High density residential (9 or more dwellings/acre).
LUCOMM		f. Commercial.
LUINDU		g. Industrial
LUCONB		h. Under construction (bare surface).
LUIDLE		i. Idle or vacant land.
LUWETL		j. Wetland.
LUPARK		k. Parkland.

(Table 1 continued)

Table 1.--Physiographic, land use and water-quality characteristics
in the Basin Characteristics File--Continued

Variable name	Numeric values	Variable identification
DETSTO	14.	Detention storage in acre feet of storage.
PERWUD	15.	Percent of watershed upstream from detention storage.
PERSEW	16.	Percent of area drained by a storm sewer system.
PERCAG	17.	Percent of streets with curb and gutter drainage.
PERDAS	18.	Percent of streets with ditch and swale drainage.
MANRNT	19.	Mean annual rainfall in inches. (long term)
TENYRR	20.	Ten year 1-hour rainfall intensity in inches per hour. (long term)
MRLxxxxx	21.	Mean annual loads of water-quality constituents in runoff, in pounds/acre. xxxxx is the 5-digit WATSTORE parameter code.
MWLxxxxx	22.	Mean annual loads of constituents in wetfall, in pounds/acre. xxxxx is as defined in item 21.
MDLxxxxx	23.	Mean annual loads of constituents in dryfall, in pounds/acre.

Table 2.--Environmental practices inventory characteristics in
the Basin Characteristics File

Variable name	Variable identification

Numeric values

AVFRSS	1.	Average frequency of street sweeping, in days.
EAFENI	2.	Estimated annual fertilizer applied to watershed, in pounds/acre of nitrogen.
EAFEPR	3.	Estimated annual fertilizer applied to watershed, in pounds/acre of phosphorous.
AVSEFF	4.	Average sewer flushing frequency, in days.
AVBCFR	5.	Average catch basin cleaning frequency, in days.
EADVTR	6.	Estimated average daily vehicle traffic, in vehicle miles per day.

Descriptive, as many as 80 characters per variable

MEFQSC	1.	Method or type of street sweeping equipment.
GAOINE	2.	Grading and agricultural ordinances in effect.
RECOLP	3.	Refuse collection practice.
SWASDP	4.	Solid waste disposal areas in watershed.
FLREFE	5.	Flood retarding features such as gravel filter strips.
LEAFDP	6.	Leaf disposal practice in watershed.
SEDSOU	7.	Identify major sediment source(s).
STPTCO	8.	Street pavement and condition.
DEICCH	9.	Deicing chemicals.

Table 3.--Characteristics related to storm events in the national
urban studies program

Variable name	Numeric values	Variable identification
SITE	1.	Site identification number which is an 8 or 15 digit station ID.
BDATE	2.	Storm begin date; year, month, day, - for example, 800601.
BTIME	3.	Storm begin time expressed in military format, 0001-2400.
EDATE	4.	Storm end date; year, month, day, - for example, 800602.
ETIME	5.	Storm end time expressed in military format, 0001-2400.
TRAINA	6.	Total rainfall, average for the basin in inches.
MAXR5	7.	Maximum 5-minute rainfall rate in inches/hour.
MAXR15	8.	Maximum 15 minute rainfall rate in inches/hour.
MAX1H	9.	Maximum 1-hour rainfall rate in inches/hour.
NDRDO2	10.	Number of dry hours prior to storm, counting backwards to storm event with rainfall greater than 0.2 inches.
DERNPD	11.	Depth of rainfall accumulated during previous 24 hours, in inches.
DERNP3	12.	Depth of rainfall accumulated during previous 72 hours, in inches.
DERNP7	13.	Depth of rainfall accumulated during previous 168 hours, in inches.
TOTRUN	14.	Total runoff, in inches, over the basin.
PEAKQ	15.	Peak discharge, in cubic feet per second.
BFLOW	16.	Base flow prior to storm, in cubic feet per second.
DURSTO	17.	Duration of storm runoff used to calculate load, in minutes.
DURRNF	18.	Duration of rainfall, in minutes.
TIMBPK	19.	Time from beginning of rainfall to hydrograph peak, in minutes.
TILASC	20.	Time since last street cleaning, in days.
RLxxxxx	21.	Storm-runoff loads of individual constituents, in pounds/acre. WATSTORE parameter codes are to be used in place of xxxxx, for example, RL01051 is the variable name for storm-runoff load for total lead.
WLxxxxx	22.	Wetfall loads of individual constituents since previous storm sampled, in pounds/acre. xxxxx is as defined in no. 21.
DLxxxxx	23.	Dryfall loads of individual constituents since previous storm in pounds/acre (interpolated from monthly dryfall rate, based on number of dry hours, for example, no. 10 above).

81

S T A T I S T I C A L   A N A L Y S I S   S Y S T E M

15:24 WEDNESDAY, APRIL 16, 1980

OBS	STATION	PARMCODE	STATCODE	DATETIME	VALUE	DATE
1	254031080191103	45	6	01OCT77:00:00	0.00	77-10-01
2	254031080191103	45	6	02OCT77:00:00	0.00	77-10-02
3	254031080191103	45	6	03OCT77:00:00	0.00	77-10-03
4	254031080191103	45	6	04OCT77:00:00	0.00	77-10-04
5	254031080191103	45	6	05OCT77:00:00	0.00	77-10-05
6	254031080191103	45	6	06OCT77:00:00	0.54	77-10-06
7	254031080191103	45	6	07OCT77:00:00	0.00	77-10-07
8	254031080191103	45	6	08OCT77:00:00	0.00	77-10-08
9	254031080191103	45	6	09OCT77:00:00	0.00	77-10-09
10	254031080191103	45	6	10OCT77:00:00	0.00	77-10-10

Figure 3.--Listing of Daily-Values Data for Parameter Code 00045

Several of the utility listing programs are:
  (1) RRLIST--Rainfall-Runoff Summary Utility Listing Program.
       This program produces a listing (figure 5) of selected
       unit rainfall and discharge information for storm
       periods. The user selects a storm and inputs a time
       period for the retrieval. The information can be
       printed out in any interval the user requests.
       Information for as many as three rain gages can be
       retrieved with output in individual and accumulated
       time intervals.
  (2) QWLIST--Water-Quality Summary Utility Listing Program.
       This program produces a tabular listing of a core
       list of water-quality constituents (figure 6). The
       information is retrieved by time periods corresponding
       to storm begin and end times and dates.
  (3) LOADS--Water-Quality Loads and Loadings Utility
       Listing Program. This program produces an individual
       basin chemical constituent load and loading listing
       for any number of constituents and storms (figure 7).
       It also produces plots of parameter values versus
       time, load versus time, and accumulated load versus
       time for each storm constituent. It will also produce
       a total loads and loading listing for each storm.

SUMMARY AND CONCLUSIONS

A data management system, the USDM system, has been created
for an urban stormwater study conducted jointly by USGS and USEPA.
The system was designed using the SAS data management concepts.
SAS was used to create all files in the system and is also used as
a manager to store, update, retrieve, and interface information to
and from files in the data base. There are many advantages to
using SAS to create and manage a data base. It is simple, easy to
use, and has a comprehensive statistical package. The most important
feature however, is the ability that SAS has to modify files.
Data base system development has progressed during the last two
decades and the data management system concepts used in this study
reflect the advancement made in computer technology during this
era. Urban stormwater data is just one application example for
which the system can be used. USGS has several other programs that
may use the data management system in the future.

20.00 CONTRIBUTING DRAINAGE AREA, SQUARE MILES
 0.60 IMPERVIOUS AREA, PERCENTAGE OF DRAINAGE AREA
 0.40 EFFECTIVE IMPERVIOUS AREA, PERCENTAGE OF DRAINAGE AREA

0.0010 AVERAGE BASIN SLOPE, FEET/MILE
0.0020 MAIN CONVEYANCE SLOPE, FEET/MILE

 0.75 PERMEABILITY OF A HORIZON OF SOIL PROFILE, INCHES/HOUR
 0.50 WATER CAPACITY, INCHES OF WATER/INCH OF SOIL
 7.20 SOIL-WATER PH OF THE A HORIZON
    1 HYDROLOGIC SOIL GROUP, SCS METHODOLOGY
      A=1, B=2, C=3, D=4

 1000 POPULATION DENSITY, PERSON/SQUARE MILE
   20 STREET DENSITY, LANES/SQUARE MILE

      LAND USE, PERCENTAGE OF DRAINAGE AREA
 0.10 RURAL AND PASTURE
 0.10 AGRICULTURAL
 0.10 LOW DENSITY RESIDENTIAL
 0.10 MEDIUM DENSITY
 0.04 HIGH DENSITY RESIDENTIAL
 0.10 COMMERCIAL
 0.10 INDUSTRIAL
 0.10 UNDER CONSTRUCTION, BARE SURFACE
 0.06 IDLE OR VACANT
 0.10 WETLAND
 0.10 PARK   ***TOTAL LAND USE = 1.00

200.00 DETENTION STORAGE, ACRE FEET OF STORAGE
 0.30 % WATERSHED UPSTREAM OF DETENTION STORAGE
 0.70 % AREA DRAINED BY A STORM SEWER SYSTEM
 0.35 % STREETS WITH CURB AND GUTTER DRAINAGE
 0.75 % STREETS WITH DITCH AND SWALE DRAINAGE

                              ***TOTAL STREET DRAINAGE = 1.10

65.00 MEAN ANNUAL RAINFALL, INCHES
 7.20 TEN-YEAR 1-HOUR RAINFALL INTENSITY, INCHES/HOUR

   30 AVERAGE FREQUENCY OF STREET SWEEPING, DAYS
  100 ESTIMATED FERTILIZER, POUNDS / ACRE OF NITROGEN
  100 ESTIMATED FERTILIZER, POUNDS / ACRE OF PHOSPHOROUS
   30 AVERAGE SEWER FLUSHING FREQUENCY, DAYS
   60 AVERAGE CATCH BASIN CLEANING FREQUENCY, DAYS
 1000 ESTIMATED AVERAGE DAILY VEHICLE TRAFFIC, VEHICLE MILES / DAY

MEFQSC - METHOD OR TYPE OF STREET SWEEPING EQUIPMENT
GAOINE - GRADING AND AGRICULTURAL ORDINANCES IN EFFECT
RECOLP - REFUSE COLLECTION PRACTICE
SWASDP - SOLID WASTE DISPOSAL AREAS IN WATERSHED
FLREFE - FLOOD RETARDING FEATURES SUCH AS GRAVEL FILTER STRIPS
LEAFDP - LEAF DISPOSAL PRACTICE IN WATERSHED
SEDSOU - IDENTIFY MAJOR SEDIMENT SOURCES
STPTCO - STREET PAVEMENT AND CONDITION
DEICCH - DEICING CHEMICALS

Figure 4.--Example of Basin Characteristics Input Program

RAINFALL AND RUNOFF DATA

DISCHARGE STATION ID: 25403108019110O
RAINFALL GAGE # 1 ID: 25403108019110.3
GAGE # 2 ID:
GAGE # 3 ID:

TIME: FROM   MAY 11, 1977 4:40:00 TO   MAY 11, 1977 6:58:00
NOTE: * * * RAINFALL AND RUNOFF IN INCHES, DISCHARGE IN CUBIC FEET PER SECOND
TIME INTERVAL OF BASIC DATA: 1.0 MINUTE(S)
PRINT INTERVAL : EVERY 1 MINUTES

TIME	DISCHARGE	ACCUMULATED RUNOFF	RAINFALL GAGE 1	ACCUMULATED RAIN 1	RAINFALL GAGE 2	ACCUMULATED RAIN 2	RAINFALL GAGE 3	ACCUMULATED RAIN 3
4:40:00	0.00	0.000	0.00	0.00	.	.	.	.
4:41:00	0.00	0.000	0.01	0.01	.	.	.	.
4:42:00	0.00	0.000	0.06	0.07	.	.	.	.
4:43:00	0.00	0.000	0.05	0.12	.	.	.	.
4:44:00	0.00	0.000	0.08	0.20	.	.	.	.
4:45:00	1.59	0.002	0.03	0.23	.	.	.	.
4:46:00	5.16	0.008	0.03	0.26	.	.	.	.
4:47:00	8.39	0.017	0.06	0.32	.	.	.	.
4:48:00	11.09	0.029	0.14	0.46	.	.	.	.
4:49:00	14.32	0.046	0.06	0.52	.	.	.	.
4:50:00	19.32	0.067	0.14	0.66	.	.	.	.

Figure 5.--Example of Rainfall-Runoff Summary Utility Listing Program

CORE LIST OF WATER QUALITY CONSTITUENTS
FOR THE NATIONAL URBAN STUDIES PROGRAM

STATION ID: 254031080191100 FROM MAY 1, 1977 7:34 TO MAY 4, 1977 13:30

D M YR	TIME	DISCHARGE (CFS)	SPECIFIC CONDUCTANCE (MICROMHOS)	PH (UNITS)	DISSOLVED SOLIDS (MG/L)	DISSOLVED NO2+NO3 AS N (MG/L)	DISSOLVED NH3 AS N (MG/L)	DISSOLVED KJELDAHL N. (MG/L)	TOTAL KJELDAHL N. (MG/L)
04MAY1977	9:27	.	.	.	.	.	.	.	1.41
04MAY1977	9:33	.	117	.	.	.	.	.	1.11
04MAY1977	9:42	.	122	.	.	.	.	.	0.83

D M YR	TIME	DISSOLVED PHOS. AS P (MG/L)	TOTAL PHOS. AS P (MG/L)	TOTAL LEAD (UG/L)	DISSOLVED ORG. CARBON (MG/L)	SUSPENDED ORG. CARBON (MG/L)	CHEMICAL OXY. DEMAND (MG/L)	FEC.COLIFORM BACTERIA (COL/100 ML)
04MAY1977	9:27	.	0.21	610.00	.	.	110.00	.
04MAY1977	9:33	.	0.19	450.00	.	.	110.00	.
04MAY1977	9:42	.	0.14	230.00	.	.	45.00	.

Figure 6.--Example of Water-Quality Summary Utility Listing Program

INDIVIDUAL BASIN CHEMICAL CONSTITUENT LOAD AND LOADINGS

STATION ID : 25403108019110   FROM 11MAY77 4:40   DRAINAGE AREA(ACRES): 14.7

CHEMICAL NAME:TOTAL NITROGEN
PARAMETER CODE: 00625

DATE	TIME	DISCHARGE (CFS)	***TIME INTERVAL DATA*** CONSTITUENT VALUE	LOAD (LBS)	LOADING (LBS/ACRE)	***ACCUMULATED DATA*** LOAD (LBS)	LOADING (LBS/ACRE)	ACCUMULATED RUNOFF(IN)
11MAY1977	4:40	0.00	1.840	0.000	0.000	0.00	0.00	0.00
11MAY1977	4:41	0.00	1.840	0.000	0.000	0.00	0.00	0.00
11MAY1977	4:42	0.00	1.840	0.000	0.000	0.00	0.00	0.00
11MAY1977	4:43	0.00	1.840	0.000	0.000	0.00	0.00	0.00
11MAY1977	4:44	0.00	1.840	0.000	0.000	0.00	0.00	0.00
11MAY1977	4:45	1.59	1.840	0.011	0.001	0.01	0.00	0.00
11MAY1977	4:46	5.16	1.840	0.036	0.002	0.05	0.00	0.01
11MAY1977	4:47	8.39	1.840	0.058	0.004	0.10	0.01	0.02
11MAY1977	4:48	11.09	1.840	0.076	0.005	0.18	0.01	0.03
11MAY1977	4:49	14.32	1.840	0.099	0.007	0.28	0.02	0.05
11MAY1977	4:50	19.32	1.840	0.133	0.009	0.41	0.03	0.07

Figure 7.--Example of Water-Quality Loads and Loadings Utility Listing Program

REFERENCES

Barr, A. J., Goodnight, J. H., Sall, J. P. and Helwig, J. T. 1979. SAS User's Guide. 1979 Edition, SAS Institute, Inc., P.O. Box 10066, Raleigh, N.C., 494 pp.

Showen, C. R. 1978. Storage and Retrieval of Water-Resources Data in Collection, Storage, Retrieval, and Publication of Water-Resources Data. U.S. Geological Survey Circular 756, pp. 20-25.

REALISTIC RAINFALL AND WATERSHED RESPONSE SIMULATIONS FOR ASSESSING WATER QUALITY IMPACTS OF LAND USE MANAGEMENT

H. E. Westerdahl
Research Ecologist
U. S. Army Engineer Waterways Experiment Station
P. O. Box 631
Vicksburg, MS 39180

J. G. Skogerboe
Physical Scientist
U. S. Army Engineer Waterways Experiment Station
P. O. Box 631
Vicksburg, MS 39180

ABSTRACT

Soil erosion and pollutant transport from watersheds occur during and immediately following a relatively few unpredictable storm events. Consequently, erosion control research has become reliant on simulated rainfall since it can be generated faster than natural rainfall. However, characteristics of natural rainfall that influence erosivity, i.e., intensity, drop-size distribution, and raindrop-fall velocity, must be accurately simulated. A modified rotating-disk rainfall simulator with programmable slit-width was developed for use in the laboratory and field locations providing a calibrated land surface area coverage of 5.5 m². A watershed modeling facility incorporating this rainfall simulator and four variable-slope and depth soil lysimeters provides the capability for specific erosion control studies. A completely self-contained 15-m trailer equipped with a 10-m boom and water supply permits the transport of the rainfall simulator to remote field sites for field application of simulated rainfall. Continuously variable rainfall intensities (up to 7 cm/hr) and duration (up to 72 hr) can be programmed to duplicate selected design storm events. Criteria for calibrating this rainfall simulator focused on achieving a drop-size distribution and impact velocity typical of natural rainfall. Moreover, uniform intensity and drop characteristics throughout the study area were equally important. By adjusting the vertical distance of drop fall, angle of the nozzle, and nozzle pressure, the optimal drop characteristics and lowest rainfall distribution coefficient of variation were determined for selected rainfall intensities and expressed as percentage opening for the programmable slit opening. Finally, those parameters which have been suggested as proportional to rainfall erosivity were compared to natural rainfall and other rainfall simulators. These parameters included: kinetic energy per unit of rainfall; momentum per unit of rainfall; total kinetic energy per unit of drop impact area; total momentum per unit of total drop impact area; kinetic energy per unit of drop impact area (by increments); and momentum per unit of drop impact area (by increments). As a pilot study, this improved rotating-disk simulator and two soil lysimeters were operated simultaneously to simulate rainfall, runoff, and soil erosion from two prototype watersheds with different land use: an overland flow wastewater treatment site in Utica, Miss., and a nearby pastureland site. A design storm with varying rainfall intensities was simulated on each soil lysimeter and comparable runoff hydrographs were obtained. The effects of vegetative cover, i.e., biomass and grass height, on runoff and soil erosion compared closely with results observed for the prototype watersheds. Results of these studies will assist in the selection of those vegetative covers that minimize soil erosion and may suggest the proper grass height to be maintained for maximizing soil erosion control.

INTRODUCTION

The storm events that produce a large percentage of the runoff and soil loss are a small percentage of the total number of storms. Over a study period of 20 years there will occur only ten or twelve storms of a two-year or greater frequency including only four or five storms of a five-year or greater frequency (Meyer, 1958). These storms may be generally distributed throughout that 20-year period or several of these storms may occur within a relatively short period. For a given land use management practice, the frequency of occurrence of these storm events is critical since the storms contribute most to annual soil losses. Rainfall simulators have assisted in more rapid evaluation of various land use management practices to minimize soil erosion. Moreover, they have been useful in clarifying the contribution of rainfall, soil, vegetative cover, and land use management on the processes of infiltration, storage, runoff, soil erosion, and, to a lesser extent, rural and urban surface runoff of pollutants.

In general, most rainfall simulators fall within two categories: drip simulators and nozzle simulators. The most common drip simulators have water container boxes with drop-formers protruding from the bottom (Chow and Harbrough, 1965; Mutchler and Moldenhauer, 1968; Black, 1970; and Romkens et. al., 1975). One major advantage of the drip simulator is its ability to produce a combination of relatively large drops at a low intensity. However, to achieve impact velocities approaching those of natural rainfall, the simulator must be placed more than 10 m above the impact surface. Consequently, it is not practical for laboratory or field investigations. The major difficulty occurs at low rainfall intensities where realistic drop sizes cannot be achieved with high impact velocity (Mutchler and Hermsmeier, 1965).

Nozzle simulators are either designed in fixed assemblies (Wilm, 1943; Bertrand and Parr, 1961; Hall, 1969; Young and Burwell 1972; and Brenner, 1973), traversing rods (Tackett and Pearson, 1965; Young and Wiersma, 1973; and Sloneker and Moldenhauer, 1974) or with rotating disk assemblies (Morin, Goldberg, and Seginer, 1967; Morin, Cluff, and Powers, 1970). When nozzles are directed downward with pressure, the impact velocity is similar to the terminal velocity of natural rain-drops. However, large orifice openings are required to obtain realistic drop sizes at high velocities, which may result in excessive application rates. To reduce the intensity from large nozzles, there have been three different approaches: (a) turn the nozzle upward and spray a large area, e.g., Type F simulator; (b) physically move the nozzle back and forth across a plot, e.g., the Rainulator developed by Meyer and McCune (1958); or (c) physically remove a portion of the water from the high capacity nozzle to obtain realistic intensities, e.g., rotating disk simulator developed by Morin, Goldberg, and Seginer (1967).

The Type F simulator and the Rainulator represent significant improvements in rainfall simulations but both have strong disadvantages. Since the drops from the Type F simulator fall from zero velocity for only 3 m, the large drops lack sufficient energy to duplicate natural rainfall. The Rainulator developed at Purdue University utilizes a 80100 Vee-jet nozzle which sprays intermittently every 5 sec to reduce the intensity to a reasonable rate. This is accomplished by moving a series of spray bars back and forth over the test area and adjusting the number of spraying nozzles to vary the intensity. Though relatively large areas can be covered, the intensity cannot be varied during a given storm event simulation. Likewise, the kinetic energy of the simulated rainfall is not as high as natural rainfall.

To circumvent the major criticisms of the rainfall simulators, Morin, Cluff, and Powers, (1970), presented substantial justification for the selection of rotating disk rainfall simulators. The rotating disk simulator utilizes a full cone spray type nozzle (Spraying Systems Co., Fulljet 1-1/2 H30) similar in principle to those used by Bertrand and Parr (1961) but much larger in capacity. The unique feature of the rotating disk simulator is implied in its name. The rotation of the disk with various size slit openings produces a variety of intensities from near zero to the full nozzle capacity. Unfortunately, the intensities can not be easily varied to simulate actual storm events since the simulator must be turned off to change from one size slit width to another.

The purpose of this paper is to describe the construction and calibration of an improved rainfall simulator using the principle of the rotating disk and to discuss the results of a pilot study whereby storm events were simulated and runoff water quality was monitored. Our objective was to construct a rainfall simulator and soil lysimeter system that could be used in a controlled environment greenhouse to quantify the runoff-infiltration processes, soil erosion, and pollutant transport from various land uses. In addition, the rainfall simulator was designed to be portable for use at remote field sites using a self-contained trailer. The rainfall simulator and soil lysimeter system was designed to assist in:

a. Assessing land use management effects on soil erosion and pollutant transport.

b. Evaluating soil stabilization techniques.

WATERSHED MODELING FACILITY

The watershed modeling facility consists of a 12- by 18-m controlled environment greenhouse. Air temperatures are maintained through the range of $20^\circ - 35^\circ C \pm 1^\circ C$. The desired temperature control is obtained using air handlers which provide about two air changes per minute with an air velocity of less than 60 m per minute. The air temperature control system is designed to have a maximum rate of rise and fall of $8^\circ C$ per hour and $6^\circ C$ per hour, respectively, under maximum solar load. Air temperature is programmed via independent cam-controllers. Simulation of light intensities ranging from beneath a forest canopy to that of full sunlight can be obtained using commercially available shading fabric and supplemental lighting. The supplemental lighting consists of 12 high intensity discharge (HID) lamp fixtures arranged as a movable and programmable light bank. The light bank will provide up to 180 watts per square meter total lighting on the soil surface within the 400- to 700-nm wave lengths. This compares favorably with the 200 watts per square meter of natural radiation in the same range (sun plus sky) at solar noon during the summer solstice (Hutchinson, 1957). Control of relative humidity from 50 to 95 percent ($\pm$5 percent) is limited to 24°C dewpoint on the high end and 13°C on the low end of the control range. This control is obtained using a system of stainless-steel air-atomizing nozzles and deionized water regulated by humidstats and programmed via independent cam-controllers. A maximum of 2 hr is required to raise or lower the relative humidity throughout its control range when the air temperature is between 20°-35°C.

As many as eight 1.2- by 4.6-m lysimeters with variable slope and depth and three programmable rotating-disk rainfall simulators have been designed for the greenhouse (Figure 1). Currently, four lysimeters and two simulators have been completed. The lysimeter, constructed of aluminum with a chemically inert interior, e.g., polyvinyl chloride (PVC) liner, is supported by four load cells capable of detecting a 1.3-kg weight change for quantifying water storage and evapotranspiration. The soil depth can be varied from 15 to 45 cm in 15-cm increments with a maximum surface slope of 20 percent (11.5°).

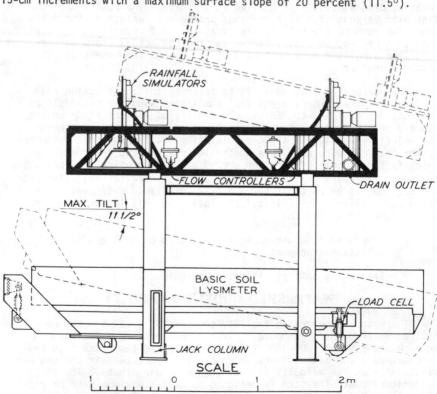

Figure 1. Rotating-disk rainfall simulator above a variable slope and depth soil lysimeter.

RAINFALL SIMULATOR DESIGN

Several criteria were established to improve the design and operation of the rotating-disk for this rainfall simulator:

a. It must possess the capability to uniformly apply realistic simulated rainfall to various land uses and soil type conditions.

b. It must possess the capability to simulate a broad range of rainfall intensities during a given storm event with a slight effect on drop-size distribution.

c. Drop size, fall velocity, and kinetic energy characteristics must be similar to natural storms.

d. It must be portable for transport to remote research sites.

e. It must possess the capability to allow varying water quality constituents of rainfall.

Some features of the U. S. Army Engineer Waterways Experiment Station (WES) rainfall simulator are similar to the one reported by Morin, Cluff, and Powers (1970); however, several important changes were made to meet the aforementioned criteria. The entire unit was constructed from stainless steel to permit manipulation of rainfall water quality, e.g., acid rainfall. Moreover, parallel attachment of two basic rainfall simulator units was performed to permit a uniform rainfall coverage area, i.e, 5.5 m^2 (Figure 2). The conical rotating disk described by Morin, Cluff, and Powers (1970) was replaced by a flat disk since the conical shape was not functional. The gear box and drive assembly were modified significantly to permit the development of a programmable slit width. The rotating disk (20.3 cm diam.) was mounted to the 200 rpm drive assembly shaft and the stainless steel 1-1/2 H30 nozzle was mounted to the 4 rpm shaft. Also, a 40-deg slit was cut radially from the center of the disk. Stepping motors were used in conjunction with the slip ring assembly to control the movement of the stainless steel cover as it progressed over the 40-deg slit opening in the rotating disk. Thus, rainfall intensity could be changed during a simulated storm event by electrical impulses to the stepping motors which would open or close the slit opening cover. This modification permitted the varying of simulated rainfall intensity to reproduce the design storm hydrograph. A Data Trak Model 5310 controller uses a programmable card on which design storm event hydrographs are plotted. The controller translates the plotted image via electrical impulses to the stepping motors whereby the slit opening cover is moved accordingly. Rainfall intensities in increments of 0.2 cm/hr up to 7.0 cm/hr and a duration of up to approximately 3 days could be reproduced precisely when the perpendicular distance from the nozzles to the ground surface was 2.0 m and the nozzles were rotating at a 4-deg angle from the perpendicular. The cocking of the nozzles to one side provided the most uniform rainfall distribution over the 5.5-m^2 target area.

Calibration of the rainfall simulator was focused on achieving a uniform drop-size distribution similar to natural rainfall with all drops falling at their terminal velocities. By adjusting the vertical distance from the nozzles to the impact surface area, the angle of the nozzles, and the nozzle pressure, the optimal drop characteristics and lowest coefficient of variation in spray distribution were determined

91

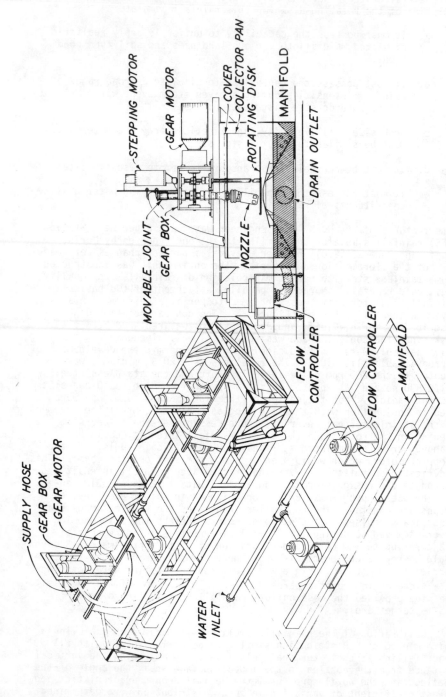

Figure 2. Schematic of the WES rotating-disk rainfall simulator.

for selected rainfall intensities and expressed as percentage opening of the programmable slit width. By changing the nozzle pressure, i.e., 50 to 85 N/m^2, only minor fluctuations were observed in the spray distribution coefficient of variation, i.e., 14 to 16 percent; however, the rainfall intensity increased approximately 40 percent. By increasing the nozzle distance from 2.0 m to 2.2 m above the impact surface area, the spray distribution coefficient of variation nearly doubled, i.e., 14 percent to 28 percent. Table 1 represents optimum design conditions whereby the perpendicular distance from the nozzle to the impact surface area was 2 m and the nozzle water pressure and angle was 65 N/m^2 and 4 degrees, respectively. The coefficient of variation over a 5.5-m^2 area for the simulated rainfall distribution was 24 percent at 25 percent slit opening and less than 14 percent at 75 percent slit opening. This was caused primarily by the greatly reduced slit-opening area toward the center of the rotating disk compared to the outer area of the slit opening, as the slit width was decreased from the 75 to 25 percent setting.

Once the optimum nozzle angle and nozzle distance above the impact area were selected to yield the lowest coefficient of variation in spray distribution over the 5.5-m^2 area, further calibration efforts were directed at achieving a drop-size distribution and fall velocity typical of natural rainfall. The drop-size distribution of simulated rainfall was determined by the basic method of Laws and Parsons (1943) which was later modified by Meyer (1958). This involved raining over pans of finely sieved flour whereby the drops were formed into flour-dough pellets of similar size to the raindrops; the pellets were oven-dried at 70°C overnight to completely dry, then sieved into appropriate size fractions.

A median drop diameter of approximately 2.0 mm was measured for each rainfall intensity from 1.7 to 7.2 cm/hr (Table 1). This was slightly less than the median of 2.5 mm for a 5-cm/hr rainfall previously reported by Laws and Parsons (1943). Through this broad range of simulated rainfall intensities, only a slight shift toward smaller drop sizes occurred; however, the percent composition of raindrops within the respective drop-size distributions of simulated and natural rainfall were similar (Figure 3). The terminal velocities versus water drop diameter for drops from the WES rainfall simulator as compared to

Table 1. Simulated rainfall intensity, median drop diameter, and rainfall distribution coefficient of variation at selected slit openings.

Slit opening (%)	Rainfall intensity (cm/hr)	Median drop diameter (mm)	Coefficient of variation (%)
25	1.7	1.9	24.0
50	3.3	1.9	13.5
75	5.4	2.0	13.8
100	7.2	2.1	14.0

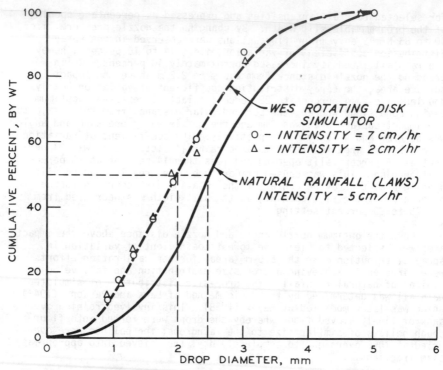

Figure 3. Drop-size distribution for the WES rotating disk rainfall
simulator using two 1-1/2 H30 nozzles at a pressure of
65 N/m² and a height of 2 m.

the maximum terminal velocities of freely falling drops is shown
in Figure 4.

COMPARISON OF RAINFALL CHARACTERISTICS WITH
OTHER SIMULATORS

Increased knowledge of raindrop characteristics and recognition of
their importance in soil erosion has resulted in the requirement that
simulated rainfall approximate not only the amount and intensity of a
design storm but also a similar drop-size distribution with all drops
falling at their terminal velocities. Unfortunately, physical limitations of all known methods have prevented this achievement. Consequently, Meyer (1965) has suggested several parameters as being proportional to rainfall erosivity which include: (a) kinetic energy
$(1/2 \ MV^2)$; (b) momentum (MV); (c) kinetic energy per unit of drop-
impact area $(1/2 \ MV^2/A_d)$; (d) momentum per unit of drop-impact area
(MV/A_d); and (e) interactions of these variables with rainfall intensity. A comparison of those simulators reviewed by Meyer (1965),
expressed as a percent of natural rainfall, is given in Table 2 along
with similar information for the WES rainfall simulator.

The energy characteristics of natural rainfall are dependent on
many factors other than intensity, e.g., wind speed, moisture content
in air, and air temperature. Morin (1970) summarized previous work
showing that kinetic energy per unit mass increases with increasing
intensity. Nozzle 5B, which is used in the Sprinkling Infiltrometer,
produces only one intensity; whereas, the 80100 nozzles of the Rainula-

Table 2. Comparison of relative erosivity ratios of selected rainfall simulators and natural rainfall. After Meyer (1965) except for the WES rainfall simulator.

Parameter	Natural Rainfall	Rainfall Simulator*				
		C	F	P	R	WES
		(percent)				
Kinetic energy per unit of rainfall ($\propto\Sigma pV^2$)	245,100 m·kg/ha·cm	44	56	26	77	95
Momentum per unit of rainfall ($\propto\Sigma pV$)	58,240 kg·sec/ha·cm	68	76	49	87	97
Total kinetic energy per unit of total drop impact area ($\propto[\Sigma D/p]$ ΣpV^2)	0.189 m·g/m	82	72	9	62	90
Total momentum per unit of total drop impact area ($\propto[\Sigma D/p][\Sigma pV]$)	0.044 kg·sec/m	126	98	18	70	93
Kinetic energy per unit of drop impact area (by increments) ($\propto\Sigma pDV^2$)	0.256 m·g/m²	65	70	10	63	71
Momentum per unit of drop impact area (by increments) ($\propto\Sigma pDV$)	0.056 kg·sec/m²	105	97	18	72	72

$\propto$ = proportionality factor

p = portion by weight in a given drop-size group

V = terminal velocity

D = drop diameter

* = Description as follows:

Type C - produces a nearly uniform drop size which is much larger than most raindrops. The drops fall from zero velocity for a distance of only 1.4 meters.

Type F - produces a drop-size distribution larger than intense rainfall. The drops fall from zero velocity for a distance of only 3 meters.

Type P (Peoples) - produces a drop-size distribution much smaller than intense natural rainfall. The drop falls from zero velocity for a distance of only 3 meters.

Type R (Rainulator) - produces a drop-size distribution slightly smaller than intense natural rainfall. Drop velocities are near terminal velocities except for large drop sizes.

WES - produces a drop-size distribution slightly smaller than intense natural rainfall. Drop velocities are near terminal velocities except for large drop sizes.

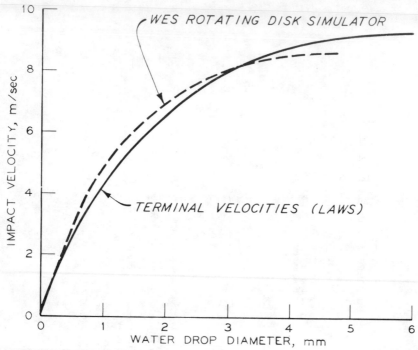

Figure 4. Impact velocity of rain drops from the WES rainfall simulator compared with maximum terminal velocities of freely falling drops.

tor requires spray overlapping to achieve two different intensities. The kinetic energy per unit of rainfall from the Rainulator would be similar at both intensities. The WES rainfall simulator has the advantage of remotely programming various intensities and thereby changing the kinetic energy per unit of rainfall during a simulated design storm by varying the slit opening on the rotating disk (Table 3). The effects on kinetic energy were similar to those reported by Wischmeier and Smith (1958) for natural rainfall. Changes in the slit opening slightly influenced the drop-size distribution and hence the kinetic energy per unit mass.

COMPARISON OF RUNOFF AND SOIL EROSION RESPONSE FROM PROTOTYPE PLOTS AND LABORATORY WATERSHED MODEL

A pilot study was conducted to test the effectivenss of the WES rainfall simulator for duplicating a design storm event on two soil lysimeters. Characteristic runoff quantity and suspended solids were compared with prototype data. Each lysimeter contained Grenada silt loam: one was obtained from an overland flow wastewater treatment site near Utica, Mississippi; and, the other was taken from a cow pasture in the same area. The soil was collected in 0.1 m^2 blocks, 15 cm deep, with the grass cover intact. The soil blocks were transported to the WES where they were carefully placed into each of the lysimeters. The two soil lysimeters were similar except for the grass coverage.

The lysimeter representing the cow pasture had a dense grass cover per unit area. The soil surface was not visible. The other lysimeter had much less grass cover with a large portion of the soil surface exposed.

Table 3. Relative erosivity ratios of WES rotating disk rainfall simulator at selected rainfall intensities.

	Rainfall intensity (cm/hr)			
	7.2	5.0	3.3	1.7
Kinetic energy per unit of rainfall (m·kg/ha·cm)	235,050	230,733	226,717	222,511
Momentum per unit of rainfall (kg·sec/ha·cm)	57,085	56,473	55,965	55,359
Total kinetic energy per unit of total drop impact area (m·kg/m²)	0.180	0.175	0.171	0.166
Total momentum per unit of total drop impact area (kg·sec/m²)	0.043	0.042	0.041	0.040
Kinetic energy per unit of drop impact area (by increments) (m·kg/m²)	0.184	0.176	0.170	0.162
Momentum per unit of drop impact area (by increments) (kg·sec/m²)	0.041	0.039	0.038	0.036

Prototype plots from the Utica, MS, overland flow wastewater treatment site were selected. These plots were 46 m in length and 4.6 m wide. Vegetative cover and plant biomass were similar to the cow pasture lysimeter. Each plot was equipped for automatic rainfall-runoff monitoring and sampling. The design storm event occurred in December of 1978 and lasted 45 min. while delivering 1.83 cm water. Runoff rates were measured using a 5-cm Parshall flume and an automatic water level recorder. Water samples were collected automatically using a refrigerated Isco sampler.

The design storm event was programmed for the WES simulator using the Data Trak (Model 5310) controller. Runoff rates from each soil lysimeter were determined from measured water sample volumes collected for 5 sec every minute during the storm simulation. Water samples were selected for suspended solids analysis from the ascending and descending portions of the hydrograph. Measurable runoff from the prototype field plots started approximately 20 min later than the two soil lysimeters. This was attributed to the fact that the field plots were ten times longer than the soil lysimeters.

The runoff hydrographs were similar from the soil lysimeter representing the cow pasture and the prototype Utica field site which had similar grass biomass (Figure 5). Though the design storm was complex, i.e., varying rainfall intensities, surface runoff from the cow pasture lysimeter and the prototype field site representing the first and second storm peaks was very similar. Slightly higher peak runoff was observed for the cow pasture lysimeter resulting from less water storage capacity of the soil compared to the prototype. However, the runoff hydrograph from the overland flow soil lysimeter, with much greater exposed soil surface area, had a predictably higher peak runoff rate that occurred earlier than either the lysimeter representing the cow pasture or the Utica field site. From Table 4, the amount of total runoff (cm) from the prototype (Utica) field site and the two lysimeters was very close. Moreover, the percentage of rainfall runoff from the prototype field site compared closely to runoff from the overland flow and cow pasture soil lysimeters. The lower plant biomass level on the overland flow soil lysimeter resulted in reduced rainfall interception and storage, thus showing a slightly higher runoff. This same effect was more noticeable when comparing the peak runoff (cm/min) from each system.

Table 4. Comparison of runoff between the Utica Field Site during a natural storm event and two soil lysimeters during simulated rainfall.

Site	Runoff (cm)	Runoff % of rainfall	Peak runoff (cm/min)	Peak runoff % natural storm event
Utica Field Site (Natural Storm Event)	1.83	87.8	0.077	--
Overland Flow Soil Lysimeter (Simulated Storm Event)	1.97	90.1	0.142	184
Cow Pasture Soil Lysimeter (Simulated Storm Event	1.91	87.2	0.088	114

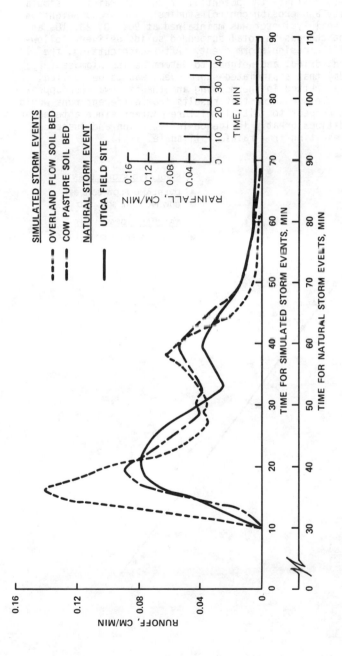

Figure 5. Comparison of simulated storm runoff hydrographs from the overland flow and cow pasture soil lysimeters with the prototype natural storm event runoff hydrograph from the Utica field site.

99

The effects of a vegetative canopy on suspended solids delivery were investigated to evaluate the potential of the WES rainfall simulator in water quality and erosion control studies. The grass height on the overland flow soil lysimeter was maintained at 90, 30, 20, 10, and 0 cm to observe the changes in total suspended solids delivery following application of the design storm event. After each cutting, the grass was collected, dried, and weighed to determine its biomass (dry weight). On the day that a simulated storm event was to be applied, the surface soil was wetted in the morning and the test was run approximately 4 hr later. This ensured that results from different runs would be comparable with respect to initial moisture content since antecedent soil moisture conditions greatly influence infiltration and suspended solids transport resulting from rainfall (Tisdale, 1951).

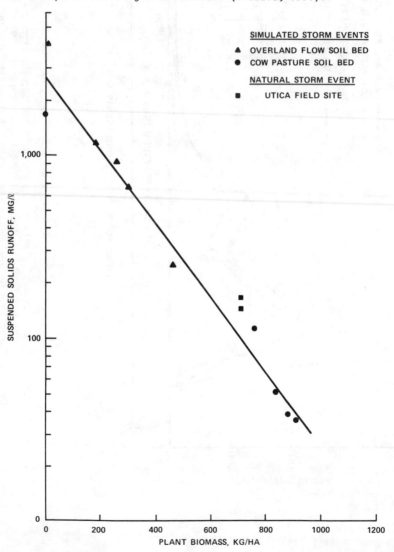

Figure 6. Relationship of plant biomass to the suspended solids load in simulated storm runoff.

The grass biomass versus the peak suspended solids concentration for each run was plotted (Figure 6) and a regression analysis of the data gave the equation:

$$Y = 10^{-0.002(X) + 3.421}$$

where

Y = suspended solids concentration in runoff, mg/ℓ
X = plant biomass, Kg/ha

The correlation coefficient (R) was -0.99. Hence, the suspended solids in runoff was very sensitive to changes in the vegetative biomass. Also plotted on the same graph are suspended solids and plant biomass data from two natural storm events that occurred at the Utica field site. Data from these natural events closely agreed with the curve derived from using the rainfall simulator/soil lysimeter system. Moreover, the overland flow soil lysimeter with less plant biomass produced more suspended solids in runoff than either the Utica field site or the cow pasture lysimeter, which both had greater and similar amounts of plant biomass.

SUMMARY

The WES rainfall simulator/soil lysimeter system has been described in terms of design criteria, construction, and performance evaluations. This rainfall simulator may be used under controlled-environment laboratory conditions or mounted on a truck and transported to remote field sites as a portable rainfall simulator. The rainfall characteristics from the WES rotating disk simulator closely duplicated natural rainfall. Moreover, the ability to program storm events with varying intensities and only slight changes in drop-size distribution is very important in the quantification of infiltration and soil erosion processes and development of improved predictive techniques.

Assessment of land use effects on runoff water quantity and quality, evaluation of soil stabilization techniques, and contaminant mobility from the land surface are other applications for which the WES rotating disk rainfall simulator is well suited. The laboratory watershed model has been shown to be a valuable tool in quickly determining the effects of various land use management decisions on runoff water quality and soil erosion.

ACKNOWLEDGEMENT

Special acknowledgement is given to J. Haskins, Mechanical Engineer at the WES, who provided much of the technical help in improving the design of the rotating disk rainfall simulator. Likewise, recognition is given to: E. Hummert, Mechanical Engineer Technician at WES, who assisted in identifying needed modifications to the rainfall simulator and lysimeters; and B. Reed, Electrical Engineer at the WES, who provided the technical help in developing the programmable controller. Also, special recognition should be given to M. Brodie and B. Gehman, cooperative students at the WES, who assisted in the calibration of the rotating disk simulator.

Appreciation is also extended to Dr. C. R. Lee, Soil Scientist at the WES, for his advice and constructive comments in the calibration of the simulator and conduct of the pilot study.

The development and calibration of the rainfall simulator as well as the pilot study described herein were supported under the Environmental Impact Research Program of the United States Army Corps of Engineers by the Waterways Experiment Station. Permission was granted by the Chief of Engineers to publish this information.

REFERENCES

Black, P. E. 1970. Runoff from Watershed Models. Water Resources Research Vol. 6, pp. 465-477.

Bertrand, A. R., and Parr, J. F. 1961. Design and Operation of the Purdue Sprinkling Infiltrometer. Purdue University Agricultural Experiment Station, Research Bulletin No. 723, 16 pp.

Brenner, R. P. 1973. A Hydrological Model Study of a Forested and a Cutover Slope. Hydrological Science Bulletin, Vol. 26, pp. 125-144.

Chow, V. T., and Harbrough, T. E. 1965. Raindrop Production for Laboratory Watershed Experimentation. Journal of Geophysical Research, Vol. 70, No. 24, pp. 6111-6119.

Hall, M. J. 1969. The Design of Nozzle Networks for the Simulation of Rainfall. Journal of Hydrological Research, Vol. 7, No. 4, pp. 449-483.

Laws, J. O., and Parsons, D. A. 1943. The Relation of Raindrop-Size to Intensity. Transactions American Geophysical Union, Vol. 24, pp. 452.

Meyer, L. D. 1958. An Investigation of Methods for Simulating Rainfall on Standard Runoff Plots and a Study of the Drop Size, Velocity, and Kinetic Energy of Selected Spray Nozzles. Purdue University, Special Report No. 81.

Meyer, L. D. 1965. Simulation of Rainfall for Soil Erosion Research. Transactions American Society of Agricultural Engineers, Vol. 8, No. 1, pp. 63-65.

Meyer, L. D, and McCune, D. L. 1958. Rainfall Simulator for Runoff Plots. Agricultural Engineer Vol. 39, No. 10, pp. 644-648.

Morin, J., Goldberg, D., and Seginer, I. 1967. A Rainfall Simulator with a Rotating Disk. Transactions American Society of Agricultural Engineers, Vol. 10, No. 1, pp. 74-79.

Morin, J., Cluff, D. B., and Powers, W. R. 1970. Realistic Rainfall Simulation for Field Investigation. Presented at 51st Meeting of American Geophysical Union in Washington, D. C., 22 pp.

Mutchler, C. K., and Moldenhauer, W. C. 1968. Applicator for Laboratory Rainfall Simulator. Transactions American Society of Agricultural Engineers, Vol. 6, pp. 220-222.

Mutchler, C. K. and Hermsmeier, T. F. 1965. A Review of Rainfall Simulators. Transactions American Society of Agricultural Engineers, Vol. 8, pp. 67-68.

Romkens, M. J. M., Glenn, T. F., Nelson, D. W., and Roth, C. B. 1975. A Laboratory Rainfall Simulator for Infiltration and Soil Development Studies. Soil Science Society of American Proceedings, Vol. 39, pp. 158-160.

Sloneker, L. L. and Moldenhauer, W. C. 1974. Effect on Varying the On-off Time of Rainfall Simulator Nozzles on Surface Sealing and Intake Rate. Soil Science Society of America Proceedings, Vol. 38, pp. 157-159.

Tackett, J. L., and Pearson, R. W. 1965. Some Characteristics of Soil Crusts Formed by Simulated Rainfall. Soil Science, Vol. 99, No. 6, pp. 407-413.

Tisdale, A. L. 1951. Antecedent Soil Moisture in Its Relation to Infiltration. Australian Journal of Agricultural Research, Vol. 2, pp. 342-348.

Wilm, H. K. 1943. The Application and Measurement of Artificial Rainfall on Types FA and F Infiltrometers. Transactions American Geophysical Union, Vol, 24, pp. 480-487.

Wischmeier, W. H., and Smith, D. D. 1958. Rainfall Energy and Its Relationship to Soil Loss. Transactions American Geophysical Union, Vol. 39, pp. 284-291.

Young, R. A., and Burwell, R. E. 1972. Prediction of Runoff and Erosion from Natural Rainfall Using a Rainfall Simulator. Soil Science Society of America Proceedings, Vol. 36, pp. 827-830.

Young, R. A., and Wiersma, J. L. 1973. The Role of Rainfall Impact in Soil Detachment and Transport. Water Resources Research, Vol. 9, No. 6, pp. 1629-1636.

INFILTRATION

RATIONAL MODELS OF INFILTRATION HYDRODYNAMICS

Roger E. Smith
Research Hydraulic Engineer
USDA-SEA, Fort Collins, Colorado

ABSTRACT

Recent advances in porous media hydrodynamics have produced an elegant comprehensive structure for the relation of rainfall pattern to the evolution of the wetting profile in a homgeneous soil, and thus to the time when runoff begins. In addition, the same basic concepts provide a basis for derivation of the infiltrability decay function. These principles of porous media hydrodynamics are stated in three infiltration theorems. Four analytically derived infiltration models, based on different assumptions concerning soil behavior but in accordance with the three theorems, are introduced and compared. Each reduces to a similar shape function in dimensionless coordinates.

The extension of homogeneous soil infiltration concepts to cases of inhomogeneity and anisotropy is explored briefly. Several cases of layered systems are examined by use of a rigorous solution of the governing Fokker-Plank equation, and the limits of simplification in general layered systems demonstrated. The effect of random spatial distribution in biasing results for "equivalent" mean conditions is also pointed out. Parameterization of the various effects of air phase flow and treatment of the physical changes to which soils are subject are two of the more important challenges facing the model representation of the infiltration phenomenon.

INTRODUCTION

From an extensive and intensive amount of research into the mechanics of flow in porous media, hydrologic engineering has progressed over the past few decades beyond the use of crude algebraic empiricisms for the description of rainfall infiltration. Although such methods as the so-called 'rational formula' are still in use as a form for estimation in the face of ignorance, or for convenience, we now have a soundly based, physically consistent and comprehensive theory of the intake of water by a soil in response to a pattern of rainfall. Moreover, with two physically based soil parameters, this theory yields rather simple functions for use in modeling or analyzing infiltration on a catchment.

In this paper, I will outline this soil infiltration model, its relation to soil water movement and its application in several cases of heterogeneous conditions such as commonly faced in catchment modeling. I will cover very little of the mathematics of porous media flow theory, which encompasses a vast body of knowledge, as has been attempted by Philip (1957) and others. Much of mathematical soil physics is of secondary interest in hydrology, because often the interest of soil physics centers on distribution and movement of water within the soil and because soil properties and boundary conditions are often chosen for mathematical facility rather than practical representativeness. This is not to criticize the contributions of soil physics, to whose scientists we in hydrology owe the bulk of the theory on which depends the infiltration model presented below. In outlining the current understanding of soil infiltration behavior it is useful to start with a brief background of porous media hydrodynamics.

BASIC CONCEPTS

Water exists in an unsaturated porous media (soil) in tiny intergranular capillary pockets and exhibits a defining relation between water content, θ, and (negative) capillary potential, Ψ. In addition, a soil exhibits a range of hydraulic conductivity, $K[L/T]$, which may be thought of as the net rate at which water will move in a soil in response to a unit potential gradient. These three properties are related monotonically with each other, as exemplified in Figure 1. Thus we may speak of $K(\Psi)$, $K(\theta)$, $\theta(K)$, . . . etc. θ_s represents soil water content at $\Psi=0$, and K_s is corresponding (maximum) conductivity. These two correspond to $\Psi = 0$.

The relations shown in Figure 1 are modified when a soil undergoes cycles of wetting and drying, forming a hysteritic relationship, but the simple relations here are adequate if we are concentrating on infiltration during rainfall periods.

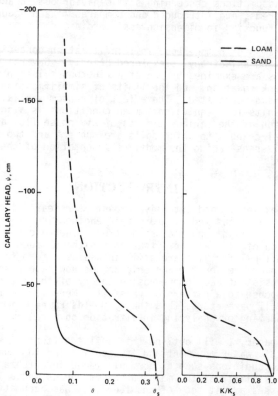

Figure 1   Examples of water content, θ, and hydraulic conductivity, K, as related to soil capillary head, Ψ, for two contrasting soil types.

Water flow in porous media is assumed to obey the following simple laws, which I will state as an initial theorem of infiltration:

Theorem I: Water moves in soil accordance with the laws of a) Darcy and b)mass conservation.

Darcy's law proposes that flow in any direction, q, is described by the negative product of the K and the hydraulic gradient of total potential, H: For flow in the z direction:

$$q = -K \frac{\partial H}{\partial z} \tag{1}$$

For z measured downward from the soil surface, unsaturated H is $\Psi - z$, and eq. (1) becomes

$$q = -K(\Psi)\left[\frac{\partial \Psi}{\partial z} - 1 \right]. \tag{2}$$

Continuity or mass balance may be expressed as

$$\frac{\partial \theta}{\partial t} + \frac{\partial q}{\partial z} = 0 \tag{3}$$

which when combined with eq. (2) becomes

$$\frac{\partial \theta}{\partial t} + \frac{\partial}{\partial z}\left(K \frac{\partial \Psi}{\partial z}\right) - \frac{\partial K}{\partial z} \tag{4}$$

which is a non-linear Fokker-Plank equation commonly referred to as Richard's equation. It ignores the complicating effect of the counter-flow of air, which must escape, move or be compressed as water is imbibed. For one-dimensional flow of the true 2-phase system two equations result. Morel-Seytoux (1973, 1976) has written extensively on the air-water system. I will concentrate here on the one-phase approximations and return to the air effects briefly later.

When water is applied at a rate r to the surface of an unsaturated soil, the mathematical model of the process is the application to eq. (4) of the boundary condition from eq. (2), i.e. the surface flux is

$$r = -K(\Psi)\left[\frac{\partial \Psi}{\partial z} - 1\right] , \quad z = 0 \tag{5}$$

Since unsaturated soil exhibits large negative values of Ψ, the water content profile adjusts at small times to satisfy eq. (5) at $z = 0$, so that $\theta(z = 0)$ and $\Psi(z = 0)$ both increase at small times to satisfy q=r. An important point from Theorem I, expressed in Eq (t), is that infiltration is fundamentally an intake of water in response to a potential gradient, and not a storage filling process as so commonly represented in conceptual models.

As t increases, for a uniform soil, two cases arise depending on the value of r. If $r < K_s$, steady r of sufficient duration will approach the asymtotic condition $\theta^S = \theta(K = r)$, or $K(\theta) = r$. If $r > K_s$, adjustment of θ and Ψ at the surface in satisfaction of eq. (5) can only continue until $\theta = \theta_s$, $K = K_s$. At this time the rate of intake at the surface begins to fall below r, rainfall excess appears, and we refer to this as the time of ponding t_p. Rainfall excess is defined as $r - q(z = 0)$, and we denote $q(z = 0)$ by $p_f(t)$. The evolution of water content profiles within the soil during this time, illustrated in Figure 2, must also follow a volume balance equation

$$\int_0^t r \, dt = \int_\infty^0 (\theta - \theta_i) dz = \int_{\theta_i}^{\theta(t)} z(\theta) d\theta. \tag{6}$$

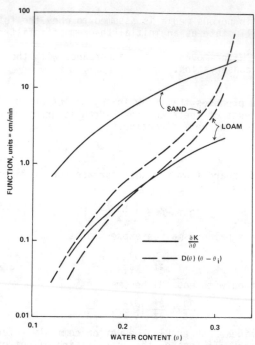

Figure 2 Water content profiles at various times for rainfall fluxes of either 4 or 8 cm/hr. Relative saturation S is θ/ϕ, and the results here are obtained by numerical solution of eq. (4).

The long-time asymptotic condition, after t_p, is $\theta = \theta_s$, and thus infiltration rate f approaches K_s in a uniform soil at large time, as $\partial\psi/\partial z$ (z = 0) approaches 0.

Infiltration to Ponding

To introduce the infiltration model produced by soil flow theory, it is useful to recast eq. (4) somewhat, in analogy to a diffusive wave equation. Introduce diffusivity, D, defined as

$$D(\theta) = K(\theta)\,\frac{d\psi}{d\theta} \tag{7}$$

so that eq. (4) becomes

$$\frac{\partial\theta}{\partial t} = \frac{\partial}{\partial z}\left(D\,\frac{\partial\theta}{\partial z}\right) - \frac{\partial K}{\partial z} \tag{8}$$

This Fokker-Plant equation would be a diffusive wave equation for constant D. Also, eq. (2) becomes

$$q = -D(\theta)\,\frac{\partial\theta}{\partial z} + K(\theta) \tag{9}$$

Solving eq. (9) for dz and substituting into eq. (6) produces

$$\int_0^t r\,dt = \int_{\theta_i}^\theta (\theta-\theta_i)\,\frac{D(\theta)}{(q-K(\theta))}\,d\theta \tag{10}$$

Different assumptions on D and q may be used (q = r at z = 0), but this expression defines a unique relation between r and t when $\theta_t = \theta_s$ at z = 0, and eq. (10) is the basis for the following; with $I \equiv \int_0^t f\,dt$:

Theorem II: For $r > K_s$, at a given initial water content θ_i, there is a unique relationship between r and infiltrated depth of water I at time of ponding, t_p.

This theorem does not contradict the principle that infiltration is related to potential gradients, but rather stresses the importance of accurate representation of the $\theta(\psi)$ relation through the function $D(\theta)$. Although the discussion above implicitly deals with uniform rainfall r, by specifying the dependence on r at time of ponding, this theorem includes the experimentally observed fact (Smith, 1971; Smith and Parlange, 1978) that the water content profile, as it developes for $t < t_p$, will adjust to moderate changes in r. This implies, however, that the theorem will not hold well for severe changes in r immediately prior to ponding. Several relations have been derived consistent with Theorem II which make somewhat different assumptions to solve eq. (10).

Assuming D and q constant within the wetted soil water profile, or D and K step functions (Smith and Parlange, 1978), eq. (10) leads to the function for ponding time obtained conceptually by Mein and Larson (1973) for their extension of the Green and Ampt model to constant rainfall infiltration:

$$I_p = rt_p = \frac{A(\theta_i) K_s}{(r - K_s)} \tag{11}$$

in which A is a soil- and θ_i-dependent parameter, and K_s is K at $\psi = 0$.

If q is assumed constant within the profile and $D(\theta - \theta_i)$ assumed closely proportional to $dK/d\theta$, the expression for ponding of Parlange and Smith (1976) is obtained:

$$I_p = \int^{t_p} r\, dt = B(\theta_i)\, \ell n\, [r_p/(r_p - K_s)] \tag{12}$$

where r_p denotes r at t_p for time varying r, and B is a parameter analogous to A. Figure 3 shows for two example soils the accuracy of the assumption on $dK/d\theta$. Neither soil is necessarily representative in the D or $\partial K/d\theta$ functions illustrated.

White et al. (1979) use eq. (10) with a similarity expression for $q(\theta, \theta_o)$, called the flux concentration relation, and numerically evaluated the evolution of $\theta(z = 0)$ for $t < t_p$ at constant r. An expression for $t_p(r)$ is not obtained, but profiles of $\theta(z)$ result from observed values of $D(\theta)$ and eq. (10).

Morel-Seytoux (1976), using an expression analogous to eq. (5) vut including parameteric treatment of the expected effects of air, plus a series of simplifying assumptions on the resulting functions, has derived an analogous expression for constant r.

$$I_p = rt_p = \frac{(\theta_s - \theta_i)H}{1 - f_i}\; [\exp(K_s/(r - K_s)) - 1] \tag{13}$$

where H is "effective capillary drive" and f_i is a function of θ_i.

Infiltration Decay

The fundamental assumptions behind Theorem II imply that for all $r > K_s$, there is a water content profile uniquely associated with each r at ponding, and in fact an association between the $\theta(z)$ profile, I and r

111

prior to ponding. Application of the same basic assumptions to the extension of the profiles, now with a fixed θ at $z = 0$, for times beyond t_p, leads to the following:

Theorem III: For $t > t_p$ during $r > K_s$ the water content profile $\theta(z)$ for infiltrability $f(f < r_p)$ is essentially the same as the $\theta(z)$ at t_p for $r_p = f$.

This theorem is rather clearly seen for step function or "piston" water profiles, which implies a universal association between infiltrated depth $I = \int_0^t f\ dt$ ($f = r$ for $t < t_p$), and f itself. But such an association is precisely what is denied in such algebraic formulae as Horton's or Kostiakov's equation, which are often used, irrespective of rainfall patterns, to "separate" rainfall excess.

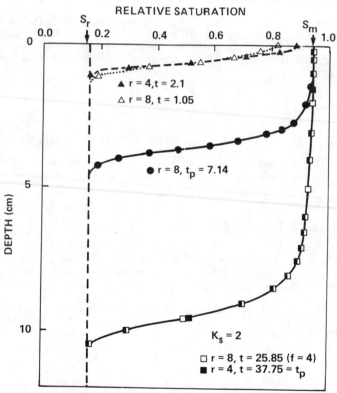

Figure 3 Comparison of the functional relations between $\partial K/\partial\theta$ and θ, and $D(\theta)(\theta-\theta_i)$ and θ. Equation (12) assumes the two functions have similar shape.

The correctness of this theorem is not so clearly seen for soil water profiles which are not step functions. As a demonstration of its validity, eq. (4) is solved numerically, subject to condition (5), for $r = 8$ cm/hr and $r = 4$ cm/hr and the results are plotted in Figure 4. The theorem says that the water content profile for $f = 4$ under $r = 8$, point (a) during the decay of infiltrability, should closely compare to the profile at $t = t_p$ for $r_p = 4$, point (b). Figure 3 includes these two profiles, which are remarkably similar considering the approximations inherent in Theorem III.

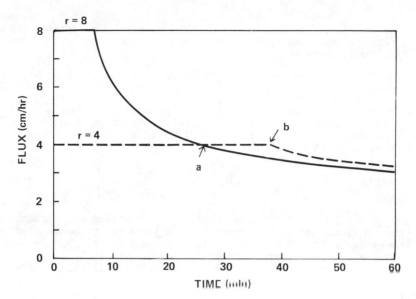

Figure 4   Infiltration flux patterns for events of r = 4 and r = 8 cm/hr
through a soil (Sand, figure 1) with K_s = 2.0.

More importantly, this theorem implies that, as indicated above, a
relation exists between I and f for all $r > K_s$ and this is a true "infiltration capacity" law, or infiltrability model.   To obtain a function for
the decay of f with time after t_p, one may express f as dI/dt, with I
generalized from I_p, by differentiation of eq. (10) replacing surface
flux r by the time variable value of f.   Thus from eq. (10), the generalized infiltration expression is

$$f = \frac{\partial}{\partial t} \left[\int_{\theta_i}^{\theta} (\theta-\theta_i) \; \frac{D(\theta)}{(f(t)-K(\theta))} \; d\theta \right] \qquad [t \geq t_p] \quad (14)$$

The different assumptions on $D(\theta)$ mentioned above produce different
expressions for the term in brackets, but for eqs. (11) and (12) we
obtain from eq. (14) an expression in f and df/dt which is rearranged and
integrated from t_p to t to find f(t) or t(f).   From eq. (12) (Smith and
Parlange, 1978) one obtains:

$$\frac{(t - t_p)}{B(\theta_i)} = \frac{1}{K_s} \; \ell n \; \frac{(r_p - K_s)f}{(f - K_s)r_p} \; - \frac{1}{f} + \frac{1}{r_p} \qquad (15)$$

Similarly operating on eq. (11), we obtain the infiltration function for
the rainfall version of the Green and Ampt expression, as given by Mein
and Larson (1973):

$$\frac{t - t_p}{A(\theta_i)} = \frac{r_p - f}{(f-K_s)(r_p-K_s)} - \frac{1}{K} \; \ell n \; \frac{(r_p - K_s)f}{(f - K_s)r_p} \qquad (16)$$

This expression becomes the "sudden ponding" or original Green and Ampt
function when $r_p \rightarrow \infty$ and $t_p \rightarrow 0$.   Both these expressions have been verified for accuracy by laboratory and/or numerical experiments (Mein and
Larson, 1976; Smith and Parlange, 1978).

113

Reasoning in reverse, one can obtain a rainfall infiltration expression from the truncated series expression for ponded infiltration from Philip (1957), which is

$$f = \tfrac{1}{2}St^{-\frac{1}{2}} + A \tag{17}$$

and was intended to describe infiltration at short times from a $\theta(z = 0) = \theta_s$ boundary condition. Integrating (17) to obtain an expression for I, we can eliminate t between the two expressions to obtain an expression analogous to (11) and (12) which is

$$I_p = rt_p = \frac{C_p}{2} \left[\left(\frac{r}{(r - K_s)} \right)^2 - 1 \right] \tag{18}$$

where $C_p = S^2/2 \, K_s$, and $S(\theta_i)$ is sorptivity, as defined and discussed by Philip (1957) and others in connection with infiltration parameters (Parlange and Smith, 1976). It should be noted that eq. (18) does not come from eq. (1) by explicit assumptions on D, q or $K(\theta)$. A similar expression was presented (but not used) by Eagleson (1978), who cited Linsley et al. (1958) as a source for the idea of an I-related f capacity relation. Interestingly, Linsley et al. (p. 179) merely state "It is often assumed that the infiltration capacity at any time during a storm is determined by the mass infiltration which has occurred up to that time". However, there is little evidence that such was done in hydrologic practice, and Linsley et al. give no citation or example procedure. Such use of an f(I) relation was suggested in the ASCE Hydrology manual in 1949, but only for estimating f during periods in a rainstorm when r < f(t). The I-dependent f relation was essentially the idea behind the "time compression" procedure of Reeves and Miller (1975), who did not actually apply it to a rainfall flux case since they forced t_p = 0 but used it to adjust f capacity for a hybrid boundary condition.

Morel-Seytoux did not extend eq. (13) to the case of $t > t_p$, but using different assumptions derived a rather awkward expression for f < r. We may however, apply theorem III to extend eq. (13) and produce an infiltration expression:

$$f = K_s [1 + 1/\ell n(I/C_H - 1)] \tag{19}$$

where $C_H = (\theta_s - \theta_i)H/(1 - f_i)$, with H and f_i), with H and f_i parameters from eq. (13).

A Comparative Look at Various Models

In terms of I and f, the Green and Ampt model from eq. (11) is

$$f = K_s \left[\frac{A(\theta_i)}{I} + 1 \right] \tag{20}$$

and the Parlange-Smith model from eq (12) implies

$$f = K_s \frac{\exp(F/B)}{\exp(F/B) - 1} \tag{21}$$

In these and in eqs (18) and (19), we have analytically derived relations between f and I, consituting expression of a "universal" relation for infiltration each with one parameter K_s, and a second parameter with units of length representing some integral expression of soil capillary properties and approximately proportional to $\theta_s - \theta_i$ (Smith and Parlange, 1978). If we define dimensionless infiltrability $f_*^1 = f/K_s$, $f_* > 1$, and dimensionless depth of infiltration $F_* = I/C$, where C is the appropriate constant in each expression, they may each be compared on a dimensionless basis. The expressions and plotted functions are shown in Figure 5.

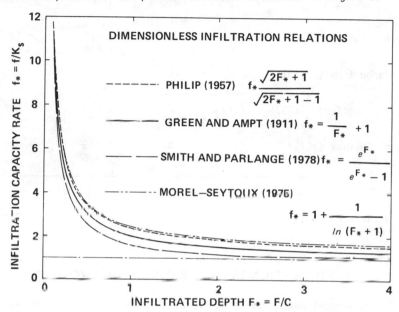

Figure 5   Dimensionless comparison of 4 physically derived infiltration models, comparing flux f_* with infiltrated depth F_*.

Only Philip's expression yields f as an explicit function of time. Each of the others may be shown by series expansion to be equivalent to $t-\frac{1}{2}$ expression to the first order at least. For example, an efficient explicit expression for use in hydrologic models may be obtained from eq. (11) after substituting f = dI/dt and integrating to get t(I):

$$K_s t = I - A(\theta_i)\, \ell n[1 + \frac{I}{A(\theta_i)}]$$ (22)

expanding the logarithmic term as a series and taking the first two terms, we use the result in an elementary difference expression to find, $(C = A(\theta_i))$:

$$\overline{f}(\Delta t) = \frac{\Delta F}{\Delta t} = [2K_s(+F) + (F/\Delta t - K_s/2)^2]^{\frac{1}{2}} - (F/\Delta t - K_s/2)$$ (23)

This expression is accurate to within a few percent, and well suited to rainfall data in pulse rate form.

To compare the theoretical accuracy of the four models discussed above, we note that if an expression properly reflects the effect of r on t_p, it should find the same value for its constant C (given a value for K_c^p) from any rainfall rate, using experimental data on infiltration to and beyond ponding. We use the solution to eq. (4) presented in Figure 4, for r = 4 and 8 cm/hr, and present calculated values of each expression's C in Table 1 for each r. Note that eq. (11) and (12) bracket the case of truly constant C, as expected from the analysis of Smith and Parlange (1978).

TABLE 1 - Theoretical consistency of the approximate ponding time expressions*

Source	Eq. no.	Rainfall Rate cm/hr	Calculated C (cm)	Relative Error $\Delta C/\bar{C}$
Mein & Larson (1973)	(11)	4 8	2.517 2.856	+ 12%
Parlange & Smith (1976)	(12)	4 8	3.631 3.309	- 9%
Philip (derived) (1957)	(18)	4 8	1.678 2.448	+ 37%
Morel-Seytoux (1976)	(13)	4 8	1.465 2.406	+ 48%

* K_s = 2.0 cm/hr

t_p (r = 8) = 7.14 min.

t_p (r = 4) = 37.75 min.

INFILTRATION INTO LAYERED SOIL PROFILES

The general analytically derived infiltration relations developed above provide an elegant and efficient model for calculating runoff or "rainfall excess" from rainfalls where $r > K_s$ on homogeneous surface soils. Heterogeneous or layered soils may be treated with the same basic principles of soil water dynamics, but the control or limiting position for creation of runoff may be anywhere in the profile. An extreme case of layered soils is when there is a layer of low permeability below a relatively pervious surface layer. This gives rise to the situation where rainfall becomes runoff when the upper layer fills from soil water "ponding" at the bottom of the upper soil and saturating the surface soil. This has been distinguished by Freeze (1980) as a different runoff generating mechanism, (the "Dunne" mechanism) (Dunne and Black, 1970) but I prefer to regard it as a special case of (subsurface) infiltration control in layered systems. In this and other layered soil cases, as well as the surface control discussed above, runoff is generated because vertical flux of a rainfall source is limited by the soil's capacity to transmit water downward. In the surface case runoff will be surface flow, and for sub-surface control, "sub-surface stormflow" may also be generated even if the whole profile depth is not saturated.

The likelihood of surface runoff in principle becomes less for soil profiles as the controlling layer for any flux is deeper in the profile. Control at depth provides a large depth of storage of water which "ponds" above, increases the opportunity for movement of this "perched" layer both through and along the controlling layer, and provides a depth through which water must move before the control layer is reached. It is a very large storm indeed which can penetrate to a depth $z = I/\theta_s$ greater than a few centimeters during the course of the storm.

Philip (1980) has treated in a quite informative sample calculation the case of ponded intake into a randomly layered system. Earlier, Bouwer (1969) treated the Green and Ampt model case of ponded flux into a layered system with monotonically decreasing K_s. Layered soils with randomly varying K_s are of limited interest in general since variation in the point of infiltration control ($r > K_s$ and "ponding" occurrence) can only shift between layers for which K_s monotonically decreases with depth. It would be of interst in "random" layering to study the distribution of the location of the control with a distribution of depth and the sequential properties of layers in conjunction with a stochastic distribution of rainfall flux.

It is inappropriate here to address the subject of layered soil infiltration in general, since many various combinations of rainfall rate and layer properties should be considered. I will limit myself to a brief consideration of two cases where a layer of more permeable soil with $K_s = K_A$ overlies a layer of less permeable soil, $K_s = K_B$, plus a short look at the case of infiltration through a surface "crust".

Decreasing Permeability; Sub-surface Control

In this example, we take $K_A > r > K_B$. It is conceptually appealing to consider application of the surface control model, described above, at the boundary between K_A and K_B. From the soil properties of soil A we could say the flow in B will asymptotically approach the case where $\theta_A = \theta(r)$, and solve a ponding equation for the formation of excess at the layer interface.

Simulation of this situation using eq. (4) illustrates why such an approach is in error. As shown in Figure 6 before ponding "should" occur at the top of layer B, soil A adjusts the rate of supply at this point increasing θ_A in the region near the boundary. The actual flux through the layer interface never exceeds approximately 85 percent of the rate of flux into the soil surface. At interface "ponding", interface flux is only 56 percent of r. Nevertheless ponding eventually occurs, with saturation on both sides of the interface, and when the surface soil layer becomes saturated, infiltration rate drops almost instantly to the rate of influx of the lower layer.

Decreasing Permeability; Surface Control

The behavior illustrated in the previous example is in marked contrast to that of the same soil layering when rainfall is sufficiently intense to cause surface control before the lower layer exerts control, i.e., $r > K_A$. Figure 7 illustrates a case where surface control is exerted just prior to the time when the evolving soil water profile intersects the layer interface. When that occurs, control shifts abruptly and dramatically to the lower soil, but infiltration flux is well-behaved and quickly resembles the curve for a soil with properties of the lower layer. This example illustrates the sensitivity of the infiltration flux to changes at the lower, steep portion of the soil water profile. Change in the rainfall rate, and thus the time of ponding $t_p(r)$, will change the relative time at which control shifts by changing the time when the lower layer is reached. This two layer case, and the alternative case with surface control for increasing permeability, has been well analyzed mathematically for the Green-Ampt model by Childs (1969), whose results are generally identical to those illustrated in Figure 7.

Increasing Permeability; Surface Crust

Consider a soil profile with a surface layer of arbitrary thickness

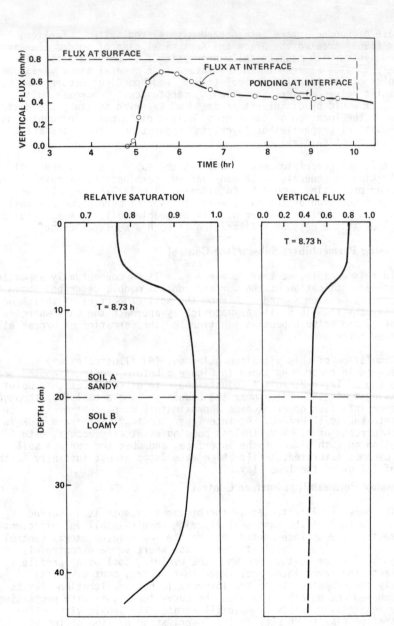

Figure 6 (a) Infiltration flux at the surface and at the soil layer
interface (20 cms) for the example with r = 1.0 for sand (K_s =
2.0) over loam (K_s = 0.2).
(b) Relative saturation profile and vertical flux profile just
prior to "ponding" at z = 20 for the layer infiltration case.
Solutions obtained by numerical treatment of eq. (4).

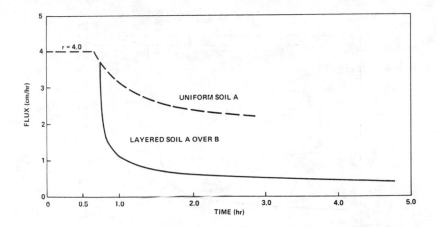

Figure 7   Infiltration pattern when control is exerted in a more permeable surface layer prior to the wetted profile encountering a less permeable lower profile. The layer interface is 10 cm below the surface.

Z_c and arbitrary saturated conductivity K_{sc}, where subscript c will refer to the crust layer, lower than the general soil below. Solutions to eq. (4) for such systems have been studied numerically by several investigators, including Whisler et al. (1978) and Curtis (1980). The general effect on gradients in the surface region is quite predictable and is exemplified by the simulation shown in Figure 8. In response to a given $r(> K_{sc})$, the potential gradient in the crust is K_s/K_{sc} greater than that in the soil below. The relative effect that the crust has on the $t_p(r)$ depends very much on the factors K_s/K_{sc}, and Z_c/I_p. I_p depends in turn on r/K_s, as seen above.

Defining the crust conductance α as K_{sc}/Z_c, the ratio of r to α is important in the relative effect of a crust. Flow through the crust must adjust to the lower K_c, and the result is a lower volume I_p as illustrated in Figure 8, as well as a lower effective K_s during $t > t_p$. The actual effective $\bar{K}_c$ is a function of the relative thickness of the soil layers having K_s and K_{sc} within the wetted zone, so that thinner crusts have much less effect, as do crusts when I_p is larger. Without a crust the rainfall rate into uniform soil A alone may cause ponding at a much larger I_p, or may not cause ponding at all. What is most needed to properly account for crust effects on infiltration is some measure of the properties crusts have in comparison to the underlying source soil, and in response to various crust forming rainfall energies.

SPATIAL VARIABILITY

By contrast to the case of series variability, several recent works have illuminated the affects of parallel or spatial variability. In this situation, infiltration rates are considered to vary in space across the surface of the catchment. Application of the results of the theory of such studies is commonly limited by lack of sufficient field information. Nevertheless, it is valuable in catchment runoff simulation to appreciate the nature of the effect of spatial variability. I will here briefly discuss the hydrologic consequences of both deterministic and random variability.

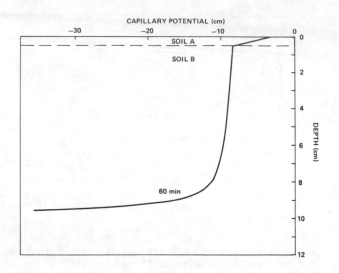

Figure 8 Water potential distribution during infiltration through a 5 mm
surface crust, represented by soil A. The gradient change at
the interface reflects directly the change in K_s.

Deterministic Spatial Variation

As demonstrated by Smith and Hebbert (1978), variation along flow
paths perpendicular to the contour lines of a catchment can have a significant effect on the catchment response due to the sensitive interaction
of successive locations. Runoff water from an upstream source will
dramatically decrease the time to ponding at a location, by increasing
the effective supplied flux. Since variation in t_p cannot conversely
effect upstream elements, the result is a severe bias in catchment
response when upslope soil properties, expecially K_s, favor early ponding, or lower F_p, in the upstream direction. Deterministic variation of
the opposite sense delays response, as would be expected, but not with
such dramatic bias. Figure 9, from Smith and Hebbert (1979) illustrates
this effect. The hydrograph defined by open circles has K_s decreasing
linearly as in the downslope direction, and the other, with open squares,
has K_s increasing downslope. The effect of upstream runoff in increasing
effective local supply rate r in the function for t_p causes significant
increase in runoff for the latter case (K_s increasing).

Random Spatial Variation

The interactive effect of soil properties along flow paths on the
catchment also affects the results of randomly varying infiltration
properties. Smith and Hebbert (1979) have demonstrated the dispersive
effect that random variation in parameters K_s and C have on the net mean
infiltration decay curve. In addition, experimental studies have shown
that K_s indeed exhibits large variance, with coefficient of variation of
the order of 1, and a skewed log-normal type distribution (Nielson,
et al., 1973).

Such distributions of infiltration properties also results in a
time-varying fraction of the catchment contributing to the runoff. Given
a distribution of the infiltration parameters of a watershed, there is a

120

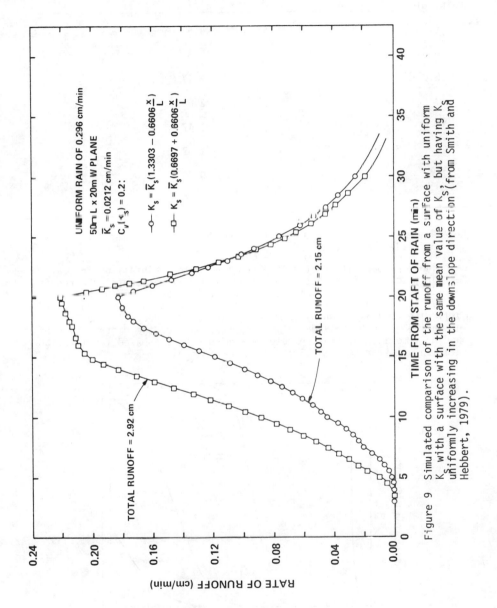

Figure 9 Simulated comparison of the runoff from a surface with uniform K_s with a surface with the same mean value of K_s, but having K_s uniformly increasing in the downslope direction (from Smith and Hebbert, 1979).

Content within the figure:

RATE OF RUNOFF (cm/min)

TIME FROM START OF RAIN (min)

UNIFORM RAIN OF 0.296 cm/min
50m L x 20m W PLANE
$\overline{K}_s = 0.0212$ cm/min
$C_v(K_s) = 0.2$:

—○— $K_s = \overline{K}_s(1.3303 - 0.6606\frac{x}{L})$

—□— $K_s = \overline{K}_s(0.6697 + 0.6606\frac{x}{L})$

TOTAL RUNOFF = 2.15 cm

TOTAL RUNOFF = 2.92 cm

TABLE 2 - Expected average non-contributing flow lengths for $r = 4K_s$

Time from start of rain, min.	Expected relative length of non-contributing upper flow path $_m{(1)}$
4.0	44.5
6.0	12.5
8.0	5.17
10	2.69
12	1.47
14	1.07
16	0.79
18	0.59

Note 1: m is the minimum distance at which soil properties are uncor-related in space.

calculable probability that at a given time after rainfall begins, a certain fraction of the catchment at its top will not contribute runoff. This is demonstrated in Table 2, reproduced from Smith and Hebbert (1979). The actual distribution of infiltration rates varies continuously with time as shown by example in Figure 10. Recently, Maller and Sharma (1980) have derived analytic expressions for the distribution of ponding time and infltration rate, assuming a Philip-based infiltration relation (similar to eq. (13)).

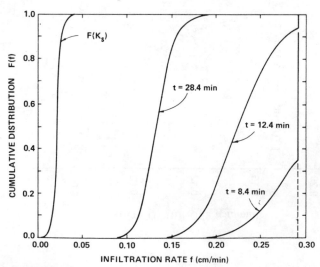

Figure 10 Time changes in net infiltration rate distribution for a soil area when K_s is log-normally distributed with mean = 1.27 cm/hr and standard derivation = 0.25 cm/hr, under r = 17 cm/hr (from Smith and Hebbert, 1979).

INFILTRATION AFFECTED BY AIR FLOW

The effect on the rate of infiltration of the movement of air during infiltration of water is subject to considerable further clarification. All work to date indicates that the effect of air is to reduce the flux of water at some stage during infiltration. Different results have been obtained by different authors depending on the assumptions made regarding air compressibility, viscosity, or manner of escape from the soil profile. In general it seems clear that infiltration of rainfall at large value of r/K_s into soils where the lower boundary is low (more than perhaps 2 metres) is little affected by air prior to ponding. Morel-Seytoux and Khanji (1976) and McWhorter (1975) have shown that f for t > t_p is affected to some extent even when air is free to escape, by virtue of its viscosity. Parlange and Hill (1979) showed that the wetting profile should steepen with air trapped ahead of a wetting front (assuming incompressibility) and Curtis and Watson (1980) indicate that the profile should be less steep, assuming air to be compressible but neglecting viscosity. In any case, the increased pressure ahead of the wetting front reduces the capillary gradient and reduces the water flux to an uncertain extent.

Morel Seytoux and Khanji (1974) have expressed the effect of air in two parameters; f_s, the "fractional flow function" which represents the relative conductivity ratio for water and air over the range of θ; and rainfall rate coefficient β, the "viscous correction factor" representing a correction when air must move ahead of the wetting front. Morel-Seytoux and Khanji suggest values for β for a few cases. It remains to determine the expected effect of air for various rainfall rates (including ∞), lower boundary conditions, and soil properties in terms of corrections to the models presented above. This will not be a simple task.

SUMMARY AND SUGGESTIONS

In the above presentation, I have outlined the derivation of several recent infiltration models which are direct descendants from porous media flow dynamics. The complete general model encompasses relations for both time to surface ponding and soil infiltrability from both flux and ponded surface conditions. Although complete expressions for some models were obtained by analogy to derivations of others (as in the extension of Philip's equation) all the models discussed have their basis in reasonable approximations to the soil physical properties rather than a conceptual model or a parameterization of an observed decay curve. I have shown also how the relation between infiltrability f and infiltrated depth I is a true "capacity" relationship, encompassing variable rainfall rates and times both before and after ponding. The physically based models now available are at least as concise in the number of descriptive parameters as any comparable empirical model.

The effect on catchment infiltration of both vertical, lateral, deterministic and random variability was outlined briefly. Much of such variation, when quantified, can be represented within the general infiltration model described.

The crusting and bulk property changes associated with agricultural practice can be treated within existing models provided such changes can be represented in terms of bulk soil hydraulic properties. The effect that rainfall rate and its temporal variation have on infiltration amounts is apparent in these physically based models, implying a strict limit on our prediction ability inherent in the quality of the available rainfall data. Much remains to be learned. The problems which I feel to be most pressing include the following:

1. In order to more easily characterize soil hydraulic properties, it
    would be useful to be able to predict changes in capillary and
    infiltration properties from variation in compaction or bulk den-
    sity. Hydraulic effects of crusting and cracking of a given soil
    are not now well understood, although we know the qualitative effect
    of compaction on the K and $\Psi(\theta)$ relations. Indeed, just a simpler,
    more reliable means to measure in situ values of K_s is always in
    demand. Hydrologists have perhaps a more pressing need than soil
    physicists to obtain more and better field measurements, as con-
    trasted with laboratory measures.

2. Most infiltrating catchment surfaces also include macroporosity,
    which modifies the simple porous media flow relations at the highest
    values of θ, or when Ψ gets very close to 0. When soils contain
    significant macroporosity, such as newly ploughed fields, cracked
    surfaces, or are densely penetrated by root or worm holes, downward
    flow from a saturated zone is significantly affected. I prefer to
    characterize such soils as composite media. A method of quantifying
    the hydraulic effect during infiltration of such composite media is
    needed. This includes the ability to quantify the often observed
    modification of infiltration properties by the presence of grass, or
    for that matter, any rooted plants.

3. Finally, though we know the general effects that the counterflow of
    air during infiltration can have, we have not yet developed our
    knowledge to the extent that we can predict when air will exhibit a
    significant effect on infiltration decay curves. Morel-Seytoux
    (1976) has treated air flow in eq. (11) by a parameter f, here
    included in $C(\theta_i)$ as well as the parameter β, here assumed to be 1,
    but their variation with initial conditions, rainfall patterns, and
    lower boundary conditions are not yet clear enough to characterize
    the importance of air counterflow in any given situation. The
    existence of macropores has been linked to the effective release of
    air by Dixon and Peterson (1971) as well as others. Any effective
    characterization of their effect should include practical field
    measurement techniques, which can be extended to predict response to
    a variety of influx conditions.

ACKNOWLEDGEMENTS

Many of the computer simulations presented in this paper were done,
during visits, through the support of the School of Australian Environmental Studies, Griffiths University, Brisbane, and the support of the
Environmental Dynamics Division, Civil Engineering Department, University
of Western Australia, Perth. I wish also to acknowledge valuable discussions with Dr. J-Y. Parlange, Griffiths University, and Dr. D. E.
Smiles, Environmental Mechanics, CSIRO, Canberra.

REFERENCES

ASCE Hydrology Committee, 1949. Hydrology Handbook. ASCE Manual of
    Engineering Practice, No. 28.

Bouwer, H., 1969. Infiltration of Water into Nonuniform Soil. Journal
    of the Irrigation and Drainage Division, American Society of Civil
    Engineers, Vol. 95, No. IR4, pp. 451-461.

Childs, E. C., 1969. An Introduction to the Physical Basis of Soil Water

Phenomena. Wiley-Interscience, London. Chapter 13, pp. 281-288.

Curtis, A. and K. K. Watson, 1980. Physical Restraints on Infiltration. Hydrology and Water Resources Symposium Papers, National Conference Publication No. 80/9. The Institution of Engineers, Australia, pp. 6-11.

Dixon, R. M. and A. E. Peterson, 1971. Water Infiltration Control: A Channel System Concept. Soil Science Society of America Proceedings, Vol. 35, pp. 968-973.

Dunne, T. and R. D. Black, 1970. Partial Area Contributions to Storm Runoff in a Small New England Watershed. Water Resources Research, Vol. 6, No. 5, pp. 1296-1311.

Eagleson, P. H., 1978. Climate, Soil and Vegetation, 5. A Derived Distribution of Storm Surface Runoff. Water Resources Research, Vol. 14, No. 5, pp. 741-748.

Freeze, R. A., 1975. A Stochastic-Conceptual Analysis of One-dimensional Flow in Nonuniform Homogeneous Media. Water Resources Research, Vol. 11, No. 5, pp. 725-741.

Freeze, R. A., 1980. A Stochastic-Conceptual Analaysis of Rainfall-Runoff Processes on a Hillslope. Water Resources Research, Vol. 16, No. 2, pp. 391-408.

Green, W. A. and G. A. Ampt., 1911. Studies on Soil Physics, I. The Flow of Air and Water Through Soils. Journal of Agricultural Science, Vol. 4, pp. 1-24.

Maller, R. A. and M. L. Sharma, 1980. Time to Ponding and Areal Infiltration from Spatially Varying Parameters. Hydrology and Water Resources Symposium. The Institution of Engineers, Australia, National Conference Publication No. 80/9, pp. 101-165.

McWhorter, D. B., 1975. Vertical Flow of Air and Water with a Flux Boundary Condition. Paper No. 75-2012, ASAE Ann. Meeting, 1973.

Mein, R. G. and C. L. Larson, 1973. Modeling Infiltration During a Steady Rain. Water Resources Research, Vol. 9, No. 2, pp. 384-394.

Morel-Seytoux, H. J., 1973. Two-Phase Flows in Porous Media. in: V. T. Chow (Editor), Advances in Hydroscience, Vol. 9, pp. 119-202, Academic Press, New York.

Morel-Seytoux, H. J., 1976. Derivation of Equations for Rainfall Infiltration. Journal of Hydrology, Vol. 31, pp. 203-219.

Morel-Seytoux, H. J. and J. Khanji, 1974. Derivation of an Equation of Infiltration. Water Resources Research, Vol. 10, pp. 795-800.

Nielson, D. R., J. W. Biggar and K. T. Erh, 1973. Spatial Variability of Field Measured Soil-water Properties. Hilgardia, Vol. 42, pp. 215-259.

Parlange, J-Y. and R. E. Smith, 1975. Ponding Time for Variable Rainfall Rates. Canadian Journal of Soil Science, Vol. 56, pp. 121-123.

Parlange, J-Y. and D. E. Hill, 1979. Air and Water Flow in a Horizontal

Column - Influences of the Air Boundary Condition. Surface and Subsurface Hydrology, H. J. Morel-Seytoux et al. eds. Water Resources Publications, Fort Collins, Colorado, pp. 374-285.

Philip, J. R., 1957. The Theory of Infiltration, 4, Sorptivity and Algebraic Infiltration Equations. Soil Science, Vol. 84, No. 3, pp. 257-264.

Philip, J. R., 1969. Theory of Infiltration. In: V. T. Chow (Editor), Advances in Hydroscience, Vol. 5, pp. 216-296, Academic Press, New York.

Philip, J. R., 1980. Field Heterogeneity: Some Basic Issues. Water Resources Research, Vol. 16, No. 2, pp. 443-448.

Reeves, M. and E. E. Miller, 1975. Estimating Infiltration for Erratic Rainfall. Water Resources Research, Vol. 11, No. 1, pp. 107-110.

Smith, R. E. and R. H. B. Hebbert, 1979. A Monte Carlo Analysis of the Hydrologic Effects of Spatial Variability of Infiltration. Water Resources Research, Vol. 15, No. 2, pp. 419-429.

Smith, R. E. and J-Y. Parlange, 1978. A Parameter-Efficient Hydrologic Infiltration Model. Water Resources Research, Vol. 14, No. 3, pp. 533-538.

Swartzendruber, D. and Hillel, 1973. The Physics of Infiltration. in: Physical Aspect of Soil, Water and Salts in the Ecosystem, A. Hadas, et al., eds. Springer-Verlag, New York, pp. 3-15.

Whisler, F. D., A. A. Curtis, A. Nikham and M. J. M. Romkens, 1977. Modeling Infiltration as Affected by Soil Crusting. Proceedings of the Third International Hydrology Symposium, Fort Collins, Colorado, pp. 400-413.

White, I., D. F. Smiles and K. M. Perroux, 1979. Absorption of Water by Soil: The Constant Flux Boundary Condition. Soil Science Society of America Journal, Vol. 43, No. 4, pp. 659-664.

PRE-PONDING CONSTANT-RATE RAINFALL INFILTRATION

I. White
Senior Research Scientist and Visiting Scientist
CSIRO, Division of Environmental Mechanics and
New Mexico Petroleum Recovery Research Center,
New Mexico Institute of Mining and Technology,
Socorro, NM 87801

B. E. Clothier
Research Scientist
CSIRO, Division of Environmental Mechanics and
DSIR, Plant Physiology Division,
Palmerston North, New Zealand

D. E. Smiles
Chief
CSIRO, Division of Environmental Mechanics
P. O. Box 821
Canberra City, ACT 2601 Australia

ABSTRACT

A summary is presented of recent developments in the theory of rainfall-infiltration. The theory, which uses the concept of approximate, but physically realistic flux-concentration relations, gives integral solutions which describe the development of profiles of soil-water content or water pressure potential during rainfall infiltration up to the time of incipient ponding. Also given are integral expressions which permit calculation of the time dependence of water-content or water-pressure potential at the soil surface during rainfall. The latter may be used to calculate the time of incipient ponding. These solutions are expressed in terms of the basic soil water properties soil-water diffusivity, $D(\theta)$, and hydraulic conductivity, $K(\theta)$, or in terms of the hydraulic conductivity-water potential, $K(\psi)$ and water potential-water content, $\psi(\theta)$.

In controlled, laboratory measurements, employing both horizontal and vertical constant-rate water application, the observed time dependence of the water content at the soil surface, the movement of the wetting front and the evolution of the water-content profiles were found to be in good agreement with predictions made using the theory. Good agreement was also found between measured and predicted times to ponding.

For field use, the time required to characterize a site dictates that the theory be written in terms of easily and rapidly determined soil-water properties. The properties chosen here were sorptivity, $S(\theta_s, \theta_n)$, and hydraulic conductivity at natural saturation, K_s. The presence of macropores greatly affects the magnitudes of $S(\theta_s, \theta_n)$ and K_s measured in the field under ponded conditions. It is shown that in order to make measurements of $S(\theta_s, \theta_n)$ and K_s which are relevant to pre-ponding rainfall infiltration, the influence of macropores must be excluded. This was achieved by using a device to measure $S(\theta_s, \theta_n)$ and K_s, which supplies water to the soil surface at a small negative pressure head.

A rainfall simulator was used to provide constant-rate rainfall at two field sites. Measured soil-water content profiles were found to be consistent with the modified theory. The shape of the field-measured profiles differed markedly from those observed in the laboratory for columns of repacked soil. It is suggested that this difference arises because of the presence of continuous, biogenically produced meso-pores which were observed in the undisturbed field soils.

The field data are also shown to be successfully described by an exact solution of a particular form of the flow equation which has $D(\theta)$ constant and $K(\theta)$ proportional to θ^2. The applicability of this form of the flow equation, known as Burgers' equation, to other field situations remains to be demonstrated. The exact solution is, however, independently of value in testing the reliability of numerical schemes for solving the flow equation.

INTRODUCTION

The prediction of water movement and distribution in soils under steady rainfall conditions is of considerable interest in both hydrology and agriculture. In the latter, the demand for efficient water use in sprinkler irrigation requires accurate assessment of both wetting front movement and the time at which surface ponding occurs. Rainfall infiltration, prior to surface ponding, differs substantially in character to that of ponded infiltration (Rubin, 1966). For that reason, the success-ful, predictive models developed for ponded infiltration (see e. g., Philip, 1969) have no direct application to rainfall infiltration.

Efforts to analyze rainfall infiltration have been based either on empirical approaches (see e.g., Johnston, Elsawy and Cochrane, 1980) or on attempts to solve the highly non-linear flow equation for flux boundary conditions. The first successful calculation of the development of soil-water content profiles during constant-rate rainfall infiltration, using basic soil-water properties, was made by Rubin and Steinhardt (1963) who used a finite difference method to solve the flow equation. Thereafter followed a number of numerical studies which examined rainfall infil-tration for a variety of initial conditions (see e.g., Whisler and Klute, 1967; Smith, 1972). The need for more easily evaluated or analytical solutions led to the use of simplified models of soil-water movement (see e.g., Mein and Larson, 1973; Braester, 1973; Swartzendruber, 1974; Ahuja and Römkens, 1974; Chu, 1978). A significant breakthrough was made by Parlange (1972) who introduced a general, but approximate, method of solving the flow equation which gave simple integral solutions. Philip and Knight (1974) showed how Parlange's method could be improved to any desired accuracy through use of a concept called the flux-concentration relation (Philip, 1973).

In recent work, a simplified form of the Philip-Knight approach has been found to predict accurately the principal features of constant-rate water application to soils both in the laboratory (Smiles, 1978; White,

Smiles and Perroux, 1979; Smiles, Perroux and Zegelin, 1980; Perroux, Smiles and White, 1981) and in the field (Clothier, White and Hamilton, 1981). In this paper, we summarize the development of the theory and its modification for field use. Examples are given of its application in predicting wetting front movement, time-dependence of the soil-water distribution, change of soil-water content at the soil surface with time and time to incipient ponding during constant-rate rainfall in both controlled laboratory experiments and in field tests.

THEORY

In the following theory of rainfall infiltration, the simplest of soil-water systems is employed. Flow is considered to be one-dimensional in the vertical direction, z, only. The soil is assumed to be uniform to a great depth and initially has a constant volumetric soil-water content, θ_n. Air ahead of the wetting front is taken to be always at atmospheric pressure. Raindrop impact effects on the soil surface (McIntyre, 1958; Morin and Benyamini, 1977) are ignored, and the soil-water flow is non-hysteretic. Initially, we take the rainfall rate, $V_0(t)$ as being time-dependent. Our neglect of hysteresis means, however, that $V_0(t)$ must be a non-decreasing function of time. For this system, the initial condition is,

$$t = 0; \quad \theta = \theta_n; \quad z > 0 \tag{1}$$

and the boundary condition in terms of the soil-water diffusivity, $D(\theta)$ and hydraulic conductivity, $K(\theta)$, is:

$$t > 0; \quad -D(\theta)\partial\theta/\partial z + K(\theta) = V_0(t); \quad \theta = \theta_0(t); \quad z = 0 \tag{2}$$

where θ is the volumetric soil-water content, $\theta_0(t)$, the water content at the soil surface, and z is taken as being positive downwards.

Our basic assumption, in common with other treatments (Rubin, 1966; Parlange, 1972), is that soil-water flow is described by Darcy's law, which in terms of $D(\theta)$ and $K(\theta)$ is:

$$v(\theta,t) = -D(\theta)\partial\theta/\partial z + K(\theta) \tag{3}$$

Here $v(\theta,t)$ is the Darcy velocity of water at a position in the soil profile where the water content is θ.

Philip (1973) introduced the flux-concentration relation, $F(\Theta,t)$:

$$F(\Theta,t) = [v(\theta,t) - K_n]/[V_0(t) - K_n] \tag{4}$$

where $K_n = K(\theta_n)$ and $\Theta = (\theta-\theta_n)/[\theta_0(t)-\theta_n]$. $F(\Theta,t)$ is a function of $\theta,\theta_0(t),\theta_n$ and soil properties, and is defined so that it monotonically increases in the region $\Theta(0,1)$ with $F(0,t) = 0$ and $F(1,t) = 1$.

Substitution of $v(\theta,t)$ from (3) into (4) and integrating gives:

$$[V_0(t)-K_n]z = \int_{\theta}^{\theta_0(t)} \frac{D(\theta')d\theta'}{F(\Theta,t)-K} \tag{5}$$

where $\kappa = [K(\theta)-K_n]/[V_0(t)-K_n]$

We now make use of the continuity statement written as:

$$\int_{\theta_n}^{\theta_0(t)} z.d\theta = \int_0^t [V_0(t)-K_n]dt \qquad (6)$$

Substitution for z from (5) in (6) and integration by parts yields:

$$[Vo(t)-K_n] \int_0^t [V_0(t)-K_n]dt = \int_{\theta_n}^{\theta_0(t)} \frac{(\theta-\theta n)D(\theta)d\theta}{F(\Theta,t)-\kappa} \qquad (7)$$

Given $D(\theta)$, $K(\theta)$, $V_0(t)$ and provided $F(\Theta,t)$ is known, the water content at the soil surface at any selected time can be calculated from (7). This value, when substituted in (5), gives the soil-water content distribution at that time. The solution is, however, not straightforward since $F(\theta,t)$ is, in general, not known a priori. Exact solutions of (5) and (7) require iterative procedures similar to those of Philip and Knight (1974). For one-dimensional constant-flux infiltration such iteration appears unnecessary, and, to a good accuracy, $F(\Theta,t)$ can be considered as time independent and having a simple Θ-dependence (Smiles, 1978; White, Smiles and Perroux, 1979; White, 1979; Perroux, Smiles and White, 1981). With this approximation, and the fact that for most practical situations, $V_0(t) >> K_n$, (5) and (7) may be written, for constant rainfall rate, $V_0(t) = V_0$, as:

$$V_0 z = \int_{\theta}^{\theta_0(t)} \frac{D(\theta')d\theta'}{F(\Theta)-\kappa} \qquad (8)$$

and

$$V_0^2 t = \int_{\theta_n}^{\theta_0(t)} \frac{(\theta-\theta_n)D(\theta)d\theta}{F(\Theta)-\kappa} \qquad (9)$$

Movement of the wetting front can be determined using (9) and (8) by calculating the position of a plane of constant moisture content, $\theta = \theta_{wf}$, where $\theta_{wf} > \theta_n$, for selected times.

The Theory in Terms of Water Pressure Potential

Smiles, Perroux and Zegelin (1980) have pointed out that $D(\theta)$ in (8) and (9) is not uniquely defined under some circumstances. For example, it is found for some soil samples that the soil becomes saturated at soil-water pressure potentials, ψ_s, which are negative (Philip, 1958; Rubin, 1966). When this occurs $D(\theta)$ is not uniquely defined in the water potential region $\psi > \psi_s$ and equations (8) and (9) are inapplicable. For this situation, the theory must be expressed in terms of the potential-water content relation, $\psi(\theta)$, and the hydraulic conductivity-potential function, $K(\psi)$ (Smiles, Perroux and Zegelin, 1980). In terms of these relations, the solutions equivalent to (8) and (9) are:

$$V_o z = \int_{\psi}^{\psi_o(t)} \frac{K(\psi')d\psi'}{F(\psi')-\kappa'} \tag{10}$$

and

$$V_o^2 t = \int_{\psi_n}^{\psi_o(t)} \frac{(\theta-\theta_n)K(\psi)d\psi}{F(\psi')-\kappa'} \tag{11}$$

where $\kappa' = K(\psi)/V_o$, ψ_n is the initial potential (at θ_n), $\psi_o(t)$ is the potential at the soil surface, $F(\psi)$, the flux-potential relation is given by (4) and is related to $F(\theta)$ through $\psi(\theta)$. When $\psi_o(t) > \psi_s$, $\theta = \theta_s$, $K(\psi) = K_s$, the saturated conductivity, $F(\psi) = 1$ and (10) and (11) become respectively:

$$V_o z = \frac{K_s[\psi_o(t)-\psi_o]}{1 - K_s/V_o} + \int_{\psi}^{\psi_s} \frac{K(\psi')d\psi'}{F(\psi')-\kappa'} \tag{12}$$

and

$$V_o^2 t = \frac{K_s[\psi_o(t)-\psi_s](\theta_s-\theta_n)}{1 - K_s/V_o} + \int_{\psi_n}^{\psi_s} \frac{(\theta-\theta_n)K(\psi)d\psi}{F(\psi) - \kappa'} \tag{13}$$

In (10), (11), (12), and (13) convenient approximations for $F(\psi)$ are made by either putting $F(\psi)=1$ (Parlange, 1972) or by taking $F(\theta)= \theta$ in which case $F(\psi)$ is just the normalized moisture characteristic.

Time to Incipient Ponding.

The time at which water ponds on the soil surface during rainfall depends on rainfall rate, soil properties and the local surface micro-relief. Because the last factor requires site-specific, geographical knowledge, the simple, deterministic theory developed here cannot be used, in general, to predict ponding time. It can, however, be used to predict a necessary precursor to ponding time; namely, the time at which the water potential at the soil surface becomes zero. This time, the time to incipient ponding, t_p (Rubin, 1966), can be found by putting $\psi_o(t)= 0$ in (13) so that:

$$V_o^2 t_p = \frac{- K_s\psi_s(\theta_s-\theta_n)}{1 - K_s/V_o} + \int_{\psi_n}^{\psi_s} \frac{(\theta-\theta_n)K(\psi)d\psi}{F(\psi) - \kappa'} \tag{14}$$

For soils with $\psi_s = 0$, incipient ponding occurs when the soil surface first becomes saturated. The time to incipient ponding for this case may be found by substituting $\theta_o(t) = \theta_s$ in (9):

$$V_o{}^2 t_p = \int_{\theta_n}^{\theta_s} \frac{(\theta - \theta_n) D(\theta) d\theta}{F(\Theta) - \kappa} \tag{15}$$

We note that for infiltration into very deep soil profiles, a necessary condition for ponding is that the rainfall rate be greater than the saturated conductivity, $V_o > K_s$. The soil surface, of course, does not pond instantaneously for $V_o > K_s$ as shown in (14) and (15).

Short-Time Approximations

For the early stages of infiltration, for systems where gravity is negligible or for $V_o >> K_s$, simplification of the solutions (8), (9), (12), (13), (14) and (15) is possible. Here the κ or κ' terms can be omitted and (8) and (9) become simply:

$$Z = V_o z = \int_{\theta}^{\theta_o(t)} \frac{D(\theta)\ d\theta}{F(\Theta)} \tag{16}$$

$$T = V_o{}^2 t = \int_{\theta_n}^{\theta_o(t)} \frac{(\theta - \theta_n) D(\theta) d\theta}{F(\Theta)} \tag{17}$$

Equations (14) and (15) similarly can be written as:

$$T_p = V_o{}^2 t_p = -K_s \psi_s (\theta_s - \theta_n) + \int_{\psi_n}^{\psi_s} \frac{(\theta - \theta_n) K(\psi)\ d\theta}{F(\psi)} \tag{18}$$

and

$$T_p = \int_{\theta_n}^{\theta_s} \frac{(\theta - \theta_n) D(\theta)\ d\theta}{F(\Theta)} \tag{19}$$

It can be seen in (16) to (19), which are the gravity-free or absorption solutions, that the process can be described in terms of the reduced variables, Z and T. This leads to solutions which are independent of the imposed rainfall rate. It follows directly that in the early stages of infiltration, the results measured for one experiment at any single water application rate can be scaled using Z and T to give results which are valid for all rainfall rates.

Practical short-time criteria for the time period over which the simpler absorption solutions are valid can be introduced. These have features in common with early time criteria for ponded infiltration (Philip, 1969). If we define a time, t_g, such that for $t < t_g$, the short time-solutions, (16) and (17), adequately represent the water-profile development, then, for a wide range of soil types, it can be shown that:

$$\text{for } V_o \geqslant K_s, \quad t_g = S^2(\theta_s, \theta_n) / 2V_o{}^2 \tag{20}$$

132

and $\qquad$ for $V_0 < K_s$ $\qquad$ $t_g = \frac{1}{3}\,[S^2(\theta_s,\theta_n)/2V_0^2](V_0/K_s)$ (21)

Here $S(\theta_s,\theta_n)$ is sorptivity (Philip, 1957). Calculations show that the use of the simpler short-time solutions up to $t = t_g$, produce errors in θ_0 of about 10% and of about 5% in the position of the wetting front.

COMPARISON OF THEORY WITH LABORATORY EXPERIMENTS

Soil-Water Profile Development

In order to test the theory exhaustively, the simplest of systems was initially chosen (White, Smiles and Perroux, 1979). Water was supplied at constant rates, ranging from 1.7mm hr.$^{-1}$ to 217mm hr.$^{-1}$, to horizontal columns of washed Bungendore fine sand. Because the columns are horizontal, gravity can be ignored, and, since $\psi_s = 0$ for the sample used, (16), (17) and (19) represent the solutions appropriate to the experiments. The only soil-water property required for these solutions is $D(\theta)$ which was measured independently. The flux concentration relation was assumed to be $F(\Theta) = \Theta.^{2-4}/\pi$ The integrations necessary to evaluate (16), (17) and (19) were performed on small programmable calculators.

The experimentally determined surface water contents for a reduced time range covering nearly 4 orders of magnitude are shown in figure 1. The measured values are compared with those calculated from (17). In figure 2 measured values of the position of the wetting front are compared with that predicted using equations (17) and (16). According to the gravity-free treatment, (16) and (17) water content profiles plotted, in terms of the reduced space variable Z, for a selected value of reduced time, T, should all fall on a single curve independent of V_0. In figure 3, water content profiles, measured at 4 values of T show that this is the case. The predicted values from (17) and (16), plotted as the solid lines in figures 1 through 3, can be seen to be in excellent agreement with the observations.

Perroux, Smiles and White (1981) also conducted constant rate rainfall experiments in vertical columns for two soils. For the vertical case with $\psi_s = 0$, the appropriate solutions are equations (8) and (9) which require that both $D(\theta)$ and $K(\theta)$ be measured. Experimentally measured water content profiles found in a silty clay loam soil are shown in figure 4. The observed values are compared with predicted profiles calculated using the full solutions (8) and (9), with $F(\Theta)=\Theta$, and using the simpler absorption solutions (16) and (17). The values of V_0 used in these experiments were 0.141, 0.336, 0.642 and 1.25 times K_s(118mm/hr.). Rainfall durations ranged from 3 to 284 minutes.

The results in figure 4 are presented in terms of the reduced variables, Z and T, which we note do not completely parametize the infiltration process. The predicted profiles using the full solutions (8) and (9) can be seen to agree well with the experimental results. The absorption solutions (16) and (17) also give an adequate description of the profile development despite the fact that (17) overestimates $\theta_0(t)$ and (16) underestimates the position of the wetting front. The duration times of all experiments in figure 4 is less than t_g given by the short time criteria (20) and (21), and so the success of the absorption solutions is expected.

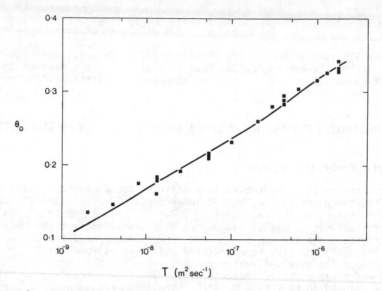

Figure 1. Water content at the soil surface as a function of reduced time during the constant rate application of water to washed Bungendore fine sand. The points are experimental values, the smooth curve was predicted using equation (17) (White, Smiles and Perroux, 1979).

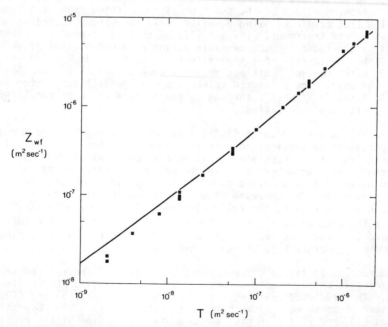

Figure 2. Movement of the reduced wetting front, $Z_{wf} = V z_{wf}$ for $\theta_{wf} = 0.03$, during constant rate water application. The experimental points are compared with the smooth curve predicted from equations (17) and (16) (White, Smiles and Perroux, 1979).

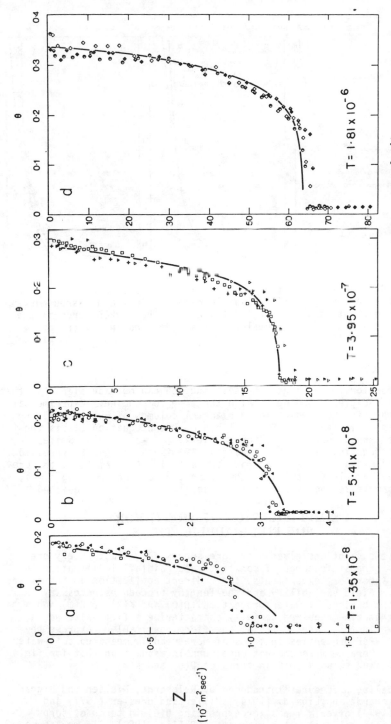

Figure 3. Reduced water content profiles at four reduced times (in $m^2 s^{-1}$) during constant rate water application to washed Bungendore fine sand. Smooth curves are calculated using equations (16) and (17). Symbols refer to different water application rates used which ranged from 3.4 to 217 mm/hr (White, Smiles and Perroux, 1979).

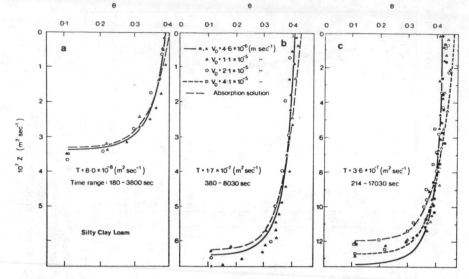

Figure 4. Reduced water content profiles for constant rate infiltration into vertical columns of a silty clay loam soil. Full and dashed curves are for the highest and lowest rainfall rates. The broken curve is the absorption solution. Experimental points are defined in (b) (Perroux, Smiles and White, 1981).

Time to Ponding

In laboratory experiments the soil surface can be made planar so that the time to incipient ponding is also the time to ponding. Experimental values of the time to ponding for horizontal columns of Bungendore fine sand are listed in Table 1. Sample 1 had $\psi_s = 0$ and equation (19) is the appropriate expression for t_p. For sample 2, $\psi_s \approx -25$ cm H_2O (Smiles, Perroux and Zegelin, 1980), and equation (18) is the relevant expression. Predicted values of t_p are also shown in Table 1. For sample 1, the predicted values are less than 14% higher than the measured ponding times, while for sample 2, the experimental values lie within the predicted range.

SIMPLIFICATION OF THE THEORY
FOR FIELD APPLICATIONS

While the solutions given above are successful in accurately predicting the general features of constant rate rainfall infiltration in controlled laboratory experiments, their direct application in the field is limited. Field variability and the lengthy procedures necessary to determine the basic soil-water properties, $D(\theta)$ and $K(\theta)$ or $K(\psi)$ and $\psi(\theta)$ mean that time required to fully characterize a field site would be excessive in most cases. It is, therefore, desirable to express the basic soil-water properties in terms of more readily measured soil-water parameters. Here we make the not unreasonable assumption that for field soils, $\psi_s = 0$, and so work only in terms of $D(\theta)$ and $K(\theta)$.

The scaling procedures introduced by Reichardt, Nielsen and Biggar (1972), Reichardt and Libardi (1974), Miller and Bresler (1977) and Brutsaert (1979) offer a means for expressing $D(\theta)$ in terms of sorptivity, an integral soil-water parameter (Philip, 1957). If we assume

Table 1: Ponding time for Bungendore fine sand.

Sample No.	ψ_s (m)	V_o (mm/hr.)	Ponding time, t_p (hr.) Observed	Predicted Eq(18)	Eq(19)	Eq(24)
		85	0.903	-	1.03	1.05±0.03
1	0	118	0.465	-	0.525	0.545±0.014
		152	0.285	-	0.315	0.328±0.008
		217	0.137	-	0.153	0.160±0.004
2	-0.25	105	0.689±0.034	0.675-0.753*	-	0.652±0.072
		209	0.166±0.006	0.168-0.188	-	0.162±0.018

* range of predicted values due to uncertainty in θ_c.

that $D(\theta)$ is an exponential function of water content, then the diffusivity can be written as (Brutsaert, 1979; White and Clothier, 1981):

$$D(\theta) = [\gamma(\beta)S^2(\theta_s,\theta_n)/(\theta_s \theta_n)^2]\exp(\beta\omega)$$

where $\gamma(\beta)$ is a known function of β (White and Clothier, 1981), $S(\theta_s,\theta_n)$ is the sorptivity measured with initial water content, θ_n, and with water supplied to the soil surface at $\psi=0$, and $\omega=(\theta-\theta_n)/\theta_s-\theta_n)$.

We here note that the exact functional form assumed for $D(\theta)$ is not critical, other forms can be used with equal success.

Sorptivity, θ_n and θ_s are readily measured in the field (Talsma, 1969; Clothier and White, 1981a), β, and hence $\gamma(\beta)$, can be determined by concurrent measurements of $S(\theta_s,\theta_n)$ and wetting front movement (Clothier and White, 1981a).

We follow Brooks and Corey (1964) and Bresler, Russo and Miller (1978) and assume that the hydraulic conductivity is given by: $K(\theta) = K_s\omega^b$.

Here K_s, the saturated hydraulic conductivity, is easily measured in the field. Correlations involving soil-type can be used to estimate b (Mualem, 1979) or it may be found by constant-rate rainfall measurements with $V_o < K_s$ (Clothier, White and Hamilton, 1981). With these expressions and using the approximation, $F(\Theta)= \Theta$, equations (8) and (9) become:

$$V_o z = \frac{\gamma(\beta)S^2(\theta_s,\theta_n)}{(\theta_s-\theta_n)^2} \int_\theta^{\theta_o(t)} \frac{\exp(\beta\omega)}{\Theta - \kappa} d\theta \qquad (22)$$

and

$$V_o^2 t = \frac{\gamma(\beta)S^2(\theta_s,\theta_n)}{(\theta_s-\theta_n)^2} \int_{\theta_n}^{\theta_o(t)} \frac{(\theta-\theta_n)\exp(\beta\omega)}{\Theta - \kappa} d\theta \qquad (23)$$

137

with $\qquad \kappa = K_s \Theta^b / V_o$

To evaluate (22) and (23), field measurements of sorptivity and wetting front position, saturated conductivity and initial and saturated volumetric water contents are required.

Even simpler expressions can be found for the time to incipient ponding (Kutílek, 1980; Perroux, Smiles and White, 1981). These are based on the recognition that, to a good approximation, twice the integral on the right hand side of equation (19) is the square of the sorptivity. So with this approximation (19) becomes:

$$t_p = S^2(\theta_s, \theta_n)/2V_o^2 \qquad (24)$$

and we expect (24) to be valid for gravity free absorption situations or when $V_o \gg K_s$. We note here that (24) ought to represent a lower limit of the time to incipient ponding. In practice, it emerges that (24) is a good approximation for rainfall rates about 5 times the saturated conductivity. Since the horizontal column experiments conform to the gravity free assumption inherent in (24), we also compare the measurements and predictions of time to ponding for the fine sand samples with that calculated using (24) in Table 1. In both cases, sorptivity was measured by supplying water to the surface at $\psi = 0$. The simple approximation (24) can be seen to be in as good agreement with the experimental results as the more precise solutions (18) or (19).

For $V_o < 5K_s$, the gravity effect must be included in any approximation. If we substitute $K = K_s F(\theta)/V_o$ in equation (15), then, to a good approximation:

$$t_p = S^2(\theta_s, \theta_n)/[2V_o(V_o-K_s)] \qquad (25)$$

and (25) represents an upper bound for t_p. We note that (25) can also be derived using the Green-Ampt model (Mein and Larson, 1972; Swartzendruber, 1974). The approximation of Parlarge and Smith (1975):

$$t_p = [S^2(\theta_s, \theta_n)/2V_oK_s] \ln[V_o/(V_o-K_s)] \qquad (26)$$

also reduces to (25) for $V_o \gg K_s$. To test the validity of equations (24) and (25) ponding times were measured in the laboratory for 5 different rainfall rates using vertical columns of a loam soil (K.M. Perroux, personal communication, 1980). The results, plotted in figure 5, are compared with the predicted values from (24) and (25). In general, as expected, (24) underestimates t_p while (25) overestimates t_p. We also show in this figure, the Parlange and Smith approximation, equation (26), which is in good agreement with the results in figure 5. We note that for V_o only slightly greater than K_s, the predictions using (25) or (26) are extremely sensitive to the measured value of K_s.

FIELD MEASUREMENT OF SOIL-WATER PARAMETERS

It is usual procedure in the field to measure both sorptivity and saturated conductivity by supplying water at a small, positive head at the soil surface. Such measurements are indeed pertinent to the pre-

138

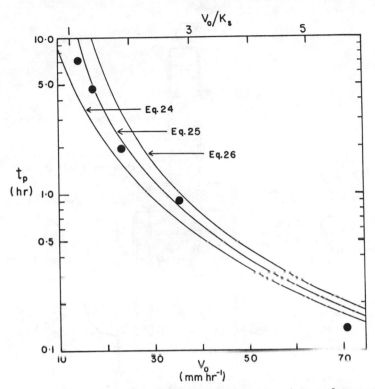

Figure 5. Ponding times for rainfall infiltration into a loam soil. The points are experimental values; the curves are predicted values.

diction of the outcome of ponded infiltration. Under rainfall infiltration, prior to ponding, it can be argued that macropores such as ant and worm holes, soil cracks and old root channels take no part in the infiltration process in light to medium textured soils (Clothier and White, 1981a). These macropores (mean diameter greater than 1mm) contribute substantially to the magnitudes of $S(\theta_s,\theta_n)$ and K_s measured under ponded conditions and, as such, produce values which have little relevance to non-ponding rainfall infiltration where water moves only through the soil matrix (taken here as the soil less macropores). For example, we have measured ponded values of $S(\theta_s,\theta_n)$ and K_s for a pastured site on a loam soil. These values, substituted in (25), gave a predicted time to incipient ponding of 180 min. for a rainfall rate of 30mm/hr. Direct measurement of the soil-water potential at the soil surface showed t_p to be between 4.5 to 6.5 min., about 1/30 of the predicted value. In this experiment, the soil surface was not affected by raindrop impact nor was compression of soil-air a problem. The soil at the field site did, however, contain a large number of worm holes and old root channels which accounted for the over-estimate of $S(\theta_s,\theta_n)$ and K_s of the soil matrix. Here incipient ponding occurred before water started to move down macropores.

In order to make measurements which are apposite to non-ponding rainfall, Clothier and White (1981a) developed a device which supplied water to the surface at a small negative head of about - 40mm H_2O. The device is shown schematically in figure 6. The actual head at which

139

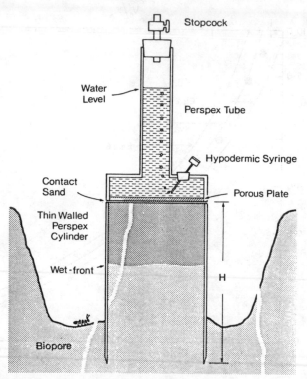

Figure 6. Device for measuring sorptivity at a slight negative water supply head. The supply head is set so that macropores, such as the ant-hole shown, play no part in the infiltration process (Clothier and White, 1981a).

water is supplied is controlled by the position and size of the hypodermic needle, shown in the figure. With this device, macropores whose average diameter exceeds 0.74mm are excluded from the flow process. To determine β simultaneously with sorptivity, a transparent ring is driven into the ground and soil excavated away from the outside of the ring. The device, placed on top of the ring, then gives values of cumulative infiltration, i, as a function of time, t, and the position of the wetting front, Z_{wf}, is recorded as a function of t. Sorptivity is found from $di/dt^{\frac{1}{2}}$ for small times. The group $S/[\lambda_{wf}(\theta_s-\theta_n)]$, where $\lambda_{wf}=Z_{wf}t^{\frac{1}{2}}$, permits calculation of β (Clothier and White, 1981a). The saturated conductivity of the soil matrix is determined by carefully removing the soil core from the profile and establishing a water table at the base of the sample. In this way hydraulic conductivities were measured using an hydraulic gradient of about 0.5.

Preliminary work indicated that these measurements, made at a small negative pressure, were relevant to rainfall infiltration up to the time of incipient ponding. For example, at the loam soil pasture site mentioned above, these negative pressure measurements gave a predicted t_p of 2-4 min., consistent with the observed values.

RAINFALL INFILTRATION IN THE FIELD

Two field sites were chosen for study (Clothier, White and Hamilton,

1981). This first was a virgin site in a 10m x 10m clearing within a dry sclerophyll forest, near Bungendore, N.S.W. The soil is a deep, apedal deposit of fine aeolian sand. The second site, in one bay of a wheat field on the Soil Conservation Service of N.S.W. Cowra Research Station, had a long history of agricultural use. The soil there was a sandy loam and infiltration measurements were made after the wheat stubble had been grazed by sheep for several months. The relevant soil properties of both sites necessary for predictions are listed in Table 2.

A rainfall simulator (Grierson and Oades, 1977) supplied constant rate rainfall to 80cm x 80cm bare soil plots. The raindrop momentum distribution and range of rainfall rates selected were typical for the areas studied. At Bungendore, rates varied between 0.15 to 0.8 of the mean saturated hydraulic conductivity and at Cowra, the rates were between 0.73 to 4.15 K_s. Moisture profile development during rainfall prior to ponding was determined by sequential gravimetric sampling of the soil profile at 1cm intervals.

Typical water-content profiles found for the Bungendore sand site, using a rainfall rate of 14mm hr.$^{-1}$ are given in figure 7. In figure 7a the results are compared with the absorption solutions (equations (22) and (23) with $K=0$), and in figure 7b with the full solutions, equations (22) and (23). We see in figure 7a that gravity starts to become noticeable after 17.5 min. The limit of application of the short time solutions, given by equation (21), is 16.1 min.

The results at the Cowra site were similar to those on the fine sand. Measured water-content profiles for a rainfall rate of 7.6mm hr.$^{-1}$ are plotted in figure 8. Also shown in the figure are predictions made using equations (22) and (23) and the soil properties in Table 2.

In order to determine time to incipient ponding, long, thin, quick-response tensiometers were placed on the soil-surface. The change of

Table 2: Soil properties at field sites.

Property	Bungendore fine sand	Cowra sandy loam
Sorptivity (m sec.$^{-\frac{1}{2}}$)	4.0×10^{-4}	2.2×10^{-4}
Conductivity at θ_s (ms^{-1})	7×10^{-6}	2.9×10^{-6}
θ_s (m^3 m^{-3})	0.335	0.33
θ_n (m^3 m^{-3})	0.03	0.029
β	3	3
b	2.4	5*
Bulk density (Kg m^{-3})	1.47	1.50

* estimated from Mualem (1979)

surface water potential measured at the Cowra site for three rainfall rates, 43.2, 20.2 and 7.56mm hr.$^{-1}$ are presented in figure 9. The lowest rate was less than the site K_s, and so the soil surface did not reach

incipient ponding as can be seen from the negative value of surface
water-potential in Figure 9.   Predicted values of t_p made using equation
(25) are indicated by the arrows on the figure.

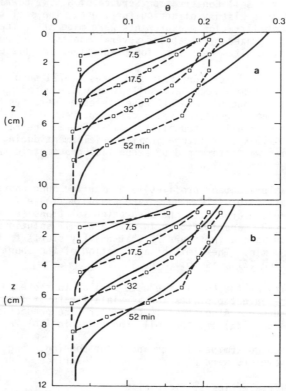

Figure 7.   Field measured water-content profiles developing in Bungendore
fine sand under a rainfall rate of 14mm/hr.   The solid lines are calculated from (a) the absorption solutions, and (b) the infiltration solutions (22) and (23) Clothier, White and Hamilton, 1981).

    For both field sites, the agreement between the predictions of the
modified theory and the measured profile development and time to incipient
ponding is good.   This is particularly so when one considers the simplistic,
uniform soil profile model used in the theory.

Shape of Field-Measured Water Content Profiles

    The almost triangular water-content profiles found at both field
sites differ sharply from the near rectangular profiles normally associated
with water movement in soils and observed in nearly all laboratory
experiments (compare the profile shapes in figures 3 and 4 with those
in figures 7 and 8).   Similar "triangular" shaped profiles have been
observed at other field sites (Hamblin and Tennant, 1981).
Such profiles indicate that the soil-water diffusivity is not strongly
dependent on water content at high water contents (Clothier and White,
1981b).   This finding is in sharp contrast to the strongly θ-dependent
$D(\theta)$ found in laboratory columns of disturbed samples of the same soils
which have been repacked to the field bulk density.   The pronounced
differences in $D(\theta)$ appear to stem from differences in pore size distributions of the field and repacked samples.

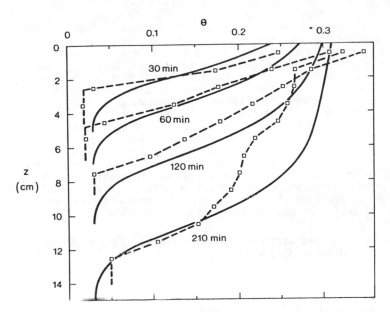

Figure 8. Field measured water-content profiles developing in Cowra sandy loam under a rainfall rate of 7.6mm/hr. The solid lines are the predicted profiles calculated using equations (22) and (23) (Clothier, White and Hamilton, 1981).

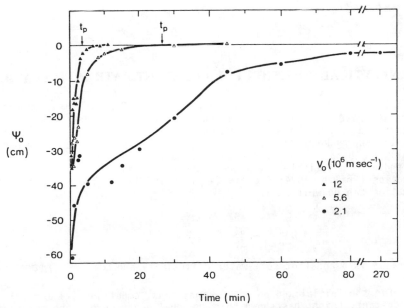

Figure 9. The increase in surface soil-water pressure potential, ψ_0, during 3 different constant-rate rainfall events. For the two ponding rates, the time to incipient ponding, t_p, calculated from equation (25) is indicated (Clothier, White and Hamilton, 1981).

A micrographic examination of thin sections taken from field cores and repacked cores of the soils revealed the presence of a greater number of large pores in the field samples than in the repacked cores. These larger pores dominate waterflow at soil-water contents close to saturation (see, e.g., Bouma et al, 1977). Such larger pores, termed mesopores, were found to be associated with biological activity such as fungal hyphae and old root channels, and tended, therefore, to be continuous and connected with the soil surface. In Table 3 a comparison is made between mesopore sizes found from point to point counts of 63cm traverses of thin sections taken from undisturbed and repacked cores of Bungendore fine sand. The tabulated results indicate that even for such an apedal soil, significant differences exist between the mesopore size distributions of undisturbed and repacked, laboratory samples. We suggest that the triangular shaped water content profiles observed in this work are due to the presence of large, continuous pores which conduct water away from the soil surface, and which is then absorbed into the soil matrix.

Table 3: Mesopore size distributions for field and repacked cores of Bungendore sand.

Pore size range (mm)	No. of pores in 63cm traverse	
	Field core	Repacked core
0.2-0.4	122	169
0.4-0.6	72	22
> 0.6	27	4

ANALYTICAL SOLUTION FOR CONSTANT-RATE INFILTRATION

For certain models of soil-water flow, analytical solutions for constant rate rainfall infiltration are available. Their use is normally restricted, since the models are considered unrealistic. The results for the Bungendore field site, where the soil-water diffusivity was only weakly θ-dependent and the hydraulic conductivity had an almost quadratic dependence on θ (Table 2), suggest the use of one such model; Burgers' equation (Philip, 1974; Clothier, Knight and White, 1981). Burgers' equation can be written as:

$$\frac{\partial \theta}{\partial t} = D\star \frac{\partial^2 \theta}{\partial z^2} - \frac{\partial (K_s \omega^2)}{\partial \theta} \frac{\partial \theta}{\partial z}$$

where $D\star$ is a constant soil-water diffusivity and $\omega = (\theta - \theta_n)/(\theta_s - \theta_n)$.

For the initial and boundary conditions, equations (1) and (2), with V_o constant, Knight has shown that Burgers' equation has the solution (Clothier, Knight and White, 1981):

$$\omega = (A-B)V_o^{\frac{1}{2}}[A+B-2f(\zeta/2\tau^{\frac{1}{2}})+2\exp(\zeta^2/4\tau)]K_s^{\frac{1}{2}}, \qquad (27)$$

where $\qquad A=f[(\zeta-2\tau)/2\tau^{\frac{1}{2}}], \quad B=f[(\zeta+2\tau)/2\tau^{\frac{1}{2}}]$

$f[\chi] = \exp(\chi^2)\text{erfc}(\chi)$ and the nondimensional space and time variables ζ and τ are $\zeta = (V_0 K_s)^{\frac{1}{2}}z/(\theta_s-\theta_n)D^*$ and

$$\tau = V_0 K_s t/(\theta_s-\theta_n)^2 D^*.$$

A simple expression for the nondimensional ponding time, $\tau_p = V_0 K_s t_p/(\theta_s-\theta_n)^2 D^*$, can be found by putting $\omega=1$ and $\zeta=0$ in (27);

$$\tau_p = \text{inverf} (K_s/V_0)^{\frac{1}{2}} \qquad (28)$$

Equation (27) can be used to calculate the development of moisture profiles, the surface soil-water content and the movement of the wetting front during constant rate infiltration, given V_0, K_s, D^* and $\theta_s-\theta_n$. As an example of the use of (27), we show in figure 10 how the depth of the wetting front changes with time at the Bungendore field site. The wetting front is given in terms of the reduced variable $z_{wf}=V_0 z_{wf}$ and time in terms of reduced time, $T=V_0^2 t$. The results are compared with values calculated using (27) for rainfall rates of $K_s/4$ and $3K_s/4$ which spans the rates used at the Bungendore site. Also shown is the simple, absorption solution. For these calculations, the linear D^* was taken to be

$$D^* = 0.5983 \, S^2(\theta_s,\theta_n)/(\theta_s-\theta_n)^2 \qquad \text{which is the mean}$$

diffusivity over the moisture range θ_s to θ_n, for the exponential diffusivity with $\beta = 3$. The wetting front has been identified with a plane of moisture content, $\theta = 0.08$. The agreement between measured values and those calculated from (27) can be seen, in Figure 10, to be excellent.

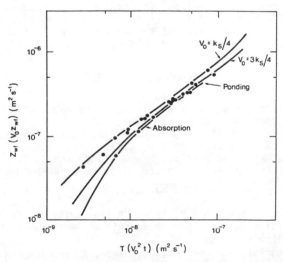

Figure 10. Reduced wet-front depth, Z_{wf}, as a function of reduced time for _in situ_ rainfall infiltration experiments in Bungendore fine sand. The curves are the predicted values from the Burger's equation solution (27) (Clothier, Knight and White, 1981).

145

Clothier, Knight and White (1981) have shown that ponding times calculated from (28) are in reasonable agreement with the numerical results for the soils studied by Parlange and Smith (1976). While this agreement and that in Figure 10 is encouraging, it has yet to be shown that Burgers' equation has wider applicability to other soils. Nonetheless, equations (27) and (28) remain useful for testing numerical models.

CONCLUSIONS

The deterministic theory of constant-rate rainfall infiltration presented here has been demonstrated to be successful in predicting the time dependence of the surface soil-water contents, the movement of the wetting fronts, the evolution of water content profiles and times to incipient ponding. This conclusion not only holds for controlled laboratory experiments, tailored to fit the theory's assumptions, but also for complex field situations. For the latter, advances in scaling theory permit the expression of the solutions in terms of easily measured soil-water properties, sorptivity and saturated conductivity. In order to determine values of these soil parameters which are relevant to non-ponding rainfall infiltration, the influence of macropores, ever present in field soils, must be excluded.

The shapes of water-content profiles in the in situ field soils examined here were profoundly different from those normally reported in laboratory experiments on repacked cores. The difference seems attributable to the presence of biologically-associated pores in the field soils. These unusually shaped profiles imply that the field soil-water diffusivity, for the sites studied, has only a weak dependence on soil water content at high water contents. For one of the sites this weak dependence meant the experimental results could be described by an analytical solution of the simplified flow equation.

In the work described here, attention has been deliberately confined to light textured, stable soils. It remains to be demonstrated that the approach adopted here is applicable to heavy clay soils. The large cracks and low hydraulic conductivity of the soil matrix of clay soils present dificulties which may be more amenable to the analysis of Hoogmoed and Bouma (1980).

ACKNOWLEDGMENTS

We wish to express our thanks to K. M. Perroux, CSIRO, Division of Environmental Mechanics, G. J. Hamilton, Soil Conservation Service of N.S.W., Cowra Research Station, J. H. Knight, CSIRO, Division of Mathematics and Statistics, J. R. Philip, CSIRO, Institute of Physical Sciences, D. R. Scotter, Massey University, New Zealand, and J. R. Sleeman, CSIRO, Division of Soils for their help and discussions throughout this work.

REFERENCES

Ahuja, L. R. and Römkens, M.J.M. 1974. A Similarity during Early Stages of Rain Infiltration. Soil Science Society of America, Proceedings, Vol. 38, pp. 541-544.

Bouma, J., Jongerius, A., Boersma, O., Jager, A. and Schoonderbeek, D. 1977. The Function of Different Types of Macropores during Saturated Flow through Four Swelling Soil Horizons. Soil Science Society of America Journal, Vol. 41, pp. 945-950.

Braester, C. 1973. Moisture Variation at the Soil Surface and the Advance of the Wetting Front during Infiltration at Constant Flux. Water Resources

Research, Vol. 9, pp. 687-694.

Bresler, E., Russo, D. and Miller, R. D. 1978. Rapid Estimate of Unsaturated Hydraulic Conductivity Function. Soil Science Society of America Journal, Vol. 42, pp. 170-172.

Brooks, R. H. and Corey, A. T. 1964. Hydraulic Properties of Porous Media. Hydrology Paper No. 3, Colorado State University, Fort Collins. March.

Brutsaert, W. 1979. Universal Constants for Scaling the Exponential Soil Water Diffusivity. Water Resources Research, Vol. 15, pp. 481-483.

Chu, S. H. 1978. Infiltration during an Unsteady Rain. Water Resources Research, Vol. 14, pp. 461-466.

Clothier, B. E., Knight, J. H. and White, I. 1981. Burgers' Equation: Application to Field Constant-Flux Infiltration. Soil Science (in Press).

Clothier, B. E. and White, I. 1981a. Measurement of Sorptivity and Soil-Water Diffusivity in the Field. Soil Science Society of America Journal, Vol. 45, pp. 241-245.

Clothier, B. E. and White, I. 1981b. Whater Diffusivity of Field Soil. Submitted to Soil Science Society of America Journal.

Clothier, B. E., White, I. and Hamilton, G. J. 1981. Constant-Rate Rainfall Infiltration: Field Experiments. Soil Science Society of America Journal, Vol. 45, pp. 245-249.

Grierson, I. T. and Oades, J. M. 1977. A Rainfall Simulator for Field Studies of Runoff and Soil Erosion. Journal of Agricultural Engineering Research, Vol. 22, pp. 37-44.

Hamblin, A. and Tennant, D. 1981. The Influence of Tillage on Soil Water Behavior. Soil Science (in Press).

Hoogmoed, W. B. and Bouma, J. 1980. A Simulation Model for Predicting Infiltration into Cracked Clay Soil. Soil Science Society of America Journal, Vol. 44, pp. 458-461.

Johnston, H. J., Elsawy, E. M. and Cochrane, S. R. 1980. Study of the Infiltration Characteristics of Undisturbed Soil under Simulated Rainfall. Earth Surface Processes, Vol. 5, pp. 159-174.

Kutilek, M. 1980. Constant-Rainfall Infiltration. Journal of Hydrology, Vol. 45, pp. 289-303.

McIntyre, D.S. 1958. Soil Splash and the Formation of Surface Crusts by Raindrop Impact. Soil Science, Vol. 85, pp. 261-266.

Mein, R.G. and Larson, C.L. 1973. Modelling Infiltration during a Steady Rain. Water Resources Research, Vol. 9, pp. 384-397.

Miller, R.D. and Bresler, E. 1977. A Quick Method for Estimating Soil-Water Diffusivity Functions. Soil Science Society of America Proceedings, Vol. 41, pp. 1021-1022.

Morin, J. and Benyamini, Y. 1977. Rainfall Infiltration into Bare Soils. Water Resources Research, Vol. 13, pp. 813-817.

Mualem, Y. 1979. A New Model for Predicting the Hydraulic Conductivity of Unsaturated Porous Media. Water Resources Research, Vol. 14, pp. 513- 522.

Parlange, J.-Y. 1972. Theory of Water Movement in Soils. 8. One Dimensional Infiltration with Constant Flux at the Surface. Soil Science, Vol. 114, pp. 1-4.

Parlange, J.-Y. and Smith, R.E. 1976. Ponding Time for Variable Rainfall Rates. Canadian Journal of Soil Science, Vol. 56, pp. 121-122.

Perroux, K. M. Smiles, D. E. and White, I. 1981. Water Movement in Uniform Soils during Constant Flux Infiltration. Submitted to Soil Science Society of America Journal.

Philip, J.R. 1957. The Theory of Infiltration: 4. Sorptivity and Algebraic Infiltration Equations. Soil Science, Vol. 84, pp. 278-286.

Philip, J.R. 1969. Theory of Infiltration. Advances in Hydroscience, Vol. 5, pp. 215-296.

Philip, J.R. 1973. On Solving the Unsaturated Flow Equation: 1. The Flux-Concentration Relation. Soil Science, Vol. 116, pp.328-335.

Philip, J.R. 1974. Recent Progress in the Solution of Non-Linear Diffusion Equations. Soil Science, Vol. 117, pp. 257-264.

Philip, J.R. and Knight, J.H. 1974. On Solving the Unsaturated Flow Equation. 3. New Quasi-Analytical Technique. Soil Science, Vol. 117, pp. 1-13

Philip, J.R. 1958. The Theory of Infiltration: 7. Soil Science, Vol. 85, pp. 333-337.

Reichardt, K. and Libardi, P.L. 1974. A New Equation to Estimate Soil-Water Diffusivity in: Isotope and Radiation Techniques in Physics and Irrigation Studies. International Atomic Energy Agency, Vienna, Austria, pp. 45-51.

Reichardt, K., Nielsen, D.R., Biggar, J.W. 1972. Scaling of Horizontal Infiltration into Homogeneous Soils. Soil Science Society of America Proceedings, Vol. 36, pp. 241-245.

Rubin, J. 1966. Theory of Rainfall Uptake by Soils Initially Drier than Their Field Capacity and Its Applications. Water Resources Research, Vol. 2, pp. 739-749.

Rubin, J. and Steinhardt, R. 1963. Soil Water Relations during Rain Infilitration: I. Theory. Soil Science Society of America Proceedings, Vol. 27, pp. 246-251.

Smiles, D. E. 1978. Constant Rate Filtration of Bentonite. Chemical Engineering Science, Vol. 33, pp. 1355-1361.

Smiles, D. E., Perroux, K.M. and Zegelin, S.J. 1980. Absorption of Water by Soil: Some Affects of a Saturated Zone. Soil Science Society of America Journal, Vol. 44, pp. 1153-1158.

Smith, R.E. 1972. The Infiltration Envelope: Results from a Theoretical Infiltrometer. Journal of Hydrology, Vol. 17, pp. 1-21.

Swartzendruber, D. 1974. Infiltration of Constant-Flux Rainfall into Soils as Analyzed by the Approach of Green and Ampt. Soil Science, Vol. 117, pp. 272-281.

Talsma, T. 1969. In situ Measurement of Sorptivity. Australian Journal of Soil Research, Vol. 7, pp. 269-276.

White, I. 1979. Measured and Approximate Flux-Concentration Relations for Absorption of Water by Soil. Soil Science Society of America Journal, Vol. 43, pp. 1074-1080.

White, I. and Clothier, B.E. 1981. Scaling Exponential Soil-Water Diffusivity: Use of the Flux-Concentration Relation. Submitted to Water Resources Research.

White, I., Smiles, D.E. and Perroux, K.M. 1979. Absorption of Water by Soil: The Constant Flux Boundary Condition. Soil Science Society of America Journal, Vol. 43, pp. 659-664.

Whisler, F. D. and Klute, A. 1967. Rainfall Infiltration into a Vertical Soil Column. Transactions of the American Society of Agricultural Engineers. Vol. 10, pp. 391-395.

USE OF SIMILAR MEDIA THEORY IN INFILTRATION AND RUNOFF RELATIONSHIPS

E. G. YOUNGS
Principal Scientific Officer
Physics Department, Rothamsted Experimental Station, Harpenden, Hertfordshire, England

ABSTRACT

In an attempt to use soil-water physics to obtain infiltration and runoff relationships for spatially varying soils, the theory of similar porous materials has been applied to dissimilar materials. In this way scaled variables, defined in terms of a microscopic characteristic length obtained from soil-water properties, have allowed the derivation of single relationships for integral soil-water behaviour such as infiltration and runoff.

For ponded infiltration into different soils at uniform initial soil-water contents, use of the similar media theory together with certain physical assumptions shows the sorptivity in Philip's two-term infiltration equation to be proportional to the fourth root of the hydraulic conductivity of the saturated soil. Thus the cumulative infiltration i is related to the time t by the equation.

$$i = \alpha K_o^{\frac{1}{4}} t^{\frac{1}{2}} + K_o t$$

where K_o is the hydraulic conductivity of the saturated soil and the constant α is proportional to the square root of the difference between the saturated (θ_o) and initial (θ_i) soil-water contents. With i measured in mm, t in s, and K_o in mm/s, the theory gives $\alpha = 6.3 \sqrt{(\theta_o - \theta_i)}$. In laboratory experiments with ten porous materials that varied in particle size and shape, α had the value 4.1 ± 0.2, which corresponds to $\theta_o - \theta_i = 0.42$.

With surface precipitation at a uniform rate F_o greater than K_o, ponding on the soil surface occurs at some time t_p, dependent on the values of F_o and K_o. Using the Green and Ampt approach in considering the soil-water movement together with similar media considerations, the scaled ponded time t_p^* is given by

$$t_p^* = \beta / \left[G(G - 1) \right]$$

where β is a constant which has the value of 0.027 when the microscopic characteristic length λ is defined as the square root of the intrinsic permeability, and G is written for the ratio F_o / K_o. t_p^* is related to t_p by

$$t_p^{\star} = \left\{ (\rho g)^2 \lambda^3 / \left[\eta \sigma (\theta_o - \theta_i) \right] \right\} t_p$$

where ρ, η and σ are the density, viscosity and surface tension of water, respectively, and g is the acceleration due to gravity. Laboratory experiments with the same materials used to test the infiltration equation agreed with this relationship for a wide range of values of G.

The Green and Ampt approach gives the cumulative runoff in terms of F_o, K_o, t_p and t.For similar porous materials, scaling gives the scaled runoff as a function of scaled time for different values of G in the form of a family of curves. Alternatively, a single relationship connecting these variables in parametric form may be used, and this was utilised in a test of theory using experimental data.

INTRODUCTION

A major difficulty that limits the application of the physics of soil-water movement in field studies of soil-water behaviour is the variation of soil-physical properties within the study area. Without a complete survey of these properties, the exact application of soil physical theory is impossible. However, such surveys are time consuming. Thus,in predictions of soil-water movement, approximations are often used in which simplifications are made while the essential physical nature of the process is retained.

In an attempt to provide a simpler basis for the physical data required for the prediction of infiltration and runoff — processes which are of concern both in hydrology and also in agricultural soil-water management — interest has recently arisen in the use of similar media theory (Miller and Miller, 1956). Several studies on catchment areas have been reported (Peck et al.,1977; Warrick et al., 1977; Sharma et al., 1980) in which use has been made of this concept, although the similar nature of the soils was not proven, and hence its application must be regarded as an approxima-tion. In laboratory studies on packed columns of dissimilar porous materials in the form of glass beads, sands, a slate dust and sieved soils, which varied both in particle size and in shape, Youngs and Price (1981) tested the applicability of similar media theory in infiltration studies. Microscopic characteristic lengths for the various porous materials were deduced from soil-physical measurements, and the infiltration data were shown to fit theoretically derived relationships when the cumulative infiltration and time were scaled in terms of reduced variables defined in terms of these lengths.

This paper reports further work using similar media theory in obtaining infiltration and runoff relationships. The theory is shown to allow the usual two-parameter infiltration equation to be replaced by a one-parameter equation. The concept is used also to deduce runoff relationships in the case where surface ponding occurs with surface precipitation. These relationships, although semi-empirical, are illustrative of the type needed in physically-based hydrological models.

A ONE-PARAMETER INFILTRATION EQUATION

Of the various equations that have been used to describe the cumulative infiltration i into a soil as a function of the time t, two that are based on physical considerations and are well accepted are that due to Green and Ampt (1911), which may be written

$$i - a \ln(1 + i/a) = bt \tag{1}$$

where a and b are constants, and Philip's (1957) two-term infiltration equation

$$i = At^{\frac{1}{2}} + Bt \tag{2}$$

where A and B are constants. These equations and the physical significance of the constants contained in them have been considered by many authors (for example, Childs, 1969, pp 275-280). In equation (1) it may be shown that the constants a and b are related to physical properties of the soil, giving

$$a = (H_o - H_f)/(\theta_o - \theta_i) \text{ and } b = K_o$$

where H_o is the depth of surface ponded water (generally taken to be negligible), H_f the head drop across the menisci at the wetting front which is the division between the soil at the initial water content θ_i and the wetting region assumed saturated at a water content θ_o, and K_o is the hydraulic conductivity of the saturated soil. The constants A and B in Philip's infiltration equation (2) may be argued to be the sorptivity and the hydraulic conductivity K_o, respectively. Swartzendruber and Youngs (1974) calculated that the two equations differ by never more than 15.1% at all times if the theoretically derived value of $a = A^2/2K_o$, as derived by Philip (1957) and also by Youngs (1968), is used. Generally Philip's infiltration equation is preferred because of its explicit form.

Using the scaled variables

$$i^* = \left\{ \rho g \lambda / \left[\sigma (\theta_o - \theta_i) \right] \right\} i \tag{3}$$

$$t^* = \left\{ (\rho g)^2 \lambda^3 / \left[\eta \sigma (\theta_o - \theta_i) \right] \right\} t \ , \tag{4}$$

as dictated by similar media theory, where λ is a microscopic characteristic length, ρ, σ and η the density, surface tension and viscosity of water, respectively, and g the acceleration due to gravity, Philip's equation may be re-written as

$$i^* = A^* t^{*\frac{1}{2}} + K_o^* t^* \tag{5}$$

where A^* and K_o^* are the scaled sorptivity and hydraulic conductivity, respectively. As considered by Youngs and Price (1981), the microscopic characteristic length λ may be defined in various ways, giving different values of the parameters A^* and K_o^*. If λ is defined as λ_k where

$$\lambda_k = (\eta K_o / \rho g)^{\frac{1}{2}} \ , \tag{6}$$

that is, as the square root of the intrinsic permeability of the porous material, then

$$i^* = 0.233 \, t^{*\frac{1}{2}} + t^*,$$

as deduced by Youngs and Price (loc. cit.).

While scaled reduced variables are useful in analysing experimental results from different soils, practical useage requires predictions of actual infiltration as a function of actual time, and the process of calculating actual variables from reduced variables is unnecessary if similar media theory is applied in the following modified form. Youngs (1968) showed that to a very good approximation, the sorptivity A is related to the soil-water parameter

$$S_o = \int_{p_i}^{p_o} K/\rho g \; dp, \tag{7}$$

where K is the hydraulic conductivity at a soil-water pressure p, and the subscripts o and i refer to the saturated and initial states of the soil, respectively, by

$$A = \left[2(\theta_o - \theta_i)S_o\right]^{\frac{1}{2}} . \tag{8}$$

Youngs and Price (1981) also defined the microscopic characteristic length λ in terms of S_o by λ_S given by

$$\lambda_S = (\eta/\sigma)S_o . \tag{9}$$

Then, combining equations (6), (7) and (8)

$$A = \left[2(\lambda_S/\lambda_K)(\theta_o - \theta_i)\sigma\right]^{\frac{1}{2}} (K_o/\rho g \eta)^{\frac{1}{4}} \tag{10}$$

Thus generally we may write equation (2) as

$$i = \alpha K_o^{\frac{1}{4}} t^{\frac{1}{2}} + K_o t . \tag{11}$$

α has dimensions $\left| L^{\frac{3}{4}} T^{-\frac{1}{4}} \right|$; with i measured in mm, t in s and K_o in mm/s, and with $\lambda_k/\lambda_s = 37$, as found by Youngs and Price (loc. cit.),

$$\alpha = 6.3(\theta_o - \theta_i)^{\frac{1}{2}} .$$

With $\theta_o - \theta_i = 0.10$, α is 2.00; with $\theta_o - \theta_i = 0.30$, α is 3.45; and with $\theta_o - \theta_i = 0.50$, α is 4.45. When the values A and K_o obtained by Youngs and Price (loc. cit.) for their range of porous materials, whose physical description is given in Table 1, are analysed, we find

$$A = (4.1 \pm 0.2)K_o^{\frac{1}{4}} . \tag{12}$$

These values of A and K_o are plotted in Fig.1; the full line is equation (12) while the symbols show experimental values. The value for α of 4.1 corresponds to $(\theta_o - \theta_i) = 0.42$, which compares with an average value of 0.38 for the saturated soil-water content of the porous materials.

Infiltration measurements (Youngs and Price, loc. cit.) for the materials listed in Table 1 are reproduced in Fig. 2. To compare these data with equation (11), the latter is rewritten in the form

$$K_o^{\frac{1}{2}} i = \alpha (K_o^{\frac{3}{2}} t)^{\frac{1}{2}} + K_o^{\frac{3}{2}} t \tag{13}$$

152

Table 1

Particle size and soil-water properties (saturated water content θ_o, hydraulic conductivity K_o and sorptivity A) and the values of the microscopic characteristic length λ_k, for the porous materials used in the experiments.

Material	Particle size	θ_o	K_o	A	λ_k
	(μm)	(mm/mm)	(mm/s)	(mm/s$^{\frac{1}{2}}$)	(μm)
Ballotini 10	210 - 325	0.384	0.115	2.52	3.34
Ballotini 12	115 - 180	0.358	0.099	2.10	3.15
Ballotini 15	60 - 95	0.370	0.066	2.40	2.57
Leighton Buzzard Sand 1	350 - 500	0.352	0.579	3.10	7.77
Leighton Buzzard Sand 2	250 - 350	0.270	0.330	3.44	7.44
Leighton Buzzard Sand 3	180 - 250	0.396	0.197	3.15	4.43
Graded Beach Sand	180 - 250	0.375	0.190	3.25	3.45
Slate Dust	40 - 125	0.431	0.00275	0.99	0.525
Woburn Sandy Loam	Mostly 60 - 350	0.392	0.00294	0.79	0.542
Rothamsted Silt Loam	Mostly 2 - 60	0.371	0.000182	0.34	0.135

This equation is shown by the full line in Fig. 3, where $K^{\frac{1}{2}}i$ is plotted as a function of $K_o t$. The dashed line results when $\theta_o - \theta_i = 0.10$ and the dotted when $\theta_o - \theta_i = 0.50$, compared with $\theta_o - \theta_i = 0.42$ for the full line. Experimental values plotted in this reduced form in Fig.3 agree with equation (13). It thus follows that the experimental data also fit equation (11).

TIME OF PONDING

If the precipitation rate on the soil surface is maintained at a rate greater than the soil's hydraulic conductivity at saturation, then ponding of the soil surface will occur at some time, with runoff subsequently taking place. This situation may be most easily considered using the Green and Ampt (1911) approach in which it is assumed that the soil-water movement takes place as a saturated plug. When the soil surface is ponded, the rate of infiltration is given by

$$\frac{di}{dt} = K_o\left[1 + H_f(\theta_o - \theta_i)/i\right] \qquad (14)$$

where i is the cumulative infiltration up to time t. If ponding occurs at time t_p, the infiltration up to this time is

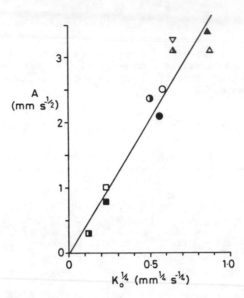

Fig. 1. Sorptivity A plotted as a function of the fourth root of saturated hydraulic conductivity. The line is $A = 4.1 \, K_o^{\frac{1}{4}}$ (equation (12)), and the symbols are experimental values (after Youngs and Price, 1981):

○ Ballotini 10, ● Ballotini 12, ◑ Ballotini 15,
△ Leighton Buzzard Sand 1, ▲ Leighton Buzzard Sand 2,
◭ Leighton Buzzard Sand 3, ▽ Graded Beach Sand,
□ Slate Dust, ■ Woburn Sandy Loam, ◧ Rothamsted Silt Loam.

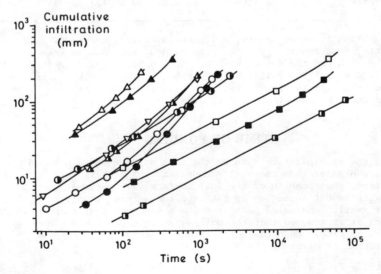

Fig. 2. Cumulative vertical downward infiltration plotted against time for the porous materials (after Youngs and Price, 1981). Symbols as in Fig. 1.

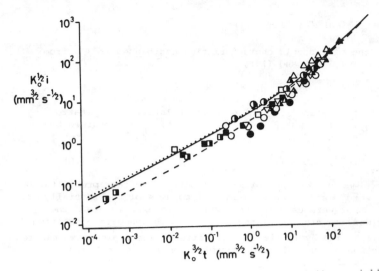

Fig. 3. The results of Fig. 2 plotted in terms of the variables $K_o^{\frac{1}{2}}i$ and $K_o^{\frac{1}{2}}t$. The full line is equation (13) with $\alpha = 4.1$ (equivalent to $(\theta_o - \theta_i) = 0.42$); the dashed line is calculated for $(\theta_o - \theta_i) = 0.10$; and the dotted for $(\theta_o - \theta_i) = 0.50$.

$$i_p = \int_o^{t_p} F \, dt,$$

where F is the flux through the soil surface at time t. Thus ponding occurs at the soil surface when

$$F > \frac{di}{dt}$$

so that, substituting for i_p in equation (14), when

$$F > K_o\left[1 + H_f(\theta_o - \theta_i)/\int_o^{t_p} F \, dt\right]. \tag{15}$$

Thus for a uniform flux F_o through the soil surface up until ponding time t_p, from equation (14),

$$F_o = K_o\left[1 + H_f(\theta_o - \theta_i)/F_o t_p\right]$$

so that

$$t_p = H_f(\theta_o - \theta_i)/\left[F_o(F_o/K_o - 1)\right]. \tag{16}$$

Now we note that

$$H_f \doteq \left[\int_p^{p_o} K/\rho g \, dp\right]/K_o \tag{17}$$

so that using this relationship (17) with equations (6) and (9) and the definition of S_o (equation (7))

$$H_f(\theta_o - \theta_i)/K_o = (\lambda_S/\lambda_K)\left\{n\sigma(\theta_o - \theta_i)/\left[(\rho g)^2 \lambda_k{}^3\right]\right\}. \tag{18}$$

The scaled ponding time t_p^* is from equation (4)

$$t^*_p = \left\{ \lambda^3 (\rho g)^2 / \left[\eta \sigma (\theta_o - \theta_i) \right] \right\} t_p \; . \tag{19}$$

Thus, if the microscopic characteristic length used is λ_k, from equations (16), (18) and (19),

$$t^*_p = \beta / \left[G(G - 1) \right]$$

where β is the ratio λ_k / λ_S and G is the relative precipitation rate, F_o / K_o. If we put $\beta = 37$, the value of λ_k / λ_S obtained by Youngs and Price (1981),

$$t^*_p = 0.027 / \left[G(G - 1) \right] \; . \tag{20}$$

The times of ponding for several uniform surface precipitation rates F_o were measured on experimental columns of the initially dry materials that were used in the infiltration experiments. The scaled times t^*_p were calculated using a λ_k obtained from equation (6) from the measured hydraulic conductivities. These are shown plotted as a function of the relative precipitation rate G in Fig. 4. The full line in this figure represents equation (20). The agreement between the experimental results and the theoretical curve indicates that the Green and Ampt approach gives a satisfactory prediction of the ponding time, and that the similar media concept again adequately encompasses the behaviour of the dissimilar porous materials used.

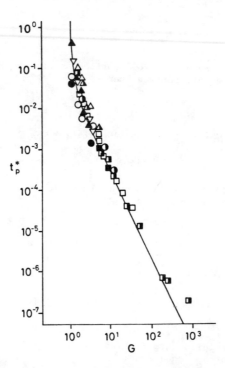

Fig. 4.   The scaled ponding time t^*_p plotted as a function of the relative precipitation rate $G \triangleq F_o / K_o$. The line is equation (20) and the symbols refer to experimental values for porous materials, identified according to the legend of Fig. 1.

RUNOFF EQUATION

The Green and Ampt approach may also be used to predict the amount of runoff that occurs after ponding with a given surface precipitation on a soil. We note that the rate of infiltration at any time is given by equation (14); that is,

$$\frac{di}{dt} = K_o\left[1 + H_f(\theta_o - \theta_i)/i\right].$$

This equation may be integrated between the time of ponding t_p, when the cumulative infiltration is i_p, to time t to give

$$K_o(t - t_p) = i - i_p$$
$$- H_f(\theta_o - \theta_i) \ln\left\{\left[H_f(\theta_o - \theta_i) + i\right]/\left[H_f(\theta_o - \theta_i) + i_p\right]\right\}. \quad (21)$$

For a constant precipitation rate F_o, noting that $i_p = F_o t_p$ and using equation (16) for t_p, the runoff $r \cong F_o t - i$ is given from equation (21) by

$$r = (F_o - K_o)(t - t_p) - F_o t_p (G - 1)\ln\left[1 + ((t/t_p - 1)/G - 1)/G\right] \quad (22)$$

In terms of scaled reduced variables t^*, t^*_p and r^*, with r^* defined by

$$r^* = \left\{\rho g \lambda/\left[\sigma(\theta_o - \theta_i)\right]\right\} r, \quad (23)$$

so that using a microscopic characteristic length λ_k defined by equation (6) in equations (4) and (23),

$$r^*/t^* = (r/t)/K_o.$$

Equation (22) may be written

$$r^* = (G - 1)(t^* - t^*_p) - G(G - 1)t^*_p \ln\left[1 + (t^*/t^*_p - 1)/G - r^*/t^*_p G^2\right] \quad (24)$$

The relationship between r^* and t^*, as predicted by equation (24), is shown in Fig. 5 for various values of G and for the corresponding values of t^*_p given by equation (20).

While equation (24) gives the predicted runoff relationship in terms of scaled variables, a more convenient form for testing the analysis against experimental results is obtained if equation (22) is recast in the form

$$(r/F_o t_p)/(G - 1) = (t/t_p - 1)/G - \ln\left[1 + (t/t_p - r/F_o t_p - 1)/G\right]$$

so that using the parameters

$$U = (t/t_p - 1)/G - (r/F_o t_p)/(G - 1) \quad (25)$$

and

$$V = (r/F_o t_p)/\left[G(G - 1)\right] \quad (26)$$

we have

$$V = \exp(U) - U - 1 \quad (27)$$

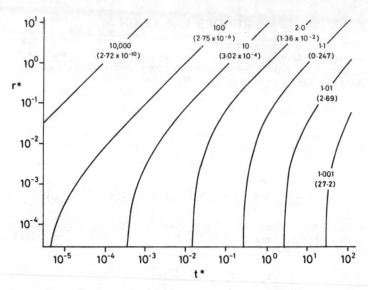

Fig. 5.   The scaled runoff r* given by equation (24), plotted
as a function of scaled time t* for various values of G
shown by the curves.   Values of t* calculated using
equation (20) are given in brackets after the G values.

Experimental runoff relationships were measured on laboratory
columns of the different porous materials listed in Table 1 for
various constant surface precipitation rates.   The results are
shown in Fig. 6 which are replotted in Fig. 7 in terms of the variables
U and V defined by equations (25) and (26).   The experimental results
are in very good agreement with equation (27) shown by the full line
in Fig. 7, and thus confirm the validity of the analysis.

DISCUSSION

The use of the theory of soil-water movement in describing
infiltration and runoff processes in catchment areas is confounded
to a large extent by the effort required to make measurements of
the physical properties when the soil varies over the area.   However,
the processes are of a physical nature and their descriptions are
therefore best embodied in physically based models.   To apply such
models, either a sufficiently large number of physical measurements
must be collected in order to describe adequately the area, or
approximations have to be made to the theory in order that fewer
data are required, while preserving at the same time the essential
physics of the problem.   In this paper we have adopted this second
course to make advances in infiltration theory through the application of similar media theory and through the making of appropriate
approximations.

The one-parameter infiltration equation in the form of
equation (11) is based on Philip's two-term infiltration equation
which itself is an approximation and which was described by
Childs (1969, p.279) as "an empirical formula to which one is led

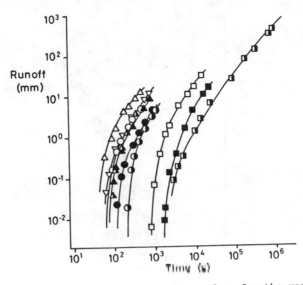

Fig. 6. Cumulative runoff from the surface for the various porous materials, identified according to the symbols used in Fig. 1, for precipitation rates:
Ballotini 10, 0.50 mm/s; Ballotini 12, 0.21 mm/s; Ballotini 15, 0.15 mm/s; Leighton Buzzard Sand 1, 1.04 mm/s; Leighton Buzzard Sand 2, 0.67 mm/s; Leighton Buzzard Sand 3, 0.50 mm/s; Graded Beach Sand, 0.44 mm/s; Slate Dust, 0.032 mm/s; Woburn Sandy Loam, 0.015 mm/s; Rothamsted Silt Loam, 0.005 mm/s.

on physical grounds". Equation (11) results with the use of an approximation for the sorptivity (equation (8)) and the application of similar media theory with empirically obtained values of the ratio of the microscopic characteristic lengths which are defined in terms of the soil-water properties (Youngs and Price, 1981); it requires only a knowledge of the saturated hydraulic conductivity when a value of $\sqrt{(\theta_o - \theta_i)}$ may be assumed. The laboratory infiltration experiments with a wide range of porous materials give confirmation to this theoretical work and show that the assumptions leading to the infiltration equation are acceptable. Applications in field hydrology depend on the ease with which the hydraulic conductivity can be measured or estimated over the given area. If the areal variability of the hydraulic conductivity is known, then the total infiltration over the catchment area R may be found by integrating equation (11):

$$\int_R i \, dR = \alpha \int_R K_o^{\frac{1}{4}} dR \, t^{\frac{1}{2}} + \int_R K_o \, dR \, t$$

The runoff relationships in the form of equation (16) for the time of ponding and of equation (22) for the cumulative runoff have been developed using the Green and Ampt (1911) approach which uses physical assumptions concerning the infiltration process. Again we have used the empirically derived value obtained for the ratio of the microscopic characteristic lengths to obtain in equation

159

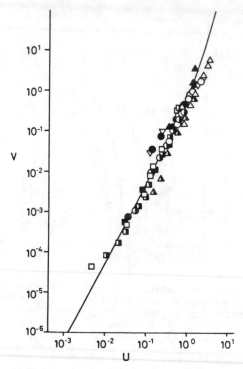

Fig. 7.   The results of Fig. 6 plotted in terms of the variables
U and V, defined by equations (25) and (26) respectively.
The line is equation (27).

(20) the scaled time of ponding. The laboratory experiments confirm
this relationship. It may be used therefore to predict, for a given
constant precipitation rate on a soil of given saturated hydraulic
conductivity, the time at which ponding occurs. The parametric form
of the runoff relationship in the form of equation (27) gives a
single relationship which includes the precipitation rate and
hydraulic conductivity instead of the family of curves given by
equation (22) (or equation (24) for scaled variables) with different
ratios of precipitation rate to the soil's hydraulic conductivity.
Equation (27) is in accord with the experimental results from the
column experiments with the various porous materials used, again
confirming the applicability of the theoretical approach to the
situation. Equations (16), (22), (24) and (27) may be regarded as
runoff relationships that have been derived from considerations of
the physics of the process and on which physically based models for
hydrological modelling may be built.

In the present consideration of the infiltration and runoff
processes, water is assumed to move one-dimensionally through the
soils in the vertical direction. No account is taken of any
lateral movement which may occur. It is thus possible to use
similar media theory to develop simple relationships for the processes
for use when there is a real variability of soil properties. A
second assumption, posing a severe limitation to the use of these
infiltration equations, is that the hydraulic conductivity is
assumed uniform with depth. The relationships are therefore

less valuable for long-term infiltration when the wetting front will penetrate to depths where there is likely to be a change of soil properties. Nevertheless, for short-term infiltration and for large precipitation rates relative to the hydraulic conductivity of the saturated soil (when ponding will occur at early times, followed by runoff), the relationships should adequately describe field processes.

REFERENCES

Childs, E.C. 1969. An Introduction to the Physical Basis of Soil Water Phenomena. John Wiley & Sons, Inc., London. pp. 493.

Green, W.H., and Ampt., G.A. 1911. Studies on Soil Physics : I. Flow of Air and Water through Soils. Journal of Agricultural Science, Vol.4, pp. 1-24.

Miller, E.E., and Miller, R.D. 1956. Physical Theory for Capillary Flow Phenomena. Journal of Applied Physics, Vol. 27, pp. 324-332.

Peck, A.J., Luxmoore, R.J., and Stolzy, J.L. 1977. Effects of Spatial Variability of Soil Hydraulic Properties in Water Budget Modeling. Water Resources Research, Vol. 13, pp. 348-354.

Philip, J.F. 1957. The Theory of Infiltration : 4. Sorptivity and Algebraic Infiltration Equations. Soil Science, Vol. 84, pp. 257-264.

Sharma, M.L., Gander, G.A., and Hunt C.G. 1980. Spatial Variability of Infiltration in a Watershed. Journal of Hydrology, Vol. 45, pp. 101-122.

Swartzendruber, D., and Youngs, E.G. 1974. A Comparison of
    Physically-Based Infiltration Equations. Soil Science,
    Vol.117, pp. 165-167.

Warrick, A.W., Mullen, G.J., and Nielsen, D.R. 1977. Scaling
    Field-Measured Soil Hydraulic Properties using a Similar
    Media Concept. Water Resources Research, Vol.13, pp.
    355-362.

Youngs, E.G. 1968. An Estimation of Sorptivity for Infiltration
    Studies from Moisture Moment Considerations. Soil Science,
    Vol.106, pp. 157-163.

Youngs, E.G., and Price, R.I. 1981. Scaling of Infiltration
    Behaviour in Dissimilar Porous Materials. Water Resources
    Research, (to be published).

UPWARD MIGRATION OF WATER UNDER COLD
BUT ABOVE FREEZING SURFACE CONDITIONS

H. J. Morel-Seytoux,
Professor of Civil Engineering
Colorado State University

E. L. Peck,
Director, Hydrologic Research Laboratory NOAA
National Weather Service

ABSTRACT

The objective of this paper is to estimate the order of magnitude
of the upward migration of water from a water table to the surface due
to a gradient of temperature directed toward the surface. The rigorous
solution would require simultaneous solution of nonlinear partial differential equations for heat and mass transport. In size the importance
of the phenomenon a very crude approach is taken in this paper which
includes many assumptions. As the mass transport is the phenomenon of
concern a steady-state temperature profile will be assumed, independent
of moisture evolution. Initially the moisture profile is assumed in
equilibrium at a uniform relatively warm temperature. Then a steady-
state temperature profile is imposed on the system with colder temperatures towards the surface. It is assumed that there is a plentiful
water supply at the water table and no evaporation from the surface.
When a thermal gradient is imposed the following transient state results
in an upward flow of water. The questions to be answered are: (1) how
fast will this migration take place, and (2) ultimately when a new
equilibrium is reached how much more water will be in storage in the
upper part of the soil.

BACKGROUND

It has been observed in the field (e.g., Ferguson et al., 1964,
Willis et al., 1964, Peck, 1974) that appreciable upward water movement·
occurs during the winter time. For example Ferguson et al. (1964,
p. 701) reports an increase of (volumetric) water content from a value
of 0.26 on December 20, 1962 to a value of 0.32 on March 20, 1963,
followed by a decrease to a value of 0.28 on April 26, 1963 at a soil
depth of 24 in. On the other hand at a depth of 42 in. the reverse
trend was noted (0.23 on December 20, 1962, 0.20 on March 20, 1963
and 0.21 on April 26, 1963). The ground was frozen to a depth of 12 in.
on January 4. Thus a relative increase of water content of the order
of 23 percent i.e., $100 \left(\dfrac{0.32 - 0.26}{0.26}\right)$ occurred right around the
depth of maximum penetration of the freezing front. The mechanism of
attraction of liquid water to a freezing front and the growth of ice
lens is well understood qualitatively and semiquantitatively
(e.g., Morel-Seytoux, 1978). Peck (1974, p. 406) reports that in the

163

spring of 1969 "the average soil moisture for the 18 stations was 44 percent" (by weight) "or approximately 12 percent greater than their average field capacity" (32 percent by weight). This corresponds to a relative increase of water content of 38 percent in the top soil. Similar water content changes were reported for the years 69-70, 70-71 and 71-72 between mid-November and February or March of the following year. However, there is a difference between these results and the spring 1969 observations. During the water years 70, 71 and 72 "frost was observed under the late season snow cover" (Peck, 1974, p. 408) whereas in "the spring of 1969 a large portion of the Rock River basin was found to be frost free." Could such a large upward water migration have been induced strictly by a temperature gradient without the added suction effect of ice?

TEMPERATURE EFFECTS

It is known that capillary pressure (and therefore field capacity) and hydraulic conductivity are temperature dependent (e.g., Klock, 1972). Klock found that the variability on (saturated) hydraulic conductivity with temperature can be completely accounted for by the dependence of hydraulic conductivity of water viscosity. A change of temperature from 25°C to 0°C will double the water viscosity and halve the hydraulic conductivity. Klock also found that warming a previously soil column from 0.3°C to 25°C would result in additional drainage of a volume equal to 12 percent of the initially drained volume. Again this effect could be completely accounted by the dependence of surface tension and temperature. In the experiment the soil column was 123 cm long and a constant water table level was maintained at the bottom of the column.

The effect of temperature on unsaturated soil properties was investigated by Jensen et al. (1970). They found similar results.

In all these investigations however comparison was made between properties at two different but both equilibrium isothermal states. It is the objective of this study to investigate the effect of a temperature gradient on the equilibrium capillary state and on the rate of migration.

TEMPERATURE DEPENDENCE

The capillary pressure is dependent upon temperature because the water-air surface tension is. Presuming that the temperature dependence is strictly reflected through the surface tension then the capillary pressure, h_c, expressed as a water height, at a uniform temperature T is deduced from that at a uniform *base* temperature T_B by the relation:

$$h_c = h_{cB}(\theta) \ \frac{374-T}{374-T_B} \tag{1}$$

where temperature is expressed in °C. The h_c curve on Figure 1 is for a uniform temperature of 15°C. For a water table 2 m below the soil surface under equilibrium conditions the moisture profile above the water table would be as shown in Figure 1 reading h_c as elevation above the water table. At the soil surface the water content would be 0.320 which is somewhat larger than field capacity estimated at 0.26,

164

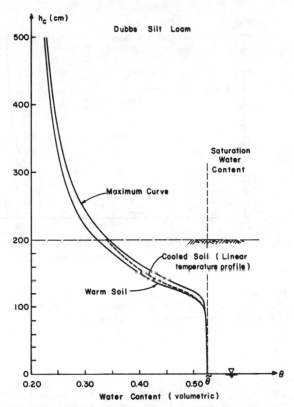

Figure 1.   Isothermal water content profiles at different
            temperatures and steady-state equilibrium profile
            for a uniform temperature gradient (data after
            Jensen et. al., 1970).

if field capacity is defined as water content at 0.3 atmosphere.
However it can be seen from Figure 2 that at a water content of 0.26
the relative permeability is very small. If a value of $k_{rw} = 0.01$
is chosen as a definition of field capacity then water content at
field capacity, θ_{fc} is more likely to be about 0.30.

STEADY STATE PROFILES

At time zero a new temperature profile is imposed on the system
and is maintained indefinitely. For simplicity it is assumed that the
temperature gradient is uniform between the soil surface and the water
table and the variation of temperature with depth is given by the
relation:

$$T = T_f + (T_B - T_f) \, z/D \tag{2}$$

where T_f is the temperature at the soil surface, T_B is the initial
uniform (base) temperature, z is the vertical coordinate oriented
positive downward with origin at the soil surface, and D is the depth
to a level where the temperature T_B is maintained.

165

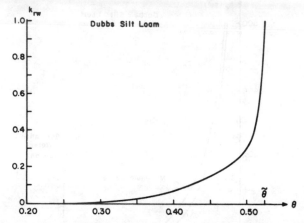

Figure 2. Relative permeability to water for Dubbs silt
           loam at 15°C (data after Jensen et. al., 1970).

Under this new maintained condition of temperature, ultimately a
new equilibrium moisture profile will develop. Under equilibrium
$dh_c + dz$ must be zero or explicitly using Eq. (1):

$$\frac{374-T}{374-T_B}\ h'_{cB}(\theta)d\theta - \frac{h_{cB}(\theta)dT}{374-T_B} + dz = 0 \tag{3}$$

where the prime indicates differentiation with respect to θ .
Expressing dz in terms of dT from Eq. (2) and rearranging one
obtains the separable differential equation:

$$\frac{h'_{cB}(\theta)}{h_{cB}(\theta) - \dfrac{D(374-T_B)}{T_B-T_f}} = -\frac{dT}{T-374} \tag{4}$$

which can be integrated to yield the equilibrium profile in the
implicit form:

$$h_{cB}(\theta) - h_{cBD} = \frac{D(374-T_B)}{(T_B-T_f)} \cdot \frac{(T_B-T)}{(374-T)} \tag{5}$$

where h_{cBD} is the capillary pressure existing at depth D or in
terms of z :

$$h_{cB}(\theta) - h_{cBD} = \frac{D(374-T_B)(D-z)}{(374-T_f)D - (T_B-T_f)z} \tag{6}$$

or in terms of elevation above the water table, h :

$$h_{cB}(\theta) - h_{cBD} = \frac{D(374-T_B)(h-h_{cBD})}{D(374-T_B) + (T_B-T_f)(h-h_{cBD})} \tag{7}$$

166

The new equilibrium profile for values of $T_B = 15°C$, $T_f = 0°C$, and $D = 200$ cm is shown on Figure 1. It was calculated from Eq. (7) rewritten in the form:

$$h - h_{cBD} = \frac{D(374-T_B)(h_{cB}-h_{cBD})}{D(374-T_B) - (T_B-T_f)(h_{cB}-h_{cBD})} \tag{8}$$

For the selected numerical values h_{cBD} is equal to zero. The difference between the two curves h_{cB} and h (linear gradient) represents the increase in moisture storage in the soil due to its cooling. In this particular instance the effect is small. The water content at the soil surface changes from 0.32 to 0.34, i.e., a relative change of 6 percent. The reduction in pore space available for infiltration is 10 percent.

Inspection of Eq. (7) rewritten for emphasis in the form:

$$h_{cB}(\theta) = h_{cDD} - \frac{h - h_{cBD}}{1 + \left(\frac{T_B-T_f}{374-T_B}\right)\left(\frac{h-h_{cBD}}{D}\right)} \tag{9}$$

indicates that the effect is more pronounced when T_B is low and $T_B - T_f$ is large, that it is maximum for $h - D$ i.e., at the point of lowest temperature and for $h_{cBD} = 0$ that is when the uniform warm base temperature region does not extend upward beyond the water table. Under these most favorable conditions the ratio $\frac{h_{cB}}{h} = \frac{374-T_B}{374-T_f}$. In other words the maximum relative decrease in capillary pressure cannot exceed $100\left(\frac{T_B-T_f}{374-T_B}\right)$ percent. Realistically T_B is at most of the order of 20°C and the difference $T_B - T_f$ might be of the order of 30°C (allowing some degree of supercooling because water is under suction and would not freeze so readily and ice is not present). Thus under these circumstances the ratio h_{cB}/h is about 0.92.

This ratio was applied at all elevations h , the corresponding $h_{cB}(\theta)$ was calculated (0.92h) and θ determined from the h_{cB} curve. Then a point of coordinate θ,h was obtained. The locus of all these points is the capillary pressure at a uniform temperature 30°C less than the original curve and is shown on Figure 1. The maximum absolute increase in water content at a given elevation (i.e., warm capillary pressure) occurs at a capillary pressure of 130 cm. The absolute change in water content is 0.03 and the relative increase in water content is 7 percent. The maximum relative increase occurs at a capillary pressure of 230 cm where it is 23 percent. The maximum absolute effect will occur at pressures where the moisture retention curve is flattest and the maximum relative effect at pressures where the moisture retention curve has highest curvature, which is usually around the field capacity value. It appears doubtful therefore that the high water contents in the Rock River basin in the spring 1969, were solely due to a water migration induced by a temperature gradient. However as much as half of it, but no more, could have been.

The calculations performed for the Dubbs silt loam were repeated for the Moody silt loam in Rock County although the capillary pressure curve is not as well defined experimentally (Figure 3). At 1/3 of an

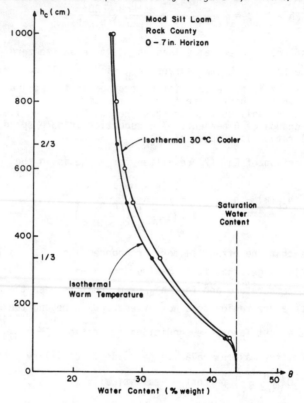

Figure 3.   Isothermal water content profiles
for Moody silt loam (data after
Jensen et. al., 1970).

atmosphere the absolute increase in water content is 1 percent by weight and the relative water content increase is 3 percent.

RATE OF MIGRATION

As has been shown previously (Morel-Seytoux and Khanji, 1974) the total velocity (i.e., algebraic sum of water and air velocities) is given by the relation:

$$v = \frac{\tilde{K} \left[\int\limits_1^2 f_w dh_c + \int\limits_1^2 f_w dz \right]}{\int\limits_1^2 \mu_{rt} \, dz} \tag{10}$$

where v is the total velocity, $\tilde{K}$ is the hydraulic conductivity at natural saturation, f_w is the *little fractional flow* function defined by the relation:

$$f_w = \frac{k_{rw}}{k_{rw} + \frac{\mu_w}{\mu_a} k_{ra}}$$ (11)

where k_{ri} and μ_i are respectively relative permeability and viscosity for phase i (water or air), μ_{rT} is the total relative viscosity defined at:

$$\mu_{rT} = \frac{f_w}{k_{rw}}$$ (12)

and where the indices 1 and 2 refer to 2 arbitrary levels in the soil. For the soil defined in Figure 1 and *assuming* a first order power curve for k_{ra} as a function of air content, the curves of k_{rw} and f_w versus h_c are shown on Figure 4. Curves of k_{ra} are rarely measured and the location of the intersection of the curves k_{rw} and f_w is somewhat uncertain. In Eq. (10) the area under the f_w curve is needed (i.e., the first integral in the numerator). For practical purposes and due to the uncertainty about the f_w curve, one actually uses the area under the k_{rw} curve. The error is not large as has been estimated for different soils in previous studies (Morel-Seytoux and Khanji, 1974). A curve of μ_{rT} versus h_c is also shown on Figure 4.

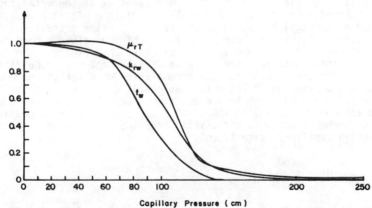

Figure 4. Hydraulic properties of Dubbs silt loam as function of capillary pressure.

Since initially the soil was in equilibrium at a temperature of 20°C it follows that under these conditions:

$$\int_0^D k_{rw} \, dh_{cB} + \int_0^D k_{rw} \, dz = 0$$ (13)

Thus the *instantaneous extra* capillary drive when the linear temperature gradient is imposed is:

$$\Delta H_c = \int_0^D k_{rw} (dh_c - dh_{cB})$$ (14)

As a rough estimate the extra capillary drive can be approximated to be:

$$\Delta H_c = \frac{1}{2} \left(\frac{T_B - T_f}{374 - T_B} \right) H_{cB}$$ (15)

where H_{cB} is the usual maximum effective capillary drive:

$$H_{cB} = \int_0^\infty k_{rw} \, dh_{cB}$$ (16)

This term can be estimated graphically as the area under the k_{rw} curve on Figure 4. Its value is about 110 cm and consequently from Eq. (15) one finds the value of the extra capillary drive:

$$\Delta H_c = \frac{1}{2} \left(\frac{20}{354} \right) 110 = 3 \text{ cm}$$ (17)

It remains to calculate the viscous resistance term in the denominator of Eq. (10). It is the area under the curve of μ_{rT} on Figure 4 in the range of h_c from 0 to 250 cm. It is of the order of 120 cm. Thus with a saturated hydraulic conductivity of the order of 3 cm/hour the total velocity v is of the order of 1.8 cm/day. Thus the sudden temperature gradient will result in a water flux at the water table of the order of 2 cm per day. Naturally as moisture migrates upward toward a new equilibrium, the extra capillary drive decreases. Also the water viscosity increases by a factor of two. Thus a more reasonable migration rate is about 0.4 cm per day. Nevertheless one can conclude that in the presence of a water table the migration rate will not be a limiting factor for the accumulation of water in the upper cold part of the soil. The water table will rarely be so close to the surface. Presuming that in the upper soil capillary pressure decreased from 250 cm ($\theta = 0.275$) to 150 cm ($\theta = 0.40$) and then remained at that value further down (a more realistic profile) then the warm isothermal effective capillary drive is only:

$$H_c = \int_{150}^{250} k_{rw} \, kh_c \simeq 1 \text{ cm} \quad ,$$

value obtained from Figure 4. The extra capillary drive due to the temperature gradient is about 0.03 cm, which is quite small. However the resistance term in the denominator has decreased considerably and from Figure 4 it can be estimated at about 1.5 cm, so that the final migration rate will be of the order of 0.3 cm per day. This value confirms the previous conclusion for the case when the water table was close to the surface.

CONCLUSIONS

The study confirms the opinion that during *severe cold* weather

170

yet *without frost* in the soil temperature gradients can produce significant upward migration of water in liquid form. The migration rate however is not limiting. What limits the water accumulation in the soil is the lack of retention capability of the soil even for a temperature drop of 30°C. Under the most favorable conditions it appears that at most half of the observed water accumulation in the spring of 1969 could be due to the phenomenon investigated. Nevertheless the magnitude of the temperature gradient induced flow is too large to be neglected in the soil moisture accounting module of existing hydrologic models such as the NWS River Forecasting System model.

Now one could speculate that the large accumulation was due to the presence of frost (ice) in the soil below the snow cover which attracted water and froze it. However since no ice was found when samples were taken in the soil during the two weeks prior to the snowmelt, the ice would have to have formed and melted before the sampling program. Even if no frost developed below the soil, there is ice (snow) right at the surface and an upward flow will be induced, not as strongly as if frost had penetrated the soil, but it will exist. In a sense the capillary pressure property in the snowpack is more akin to that of a coarse sand rather than to that of a silty loam. The hypothesis is particularly plausible because during the winter the snowpack was very dry.

The problem of this spring 1969 large water accumulation has only been partially resolved. Further studies are needed and will be done first on the basis of the hypothesis of no frost penetration.

ACKNOWLEDGMENTS

The work upon which this paper is based was supported in part by funds provided by the National Oceanic and Atmospheric Administration, National Weather Service, Hydrologic Research Laboratory. The senior author wishes to thank Dr. Richard Farnsworth and Dr. Eric Anderson, both of the Laboratory, for useful discussions on this subject and on the NWS River Forecasting System in general, during his sabbatical leave with the Laboratory during the 1979 summer.

REFERENCES

Ferguson, H., P. L. Brown and D. D. Dickey. 1964. Water Movement and Loss Under Frozen Soil Conditions, Soil Science Society of America Proceedings, Vol. 28, No. 4, pp. 700-703.

Jensen, R. D., M. Haridasan and G. S. Rahi. 1970. The Effect of Temperature on Water Flow in Soils, Report, Water Resources Research Institute, Mississippi State University, State College, Mississippi.

Klock, G. O. 1972. Snowmelt Temperature Influence on Infiltration and Soil Water Retention, Journal of Soil and Water Conservation, Vol. 27, No. 1, pp. 12-14.

Morel-Seytoux, H. J. 1978. Upward Migration of Moisture and Heaving in Frozen Soils, Proceedings, Modeling of Snow Cover Runoff, S. C. Colbeck and M. Ray, editors, U. S. Army Cold Regions Research and Engineering Laboratory, Hanover, New Hampshire, 26-28 September 1978, pp. 328-335.

Morel-Seytoux, H. J. and J. Khanji. 1974. Derivation of Equation of Infiltration, Water Resources Research Journal, Vol. 10, No. 4, August 1974, pp. 795-800.

Peck, E. L. 1974. Effect of Snow Cover on Upward Movement of Soil Moisture, Proceedings, Modeling of Snow Cover Runoff, S. C. Colbeck and M. Ray, editors, U. S. Army Cold Regions Research and Engineering Laboratory, Hanover, New Hampshire, 26-28 September 1978, pp. 328-335.

Willis, W. O. et al. 1964. Water Table Changes and Soil Moisture Loss Under Frozen Conditions, Soil Science, Vol. 98, No. 4, October 1964.

THE ESTIMATION OF SOIL HYDRAULIC PROPERTIES FROM INFLOW DATA

D. W. Zachmann, Associate Professor
Colorado State University

P. C. DuChateau, Professor
Colorado State University

A. Klute, Professor
Colorado State University
and USDA-ARS

ABSTRACT

Mathematical simulation of unsaturated soil-water flow requires knowledge of the soil's hydraulic properties. If the pressure head formulation of the flow equation is used to simulate nonhysteretic flow, the soil's hydraulic properties can be described by the functions $\theta = \theta(h)$ and $K = K(h)$ where θ, K and h represent volumetric water content, hydraulic conductivity and capillary pressure head, respectively. This paper examines the feasibility of estimating these hydraulic functions from inflow data.

Suppose the functions $\theta^*(h)$ and $K^*(h)$ characterize infiltration into a soil column with an initially, nearly uniform water content. If water is applied to the upper surface of this column at a sufficiently high rate and an outlet is provided to limit ponding, then runoff occurs. For a given application rate, the inflow rate is determined by the soil's hydraulic functions $\theta^*(h)$ and $K^*(h)$.

The inverse problem of simultaneously estimating $\theta^*(h)$ and $K^*(h)$ from runoff data is formulated as a parameter identification problem. The functions $\theta^*(h)$ and $K^*(h)$ are estimated by $\theta(h; \alpha, \beta)$ and $K(h; \alpha, \beta)$, respectively. The functional dependence of $\theta(h; \alpha, \beta)$ and $K(h; \alpha, \beta)$ is known and the parameter values α and β are obtained by an optimization algorithm which seeks to match the inflow data.

INTRODUCTION

Any quantitative analysis of fluid flow through an unsaturated soil requires knowledge of the soil's hydarulic properties. If volumetric water content, $\theta = \theta(h)$, and hydraulic conductivity, $K = K(h)$ are assumed to be single-valued functions of capillary pressure head, h, then the soil's hydraulic properties are embodied in the functions θ and K. Our goal in this paper is to study the feasibility of using a parameter identification method to simultaneously approximate $\theta(h)$ and $K(h)$ from runoff or, equivalently, cumulative inflow data. In real flow systems it has been observed that, due to hysteresis, $\theta(h)$ and $K(h)$ are generally not single valued. Later we will indicate how hysteresis restricts the scope of application of our parameter identification method.

Recently we have used parameter identification techniques to approximate a soil's hydraulic functions during drainage, DuChateau, Klute and Zachmann (1980), Zachmann, DuChateau and Klute (1980 a,b). For draining columns earlier similar works include Whisler and Watson (1968) and Skaggs, Monke and Huggins (1971, 1973). This paper represents our first efforts at using parameter identification to simultaneously esti- mate volumetric water content and hydraulic conductivity as functions of capillary pressure head during infiltration.

Consider vertical flow in a homogeneous soil column extending from $z = 0$ at the top to $z = L$ at the bottom. The continuity equation and Darcy's law combine to yield the pressure head form of the one-dimensional flow equation,

$$C(h)h_t = (K(h)(h_z - 1))_z, \quad 0 < z < L, \ t > 0 \tag{1.1}$$

where $h = h(z, t)$, $C(h) = \theta'(h)$ and subscripts denote partial differen- tiation. We will consider equation (1.1) subject to the initial and boundary conditions

$$h(z, 0) = z - h_0, \qquad 0 < z < L, \ h_0 > 0 \tag{1.2}$$

$$h(0, t) = \varepsilon, \qquad t > 0, \ 0 \le \varepsilon \ll 1 \tag{1.3}$$

$$h_z(L, t) = 1, \qquad t > 0. \tag{1.4}$$

The initial condition equation (1.2) represents, for any constant h_0, an equilibrium solution of equation (1.1). In an attempt to avoid a highly hysteretic ensuing flow we will take h_0 sufficiently large so that the column has a very low, nearly uniform volumetric water content. The boundary condition (1.3) can be obtained by supplying water to the upper surface of the column at a rate greater than the soil's saturated conduc- tivity while allowing all but a small, fixed amount of ponded water to escape. Equation (1.4) implies no flow across the lower surface of the column. In many of the applications we envision, the flow system will be considered for a short enough period of time and large enough L that the column can be viewed as semi-infinite in which case the boundary condi- tion (1.4) can be ignored. In summary, equations (1.1)-(1.4) govern infiltration into an initially, relatively dry soil on which from time zero on, a small fixed amount of water is ponded.

If the functions $\theta = \theta(h)$ and $K = K(h)$ are known, then constructing a numerical solution to the initial boundary value problem (IBVP) (1.1)- (1.4) is straightforward. Having obtained a numerical solution to the IBVP, we could then proceed to calculate the cumulative inflow per unit of cross-sectional area, $I(t)$, as follows.

$$I(t) = \int_0^L \theta(h(z, t))dz - \theta_0 \tag{1.5}$$

where θ_0 denotes the, assumed known, initial water content of the column. Note that for the above IBVP, $I(t)$ is observable, since we control the rate of application and can observe the runoff rate.

In this paper we adopt the point of view that $\theta(h)$ and $K(h)$ are unknown functions of h and we will consider the following inverse problem.

Given $I(t)$, determine $\theta(h)$ and $K(h)$ satisfying equations (1.1)-(1.5). Our approach to this inverse problem is to assume that $\theta(h)$ and $K(h)$ each depend on h in a known manner and that $\theta(h)$ and $K(h)$ each contain a (small) number of free parameters. Any choice of parameters fixes an estimate of $\theta(h)$ and $K(h)$ which can be used to numerically solve the IBVP (1.1)-(1.4). At this point the right side of equation (1.5) can be computed and compared with the known left side, $I(t)$. From a number of choices of parameters, we view the "best" choice as the one which provides the "best" agreement between $I(t)$ and the right side of equation (1.5). We use an optimization routine to seek the "best" choice of parameter values. The notion of "best" fit is dependent on the times at which we choose to compare the right side of equation (1.5) with the inflow data $I(t)$. More details of the parameter identification method are given in Zachmann, DuChateau and Klute (1980 a,b). In the next section we apply the identification method to some examples of numerical simulations.

SIMULATIONS

In this section we present some numerical simulations which indicate that it is possible to use the parameter identification method to approximate $\theta(h)$ and $K(h)$ during infiltration. These numerical simulations also provide valuable insights regarding applications of the identification method to real flow systems. For the simulations we have selected we require two empirical forms of each $\theta(h)$ and $K(h)$. Let the Form 1 $\theta(h)$ and $K(h)$ be defined by

$$\theta(h) = \begin{cases} \phi, & h \geq 0 \\ \phi/(1 + \beta h^2), & h < 0 \end{cases} \tag{2.1}$$

$$K(h) = \begin{cases} K_s, & h \geq 0 \\ K_s(\theta(h)/\phi)^\alpha, & h < 0 \end{cases} \tag{2.2}$$

and the Form 2 $\theta(h)$ and $K(h)$ be defined by

$$\theta(h) = \begin{cases} \phi, & h \geq 0 \\ \phi e^{\beta h}(1 - \beta h), & h < 0 \end{cases} \tag{2.3}$$

$$K(h) = \begin{cases} K_s, & h \geq 0 \\ K_s(\theta(h)/\phi)^\alpha, & h < 0 \end{cases} \tag{2.4}$$

where ϕ denotes porosity and K_s represents saturated conductivity. The choices of the empirical forms (2.1) and (2.3) for the volumetric water content were motivated by the qualitative knowledge that for a real soil

$$\lim_{h \to 0^-} \theta(h) \leq \phi$$

and

$$\theta'(h) \geq 0 \quad \text{for } h \leq 0.$$

The Form 1 $\theta(h)$ function is a special case of that considered by Ahuja and Swarzendruber (1972) and Haverkamp et al (1977). The choice of the form for hydraulic conductivity $K(h)$ is motivated by empirical evidence, Mualem (1976), that for some soils K is a power function of θ over a certain range of moisture content.

At this point Forms 1 and 2 each contain the four parameters, α, β, ϕ and K_s. In general a soil's porosity and saturated conductivity can

175

be determined in a fairly straightforward manner. Therefore, in both Forms 1 and 2 we view ϕ and K_s as known and α and β as parameters to be identified. To be specific, in each of the simulations to follow we will take ϕ = .5 and K_s = .3m/hr. We now proceed with our numerical simulations with which we examine some aspects of the problem of estimating the parameters α and β from cumulative inflow data.

Simulation 1 deals only with the Form 1 $\theta(h)$ and $K(h)$ and in the IBVP (1.1)-(1.4) we take h_0 = 1.m., ε = 0 m. and L = .25m. First note that if Form 1 $\theta(h)$ and $K(h)$ are used to numerically solve the IBVP (1.1)-(1.4), then the resulting solution h is dependent not only on z and t but also on the choices of α and β. To emphasize this dependence we write $h(z, t; \alpha, \beta)$. Since h is dependent on the choice of α and β, it follows that the cumulative inflow, I, is also α,β-dependent and therefore we write $I(t; \alpha, \beta)$.

Suppose that during infiltration a "soil" is characterized by the Form 1 hydraulic functions $\theta^*(h) = \theta(h; \alpha^*, \beta^*)$ and $K^*(h) = K(h; \alpha^*, \beta^*)$ where α^* = 3. and β^* = 10.m^{-2}. That is, $\theta^*(h)$ and $K^*(h)$ are respectively given by equations (2.1) and (2.2) with $\alpha = \alpha^*$ and $\beta = \beta^*$. The cumulative inflow into the "soil" column can be numerically calculated using equation (1.5) and is denoted by $I^*(t) = I(t; \alpha^*, \beta^*)$.

At this point we adopt the view that α^* and β^* are fixed but unknown to us; however, the cumulative inflow function $I^*(t)$ is given to us. In particular this means we should view the inflow values in the first row of Table 1 as data values of inflow. To simultaneously estimate the "soil's" hydraulic functions we proceed as follows.

i)   Assume $\theta(h; \alpha, \beta)$ and $K(h; \alpha, \beta)$ are of a specific functional form in h. In Simulation 1 we take θ and K to be of Form 1 while for Simulation 2 θ and K will be of Form 2.

ii)   Select some times, say t_j = .01j hr., j = 1, 2, ..., 5, at which the estimated and actual inflows are to be compared.

iii) Define an error functional
$$E(\alpha, \beta) = \sum_{j=1}^{5} [I^*(t_j) - I(t_j; \alpha, \beta)]^2 \tag{2.5}$$

where $I(t_j; \alpha, \beta)$ is the estimated inflow at time t_j obtained by using (i) and equations (1.1)-(1.5).

iv)   Systematically adjust α and β with the goal of minimizing $E(\alpha, \beta)$.

Some remarks concerning the identification procedure (i)-(iv) are in order. First, we have adopted (i) to simplify this first simulation. Having been given only $I^*(t)$ we would be very fortunate to select θ and K which have the same functional h-dependence as their counterparts which correspond to the "soil". In (ii) the choice of data times t_j was somewhat arbitrary; however, our experience with the identification procedure applied to draining columns indicates that, as long as the data times are fairly spread out, no advantage is gained by using a very large number of data times. Note that in (iii) the calculation of $E(\alpha, \beta)$ requires that, for each choice of α and β, the IBVP (1.1)-(1.4) be numerically solved and then the integral in equation (1.5) be numerically approximated. We have used standard finite difference methods to solve the IBVP and a composite Simpson's rule to evaluate E. For the questions we wanted to examine in this paper, we found it was sufficient to consider only the very early stages of infiltration and therefore in our simulations we solved the IBVP over only a small time interval. In applications to real flow systems we anticipate considering much larger time intervals. Finally, we remark that in this first simulation step (iv) is unnecessary since we have just adopted the view that α^* and β^* are unknown (we really know them).

From (i)-(iv) and equation (2.5) it follows that $E(\alpha, \beta) \geq 0$ and $E(\alpha^*, \beta^*) = 0$. Thus, in this first simulation the only way the identification procedure could fail is if the parameters α^*, β^* are not the only values for which $E(\alpha, \beta) = 0$. For example, if there were a curve in the $\alpha\beta$-plane along which $E(\alpha, \beta) = 0$, then we would have infinitely many equally "good" Form 1 approximations to $\theta^*(h)$ and $K^*(h)$. Since it is impractical to search through all positive values of α and β for zeros of $E(\alpha, \beta)$, we consider only the behavior of $E(\alpha, \beta)$ near (α^*, β^*). If we take $\Delta\alpha = .3$ and $\Delta\beta = 1.\mathrm{cm}^{-2}$ then $\Delta\alpha$ and $\Delta\beta$ represent 10% perturbations of α^* and β^*. In Table 1 we have displayed $I(t_j, \alpha^*, \beta^*)$ and $I(t_j, \alpha^* + r\Delta\alpha, \beta^* + s\Delta\beta)$ for all combinations of r and s as they take on the values 0 or ± 1. Note that r = s = 0, the I-values in the first row of Table 1, corresponds to the "soil's" cumulative inflow I^* at the times t_j.

Table 1 shows that

$$I(t_j; \alpha^* + \Delta\alpha, \beta^*) < I^*(t_j) < I(t_j; \alpha^* + \Delta\alpha, \beta^* - \Delta\beta), \quad j = 1, 2, \ldots, 5.$$

In view of the above inequalities, it appears that it is possible for the five equations

Table 1. Form 1 simulated cumulative inflow, I, and error functiona, E, at (α^*, β^*) and neighboring (α, β). $\Delta\alpha = .3$, $\Delta\beta = 1.\mathrm{m}^{-2}$.

$t_j(.01$ hr)	1	2	3	4	5	$100\ E(\alpha^*+r\Delta\alpha,\ \beta^*+s\Delta\beta)(\mathrm{cm}^2)$
(r,s)	\multicolumn{5}{l}{$I(t_j;\ \alpha^* + r\Delta\alpha,\ \beta^* + s\Delta\beta)(\mathrm{cm})$}	
(0,0)	2.51	3.57	4.42	5.17	5.84	0.00
(1,0)	2.46	3.52	4.34	5.09	5.73	2.99
(1,1)	2.31	3.35	4.16	4.90	5.54	31.89
(0,1)	2.35	3.40	4.24	4.98	5.63	16.71
(-1,1)	2.40	3.47	4.33	5.08	5.75	4.64
(-1,0)	2.56	3.64	4.51	5.27	5.96	3.99
(-1,-1)	2.74	3.84	4.73	5.49	6.20	45.39
(0,-1)	2.69	3.77	4.64	5.39	6.07	22.21
(1,-1)	2.64	3.71	4.55	5.30	5.96	8.47

$$I^*(t_j) = I(t_j; \alpha^* + \Delta\alpha, \beta^* - \tau\Delta\beta), \ j = 1, 2, \ldots, 5 \qquad (2,6)$$

be simultaneously satisfied for some τ, $0 < \tau < 1$, in which case $E(\alpha, \beta)$ would be zero for some (α, β) other than (α^*, β^*). To investigate equations (2.6) we allowed τ to vary from zero to one in small increments and found that each of the five equations (2.6) was satisfied exactly once, but for no τ were even two of equations (2.5) satisfied simultaneously. Thus $E(\alpha^* + \Delta\alpha, \beta^* - \tau\Delta\beta) \neq 0$ for any τ between zero and one.

To further study the behavior of $E(\alpha, \beta)$ near (α^*, β^*) we can employ the theorem (Taylor, 1955; p. 232) which asserts that if at (α^*, β^*) $E_\alpha = 0$, $E_\beta = 0$, $E_{\alpha\alpha} > 0$ and $E_{\alpha\beta}^2 - E_{\alpha\alpha}E_{\beta\beta} < 0$, then E has a strong local minimum at (α^*, β^*). Using the last column of Table 1 and centered finite differences to approximate the first and second partial derivatives of E at (α^*, β^*), we find that, to within the order of the discretization errors, the hypotheses of the above theorem are satisfied. All of the evidence that we have considered indicates that $E(\alpha, \beta)$ has a strong local minimum at (α^*, β^*). Thus it appears that when Form 1 $\theta(h)$ and $K(h)$ are used to approximate a "soil's" Form 1 hydraulic functions, $\theta^*(h)$ and $K(h)$, then the identification problem (i)-(iv) is locally, uniquely solvable.

In Simulation 2 we again assume the "soil" is characterized by the Form 1 hydraulic functions $\theta^*(h)$ and $K^*(h)$ defined above. We also assume that infiltration into the "soil" column is governed by the same IBVP considered in Simulation 1. Specifically, we again take $h_0 = 1.m$, $\varepsilon = 0.m$. and $L = .25m$ in equations (1.1)-(1.4). The purpose of this second simulation is to show that the identification procedure (i)-(iv) can be used to fit Form 2 $\theta(h; \alpha, \beta)$ and $K(h; \alpha, \beta)$ to the "soil's" Form 1 hydraulic functions $\theta^*(h)$ and $K^*(h)$.

In applying the identification procedure here we will use the same data times t_j as in the first simulation. The values $I^*(t_j)$, j=1,2,...,5 , found in the first row of Table 1 are to be treated as data values of cumulative inflow. Note that in this second simulation each evaluation of the error functional $E(\alpha, \beta)$ requires that the equations (1.1)-(1.5) be solved using Form 2 θ and K. We used the same finite difference method to solve equations (1.1)-(1.4) and the same quadrature rule to approximate the integral in equation (1.5) as we did in the first simulation. Step (iv) of the identification procedure, the adjustment of the parameters α and β, was carried out using an optimization routine based on the method of steepest descent and in this case required less than twenty evaluations of E to find the values $\hat{\alpha} = 3.5$, $\hat{\beta} = 4.5m^{-1}$ which approximately minimize E.

Table 2 shows the values of Form 2 $I(t_j; \hat{\alpha} + r\Delta\alpha, \hat{\beta} + s\Delta\beta)$ for r, s = 0, ± 1 and $\Delta\alpha = .5$, $\Delta\beta = .5m^{-1}$. We emphasize that in Table 2 all values of cumulative inflow except those in the last row are computed using Form 2 $\theta(h; \alpha, \beta)$ and $K(h; \alpha, \beta)$. An examination of Table 2, and the fact that a finite difference approximation of $E_{\alpha\beta}^2 - E_{\alpha\alpha}E_{\beta\beta}$ is negative at $(\hat{\alpha}, \hat{\beta})$ indicates that, again, $E(\alpha, \beta)$ has a strong local minimum at or near $(\hat{\alpha}, \hat{\beta})$. Table 3 shows a comparison of the "soil's" Form 1 volumetric water content, $\theta^*(h)$, and the Form 2 $\theta(h; \hat{\alpha}, \hat{\beta})$ obtained by parameter identification from only a knowledge of cumulative inflow into the "soil" column.

Table 2. Form 2 simulated cumulative inflow, I, and error functional, E, at $(\hat{\alpha}, \hat{\beta})$ and neighboring (α, β). $\Delta\alpha = .5$, $\Delta\beta = .5m^{-1}$.

$t_j(.01\ hr)$ (r,s)	1	2	3	4	5	$100\ E(\hat{\alpha}+r\Delta\alpha,\ \hat{\beta}+s\Delta\beta)(cm^2)$
	$I(t_j;\ \hat{\alpha} + r\Delta\alpha,\ \hat{\beta} + s\Delta\beta)(cm)$					
(0,0)	2.40	3.53	4.38	5.18	5.84	1.54
(1,0)	2.33	3.46	4.26	5.06	5.70	10.18
(1,1)	1.90	3.01	3.77	4.53	5.20	192.74
(0,1)	1.96	3.08	3.88	4.67	5.32	135.46
(-1,1)	2.04	3.16	4.02	4.82	5.48	80.11
(-1,0)	2.48	3.62	4.54	5.32	6.03	7.64
(-1,-1)	3.06	4.22	5.18	5.97	6.72	271.70
(0,-1)	2.98	4.11	5.04	5.81	6.52	176.89
(1,-1)	2.91	4.03	4.90	5.69	6.34	112.24

Table 3. Comparison of the Form 2 approximation, $\theta(h;\ \hat{\alpha},\ \hat{\beta})$, to the "soil's" Form 1 water content function, $\theta(h;\ \alpha^*,\ \beta^*)$.

-h(m)	0.	.2	.4	.6	.8	1.
Form 1 $\theta(h;\alpha^*,\beta^*)/\phi$	1.	.714	.385	.217	.135	.091
Form 2 $\theta(h;\alpha,\beta)/\phi$	1.	.772	.463	.249	.126	.061

SUMMARY

We have used numerical simulations to examine the problem of estimating a hypothetical soil's hydraulic functions during infiltration. The two simulations presented in the previous section indicate that it is feasible to use a parameter identification technique to approximate volumetric water content and hydraulic conductivity as functions of capillary pressure from a knowledge of cumulative inflow into a soil column.

When we apply our methods to real flow systems we will have to deal with several problems which we were able to avoid or minimize in these idealized numerical simulations. One of these problems is the selection of the functional dependence of θ and K on h. For example, if we are given a real flow system, should we use Form 1 or Form 2 to approximate θ and K, or is neither form appropriate? For the case of draining columns we have considered this question in Zachmann, DuChateau and Klute (1980a). Our argument there was that if a certain form is used in the identification procedure, then small changes in the geometry of the flow system should not produce large changes in the values of the identified parameters. We intend to explore similar considerations for the case of infiltration. A second problem which must be confronted is hysteresis. In our physical experiments we intend to attempt to minimize hysteresis effects by considering inflow systems which begin at a fairly uniform moisture content. An example of an alternative to the "flow" system we used in our simulations would be one in which flow equation (1.1) and the initial condition, equation (1.2) are unchanged, but the roles of the boundary conditions, equations (1.3) and (1.4) are reversed. This would cause water to be drawn up into the column from the lower end, and would produce, for all times, a smaller derivative spatial gradient of water content with respect to depth than did the case we considered in Section 2. We are also investigating the use of a pressure cell to avoid hysteresis effects.

ACKNOWLEDGEMENTS

This work was supported by the National Science Foundation Grant ENG-7803227.

REFERENCES

Ahuja, L.R., and Swartzendruber, D. 1972. An improved form of soil-water diffusivity function. Soil Sci. Soc. Am. Proc. 36:9-14.

DuChateau, P.C., Klute, A. and Zachmann, D.W. 1980. Identification of soil hydraulic properties from experimental discharge data. Submitted to Soil Sci. Soc. Am. J.

Haverkamp, R., Vauclin, M., Touma, J., Wierenca, P.J., and Vachaud, G. 1977. A comparison of numerical simulation models for one-dimensional infiltration. Soil Sci. Soc. Am. J. 41:285-294.

Mualem, Y. 1976. A new model for predicting the hydraulic conductivity of unsaturated porous media. Water Res. Res. Vol. 12 No. 3:513-522.

Skaggs, W.R., Monke, E.J., and Huggins, L.F. 1971. An approximate method for defining the hydarulic conductivity and pressure potential relationship for soils. Trans. ASAE: 130-133.

Skaggs, W.R., Monke, E.J., and Huggins, L.F. 1973. Experimental evaluation of a method for determining unsaturated hydarulic conductivity. Trans. ASAE: 85-88.

Taylor, A.E. 1955. Advanced Calculus, Ginn and Co., Boston, p. 232.

Whisler, F.D., Watson, K.K. 1968. One-dimensional gravity drainage of uniform columns of porous materials. J. of Hydrol. 6:277-296.

Zachmann, D.W., DuChateau, P.C., and Klute, A. 1980. Simultaneous approximation of water capacity and soil hydraulic conductivity by parameter identification. To appear in Soil Science.

Zachmann, D.W., DuChateau, P.C. and Klute, A. 1980. The calibration of the Richards flow equation by parameter identification. To appear in Soil Sci. Soc. Am. J.

Section 3
EVAPOTRANSPIRATION

MATHEMATICAL MODELING OF EVAPOTRANSPIRATION ON AGRICULTURAL WATERSHEDS

Keith E. Saxton, Hydrologist
U.S. Department of Agricultural, Science and Education Administration
Agricultural Research

ABSTRACT

Evapotranspiration (ET) is one of the principle hydrologic processes on agricultural landscapes which determines the soil water time and depth distribution, crop production, and antecedent hydrologic conditions. It is a complicated process which involves several variables of the atmospheric boundary layer, soil profiles, and the biological growth and control by plant communities.

A common approach to predict ET is to estimate the potential atmospheric energy available at the plant and soil surfaces, then determine the proportion of this energy utilized for conversion of liquid water to vapor depending on the water availability or rate of transmission from within the soil profile. Several methods to estimate the potential ET have been developed and tested. They contain one or more atmospheric variables, or an indirect measurement of them, which are often combined with a representation of the surface conditions and interactions. Some, like the Penman equation, are based on the physics of combining the vertical radiation budget with turbulent boundary flow over the land surface. More empirical methods based on solar radiation, air temperature, air humidity, pan evaporation, or some combination of these have proven to be practical.

The availability of water to the plant and soil surfaces is a dynamic and complex phenomena of biological and physical controls. Some precipitation is intercepted on the plant, soil, and residue surfaces and is readily evaporated and returned to the atmosphere. Water within the soil profile is returned to the surface through plant roots or capillary soil water movement. Exposed soil surfaces soon dry to the stage that the rate of soil water movement limits water availability. Crop canopy and root distribution play a dynamic role in connecting soil water to the atmosphere. Phenologic development and stomatal resistance due to water stress are significant factors in all plants and especially so for annual agricultural crops.

Several models of the complete dynamic ET process have been developed in recent years which vary considerably in complexity from single equations with empirical coefficients to very detailed physical representations. The utility of each method depends upon its requirements for input data, location calibration, and expected accuracy.

---

[1]Contribution from the USDA-SEA-AR in cooperation with Washington State University, Pullman, WA.

INTRODUCTION

Evapotranspiration (ET) on natural landscapes is a process that moves vast amounts of water and consumes the bulk of the solar energy reaching the earth's surface. In its most simple form, ET is the conversion of liquid water to vapor at an evaporating surface and the vertical transport of this water vapor upward into the atmospheric boundary layer. Within the hydrologic cycle, ET is next to precipitation in the quantity of water involved, and in fact they become nearly identical in semi-arid and arid climates. And of the total solar energy available at the soil or plant surfaces, some 80 to 90 percent often is utilized to provide energy for the liquid-to-vapor conversion of water if water is readily available (Priestley and Taylor, 1972). In this wet case, lesser amounts of solar energy are used for sensible and soil heating with a relatively minor amount utilized by the photosynthetic process.

Through its magnitude and close alliance with plant growth, ET is a highly influential process to be considered and predicted for estimates related to hydrology, erosion, and crop production. The status of soil water is primarily dependent on time distributions of precipitation and ET. And in return, soil water significantly influences runoff generation, soil erodibility, and crop growth and yield.

Provisions for simultaneous and continuous energy and water availability at a common surface, plus the transport process, involves many complex phenomena within the subject areas of meteorology, botany, and soil physics. In addition, there are many interactions with variables such as topography, latitude, and elevation. Particularly large variations of ET occur on agricultural watersheds because of farming operations and crop selection. This complexity of the ET system which involves driving and controlling factors by the climate, soil, and vegetation makes it one of the more difficult processes within the hydrologic cycle to measure, understand, and predict.

To model ET for numeric predictions causes us to examine the entire ET system and to define the principle cause-and-effect relationships. The required accuracy for the objectives at hand and data availability must be decided. The temporal and spatial delineations must be considered. While ET proceeds continuously in strong diurnal fashion, integrated values of daily or weekly values are often most useful and expedient. And similarly, regardless that each small unit of the surface is unique and provides an ET rate different from others, there is much commonality across fields and watersheds in both conservative atmospheric inputs and land surface response. Therefore, single ET estimates are frequently applied to fairly large areas with useful results.

Thus, it is from this view that, although the ET process is complex, it can be defined, described, and predicted through our current understanding with an accuracy which will provide useful estimates and guidance for research. It is the purpose of this paper to describe the major system processes which determine ET from vegetated surfaces as viewed by contemporary scientists. Mathematical representations and numeric quantification are intentionally held to a minimum, but appropriate references to published results and models are provided to guide the interested reader.

AN OVERVIEW OF ET PRINCIPLES

Evapotranspiration from vegetated surfaces requires energy inputs, water availability, and transport processes from the surface into the atmosphere. The flux of water vapor is largely limited by one or more of these requirements (Wiegand and Taylor, 1961). Several researchers have provided good descriptions of these primary variables which determine ET rates (Tanner, 1957; Goodell, 1966; Penman et al., 1967; Gray, 1970; Campbell, 1977). Eagleson (1978) provided an extensive mathematical review which integrated the principles of ET into hydrologic predictions.

Because ET is a phase change of water, large energy inputs are required. At a nominal value of 580 cal/g, a daily ET of 5 mm (0.2 in) will require the equivalent of 4480 kg/ha (2 t/a) coal. Solar radiation usually supplies 80 to 100 percent of this energy and is often the factor limiting ET.

For nonirrigated agriculture, water availability to the evaporating plant and soil surfaces also often limits ET. Thus, the rate of ET is limited to the conduction rate of soil water to the soil surface and plant roots, and through the plant system. In these circumstances, absorbed radiant energy (incoming minus reflected) in excess of that required to transform the available water is dissipated primarily by an increase of sensible heat in the air, soil, and plant canopy.

The aerodynamic transport process of water vapor upward from the evaporating surface for most vegetated situations does not often significantly limit the ET process. Although molecular diffusion is involved at the soil and plant surfaces, turbulent diffusion dominates. This is caused mostly by wind shear but also by thermal convection under calmer conditions. The diffusion of water vapor from the soil and plant surfaces and within their structures is highly complex when examined in detail, but essential to the ET mechanism (Sellers, 1965).

The horizontal advection of sensible heat from areas of excess energy to areas of limited energy is also important for ET. This is often called the "clothes line" or "oasis" effect and is best exemplified by a wet vegetated field downwind of a dry desert area, like that reported by Davenport and Hudson (1967). Significant advection effects are often encountered in much less obvious circumstances and over large areas (Rosenberg, 1969).

Evapotranspiration varies spatially as a result of variations in climate, crops, or soils. Climatic variables related to ET tend to be conservative and often do not change rapidly or significantly over considerable distance. However, we cannot make generalizations because local elevations, aspects, orographic effects, and cropping patterns can cause large ET changes. The variation of crops and soils over a region in question will need to be treated either by separate considerations of major combinations or broad scale averages. Some spatial averaging is implicit in every ET estimate and the user must acknowledge and quantify the effects with respect to the application.

The soil-plant-atmosphere system may be represented schematically as shown in Fig. 1. Of the water budget components, ET is usually the largest after precipitation. The interaction of ET with the other components, like rooting and soil moisture profiles, and the dynamic nature of these many components with time becomes readily apparent as the water budget of this system is computed.

Many methods of estimating ET, whether for hydrologic models or

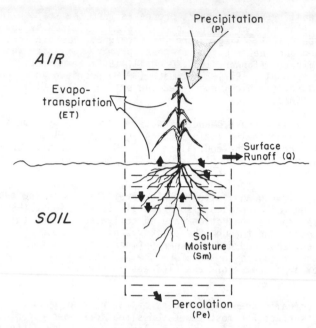

Figure 1. Schematic of the control volume for water budgets of the soil-
plant-air system (Saxton et al., 1974b).

irrigation scheduling, follow a concept of a vertical water budget
within a system like that in Fig. 1. In general, the procedure is to
first estimate or measure a potential for ET based on meteorological
factors, then compute the amount of that potential that is utilized by
the actual ET processes, given the current status of the plant and
soil water related characteristics. To apply such a procedure
requires variables that describe: (a) the potential ET; (b) plant
water relationships; and (c) soil water evaporation relationships.
Each of these is the subject of the next three sections, followed by a
brief review of several procedures where the concepts are combined to
predict actual ET.

ESTIMATING POTENTIAL ET

The potential for ET (PET) is most often defined as an atmospheric
determined quantity. This assumes that the ET flux will not exceed
the available energy from both radiant and convective sources. This
is generally a workable assumption for most predictive methods and
allows considering the atmospheric variables apart from the plant and
soil effects. However, interactions and feedback from the plant and
soil to the atmosphere do not allow complete isolation. The definition of PET for predicting ET from irrigated fields has often been
that from a well-watered reference crop. In this approach, some crop
and soil variables become partially integrated into the PET values
which make them less generally applicable than the PET values derived
from atmospheric variables.

For most hydrologic applications, estimates of PET are based on
daily, weekly, or sometimes even monthly measurements. These may
include one or more atmospheric variables, such as solar or net

radiation, air temperature and humidity, or some related measurement such as pan evaporation. Direct measurement or estimation of variables such as vapor or heat flux is difficult and not yet practical for most applications. Others, like radiation, have only been routinely measured for a relatively few years, and even now are not commonly available. As a result, most procedures for estimating PET are empirically based on atmospheric-related variables or methods. Fortunately, the primary causative atmospheric variables are relatively conservative in space where strong topographic or cultural changes are absent, thus PET estimates can often be transferred some distance with minimal error. For hydrologic applications, transfer is often necessary because data are not available on the area where needed.

There are numerous reported methods for estimating PET, each unique because of its basis, data requirements, area of application, and accuracy. The following are a few of the methods which are often applied in hydrologic predictions. Should the reader elect to apply one or more of these methods, he is admonished to study the cited references and methodology carefully to fully understand the assumptions and limitations.

Pan Evaporation

Measured evaporation from a shallow pan of water is one of the oldest and most common methods of estimating PET. It is an indirect integration of the principal atmospheric variables related to ET. Given some standardization of pan shape, environmental setting, and operation, good correlations have been developed between pan evaporation, E_p, and PET by a simple relation

$$PET = C_{ET} E_p \qquad [1]$$

where C_{ET} is a coefficient.

Pan-to-PET coefficients (C_{ET}) are necessary because evaporation from a pan is generally more than that from a well-wetted vegetated surface, or even a pond, due to the pan's excessive exposure and lower reflectance of solar radiation. The U.S. Weather Bureau Class A pan is a metal pan 122 cm in diameter, 25 cm high, and mounted with its bottom about 1 cm above the surrounding soil. Thus, it is capable of receiving and utilizing more atmospheric energy than larger, less exposed surfaces and the pan water often becomes quite warm. These coefficients are influenced by the pan surroundings, fetch, relative humidity, and wind speed (Jensen, 1973, pp. 74; Hanson and Rauzi, 1977). As these variables change, values of C_{ET} can range from 0.5 to 0.8. However, over several days and for other than extreme or unusual conditions, a much more stable value will prevail. Many examples of pan-to-PET coefficients can be found in texts and publications (Hargreaves, 1966; Richardson and Ritchie, 1973; Saxton et al., 1974a).

Although specific coefficient values for application to any given situation or pan may have to be found by calibration, representative values from other studies will provide good guidance. Mean monthly coefficient values are graphically shown by Jensen (1973, p. 79) for 10 widely separate locations over the world. For the eight locations with a uniform grass cover, the mean annual coefficients varied from 0.72 to 0.83 and averaged 0.77.

Measurements or estimates of pan evaporation are available from many U.S. Weather Bureau reporting stations, research stations, and meteorologic offices of other agencies. Most measurements are from the standardized U.S. Weather Bureau Class A pan, although many other pan types, like sunken, screened, floating, and insulated pans, have been used. Data from each of these pans are unique and will relate to PET by a different set of coefficients. Generalized maps of Class A pan and pond evaporation are available to estimate average conditions. Kohler et al. (1959) and Nordenson (1962) provided maps and seasonal distributions. Morton (1979) provides an estimation method of lake evaporation based on temperature, humidity, and radiation.

Energy Budget

Methods of estimating PET based on the vertical energy budget of a vegetated surface have a physical basis because energy limits ET where moisture is readily available and the necessary vapor transport occurs. Figure 2 shows the major components of the energy budget which form the basis for the several methods that use this approach. Except for cases of significant advection, like field edges and oasis effects, the horizontal components are usually negligible. The budget (in cal/cm^2/min, except as noted) of the major vertical components may be expressed as

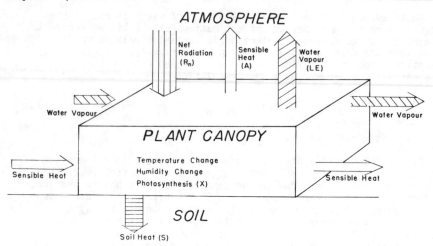

Figure 2.   Energy budget of a vegetated surface (Gray, 1970).

$$R_n = A + LE + S + X \qquad [2a]$$

and

$$R_n = R_s - aR_s + R_1 - R_{1r} \qquad [2b]$$

where
    R_n = net radiation
    R_s = incoming solar radiation (short-wave)
    aR_s = solar radiation reflected
    R_1 = incoming radiation (long wave)
    R_{1r} = emitted long wave radiation
    A = sensible heat of air

LE = latent heat of water vapor
   L = latent heat of vaporization cal/cm^3 or cal/g, and
   E = depth of evaporative water, cm^3/cm^2/min
S = soil heat
X = miscellaneous heat sinks, like plant and air heat storage,
   photosynthesis, $\cdots$

Tanner (1957) summarized the heat budget for agricultural surfaces, and Brown and Covey (1966) demonstrated the use of this approach for a cornfield.

Recognizing that A and LE terms are of much greater magnitude than S and X, Bowen (1926) proposed using the ratio of sensible to latent heat

$$\beta = A/(LE) \qquad\qquad [3]$$

where β is commonly referred to as the Bowen ratio. Equation [2a] with S and X neglected and equation [3] substituted becomes

$$LE = R_n/(1+\beta) \qquad\qquad [4]$$

The value of β can be calculated from gradients of air temperature and vapor pressure above the evaporating surface, which is a relatively difficult measurement usually made only at research installations (Fritschen, 1966). Typical β values range from 0.1 to 0.3 for moist conditions (Priestly and Taylor, 1972).

Radiation Methods

Because solar radiation provides the required energy for the phase change of water and often limits the ET process where water is readily available, any number of methods have been developed to estimate PET using a radiation data base. Most often, direct solar radiation is used because this is frequently measured by national weather networks, and air temperature has been found through correlation to be a useful second variable. Two such methods will be discussed.

For estimates of 5 days or longer, Jensen and Haise (1963) developed the following prediction equation:

$$PET = (0.025 T + 0.078)R_s \qquad\qquad [5]$$

where
PET = potential ET (30-50 cm alfalfa)(cm/day)
T   = mean air temperature ($^\circ$C), and
R_s = solar radiation (cm/day).

The Jensen-Haise method was related to PET for a full cover crop, like alfalfa, and other crops by crop coefficient curves like those developed by the U.S. Bureau of Reclamation for irrigation designs (Robb, 1966). Other applications have been made for irrigation scheduling (Jensen, 1973).

Turc (1961) proposed a similar empirical equation for PET based on radiation and temperature of:

$$PET = 0.40 T (R_s + 50)/(T + 15) \qquad\qquad [6]$$

where T is monthly mean air temperature, $^\circ$C, R_s is monthly radiation, ly, and PET is potential ET mm/month.

Priestly and Taylor (1972) showed that for a horizontally uniform saturated surface the equation

$$LE = \frac{\alpha \Delta/\lambda \; RN}{\Delta/\lambda} \qquad\qquad [7]$$

would provide good results. The coefficient α had values of about 1.30 ± 0.05. It is important to note the limitation that this is only applicable where advection (as represented by Penman's second term) is neglibible. The authors also note the correspondence of this method with the Bowen ratio method.

Obtaining reliable radiation values by either measurement or estimation often becomes the key to successful applications of energy budget methods. Direct radiation measurements are usually one of two types: (a) total incoming solar, R_s, (e.g., by Epply pyrheliometer) or (b) all wave net radiation, R_n (e.g., by Fritschen type net radiometer). Net radiation can be used directly to predict PET. Solar radiation, which is the most common measurement at meteorologic stations, can be used to estimate R_n taking into account the albedo and heating coefficients of the plant and soil surfaces. Albedo and emittance vary with stage of plant growth, soil color, degree of wetness, crop and soil tempera- ture, sun angle, and other factors. Davies and Buttimor (1969) pooled data for many crops and world locations and concluded that many sur- faces are similar enough that mean values can be quite useful. The mean of their relationships is

$$R_n = 0.63 \; R_s - 48 \qquad\qquad [8]$$

Jensen (1973), in his Table B1, also summarizes radiation measurements from 28 diverse locations and sites. Almost all results had a correlation coefficient greater than 0.90, many above 0.95. The means for these data provide the relationship

$$R_n = 0.65 \; R_s - 45 \qquad\qquad [9]$$

which is almost identical with equation [8]. Saxton (1972) reported a relationship very similar to those of equations [8] and [9]; however, he further showed a seasonal trend in the R_n/R_s ratio, which ranged from about 0.40 at mid-March and November to about 0.55 in mid-July. Other data and summaries relating net and solar radiation have been reported by Reifsnyder and Lull (1965), Stanhill et al. (1966), Fritschen (1967), Linacre (1968), and Coulson (1975).

The slope and aspect of a plane at the Earth's surface may introduce considerable variation to the incident radiation as compared with a horizontal plane. A watershed is composed of a multitude of individual facets, but the average slope-aspect effect can be determined from the view that all incoming radiation must pass through a plane defined by points on the watershed boundary, thus a single plane can be used for some objectives. The slope and orientation will be particularly important for relatively small, steep watersheds. Many calculations of slope-aspect effects on solar radiation have been published (Frank and Lee, 1966; Buffo et al., 1972).

Aerodynamic Profile Methods

The measurement of water vapor, as it is transported away from an evaporating surface, offers the potential of the most direct measurement of ET. The approach usually involves measuring temperature and vapor pressure of the air at two or more heights above the evaporating crop and a profile of wind velocities to define moisture and temperature gradients and fluctuations of wind velocity and humidity at a single height. The measurements are all quite sensitive and the amount of required data is voluminous unless processed internally by modern electronics.

Considerable research has been conducted with sophisticated instrumentation, like that described by Dyer (1961) for the mass transfer-eddy flux method or that of Parmele and Jacoby (1975) for the Bowen ratio measurements. Good results have been obtained as compared with lysimeters, but instrumentation and techniques for hydrologic measurements or predictions are not yet developed to the point that these methods can be routinely applied.

Combination Method

Neither the vertical energy budget nor the aerodynamic methods are capable of predicting PET without assumptions and limitations. Penman (1948, 1956) developed a method to combine these two theories which removed some of the limitations, and his equation is widely used. With refinement and testing (Businger, 1956; Tanner and Pelton, 1960; van Bavel, 1966) it now represents one of the more reliable techniques for predicting PET from climatic data.

The complete derivation of the combination equation is quite lengthy and involves many micrometeorologic concepts. The derivation can be divided into the following steps: (a) define the vertical energy budget of the soil or plant surface, (b) apply the Dalton-type transport function to obtain Bowen's ratio, (c) apply Penman's psychrometric simplification to eliminate the need for surface temperature, and (d) apply the vertical transport equation obtained from turbulent transport theory. A complete development for each of these steps is given by Saxton (1972, Appendix A) and abbreviated derivations are found in many sources (e.g., Jensen, 1973, p. 70). Several assumptions are made in the course of the derivation, like air thermal stability and equal transport coefficients of momentum and vapor, but these seem to have negligible effects for most applications.

The combination equation may be written:

$$LE = \frac{\Delta/\gamma \; R_n + (K \; L \; d_a \; u_a)/[\ln(z_a-d/z_0)]^2}{1 + (\Delta/\gamma)} \qquad [10]$$

and

$$K = \frac{\rho \; k^2 \; \varepsilon}{P} \qquad [11]$$

where
  E = potential evapotranspiration rate (cm/day)
  Δ = slope of psychrometric saturation line (mbars/$^\circ$C)
  γ = psychrometric constant (mbars/$^\circ$C)
  R_n = net radiation flux (cal/cm^2/day)
  L = latent heat of vaporization (cal/g)

d_a = saturation vapor pressure deficit of air (mbars)
u_a = windspeed at elevation z_a (m/day)
z_a = anemometer height above soil (cm)
d = wind profile displacement height (cm)
z_0 = wind profile roughness height (cm)
ρ = air density (g/cm^3)
k = von Karman coefficient (0.41)
ε = water/air molecular ratio (0.622), and
p = ambient air pressure (mbars).
All terms of K and the value of L are treated as constants in most applications. Care must be exercised to maintain consistent units throughout.

Application of the combination equation [10] requires measurements or estimates of four variables--net radiation, air temperature, air humidity, and horizontal wind movement--plus appropriate values for the other parameters which can usually be treated as constants for a given site. Net radiation can be assessed the same as for the energy budget approach. The (Δ/γ) term is a function of air temperature and tabled values are available (van Bavel, 1966).

Temperature-Based Methods

Air temperature is largely a function of incoming solar radiation as is PET, thus there is some basis for the several methods which have used this readily available data. Pelton et al. (1960) provided an evaluation of temperature-based methods. Because of several inter- ferring factors such as heat storage, latent and sensible heat division, and boundary-layer turbulence, experience has shown that temperature-based methods are only reasonably accurate for periods of weeks or months to obtain the benefit of natural averaging and inte- gration. They are quite useful for broad-scale planning and PET esti- mates in regions of limited data.

Comparison of Methods

The selection of a method for PET estimates depends on several criteria. Data availability often dominates. Accuracy required and time available to develop accurate estimates from available data sources are important. Whether the estimates are in retrospect, current time or projections will often dictate time and data avail- ability. Studies comparing the results of several methods were reported by McGuinness and Bordne (1972); and Parmele and McGuinness (1974), Doorenbos and Pruitt (1975), and Burman (1976) showed similar comparisons for a variety of stations.

In a review of 15 methods for estimating PET, including those just discussed, Jensen (1973, Table 7.3) showed that only the combination equations of Penman or its modifications by van Bavel and others and the R_n-based method of Jensen-Haise would be recommended for periods of 5 days or less. He noted that the availability of meteorological data alone should not be the sole criterion in selecting a method, since some of the needed data can be estimated with sufficient accuracy to permit using one of the better methods. Estimating proce- dures such as those of Nicks and Harp (1980) could be applied. He concluded that, in general, energy balance or energy balance-aerodynamic equations will provide the most accurate results of the various meteorological methods because they are based on physical laws and rational relation- ships.

PLANT TRANSPIRATION

Plants play a significant and dynamic role in the evapotranspiration process, particularly on areas largely used for agricultural crops, which are sown, harvested, and cultivated with considerable variation in both time and space. For well-vegetated surfaces, most of the radiant energy is absorbed and dissipated by the leaf surfaces. Thus, the plants have primary control of the amount utilized for latent heat through stomatal control, water availability, and root proliferation. Plant effects have received more emphasis in recent years. Federer (1975) noted a trend of recent research from ET as a physically controlled process to ET as physiologically controlled.

For agricultural applications, it is quite important to represent the plant functions because vegetation in many climates seldom transpires at a potential rate. This may either be the result of plant control or water availability. Recently developed ET methods have combined these plant effects and descriptions through system models of ET from crops with incomplete canopies and limited water where transpiration and soil evaporation are calculated separately (Ritchie, 1972; Saxton et al., 1974b; Tanner and Jury, 1976).

Plant characteristics for ET methods can be divided into the main categories of (a) canopy cover, (b) phenology, (c) roots, and (d) water stress. While there are many interactions among these categories, they represent major considerations for computational purposes and provide a useful framework for discussion and model representations.

Canopy

The dynamic development, maturation, and decay of crop canopies significantly influence plant transpiration effects. For annual crops, like corn or cotton, the canopy very rapidly develops from nothing to nearly full soil cover and then matures and is harvested. The canopy of any particular day largely determines the amount of intercepted solar radiation or adsorbed advection utilized for plant transpiration, thus agricultural hydrologic models must provide a representation of this dynamic plant behavior.

A direct approach is to graph the canopy growth curve versus time to represent the percent of ground shading throughout the year. To define crop canopy curves requires knowledge or observations of normal planting dates, emergence times, rate of development, tasseling or blooming dates, harvest dates, regrowth or residue conditions. To represent canopy as average daily soil shading primarily is to partition the radiant energy between plant and soil, thus modifications need to be considered if advection is expected to play a significant role. Although not highly accurate, an empirical canopy curve based on local knowledge of crop growth will often adequately represent crop canopy effects.

Recent research on crop effects has used the ratio of leaf area divided by soil surface area as a leaf area index (LAI) to relate measured ET to effective canopy. This measure compares canopy effects of different crops, although it has not been entirely satisfactory among crops of widely differing canopy architecture. Ritchie and Burnett (1971) and Kristensen (1974) showed very similar curves for barley, sugarbeets, long and short grass, corn, and sorghum. For almost all cases, the actual-to-potential ET (AET/PET) ratio approached 1.0 as the LAI approached 3.0. Although LAI values relate closely to the AET/PET ratio and provide a direct measure of crop canopy, they are difficult to predict, and estimates of canopy cover as a percent of the soil shading may yet be the most practical (Adams et al., 1976).

Phenology

The phenological development of plants often modifies a plant's ability to transpire. As a crop matures, its need for water and ability to transpire are diminished (Salisbury and Ross, 1978). Because phenological changes may occur independent of the crop canopy present, this effect must be introduced as a modifier.

As with canopy, a time distribution graph of the relative ability of a plant canopy to transpire will often be adequate for agricultural hydrology models. The effect being represented is the transpiration ability of the canopy existing at any time as compared with that of a fully transpiring, equal canopy. Crop maturation is the principle cause for loss of ability, but drying of leaves from stress of heat, moisture, or insects may also cause modifications. For example, cool season grasses may mature and become somewhat dormant during mid-season, then recover as fall cooling begins.

Crop residues pose a special canopy situation since they intercept radiation but have no ability to utilize that energy for evaporation unless intercepted precipitation is present. This can be considered a special case of crop phenology where the crop has lost all ability to transpire, or the residues may be considered as part of the soil eva-poration process. Nevertheless, the residues change characteristics over time through decay and destruction by tillage and must be dynam-ically represented.

The combined effects of plant canopy and phenology have often been represented by crop coefficient curves, particularly for irrigation ET estimates. To use this approach, we must carefully note the deri-vation and intended application of these coefficient curves because they are empirically determined and include specific conditions and assumptions. Typical crop coefficient curves are reported by the USDA Soil Conservation Service (1967) and Jensen (1973). Doorenbos and Pruitt (1975) provided a method to develop curves for many crops under a wide range of growing season lengths and climates.

Roots

Crop roots are as important as canopy in the process of connecting soil water with atmospheric energy and the resulting transpiration. However, root distribution and their effectiveness are more difficult to study and quantify. Much work has been done to quantify the root development of major crops. Corn roots were studied by Taylor and Klepper (1973) and Mengel and Barker (1974). Soybean data were re-ported by Stone et al. (1976). Some information is available on most crops and basic relationships are presented in texts like that of Whittington (1968). Taylor and Klepper (1978) provide an extensive review of rooting effects on plant-water relationships.

Many ET models have simply considered depth of maximum rooting as a predetermined parameter or fitted coefficient. For more physically related modeling, the time and depth of rooting density is required, especially for annual crops which establish a complete new root system each year. Only in this way can soil water profiles and their inter-action with the root profiles be modeled. This becomes very important when large differences in water contents exist with soil depth. Crop transpiration may be severely limited if the rooting patterns do not coincide with available moisture in the soil profile.

The water uptake by plants and the mathematical representation of this phenomenon have received considerable attention in recent years (Feddes et al., 1976; Hillel and Talpaz, 1976). In addition to representing time and depth of rooting densities, several approaches for calculating the resistance of water flow to the root, then through the root, stems, and leaves have been proposed. Root age and location seem to be quite important. Taylor and Klepper (1973) speculated that the deeper, less dense roots may be more effective for water uptake because they are younger and usually in wetter soil. Studies in this level of detail have not yet provided additional prediction capability; thus, including basic root-density dynamics for interaction with available soil water may be the most sophistication now warranted.

Most crops have a genetic rooting characteristic that will provide estimates of root density distributions with depth and time; and, in turn, estimates of water extractions. Older roots are less efficient in water extraction, thus root quantity distributions must be modified. Characteristic rooting patterns can be significantly modified by the soil root environment like dense layers, poor aeration, very dry or wet, and chemicals. Some deep-rooted plants, like alfalfa or trees, may extract water from deep, wet layers or shallow groundwater and pose special modeling problems.

Water Stress

The lack of available plant water and/or high evaporative demands will cause most plants to biologically react by closing stoma, reducing transpiration, and reducing assimilation and metabolic reactions. Continued stress results in leaf drop and tissue death. While this process has long been observed, there is yet considerable controversy and lack of definitive predictive relationships. Many studies have been conducted and reported, but the results are quite variable and contradictory, probably because of crop, soil, climatic, or technique differences.

Mustonen and McGuinness (1968) and Baier (1969) summarized several relationships between plant-available soil water and actual/potential transpiration ratio. There were wide differences of opinion. Some of these relationships are derived for unusual conditions, like deep-rooted crops in sandy soil. Other relationships are simple mathematical expressions for expediency. Most data have considerable variation. The variation of soil depth used to define the quantity of available water also caused some differences.

Denmead and Shaw (1962) developed basic plant-stress data using a large-container study. They later used these results for predicting transpiration from corn and meadow. Holmes and Robertson (1963) showed similar relationships. Ritchie (1973) did not obtain similar results, but he defined available water using the entire soil profile--not that related only to current plant roots--which thus precluded a direct comparison.

It is generally agreed that both plant-available soil moisture (or soil water pressure) and the atmospheric demand determine that proportion of potential transpiration a plant will achieve, i.e., the actual/potential ratio. Given a moderate available soil water status, a plant under low atmospheric demand may achieve nearly all of that demand, but the same moisture level and a high atmospheric demand may result in moisture stress and a significant reduction of transpiration from the potential.

Water stress relationships will require calibraton for each soil depth used to define available soil water, each crop, and perhaps for each soil. Soil water pressure (capillary suction) may be a better measure than plant available water. Additional effects that have been investigated, like soil conductivity near the roots and plant water pressure, may eventually reduce the need for calibration. Some recent simulations have attempted to treat the movement of water through the soil to the roots and through the roots and canopy as a series of conduits with internal and boundary resistances, but this approach will require further development and testing before it can be readily applied.

SOIL WATER EVAPORATION

The evaporation of water from the soil surface involves the same basic physics as any other evaporation, i.e., a liquid-to-vapor phase change—limited by the availability of either water or energy—and upward vapor transport. For bare soil or fields with incomplete canopies, this component of the evapotranspiration process is highly important and can involve significant quantities of water. Given that bare soil is readily exposed to the atmosphere due to lack of vegetative cover, water availability frequently limits the soil water evaporation rate since it can only be supplied by recent rainfall or upward conductivity from within the soil.

Soil water evaporation is often described as occurring at three separate stages beginning with wet soil (Gardner and Hillel, 1962). In the first stage, the drying rate is limited by and equals the evaporative demand (available energy). During the second stage, water availability progressively becomes more limiting. The third stage is described as an extension of the second stage but is limited to a more constant rate. Studies utilizing the concept of concurrent flow of heat and water by van Bavel and Hillel (1976) clearly demonstrated the first two stages but did not support the notion of a third stage.

The approach by van Bavel and Hillel (1976) of simultaneous heat and water flux at the soil surface and within the soil profile provides a more detailed and accurate prediction of soil evaporation but requires significantly more data input and computational time. An intermediate, and perhaps more feasible approach, may be to consider intercepted and ponded water at or near the soil surface for stage one drying and upward unsaturated flow for stage two.

Considerable effort has been made to explain many influences on soil evaporation, like mulches, residues, wetting methods, crust formation, and tillage (Unger and Parker, 1976; Adams et al. 1976). Each of these effects can be significant, but the cause can usually be attributed to one of the physical limitations. Thus, for hydrologic predictions, separating the atmospheric potential reaching the soil surface from that going to the vegetation and accounting for the moisture availability limitations will provide first estimates of soil evaporation with some calibration yet necessary.

COMPUTING ACTUAL ET

To compute an estimate of the actual ET emanating from a particular agricultural field or vegetated surface requires a procedure that considers and incorporates the several major processes just discussed. The objective for each ET estimate must be clear because this will suggest the parameters of space and time incrementation and expected accuracy. Even then, considerable averaging of inputs and effects will be needed depending on information and data available.

There has been a wide variety of actual ET prediction methods developed and reported in recent years. These range from short single equations to sets of equations in a decision logic to quite complex computer models. Each is unique in its objective, accuracy, and data required. The following are a few examples across this range of complexity.

Agricultural hydrologists have traditionally estimated the antecedent soil moisture status for events by simple water budget-ET equations such as the antecedent retention index reported by Saxton and Lenz (1967). More recently, slightly enhanced water budget methods have been applied to hydrology analyses and predictive models (Haan, 1972; Holtan et al., 1975). Actual ET estimates for irrigation situations have become more complex as reported by Jensen (1973), and with the exception that these methods usually do not consider water stress, they have good application potential to a variety of other agricultural situations.

Ritchie (1972) composed a model to incorporate most basic considerations through a series of interactive equations. Potential for ET was based on the Penman method, soil evaporation and transpiration were computed separately, a leaf-area index (LAI) described plant growth, and the soil water for the entire profile was budgeted, all on a daily time sequence.

A comprehensive model to compute daily actual ET from small watersheds was developed and reported by Saxton et al. (1974b). This model, shown schematically in Fig. 3 separates the major climatic, crop, and soil effects into a calculation procedure with emphasis on graphical representation of principle relationships. Calculated amounts of interception evaporation, soil evaporation, and plant transpiration are combined to provide daily actual ET estimates.

Beginning at the top of Fig. 3, a daily potential ET is computed by any one of the several methods previously discussed. Intercepted water at the plant and soil surfaces is then considered to have first use of the potential ET energy, and no limits are imposed. Remaining potential ET is divided between soil water evaporation or plant transpiration according to plant canopy present. Actual soil evaporation is the potential limited by soil water content at the surface, except in the very wet range, thus representing the traditional two-stage drying sequence. For dry soil with a plant canopy, a percent of the unused soil evaporaton potential is returned to the plant transpiration potential to account for radiated and convected energy from the heated soil and air. Actual transpiration is computed through sequential consideration of: (1) plant phenology to describe the transpirability of the existing canopy, (2) a root distribution to reflect where in the soil profile the plant is attempting to obtain water, and (3) a water stress relationship which is applied to each soil layer and is a function of the plant available water of that soil layer and the atmospheric demand on the plant. The soil water is adjusted by abstracting the daily actual ET from each soil layer with roots, adding daily infiltration computed from daily precipitation minus measured or estimated runoff, and estimating soil water redistribution and percolation by a Darcy-type unsaturated flow computation.

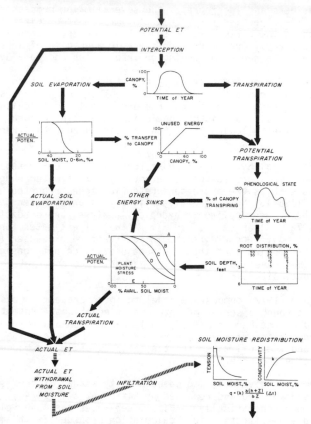

Figure 3. Calculation flow chart for computing daily actual ET and soil
water profiles by the SPAW model (Saxton et al., 1974b).

Several methods have been reported which apply best to basin
hydrology or specific land uses. The ET method by Crawford and
Linsley (1966) in their Stanford Watershed Model IV considers the
effect of areal variation through a coefficient which estimates percent of area which attains varying percentages of an evaporative
opportunity, which is in turn a function of a time-dependent coefficient to represent crop growth. Using pan evaporation as a potential ET, this model combines estimated actual ET from interception,
two soil zones, and groundwater to estimate a total daily actual ET.
No attempt is made to separate soil evaporation from transpiration,
and calibration parameters are obtained by fitting to observed watershed data.

Morton (1976) presented a method based on regional climatic data
and showed its application to many large basins in the United States
and Canada. Hanson (1976) and Aase et al. (1973) have developed ET
prediction equations for native rangeland of the western U.S. Federer
(1979) described a detailed model of transpiration that simulates both
diurnal and growing season behavior of a hardwood forest.

Other methods have restricted their data inputs to readily available climatologic data which often makes them more practical than more
sophisticated methods. Eagleman (1967) described a method based on

air temperature and humidity. Kanemasu et al. (1976) and Rosenthal et al (1977) described methods using air temperature, R_n, and LAI.

Several methods have been developed which describe the ET processes within the soil-plant-atmosphere system. The soil-plant-atmosphere model (SPAM) described by Shawcroft et al. (1974) treats the ET and plant growth characteristics in detail. A similar model reported by Hanks et al. (1969) concentrates more on the soil water and its plant interaction.

Even more sophisticated models are being developed as improved computing capacity and ease of programming become available. Kristensen and Jensen (1975) applied a detailed ET model to the crops of barley, sugarbeets, and grass over a 4-yr period with reported accuracy of 10 percent, and Wright and Jensen (1978) reported similar success for irrigated conditions with a different model. van Bavel and Ahmed (1976) described a model of the soil water flow and root uptake programmed in CSMP. van Keulen (1975) presented a model programmed in CSMP with all details well described.

Not all recent methods of modeling ET have been cited. Those that are presented vary widely in their objectives, the amount of detail required, computation time, and required data. Comparison of the several methods shows that the best results are obtained when a potential ET is derived from an energy balance-aerodynamic method and the soil-plant-atmosphere system is represented by dynamic simulations of the major processes which determine actual ET. Performance of a water yield model will be quite sensitive to the ET method used, whereas a flood peak model may be largely indifferent to the ET method used. Thus, each ET application must match the prediction method to the objectives.

REFERENCES

Aase, J.K., Wight, J.R., and Siddoway, F.H. 1973. Estimating Soil Water Content on Native Rangeland. Agricultural Meteorology, Vol. 12, No. 2, pp. 185-191.
Adams, J.E., Arkin, G.F., and Ritchie, J.T. 1976. Influence of Row Spacing and Straw Mulch on First Stage Drying. Soil Science Society of America Journal, Vol. 40, No. 3, pp. 436-442.

Baier, W. 1969. Concepts of Soil Moisture Availability and Their Effect on Soil Moisture Estimates from a Meteorological Budget. Agricultural Meteorology, Vol. 6, pp. 165-178.

Bowen, I.S. 1926. The Ratio of Heat Losses by Conduction and by Evapo ration from Any Water Surface. Physical Review, Vol. 27, pp. 779-787.

Brown, K.W. and Covey, W. 1966. The Energy-Budget Evaluation of the Micro-Meteorological Transfer Process within a Cornfield. Agricultural Meteorology, Vol. 3, pp. 73-96.

Buffo, J., Fritschen, L., and Murphy, J. 1972. Direct Solar Radiation on Various Slopes from 0 to 60 Degrees North Latitude. USDA Forest Service Research Paper PNW-142, 74 pp.

Businger, J.A. 1956. Some Remarks on Penman's Equation for the Evapotranspiration. Netherlands Journal of Agricultural Science, Vol. 4, pp. 77-80.

Burman, R.D. 1976. Intercontinental Comparison of Evaporation Estimates. American Society of Civil Engineering Proceedings, Vol. 102, pp. 109-118.

Campbell, G.S. 1977. An Introduction to Environmental Biophysics, Springer-Verlag, New York.

Coulson, K.L. 1975. Solar and Terrestrial Radiation, Academic Press, New York.

Crawford, N.H. and Linsley, R.K. 1966. Digital Simulation in Hydrology: Stanford Watershed Model IV. Technical Report No. 39, Department of Civil Engineering, Stanford University, Palo Alto, California.

Davenport, D.C. and Hudson, J.P. 1967. Changes in Evaporation Rates Along a 17-km Transect in the Sudan Gezira. Agricultural Meteorology, Vol. 4, pp. 339-352.

Davies, J.A. and Buttimor, P.H. 1969. Reflection Coefficients, Heating Coefficients, and Net Radiation at Simcoe, Southern Ontario. Agricultural Meteorology, Vol. 6, pp. 373-386.

Denmead, O.T. and Shaw, R.H. 1962. Availability of Soil Water to Plants as Affected by Soil Moisture Content and Meteorological Conditions. Agronomy Journal, Vol. 45, pp. 385-390.

Doorenbos, J. and Pruitt, W.D. 1975. Crop Water Requirements, Irriga- tion and Drainage, Paper No. 24, 179 pp., FAO, Rome. (Revised 1977, 144 pp.)

Dyer, A.J. 1961. Measurements of Evaporation and Heat Transfer in the Lower Atmosphere by an Automatic Eddy-Convection Technique. Quarterly Journal of the Meteorological Society, Vol. 87, pp. 401-412.

Eagleman, J.R. 1967. Pan Evaporation, Potential, and Actual Evapotrans- piration. Journal of Applied Meteorology, Vol. 6, pp. 482-488.

Eagleson, P.S. 1978. Climate, Soil, and Vegetation. Water Resources Research, Vol. 15, pp. 705-776.

Feddes, R.A., Kowalik, P., Kolinska-Malinka, K., and Zaradny, H. 1976. Simulation of Field Water Uptake by Plants Using a Soil Water Dependent Root Extraction Function. Journal of Hydrology (Amsterdam), Vol. 31, pp. 13-26.

Federer, C.A. 1975. Evapotranspiration (Literature Review 1971-1974). Reviews of Geophysics and Space Physics, Vol. 13, No. 3, pp. 442-445

Federer, C.A. 1979. A Soil-Plant-Atmosphere Model for Transpiration and Availability of Soil Water. Water Resources Research, Vol. 15, pp. 555-562.

Frank, E.C. and Lee, R. 1966. Potential Solar Beam Irradiation on Slopes: Tables for 30° to 50° latitude. U.S. Forest Service Research Paper RM-18, Rocky Mountain Forest and Range Experiment Station, Fort Collins, Colorado.

Fritschen, L.J. 1966. Evapotranspiration Rates of Field Crops Deter- mined by the Bowen Ratio Method. Agronomy Journal, Vol. 58, pp. 339-342.

Fritschen, L.J. 1967. Net and Solar Radiation Relations over Irrigated Field Crops. Agricultural Meteorology, Vol. 4, pp. 55–62.

Gardner, W.R. and Hillel, D.I. 1962. The Relation of External Evaporative Conditions to the Drying of Soils. Journal of Geophysical Research, Vol. 67, pp. 4319–4325.

Goodell, B.C. 1966. Watershed Treatment Effects on Evapotranspiration. International Symposium on Forest Hydrology, Pennsylvania State University, Aug. 29–Sept. 10, 1965, Pergaman Press, New York.

Gray, D.M. (Ed.). 1970. Handbook on the Principles of Hydrology. Natural Research Council of Canada, Water Information Center, Port Washington, New York.

Haan, C.T. 1972. A Water Yield Model for Small Watersheds. Water Resources Research, Vol. 8, pp. 58–69.

Hanks, R.J., Klute, A., and Bresler, E. 1969. A Numeric Method for Estimating Infiltration, Redistribution, Drainage, and Evaporation of Water from Soil. Water Resources Research, Vol. 5, pp. 1064–1069.

Hanson, C.L. 1976. Model for Predicting Evapotranspiration From Native Rangelands in the Northern Great Plains. Transactions of the American Society of Agricultural Engineers, Vol. 19, pp. 471–477.

Hanson, C.L. and Rauzi, F. 1977. Class A Pan Evaporation as Affected by Shelter, and a Daily Prediction Equation. Agricultural Meteorology. Vol. 10, pp. 27–35.

Hargreaves, G.H. 1966. Consumptive Use Computations from Evaporation Pan Data, pp. 35–62, Irrigation and Drainage Special Conference Proceedings, Nov. 2–4, American Society of Civil Engineers, New York.

Hillel, D. and Talpaz, H. 1976. Simulation of Root Growth and Its Effect on the Pattern of Soil Water Uptake by a Nonuniform Root System. Soil Science, Vol. 121, pp. 307–312.

Holmes, R.M. and Robertson, G.W. 1963. Application of the Relationship Between Actual and Potential Evapotranspiration in Dry Land Agriculture. Transactions of the American Society of Agricultural Engineers, Vol. 6, pp. 65–67.

Holtan, H.N., Stiltner, G.J., Hensen, W.H., and Lopez, N.C. 1975. USDAHL-74 Revised Model of Watershed Hydrology. U.S. Department of Agriculture Technical Bulletin 1518, 99 pp.

Jensen, M.E. (Ed.). 1973. Consumptive Use of Water and Irrigation Water Requirements, 215 pp. American Society of Civil Engineers, New York.

Jensen, M.E. and Haise, R.H. 1963. Estimating Evapotranspiration from Solar Radiation. American Society of Civil Engineers Proceedings, Vol. 89, pp. 15–41.

Kanemasu, E.T., Stone, L.R. and Powers, W.L. 1976. Evapotranspiration Model Tested for Soybean and Sorghum. Agronomy Journal, Vol. 68, pp. 569–572.

Kohler, M.A., Nordenson, T.J., and Baker, D.R. 1959. Evaporation Maps for the United States. Technical Paper No. 37, U.S. Weather Bureau, Washington, D.C.

Kristensen, K.J. 1974. Actual Evapotranspiration in Relation to Leaf Area. Nordic Hydrology, Vol. 5, pp. 173–182.

Kristensen, K.J. and Jensen, S.E. 1975. A Model for Estimating Actual Evapotranspiration from Potential Evapotranspiration. Nordic Hydrology, Vol. 6, pp. 170–188.

Linacre, E.T. 1968. Estimating the Net Radiation Flux. Agricultural Meteorology, Vol. 5, pp. 49–63.

McGuinness, J.L. and Bordne, E.F. 1972. A Comparison of Lysimeter-Derived Potential Evapotranspiration with Computed Values. U.S. Department of Agriculture Technical Bulletin No. 1452.

Mengel, D.B. and Barker, S.A. 1974. Development and Distribution of the Corn Root System under Field Conditions. Agronomy Journal, Vol. 66, pp. 341–344.

Morton, F.I. 1976. Climatological Estimates of Evapotranspiration. American Society of Civil Engineers Proceedings, Vol. 102, pp. 275-291.
Morton, F.I. 1979. Climatological Estimates of Lake Evaporation. Water Resources Research, Vol. 15, pp. 64-76.

Mustonen, S.E. and McGuinness, J.L. 1968. Estimating Evapotranspiration in a Humid Region. U.S. Department of Agriculture Technical Bulletin No. 1389, 123 pp.
Nicks, A.D. and Harp, J.F. 1980. Stochastic Generation of Temperatures and Solar Radiation Data. Journal of Hydrology, Vol. 48, pp. 1-17.
Nordenson, T.J. 1962. Evaporation from the 17 Western States. Geological Survey Professional Paper No. 272-D, U.S. Government Printing Office, Washington, D.C.
Parmele, L.H. and Jacoby, E.L. 1975. Estimating Evapotranspiration under Nonhomogeneous Field Conditions. U.S. Department of Agriculture, Agriculture Research Service, Report No. ARS-NE-51.
Parmele, L.H. and McGuinness, J.L. 1974. Comparisons of Measured and Estimated Daily Potential Evapotranspiration in a Humid Region. Journal of Hydrology, Vol. 22, pp. 239-251.
Pelton, W.L., King, K.M., and Tanner, C.B. 1960. An Evaluation of the Thornthwaite and Mean Temperature Method for Determining Potential ET. Agronomy Journal, Vol. 52, pp. 387-395.
Penman, H.L. 1948. Natural Evaporation from Open Water, Bare Soil, and Grass. Proceedings of Royal Society, Serial A, No. 1032, Vol. 193, pp. 120-145.
Penman, H.L. 1956. Estimating Evaporation. American Geophysical Union, Vol. 4, pp. 9-29.
Penman, H.L., Angus, D.E., and van Bavel, C.H.M. 1967. Microclimatic Factors Affecting Evaporation and Transpiration. in: R.M. Hagen, H.R. Haise, and T.W. Edminster (Editors), Chapter 26, Irrigation of Agricultural Lands, Special Publication No. 11 American Society of Agronomy, Madison, Wisconsin.
Priestley, C.H.B., and Taylor, R.J. 1972. On the Assessment of Surface Heat Flux and Evaporation Using Large-Scale Parameters. Monthly Weather Review, Vol. 100, No. 2, pp. 81-92.
Reifsnyder, W.E. and Lull, H.W. 1965. Radiant Energy in Relation to Forests. U.S. Department of Agriculture Technical Bulletin No. 1344, 111 pp.

Richardson, C.W. and Ritchie, J.T. 1973. Soil Water Balance for Small Watersheds. Transactions of the American Society of Agricultural Engineers, Vol. 16, pp. 72-77.
Ritchie, J.T. 1972. Model for Predicting Evaporation from a Row Crop with Incomplete Cover. Water Resources Research, Vol. 8, pp. 1204-1213.
Ritchie, J.T. 1973. Influence of Soil Water Status and Meteorological Conditions on Evaporation from a Corn Canopy. Agronomy Journal, Vol. 65, pp. 893-897.
Ritchie, J.R. and Burnett, E. 1971. Dryland Evaporative Flux in a Subhumid Climate: II. Plant Influences. Agronomy Journal, Vol. 63, pp. 56-62.
Robb, D.C.N. 1966. Consumptive Use Estimates from Solar Radiation and Temperature. in: Methods for Estimating Evaporation, pp. 169-191, Irrigation and Drainage Specialty Conference, Las Vegas, Nevada, Nov. 2-4, American Society of Civil Engineering, New York.
Rosenberg, N.J. 1969. Advective Contribution of Energy Utilized in Evapotranspiration by Alfalfa in the East Central Great Plains. Agricultural Meteorology, Vol. 6, pp. 179-184.
Rosenthal, W.D., Kanemasu, E.T., Raney, R.J., and Stone, L.R. 1977. Evaluation of an Evapotranspiration Model for Corn. Agronomy Journal, Vol. 69, pp. 461-464.

Salisbury, F.B. and Ross, C.W. 1978. Plant Physiology. Wadsworth Publishing Co., Belmont, CA. 422 pp.

Saxton, K.E. 1972. Watershed Evapotranspiration by the Combination Method. Unpublished Ph.D. Thesis, Iowa State University Library, Ames, Iowa (Dissertation Abstract International, Vol. 33, pp. 1514b-1515b, 1972).

Saxton, K.E. and Lenz, A.T. 1967. Antecedent Retention Indexes Predict Soil Moisture. American Society of Civil Engineers Proceedings, Vol. 93, pp. 223-241.

Saxton, K.E., Johnson, H.P., and Shaw, R.H. 1974a. Watershed Evapotranspiration Estimated by the Combination Method. Transactions of the American Society of Agricultural Engineers, Vol. 17, pp. 668-672.

Saxton, K.E., Johnson, H.P., and Shaw, R.H. 1974b. Modeling Evapotranspiration and Soil Moisture. Transactions of the American Society of Agricultural Engineers, Vol. 17, pp. 673-677.

Sellers, W. D. 1965. Physical Climatology. The Univ. of Chicago Press. 272 pp.

Shawcroft, R.W., Lemon, E.R., Allen, L.H., Stewart, D.W., and Jensen, S.E. 1974. The Soil-Plant-Atmosphere Model and Some of Its Predictions. Agricultural Meteorology, Vol. 14, pp. 287-307.

Stanhill, G., Hofstede, G.J., and Kalma, J.D. 1966. Radiation Balance of Natural and Agricultural Vegetation Quarterly Journal of Royal Meteorological Society, Vol. 92, pp. 128-140.

Stone, L.R., Teare, I.D., Nickell, C.D., and Mayaki, W.C. 1976. Soybean Root Development and Soil Water Depletion. Agronomy Journal, Vol. 68, pp. 677-680.

Tanner, C.B. 1957. Factors Affecting Evaporation from Plants and Soils. Journal of Soil and Water Conservation, Vol. 12, pp. 221-227.

Tanner, C.B. and Jury, W.A. 1976. Estimating Evaporation and Transpiration from a Row Crop During Incomplete Cover. Agronomy Journal, Vol. 68, pp. 239-243.

Tanner, C.B. and Pelton, W.L. 1960. Potential Evapotranspiration Estimates by the Approximate Energy Balance Method of Penman. Journal of Geophysical Research, Vol. 63, pp. 3391-3413.

Taylor, H.M. and Klepper, B. 1973. Rooting Density and Water Extraction Patterns for Corn (Zea mays L.). Agronomy Journal, Vol. 65, pp. 965-968.

Taylor, H.M. and Klepper, B. 1978. The Role of Rooting Characteristics in the Supply of Water to Plants. Advances in Agronomy, Vol. 30, pp. 99-128.

Turc, L. 1961. Evaluations Des Besoins en Eau d'Irrigation, Evapotranspiration Potentielle. Annales Agronomiques, Vol. 12, pp. 13-49.

Unger, P.W. and Parker, J.J. 1976. Evaporation Reduction from Soil with Wheat, Sorghum, and Cotton Residues. Soil Science Society of America Journal, Vol. 40, pp. 938-942.

USDA, Soil Conservation Service. 1967. Irrigation Water Requirements. Technical Release No. 21, 88 pp.

van Bavel, C.H.M. 1966. Potential Evaporation: The Combination Concept and Its Experimental Verification. Water Resources Research, Vol. 12, pp. 455-467.

van Bavel, C.H.M. and Ahmed, J. 1976. Dynamic Simulation of Water Depletion in the Root Zone. Ecological Modelling, Vol. 2, pp. 189-212.

van Bavel, C.H.M. and Hillel, D.I. 1976. Calculating Potential and Actual Evapotranspiration from a Bare Soil Surface by Simulation of Concurrent Flow of Water and Heat. Agricultural Meteorology, Vol. 17, pp. 453-476.

van Keulen, H. 1975. Simulation of Water Use and Herbage Growth in Arid Regions. Simulation Monograph, Pudoc, Wageningen, The Netherlands.

Weigand, C.L. and Taylor, S.A. 1961. Evaporative Drying of Porous Media. Utah State University Agricultural Experiment Station Special Report 15, Logan, Utah.

Whittington, W.J. 1968. Root Growth. Plenum Press, New York.

Wright, J.L. and Jensen, M.E. 1978. Development and Evaluation of Evapo-transpiration Models for Irrigation Scheduling. Transactions of the American Society of Agricultural Engineering, Vol. 21, pp. 88-91 and 96.

DISTRIBUTED WATERSHED EVAPOTRANSPIRATION MODELING IN HILLY TERRAIN

Roland E. Schulze
Associate Research Professor
Department of Agricultural Engineering
University of Natal
Pietermaritzburg, 3201 South Africa

ABSTRACT

In a spatial study of evapotranspiration (ET), particularly in hilly terrain, the nature of the surface exerts a strong modifying influence on ET. Variations in ET may thus be divided into temporal, climatological variations and spatial variations due to slope, aspect and differences in soil moisture as well as to transpirational characteristics of vegetation. A distributed watershed model for actual ET is developed by simulating ET at intersections of a square grid superimposed over four small, steep experimental rangeland watersheds at Cathedral Peak in South Africa. At each intersection at daily intervals, the net radiation load is calculated from slope, aspect and climatological inputs and used in the van Bavel potential evapotranspiration (PE) equation, which accounts also for aerodynamic differences associated with vegetation type and treatment. Daily actual evapotranspiration (AE) is estimated at each of 173 grid intersections for a three-layered soil profile using the Baier and Robertson AE equation which includes as temporal and spatial variables PE, as well as rooting, water holding and moisture depletion characteristics for each layer. Soil moisture is redistributed by accounting for precipitation, surface runoff, saturated/unsaturated soil moisture movement and drainage.

Integrated watershed AE by the model is tested against "control" AE, which is derived for the gaged watersheds by the hydrologic equation. For 91 test periods of short and long duration in all seasons and at all levels of watershed moisture status r for simulated vs control AE is 0.98 with the slope of the regression equation at 0.99. The model is shown to be sensitive particularly to PE and to moisture holding capacity of the topsoil.

BACKGROUND

Evapotranspiration is a complex vapor loss process involving water movement through the soil system, the root and the plant system to the plant/atmosphere interface. This vapor loss phenomenon is of immense importance in South Africa where economic development is linked closely to a stable water supply. The country is generally not endowed with an abundance of water, with a mean annual precipitation of less than 500 mm, and an estimated 91 percent is lost again to the atmosphere by evapotranspiration (Whitmore, 1971).

In the Drakensberg of South Africa an interest in water production and thus modeling of evapotranspiration (ET) is seen to be of special significance, for this mountain range and its foothills constitute the major inland source area of the country's water supply with annual and seasonal supplies of water available to augment deficits elsewhere.

However, modeling spatial differences in ET, particularly in hilly terrain, remains one of the vexing problems in hydrology. While ET may be seen as a response function resulting from two co-existing "forcing functions", namely a supply of water and solar radiation (Lettau, 1969), in a spatial study the nature of the surface may exert a strong modifying influence on ET. Thus, variations in ET may be subdivided into temporal (climatological) variations and spatial variations due to slope/aspect effects on the radiation load and differences in soil moisture as well as in transpirational characteristics of vegetation.

AIMS

Evapotranspiration from a watershed involves the integrated evaporative losses from an infinite number of diverse surfaces making up the watershed, all transpiring at different rates. Using the concepts outlined above, this paper aims at modeling watershed ET on four selected rangeland watersheds at Cathedral Peak in the Drakensberg mountain range (29°00'S, long. 29°15'E, Figure 1). These instrumented watersheds at an elevation around 2000 m vary in area from 0.49 to 0.73 km^2, have different steepnesses and predominant aspects as well as different treatments in regard to range burning, grazing and fire protection (Table 1). Mean annual precipitation is 1300-1500 mm of which 80% falls in the summer months, October to March (Schulze, 1979).

The model treats ET as a distributed variable by estimating it

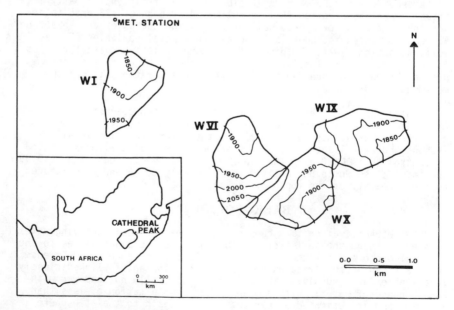

Figure 1.   Cathedral Peak Watersheds   I, VI, IX and X.

Table 1. Watershed characteristics, Cathedral Peak

Watershed	Area (km^2)	Aspect	Relief (m)	Slope (%)	Range Treatment
I	0.49	NNE	1829-2003	19	Rangeland; two camps, spring burned in alternate years; grazed
VI	0.68	N	1847-2076	27	Rangeland, spring burned in alternate years; no grazing
IX	0.65	ESE	1820-1984	22	Protected from fire and grazing to determine vegetation succession effects on hydrology
X	0.73	SE	1838-2079	26	Rangeland, spring burned in alternate years; no grazing

at intersections of a square grid of side 150 m superimposed over
the watersheds. At each intersection, at daily intervals, the
solar radiation load is calculated from slope, aspect and climatological inputs and used in the Penman-type potential evapotranspiration (PE) equation as modified by van Bavel (1966), to account
for aerodynamic differences associated with vegetation type and
treatment. Daily actual evapotranspiration (AE) is then estimated
at each grid intersection using the equation of Baier and Robertson
(1966), which incorporates PE, plant available moisture, moisture
depletion and root extraction characteristics. Saturated/unsaturated
soil moisture movement, effects of plant stress and surface runoff
routing are incorporated into the model. In order to test the model,
the integrated AE losses from all grid points in a watershed are regressed against "control" AE determined for that gaged watershed by
the hydrologic equation. The "control" AE is the residual between
measured precipitation and streamflow for selected time intervals
for which the water storage variable of the equation is cancelled
out, this being assumed to exist at equal heights of the hydrograph
of a perennial stream during periods of baseflow recession.

THE BAIER AND ROBERTSON EQUATION FOR ACTUAL EVAPOTRANSPIRATION

For the purpose of daily estimations of AE the Baier and Robertson (1966) equation was selected because it facilitates a multi-soil
zone approach to budgeting of soil moisture, SM. The basic concept
of the equation is that of sub-dividing the soil (to rooting depth)
into zones of varying plant available moisture from which water is
evapotranspired simultaneously, but in different proportions, related to the plant rooting distribution. Evapotranspiration takes
place according to the daily rate of moisture demand and the available SM in each zone.

As used for a three-zoned soil profile, the Baier and Robertson
equation may be given as :

$$AE_i = \sum_{j=1}^{3} k_j \cdot (SM_{j(i-1)}/PAM_j) \cdot Z_j \cdot PE_i$$

207

where

AE_i = AE in mm for day i

k_j^i = coefficient representing the moisture extraction from that zone by the rooting system

$SM_{j(i-1)}$ = available SM in mm in the jth zone at the end of day(i-1). Its value is subject to moisture loss by AE from that zone on day (i-1), adjusted by possible gains from precipitation and/or capillary action and by possible losses from percolation or unsaturated flow

Z_j = a variable which allows depletion of SM to proceed at different rates according to soil properties and plant stress conditions

and PE_i = potential evapotranspiration in mm for day i, determined for varying vegetation, meteorological and topographic conditions.

Components of the above equation are now discussed in more detail with emphasis on their spatial and/or temporal variability.

THE ROOTING COEFFICIENT, k_j

The rooting coefficient, k_j, expresses the fraction of soil water withdrawn simultaneously by the plant roots from each of the three soil zones. The k_j values have to be estimated to resemble the most probable rooting pattern, therefore having to account for genetic and environmental factors on transpiration such as winter dormancy, root development in spring and decay in autumn, yet allowing for some transpiration to take place during senescence or under conditions of limited moisture. From k_j values given by Baier (1969), suitably adapted to the three-layered soil profile used at Cathedral Peak, and from observations in situ (J.E. Granger, personal communication, 1976), monthly values were assigned for the growth patterns of the dominant range grass species of the four watersheds, namely Themeda triandra.

Table 2. Monthly values of the rooting coefficient k_j for Themeda triandra at Cathedral Peak

Water-shed	Soil Horizon	Jan	Feb	Mar	Apr	May	Jun	Jul	Aug	Sep	Oct	Nov	Dec
I, VI and X	A	.90	.90	.90	.94	.94	.94	.94	.94	.92	.92	.90	.90
	B1	.07	.07	.07	.04	.04	.04	.04	.04	.06	.06	.07	.07
	B2	.03	.03	.03	.02	.02	.02	.02	.02	.02	.02	.03	.03
IX	A	.90	.90	.90	.90	.94	.94	.94	.94	.94	.92	.92	.90
	B1	.07	.07	.07	.07	.04	.04	.04	.04	.04	.06	.06	.07
	B2	.03	.03	.03	.03	.02	.02	.02	.02	.02	.02	.02	.03

TOTAL PLANT AVAILABLE SOIL MOISTURE, PAM_j

In an assessment of watershed ET a vital role is played by the soil, for it is the capacity of the soil to absorb, retain and release water that is prime regulator of the ET response to the watershed (England and Stephenson, 1970).

Soils data are often used in ET computations by "lumping" the characteristics of many soils to derive an average areal parameter. A watershed is, however, not a "lumped" system in regard to soils and pronounced differences in magnitude and sequence of hydrological processes affecting AE have been observed in soil units within a catchment. The delineation of soil units which are relatively homogeneous with respect to hydrologic response is thus critical. The soil properties more closely related to infiltration and ET than any other easily measured site variables have been found by many researchers to be soil thickness and texture, which, together can be used to estimate water holding capacity (for example, Rauzi and Kuhlman, 1961; England, 1970).

Soils in the four selected watersheds were therefore mapped and classified into recognizable and identifiable units within which soil moisture properties and behavior would vary within defined and relatively narrow limits. Values of total plant available moisture, PAM_j, for the three horizons of each soil mapping unit were then determined. Soil moisture constants (i.e. plant available and gravitational water as percentages by volume) as given by England (1970) and Wilkinson (1974) were assigned according to each mapping unit's and horizon's soil textural class (mostly silty clays), these having been determined by mechanical analyses. Used together with soil depths, the moisture constants gave values of PAM_j, which were then applied as storage factors at the grid of points selected for ET computations. The derivation of PAM_j is illustrated in Table 3 using the soil mapping units from Watershed IX at Cathedral Peak by way of example.

Table 3   Derivation of PAM_j: an example from Cathedral Peak Watershed IX

Soil Mapping Unit	Horizon	Depth (mm)	PAM by Volume (%)	Gravitational Water (%)	Total PAM_j (mm)
Mispah	A	41.6	12.3	88	5.1
	B1	20.8	12.3	88	2.6
	B2	20.8	12.3	88	2.6
Shallow	A	198.1	12.3	88	24.4
Clovelly	B1	76.2	13.3	81	10.1
	B2	167.2	13.8	78	23.1
Shallow	A	121.8	12.3	88	15.0
Griffin	B1	168.7	13.3	81	22.4
	B2	229.5	13.8	78	31.7
Griffin	A	205.5	12.3	88	23.3
	B1	171.1	13.3	81	22.6
	B2	526.4	13.8	78	72.6
Hutton	A	226.1	12.3	88	27.8
	B1	170.2	13.3	81	22.6
	B2	518.1	13.8	78	71.5
Katspruit	A	278.3	12.3	88	34.2
	B1	150.0	13.3	81	20.0
	B2	345.6	13.8	78	47.7

THE SOIL MOISTURE DEPLETION RATE, Z_j

The soil moisture depletion rate, a manifestation of moisture

stress expressed as the decline of AE relative to PE with SM depletion, has been shown in a vast literature (for example, Hounam, 1971;
Slabbers, 1980) to vary with one or more of soil properties, vegetation characteristics and evaporative demand of the atmosphere.
Probably the most realistic type of depletion curve is exemplified
by the Denmead and Shaw (1962) relationship which allows the curve to
change in its accounting for plant physiological responses to atmospheric demand. While this method was tested on Cathedral Peak data,
the most consistent results there were obtained using the Thornthwaite
and Mather (1954) assumption of AE/PE being linearly related to the
relative soil moisture status, i.e. SM/PAM. This assumption is
also considered realistic for use with fine-textured soils (Salter
and Williams, 1965) such as the silty clays at Cathedral Peak, and
it has been used with success in short grassland studies elsewhere
(Baier, 1969).

POTENTIAL EVAPOTRANSPIRATION, PE_i

Research dating back to the 1940's (for example, Croft, 1944;
Lee, 1964; Rouse and Wilson, 1969) illustrates that spatial variations in the water balance exist that are too large to be ignored
and which result from radiant energy regimes of different slopes and
aspects. Such variations in hydrologic output are caused largely
by soil moisture losses through ET being affected by topography. A
necessary step toward estimations of watershed AE with the Baier and
Robertson equation is therefore the estimation of PE for sloping
terrain.

The idea of evaluating PE through the evaluation of energy exchange at the non-horizontal radiating surface has frequency been
suggested in the past two decades (for instance Lee, 1964; Garnier,
1970) especially as a major driving force of the ET process is
supplied by net radiation. Rouse (1970) has shown conclusively that
PE on slopes can be estimated to a high degree of accuracy from
radiation variations alone while Reid (1973) maintains that differences in PE on slopes, even if small over short-term periods, become
significant in "altering the gross hydrological balance" over extended
periods.

In this paper the interception of radiant energy appropriate to
slope, aspect and albedo in mountainous terrain was considered one
important variable in the estimation of PE at Cathedral Peak. A
second important variable, the result of differences in watershed
treatment on vegetation (burning, grazing, protection) was the influence of surface roughness of the vegetative cover on turbulence
and the aerodynamic factor in ET. In order to accommodate these
two major variables encountered in a spatial study, the van Bavel
(1966) modification of the Penman (1948) equation for PE was selected.
In this equation

$$PE(mm\ day^{-1}) = (\Delta/\gamma\ R_n + B_v \cdot P_d) / (\Delta/\gamma + 1)$$

where

Δ = slope of the saturated vapor pressure curve
at air temperature

γ = psychrometric constant

R_n = daily net radiation fluxes in mm equivalent
of water

P_d = saturation deficit in mb and

B_v = aerodynamically defined wind function.

In a form similar to that used by Bordne and McGuinness (1973) the aerodynamic function may be given as

$$B_v (\text{mm mb}^{-1}) = 0.1222u(298.16/t_a)(1013.25/p) \,/\, \left[\ln(z/z_0) \right]^2$$

in which

u = windrun in km day^{-1} at 2 m
t_a = air temperature in $^\circ$K
p = atmospheric pressure in mb
z = elevation, by convention in cm, at which wind velocity is recorded (i.e. 200 cm) and
z_0 = vegetation roughness parameter, also conventionally given in cm, for different vegetation heights.

The meteorological variables in the above equations were obtained, by direct measurement or by derivation, from the meteorological station at the Cathedral Peak Watersheds (Figure 1). Two components of the above equations do, however, require amplification, viz., the derivations of the vegetation roughness parameter and of net radiation fluxes on hilly terrain.

The Vegetation Roughness Parameter, z_0

With the vegetation in the selected watersheds grazed, burnt or protected, considerable temporal and spatial variation of z_0 occurs. For this research the equation for z_0 by Szeicz, Enrödi and Tajchman (1969) was selected because it was derived for vegetation with small leaf areas such as grasses. In this equation

$$\log_{10} z_0 = \log_{10} H - 0.98 \text{ cm}$$

where

H = vegetation height in cm.

A detailed functional vegetation survey, i.e., with regard to height, recovery after burn or growth stages, was therefore undertaken in the watersheds to obtain representative z_0 values which would vary in space and with season or treatment.

Net Radiation Fluxes on Hilly Terrain

The net radiation factor has been termed the "forcing function" (Lettau, 1969) or the "main driving force" (Reid, 1973) in relation to ET and it is usual that, in the absence of advection, about 80% of the variance of PE may be explained by variance of net radiation. Referring to the role of radiant energy in the spatial variation of ET, Hounam (1971) underlines first the topographic significance of watersheds in determining the interception of incoming radiation through differences in gradient, aspect and elevation; secondly, the ability of the hilly terrain to reflect radiation at different rates through variations of vegetation and topography and thirdly, the reradiation of longwave energy according to the temperature of the surface, which again is largely dependent on topography - all three factors resulting in inter- and intra-watershed variations of potential water loss to the atmosphere.

Details of partitioning incoming and outgoing radiation fluxes from meteorological station measurements to obtain radiation loads on sloping terrain have been previously described by the author (Schulze, 1975; 1976a). A summary of the significance of terrain on incoming radiation loads and resultant PE at Cathedral Peak is afforded by Table 4, showing that particularly in the equinoctal and winter months significant intra- and inter-watershed variations in PE may occur as a result of topographic influences.

Table 4. Seasonal inter- and intra-watershed differences in incoming radiation loads and PE at Cathedral Peak

Season and Watershed		Incoming Radiation Load ($10^6 Jm^{-2}day^{-1}$) Means	Range	Potential Evapotranspiration (mm) Means	Range
January	I	19.9	18.6 - 20.7	4.3	4.0 - 5.0
(Mid-	VI	19.4	17.7 - 20.7	4.5	3.9 - 5.1
Summer)	IX	19.9	17.7 - 21.0	5.5	4.7 - 7.5
	X	19.6	16.8 - 20.5	4.8	1.1 - 5.1
April	I	21.7	17.4 - 22.2	5.0	3.9 - 5.5
(Autumn)	VI	22.7	20.4 - 25.2	5.6	4.9 - 6.1
	IX	19.5	12.0 - 22.2	5.8	3.0 - 7.9
	X	18.3	12.0 - 22.2	4.8	2.9 - 7.1
July	I	17.7	14.4 - 19.2	6.2	5.8 - 6.3
(Mid-	VI	19.3	15.0 - 21.6	6.7	5.8 - 8.0
Winter)	IX	14.6	8.4 - 18.6	7.4	3.9 - 9.0
	X	13.2	3.0 - 18.6	5.8	3.0 - 9.0

AE MODELING PROCEDURES

At each sampling point, i.e., at the intersections of the square grid of side 150 m superimposed over the watersheds, a permanent data file for spatial variables was set up with relevant information on the grid points'

(1) location - for computer mapping and surface runoff routing procedures
(2) elevation - for temperature estimations in net radiation determinations
(3) slope - for modeling radiant energy loads and depression storage
(4) aspect - for modeling radiant energy loads
(5) total plant available moisture - for a three-layered soil profile used in the SM and AE modeling, and
(6) vegetation type/treatment - for calculations of albedo and aerodynamic resistance in PE modeling.

The number of grid points for Watersheds I, VI, IX and X were, respectively, 38, 46, 42 and 47. Relevant meteorological information at the daily level for the period of investigation April 1969 to September 1971 constituted the temporal data file. The study period was selected because of ready availability and high quality of the meteorological and hydrological data.

The computational sequence for daily estimations of AE at each grid point is listed below :

(1) Instructions regarding the selection of rooting coefficients are read, selections being by month and by watershed.
(2) Instructions regarding dates of range burning are read.
(3) PE at each grid point is estimated for the day in a separate

routine using meteorological, vegetation and topographic
information, and the value of PE then read.

(4) Similarly, daily precipitation is estimated separately for
each grid point and read. Being mountainous terrain with
highly localized rainfall, a four-dimensional/nine-
variable trend surface analysis was used for areal precipitation determinations (Schulze, 1976b; 1979).

(5) An atmospheric stress factor is calculated for use with the
Denmead and Shaw (1962) option of the soil moisture
depletion rate.

(6) Soil moisture content for each of the three soil zones is
fixed from the previous day's conditions.

(7) AE for the day is calculated for each soil zone by the
Baier and Robertson equation already outlined.

(8) Thereupon SM contents for each soil zone are readjusted
using the following criteria :
   (a) If precipitation occurs on a day, surface abstractions
       are first determined. These comprise interception at
       a maximum 1.5 mm per day (Whitmore, 1971) and depression
       storage, expressed for the grass vegetation at Cathedral
       Peak as a function of slope, the expression having been
       derived from data presented by Musgrave and Holtan (1964).
       The precipitation remaining after surface abstractions is
       distributed in the soil profile from the top down.
   (b) Surface flow is assumed to occur when the soil profile
       reaches field capacity and the A horizon reaches satura-
       tion (gravitational water content calculated from Table 3).
   (c) Surface flow is routed successively downslope from one
       grid point to another predetermined grid point to be added
       there to the value of precipitation. At "terminal" grid
       points, i.e. ones with a stream flowing within the grid's
       area of interest, the quickflow is "lost" to possible
       infiltration and subsequent ET as streamflow.

(9) If present, any gravitational water, i.e. soil water in
excess of field capacity, is percolated down to the next
soil zone at an exponential rate as determined by Hill (1965)
for South African soils with textures similar to those found
in the watersheds.

(10) Finally, redistributions of soil moisture, either upwards
or downwards, are accommodated for unsaturated conditions.
Daily rates of redistribution, dependent on relative moisture
statuses of the respective soil zones, were taken from research
by Kniesel, Baird and Hartman (1969).

(11) A final soil moisture content for each soil zone is thus obtained and stored for a repetition of the computational
sequence the following day.

RESULTS

Actual evapotranspiration by the procedures outlined above was
computed on a daily basis at each of the grid points, thereby facilitating a spatially variable but integrated watershed AE to be simulated
for any length of time. In order to set realistic starting values
for the SM component of the Baier and Robertson AE equation, an
initial starting date was selected where, after prolonged and consistent rains, the SM of all three zones was assumed to equal field
capacity, i.e. PAM_j.

Simulated watershed AE, SWAE, was regressed against "control"

watershed AE to test the model. For the "control" or observed watershed AE, OWAE, the simple hydrologic equation was used in the form

         OWAE = P - Q ± ΔS
where
         P = watershed precipitation in mm
         Q = streamflow in mm and
         S = watershed moisture storages in mm.

     With a dense and efficient network of raingages in the watersheds
(Schulze, 1976b) and accurate measurement of streamflow it remained
for ΔS to be assessed in order to solve for OWAE.   The ΔS term was
cancelled out of the equation by selecting time periods such that the
same amounts of soil and ground moisture could be assumed stored in
the watershed at the commencing and ending dates of a test period.
This requirement is assumed to exist for equal heights of the hydrograph of perennial streams during periods of baseflow recession,
as stated by Linsley, Kohler and Paulhus (1949) and Lambert (1969).
OWAE estimated by the hydrologic equation under the above conditions,
especially over long periods, has the advantage that it "itegrates
all spatial variations over a watershed without the need to know
details of the variations" (Hounam, 1971).   Consequently, this
method was used as the proposed SWAE experimental "control" at Cathedral Peak.

Table 5. Examples of observed and estimated watershed AE for Watershed
          IX at Cathedral Peak

Test No.	Test Period Yr Mo Dy- Yr Mo Dy	Description	Flow Rate (m^{3}day^{-1})	Precip- itation (mm)	Stream- flow (mm)	OWAE (mm)	SWAE (mm)	SWAE/ OWAE (%)
1	69.07.12- 71.07.22	2 years - winter	500	2405.1	880.0	1517.1	1576.6	103.9
7	70.03.14- 71.04.28	13 months - autumn	1150	1307.5	473.0	834.5	857.4	102.7
10	69.04.18- 70.03.04	11 months - autumn	1285	1057.8	411.3	646.5	705.9	109.2
13	69.10.20- 70.07.27	8 months - spr. to winter	255	968.8	362.0	606.8	646.5	106.5
18	69.06.07- 69.10.25	5 months - winter to spr.	750	281.9	97.6	184.3	176.8	95.9
20	69.12.09- 70.03.29	3 months - summer	850	594.9	211.0	383.9	374.5	97.6
23	70.01.23- 70.03.14	2 months - summer	1060	282.6	112.0	170.6	162.8	95.4
26	70.10.20 - 70.11.19	1 month - spring	640	88.3	24.7	63.6	85.1	133.8
29	69.08.31 - 69.09.20	3 weeks - winter	340	72.3	13.6	58.7	56.6	96.4

     A total of 91 test periods, for which ΔS was taken as zero,
were selected from hydrographs of the four watersheds.   These periods

range from long-duration two-year to short-duration three-week periods under conditions of both high and low baseflow in all seasons. A sample of results of OWAE and SWAE for Watershed IX is given in Table 5 to illustrate procedures described above. The 91 values for SWAE were then regressed against OWAE values to test the usefulness of the model. Results are shown in Figure 2.

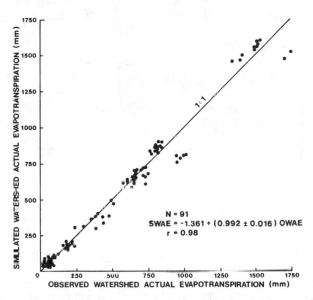

Figure 2. Simulated versus observed watershed actual evapotranspiration.

Considering the complexity of the terrain at Cathedral Peak, with marked inter- and intra-watershed variation in exposure, soil moisture capacities and vegetation treatment, the overall estimation of SWAE is highly satisfactory, correlating at $r = 0.98$ and with a slope of 0.99. For the individual watersheds correlation coefficients were all above 0.99 but the slopes of the regression equations ranged from 0.85 in Watershed I to 1.05 in Watershed X. In all watersheds the longer duration comparisons could be simulated with less relative error and the most variable results occurred in short duration tests, particularly in the transitional periods between summer and winter. This may be ascribed to the omission of the soil heat flux in the model. It is thought that the underestimation in Watershed I, which was grazed, may have occurred by the van Bavel PE equation's being oversensitive to vegetation height. The same conclusion was reached at individual grid points in all four watersheds where PE estimates were considered unrealistically high for vegetation heights in excess of one meter.

With the simulation of AE by a distributed model shown to be realistic when all the sample points within a watershed are integrated to give a lumped watershed value for AE, the model can be used to display spatial variability of AE within watersheds. An example of this is given by Figure 3, where accumulated AE for the period October 1, 1969 to September 30, 1971 was plotted. Inter-watershed variation of AE ranges from 1400 - 1600 mm in Watershed I, from 1000 - 2000 mm in VI and IX and from 1000 - 2200 mm in X. While

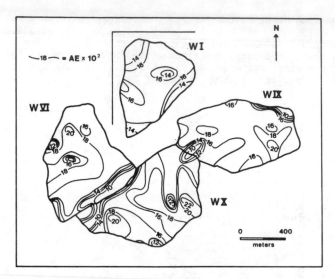

Figure 3. Spatial variation of simulated actual
evapotranspiration for the period October 1,1969
to September 30, 1971.

the detailed soil map of the watersheds is not presented in this
paper, the spatial variation of AE, in these watersheds with predominantly shallow-rooted vegetation (Table 2), may be shown to be
closely associated with depth of the A soil horizon.

Sensitivity analyses were run to test the model's response to
major input variables such as the rooting coefficient, unsaturated
soil moisture redistribution and PE. Of these, the model was most
sensitive to PE. Parmele (1972), for example, has shown that
systematic biases in PE input have significant cumulative effects on
outputs of hydrologic models.   In the present model systematic PE
biases of + and - 20 percent generally yielded errors of 10 percent
in times when water was not a limiting factor to ET, but these
errors decreased to two percent in the dry winters during senescence.

CONCLUSIONS

Because evapotranspiration is a major response component in the
hydrologic cycle, any minor change in AE can cause a pronounced relative effect on smaller, but perhaps more tangible, components of

the cycle like runoff.   The necessity for simulation of AE is thus of great importance.   Recognizing not only temporal climatological differences, but also spatial differences, particularly in regard to soil water holding capacities and the PE input (where the influence of slope and aspect on radiant energy loading and the effect of aerodynamic roughness of vegetation were considered), watershed AE was realistically simulated in a complex natural environment.   The model at this stage is not yet user oriented.   The major components requiring simplification are the estimation of topographically induced differences in radiant energy loading, while the highly sensitive aerodynamic roughness factor in the PE equation may well be replaced by (say) leaf area index.   A distributed modeling approach to AE is, however, seen as an important step in improvements to simple and more complex models of water yield, crop yield or stormflow estimation, because for all those cases the watershed's soil moisture status, as influenced largely by AE, is of vital significance to crop or hydrologic response.

ACKNOWLEDGEMENTS

This research was supported by grants from the South African Council for Scientific and Industrial Research and the University of Natal Research Fund.   Meteorological and streamflow data were kindly made available to the author by the State Department of Water Affairs, Forestry and Environmental Conservation.

REFERENCES

Baier, W. 1969. Concepts of soil moisture availability and their effect on soil moisture estimates from a meteorological budget. Agricultural Meteorology, Vol. 6, pp. 165-178.

Baier, W. and Robertson, G.W. 1966. A new versatile soil moisture budget. Canadian Journal of Plant Science, Vol. 46, pp. 299-315.

Bordne, E.F. and McGuinness, J.L. 1973. Some procedures for calculating potential evapotranspiration. Professional Geographer, Vol. 25, pp. 22-28.

Croft, A.R. 1944. Some recharge, discharge phenomena of north-and south-facing watershed-lands in the Wasatch mountains. Transactions of the American Geophysical Union, Vol. 25, pp. 881-889.

Denmead, O.T. and Shaw, R.H. 1962. Availability of soil water to plants as affected by soil moisture content and meteorological conditions. Agronomy Journal, Vol. 54, pp. 385-390.

England, C.B. 1970. Land capability classes as hydrologic response units in watersheds. Proc. Symp. on Inter-disciplinary Aspects of Watershed Management, Bozeman, Montana. American Society of Civil Engineers, New York, pp. 53-64.

England, C.B. and Stephenson, G.R. 1970. Response units for evaluating the hydrologic performance of rangeland watersheds. Journal of Hydrology, Vol. 11, pp. 89-97.

Garnier, B.J. 1970. Some thoughts on evaluating the distribution of potential evapotranspiration. Climatology Bulletin, Department of Geography, McGill University, Montreal, Vol. 7, pp. 1-7.

Granger, J.E. 1976. Department of Botany, University of Transkei, Umtata. Personal communication.

Hill, J.N.S. 1965. Moisture characteristics of some Natal Sugarbelt soils. South African Journal of Agricultural Science, Vol. 8, pp. 767-774.

Hounam, C.E. 1971. Problems of Evaporation Assessment in the Water Balance. Report No. 13, WMO/IHD Projects, Geneva.

Kniesel, W.B., Baird, R.W. and Hartman, M.A. 1969. Runoff volume prediction from climatic data. Water Resources Research, Vol. 5, pp. 84-94.

Lambert, A.E. 1969. A comprehensive rainfall/runoff model for an upland catchment area. Journal of the Institute of Water Engineers, Vol. 23, pp. 231-238.

Lee, R. 1964. Potential insolation as a topoclimatic characteristic of drainage basins. Bulletin International Association of Hydrological Sciences, Vol. 9, pp. 27-41.

Lettau, H. 1969. Evapotranspiration climatonomy: 1. A new approach to numerical prediction of monthly evapotranspiration, runoff, and soil moisture storage. Monthly Weather Review, Vol. 97, pp. 691-699.

Linsley, R.K., Kohler, M.A. and Paulhus, J.L.H. 1949. Applied Hydrology. McGraw-Hill, New York.

Musgrave, G.W. and Holtan, H.N. 1964. Infiltration. In V.T. Chow (Editor), Handbook of Applied Hydrology. McGraw-Hill, New York. Section 12.

Parmele, L.H. 1972. Errors in output of hydrologic models due to errors in input potential evapotranspiration. Water Resources Research, Vol. 8, pp. 348-359.

Penman, H.L. 1948. Natural evaporation from open water, bare soil and grass. Proceedings of the Royal Society London, Vol. A 193, pp. 120-146.

Rauzi, F. and Kuhlman, A.R. 1961. Water intake by soil and vegetation on certain western South Dakota rangelands. Journal of Range Management, Vol. 14, pp. 267-271.

Reid, I. 1973. The influence of slope orientation upon the soil moisture regime, and its hydrolomorphological significance. Journal of Hydrology, Vol. 19, pp. 309-321.

Rouse, W.R. 1970. Relations between radiant energy supply and evapotranspiration from sloping terrain: An example. Canadian Geographer, Vol. 14, pp. 27-37.

Rouse, W.R. and Wilson, R.G. 1969. Time and space variations in the radiant energy fluxes over sloping forested terrain and their influence on seasonal heat and water balances at a middle latitude site. Geographiska Annaler, Vol. 51A, pp. 160-175.

Salter, P.J. and Williams, J.B. 1965. The influence of texture on the moisture characteristics of soils: Available water capacity and moisture release characteristics. Journal of Soil Science, Vol. 16, pp. 310-317.

Schulze, R.E. 1975. Mapping potential evapotranspiration in hilly terrain. South African Geographical Journal, Vol. 57, pp. 26-35.

Schulze, R.E. 1976a. A physically based method of estimating solar radiation from suncards. Agricultural Meteorology, Vol. 16, pp. 85-101.

Schulze, R.E. 1976b. On the application of trend surfaces of precipitation to mountainous areas. Water South Africa, Vol. 2, pp. 110-118.

Schulze, R.E. 1979. Hydrology and Water Resources of the Drakensberg. Natal Town and Regional Planning Commission, Pietermaritzburg. 179 pp.

Slabbers, P.J. 1980. Practical prediction of actual evapotranspiration. Irrigation Science, Vol. 1, pp. 185-196.

Szeicz, G., Enrödi, G. and Tajchman, S. 1969. Aerodynamic and surface factors in evaporation. Water Resources Research, Vol. 5, pp. 380-394.

Thornthwaite, C.W. and Mather, J.R. 1954. The computation of soil moisture. Drexel Institute of Technology, Publications in Climatology, Vol. 7, pp. 397-402.

Van Bavel, C.H.M. 1966. Potential evapotranspiration: The combination concept and its experimental verification. Water Resources Research, Vol. 2, pp. 455-467.

Whitmore, J.S. 1971. South Africa's water budget. South African Journal of Science, Vol. 67, pp. 166-176.

Wilkinson, G. 1974. Department of Soil Science and Agrometeorology, University of Natal, Pietermaritzburg. Personal communication.

MODELING EFFECTS OF SURFACE CONFIGURATION ON SOIL MOISTURE STORAGE, EVAPOTRANSPIRATION AND RUNOFF

Francis I. Idike, Teaching Assistant in Agricultural Engineering

Curtis L. Larson, Professor of Agricultural Engineering

Donald C. Slack, Associate Professor of Agricultural Engineering
University of Minnesota, St. Paul, Minnesota 55108

ABSTRACT

A computer model of the soil-water balance has been developed. Included are submodels for surface storage, infiltration, soil-water redistribution and evapotranspiration (ET). Soil-water storage is modeled by 15-30 centimeter layers up to 152 cm (6 feet). Moisture movement between layers (downward or upward) is calculated hourly or at two hour intervals from soil moisture gradients between layers, using Darcy's law applied to unsaturated porous media.

Soil moisture data under corn for a field location in southwest Minnesota were analyzed to develop relationships for (1) ET as a function of soil moisture content, crop stage and weather and (2) moisture extraction by layers from the soil profile for corn. Infiltration is calculated by periods of one hour or less using the Green and Ampt method as modified by Mein and Larson. Input data for the model include hourly rainfall values, daily pan evaporation and soil hydraulic characteristics.

The model was tested against observed moisture contents in various soil layers at the Minnesota location for a period of seven years, not including the years used in developing the model or evaluating parameters. The model performed well for most years with the worst predicted moisture content values differing by about 0.10 (vol./vol.) from the observed values.

In addition to conventional tillage, basin tillage (contour ridging) was used with the model to make projections of its effect on soil moisture levels under corn production at Lamberton in southwest Minnesota, and Del Rio, in south central Texas. The model results for both locations indicated that basin tillage significantly increases soil moisture levels and reduces crop stress, but only if the basins are in place throughout the year.

INTRODUCTION

Agriculture is probably the segment of society which depends most on water resources of sufficient quantity and acceptable quality. Unfortunately, the available water is not always sufficient for full agricultural production. Various studies have shown that evapotranspiration (ET) in many areas exceeds the average total rainfall during the crop growing season. Irrigation is, of course, a possible alternative but this requires additional sources of water and energy and may not be economical in areas having marginal rainfall during the growing season.

In such areas, consideration should be given to production-management systems that reduce surface runoff and thereby use the rainfall more efficiently. Reduction in erosion would be an added benefit.

Among the methods used to reduce runoff are level borders, terraces and farming on the contour. These methods are effective but they have certain disadvantages such as "point rows," cost of land formation and maintenance. Under many conditions, furrow diking or basin tillage can control runoff more effectively and less expensively than any of the above methods (Bilbro and Hudspeth, 1977; Hudspeth, 1978). Hudspeth (1978) describes basin tillage as a method of tillage in which (deep) furrows are formed between the rows and dammed at short intervals (e.g., 4 meters) forming small basins in which rainfall is impounded for subsequent infiltration.

Whether or not added infiltration is beneficial to crops depends on the daily soil moisture content, which can be estimated by the water balance approach. The soil-water balance involves a systematic evaluation (usually daily) of the gains to and losses from the soil moisture reservoir. Gains occur mainly in the form of infiltration from rainfall, snowmelt or irrigation. The principal losses are evapotranspiration and surface runoff, augmented at times by percolation beyond the root zone.

Van Bavel (1953, 1956) used the water balance principle to estimate water surplus and predict drought, with ET estimated by the Thornthwaite method. Shanholtz and Lillard (1970) developed a mathematical model to simulate the soil-water status under two contrasting tillage systems. Their model represented interception as a function of crop height and vegetative development while evapotranspiration after crop emergence was estimated using a family of curves. The soil-water balance model developed by Richardson and Ritchie (1973) uses pan evaporation (or net radiation), leaf area and several soil parameters to predict daily soil evaporation, transpiration, percolation and soil-water content. Other investigators, e.g. Saxton et al. (1974), have used net radiation and the energy balance approach to more accurately calculate daily ET and soil-water content.

Evapotranspiration varies greatly with the vegetation and with meteorologic conditions. At times, it is limited by soil moisture availability. Denmead and Shaw (1962) and later Holmes and Robertson (1963) studied ET in detail and found that, as the soil moisture decreases, evapotranspiration rate declines and that this decline begins at a higher moisture content when the atmospheric evaporative demand rate is high. The vegetative growth stage of the crop influences evapotranspiration with ET increasing as the crop approaches maturity and then decreasing for some crops and remaining constant for others (Stegman et al., 1977).

The overall objective of this study was to determine whether increasing surface storage (by basin tillage, for example) can increase soil water availability during the growing season in selected areas. To accomplish this, a soil-water model incorporating various features of existing models (for the most part) was developed and tested. Model features desired were: (1) uses widely available weather data, (2) uses physically based, measureable soil parameters in calculating infiltration and soil-water movement, (3) considers effects of crop, crop stage and limited soil-water, (4) adequately represents surface storage effects, (5) represents effects of off-season precipitation, and (6) is relatively simple but sufficiently accurate for the intended use.

ET RELATIONSHIPS

Evapotranspiration (ET) can be determined from soil moisture measurements by applying the water balance equation over a given time interval. Letting S_1 and S_2 equal total soil moisture stored at the beginning and end of the time interval, respectively,

$$ET = S_1 - S_2 + P - Q - DP \tag{1}$$

where P is the amount of rainfall, irrigation or snowmelt, Q is the runoff amount and DP is the amount of deep percolation, if any. In soil moisture accounting, ET is estimated each day and the new soil moisture storage (S_2) is calculated by Eq. (1).

Potential ET (PET) is often calculated in hydrologic models from pan evaporation data because daily values of pan evaporation (E_p) are widely available. Using the "pan model,"

$$PET = K_c E_p \tag{2}$$

where K_c is the combined crop-pan coefficient, referred to herein as the crop coefficient. Thus, K_c is less than 1.0 and varies with the crop and with the crop stage. When the soil moisture supply is inadequate, the actual ET is less than PET and can be expressed as

$$ET = K_w K_c E_p \tag{3}$$

where K_w is the "soil moisture factor" and represents the effect of inadequate soil-water on ET. Comparing Eqs. (2) and (3), one sees that K_w is by definition equal to ET/PET and may have values in the range from zero to 1.0. Using measured ET values and pan evaporation data at the same site, one can evaluate the terms K_c and K_w in Eqs. (2) and (3) for different crop and soil moisture conditions.

The data for this study were collected at the University of Minnesota Agricultural Experiment Station at Lamberton in southwest Minnesota. Soil moisture data for a field under continuous corn and at various soil depths up to 152 cm (6 feet) were available for over fifteen years between 1960 and 1979 (generally at biweekly intervals after 1970). The soil type at the Lamberton Experiment Station is the Nicollet-Webster soil association which is mostly clay loam. The topography is nearly level to undulating. The average annual temperature is 8°C (46°F) and the seasons of spring, summer and fall average about 7°C, 22°C, and 9°C, respectively. Annual precipitation averages about 630 mm (24.8 in) and nearly 42% of this total falls in June, July, and August.

Values of actual ET were obtained by Eq. (1) applied to intervals of two weeks or less for a total of nine years except for periods where runoff or deep percolation could have occurred. Runoff was assumed possible when the day's rainfall (or any day's rainfall within the time interval under consideration) exceeded 1.02 cm (0.4 in) and deep percolation when the moisture content of the lowest soil zone with the crop's rooting depth exceeded field capacity. During these periods, the observed data were not used.

EVALUATION OF K_c and K_w

During periods when soil moisture is not limiting (ET = PET), Eq. (2) can be used to determine values of K_c for various stages of crop growth.

Crop growth stage was expressed as DAE (days after emergence of the crop). Figure 1 shows the curve of best fit of the data points of K_c as a function of DAE. A second order polynomial equation was fitted to the data set, (R^2 = 0.76), giving the following equation:

$$K_c = 0.152 + 0.0164(DAE) - 0.00012(DAE)^2 \qquad (4)$$

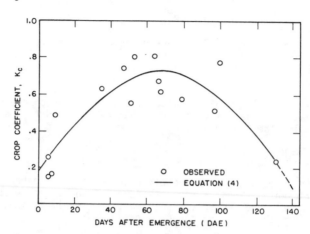

Figure 1.   Crop-Pan Coefficient as a Function of Growth Stage for Corn (Lamberton, Minnesota).

At corn emergence, the value of K_c (about 0.20) represents mainly soil evaporation. As the corn grows to full height, K_c increases rapidly to a maximum (about 0.72) and then declines. Denmead and Shaw (1959) obtained a similar curve, but with K_c varying from a minimum of 0.35 to a maximum of 0.82.

For the evaluation of K_w (Eq. 3), the observed data were arbitrarily divided into three levels of evaporative demand rate (Denmead and Shaw, 1962). For this purpose, evaporative demand is defined as the measured pan evaporation, E_p, for a single day or the average value over a given time interval. The three categories used were as follows:

(1) High demand rate; E_p > 0.762 cm (0.30 in) per day,

(2) Moderate demand rate; $0.584 \leq E_p \leq 0.762$ cm per day, (0.23 to 0.30 in per day),

(3) Low demand rate; E_p > 0.584 cm (0.23 in) per day.

Nine years of data were used for this analysis. Knowing values of K_c (Eq. 4), K_w values were then calculated for each demand rate level by Eq. (3). For each K_w value, the ratio of the total profile soil moisture content to the profile moisture content at field capacity (this ratio is designated as PM) was determined.

Figure 2 is a plot of K_w against PM for the three evaporative demand categories. For each category, the data points were fitted to both polynomial and linear equations using regression analysis procedure. The linear relationships were simpler and statistically almost as good as the polynomial equation and were adopted for this study. The general form of this relationship is:

$$K_w = a + b(PM) \tag{5}$$

where "a" and "b" are constants (Table 1) and $0.0 \leq K_w \leq 1.0$.

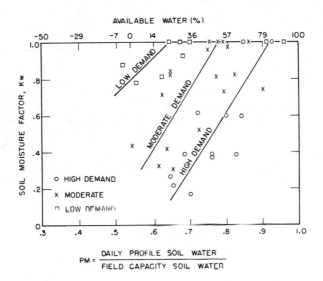

Fig. 2.   Soil Moisture Factor, K_w, as a Function of Soil-Water Content and Evaporative Demand (Corn at Lamberton, MN).

Table 1.   Values of the Constants a and b in Equation (5) for Corn Based on Data from Lamberton, MN.

Evaporative Demand Category	a	b	Number of Data Points	R^2
High	-1.876	3.15	12	0.74
Moderate	-1.530	3.28	17	0.63
Low	-0.256	1.98	9	0.76

The results shown in Fig. 2 are similar to those obtained by Denmead and Shaw (1962). In this study, 0.23 in per day was used as the upper limit of low demand to more equally divide the data points, whereas Denmead and Shaw used 0.20 in. per day.

ET EXTRACTION PATTERN

The ET extraction pattern was determined from the analysis of the data on soil moisture depletion at various depths throughout the growing season. To determine the soil depth contributing moisture for ET, the ratio of moisture extracted from each depth increment (ET_z) to the total ET (extraction ratio) was calculated over the entire root zone at various growth stages. Average values of the extraction ratio (Table 2) show that moisture extraction extends deeper into the root zone as the growing season progresses, as expected.

The soil depth contributing ET moisture at a given growth stage (Table 2) determines the number of soil zones being utilized. Therefore, the analysis to develop Table 2 was carried out using soil moisture data over only those time intervals when the total soil moisture within the root zone was considered non-limiting (i.e., K_w = 1.0). The soil moisture loss from each soil zone is the fraction of the total ET extracted from the soil profile over the time interval analyzed at the prevailing growth stage (Table 2). Full details on the determination of ET extraction pattern as well as the whole ET relationship has been given by Idike (1981).

Table 2 is similar to one developed by Saxton et al. (1974), using data presented by several previous investigators. Although the latter was in terms of calendar days, the two tables differ by small amounts, usually less than 0.05.

Table 2.   Fractions Extracted from Various Soil Zones During the Growth Period for Corn at Lamberton, Minnesota. (DAE - days after emergence).

	FRACTION EXTRACTED:						
Soil Layer (cm)	DAE 1-10	DAE 11-20	DAE 21-30	DAE 31-40	DAE 41-55	DAE 56-70	DAE > 70
0-15	1.00	.74	.43	.37	.23	.26	.23
15-30		.26	.29	.24	.21	.21	.14
30-46			.28	.25	.21	.13	.10
46-61				.14	.18	.12	.09
61-91					.17	.17	.15
91-122						.11	.16
122-152							.13

THE PROPOSED SOIL-WATER MODEL

The model developed in this study consists of three major components:

(a)   The infiltration and surface storage submodel which determines the amount of recharge entering the soil profile.

(b)   The redistribution component which redistributes the recharge within the soil profile.

(c)   The evapotranspiration routine which extracts the evapotranspiration requirement from the soil profile.

The 152 cm (6 feet) soil profile is divided into seven soil zones (Table 2), with each zone assumed homogeneous. The underlying subsoil (beyond 152 cm) is assumed to be at a constant moisture content throughout the growing season. Input data to the model include climatological data (hourly precipitation and daily pan evaporation), soil hydraulic characteristics and soil moisture data.

The infiltration process is modeled using the Green and Ampt approach as modified by Mein and Larson (1971) (the GAML model). Eq. (6) gives the depth of infiltration, F_s, at the time of surface ponding. Eq. (7), the integrated form of the GAML model, gives the cumulative infiltration, F, to any time, t, after surface ponding.

$$F_s = \frac{S_{av}(IMD)}{I/K_s - 1} \tag{6}$$

$$K_s(t - t_p + t_p') = F - (IMD)S_{av}\log_e\left(\frac{F}{(IMD)S_{av}} + 1\right) \tag{7}$$

where

I = rainfall rate (depth/time)

t_p = time to surface ponding (time)

t_p' = the equivalent time to infiltrate F_p under initially ponded conditions. t_p' is calculated by the original Green and Ampt equation, or by Eq. (7) with $t_p = t_p' = 0.0$.

IMD = initial moisture deficit in the topsoil zone (vol./vol.)

S_{av} = average capillary suction at the wetting front (length)

K_s = hydraulic conductivity in the topsoil zone at field saturation (depth/time)

Field saturation is estimated from the fillable porosity (Slack, 1980) which, for an agricultural soil, was estimated as 90 percent of the volumetric moisture content (Moore et al, 1981). The average capillary suction at the wetting front, S_{av}, was determined using the method of Clapp and Hornberger (1978) and Brakensiek (1979). The model was used with a one-hour time step to model infiltration during a rainfall event and a ten-minute time step to model infiltration from surface storage after the rainfall event ceases. Infiltration occurs only into the topsoil zone from which further soil moisture movement occurs by redistribution.

Interception values were estimated by considering both the rainfall amount and the vegetative growth stage of the crop using information contained in the literature (Shanholtz and Lillard, 1970; Moore and Larson, 1979; Saxton et al., 1974). The initial available surface storage resulting from conventional tillage was estimated from the results of the study of Allmaras et al. (1967) while the initial available storage provided by basin tillage treatment was assumed to be 7.6 cm (3 in) based on the initial results of the study by Kolstad (1981). The change in the surface depression storage resulting from precipitation impact effects is represented and evaluated using the procedure and relationships developed by Mitchell and Jones (1976, 1978).

The redistribution routine calculates vertical soil moisture movement (upward or downward) using the one dimensional Darcy's equation for unsaturated flow. Hydraulic conductivities are determined using the method of Campbell (1974). Redistribution starts from the plow layer (topsoil zone) and continues down to the seventh and bottom soil zone. Movement of soil moisture below the seventh soil zone is considered as deep percolation. Calculations are made hourly on days with rain and every two hours on rainless days and are based on the soil moisture at the beginning of each time increment.

The evapotranspiration routines uses the ET relationships described earlier. The submodel operates on a daily basis and identifies or determines:

(a) The type of evaporative demand day from the pan evaporation for each day,

(b) The crop growth stage (DAE) and

(c) The total profile soil moisture content.

With this information, the actual ET for the day is computed using Eqs. (3), (4), and (5). The computed ET is extracted from the soil profile; its proportions from the various contributing soil levels being determined from Table 2. The estimation of direct soil surface evaporation as well as further details on all aspects of the soil-water model and its development was presented in considerably more detail by Idike (1981).

MODEL OPERATION AND TESTS

In general, the model operates on a daily basis with inputs of daily measurements (pan evaporation and hourly precipitation). Initial conditions consist of observed soil moisture contents for the various soil zones at the beginning of the growing season.

The model was tested against observed moisture contents in various soil layers at the Minnesota location of Lamberton for a period of seven years, not including the years used in developing the model or evaluating parameters. Plots of predicted and observed soil-water values for three years at all depths (Figs. 3 and 4) show good agreement most of the time, with the exception of 1979. Trends in soil-water content are satisfactorily predicted by the model, but short-term fluctuations were sometimes missed, especially in the lower zones. For the remaining four years (not shown), similar results were obtained, with relatively poor predictions in one year (1972).

Table 3.   Root Mean Square Errors and Differences in Predicted and Observed Soil Moisture Content Values (vol./vol.), Lamberton, Minnesota (7-year mean values).

Soil Zone	Mean Square Error of Prediction	Mean Albegraic Error of Predicted Values (Predicted - Observed)	Base Line Root Mean Square Error
1	0.044	-0.012	0.056
2	0.064	-0.050	0.066
3	0.054	-0.035	0.070
4	0.046	-0.028	0.065
5	0.036	-0.018	0.055
6	0.036	-0.015	0.040
7	0.029	-0.001	0.034
Pooled Data	0.045	-0.023	0.057

For the entire seven years of model tests, mean square errors of prediction were calculated by soil zones (Table 3). Overall, the mean error was

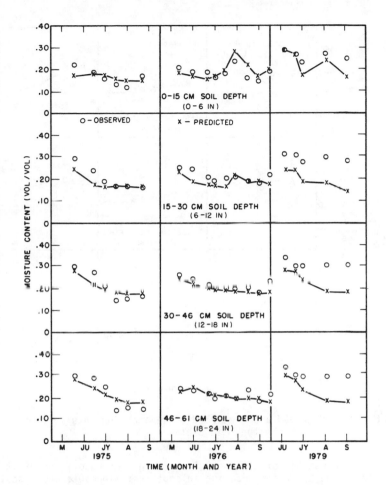

Fig. 3.   Observed and Predicted Soil-Water Contents for Soil Zones
            1-4, Lamberton, Minnesota.

0.045 (vol./vol.). The predictions are seen to be least accurate in Zones
2 and 3 and most accurate in Zones 5 to 7. At Lamberton, Minnesota, the
model underpredicted soil-water contents a small amount (on the average
0.023), as indicated by the mean algebraic errors in Table 3.

    These errors seem to be reasonable and acceptable. Thus, the authors
believe that the model predicts soil-water content with sufficient accuracy
for the main objective of this study, i.e. for comparing soil-water contents
during the growing season for various surface configurations.

PROJECTIONS

    Two variations of basin tillage (contour ridging), in addition to con-
ventional tillage were incorporated into the model to make predictions of
their effects on soil moisture levels under corn. The first is with the
ridges and basins in place during the growing season only and the other is
with the basins in place all year-round. In both cases, the basins were

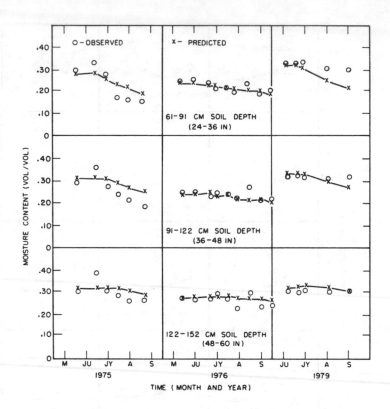

Fig. 4. Observed and Predicted Soil-Water Contents for Soil Zones
5-7, Lamberton, Minnesota.

assumed to be capable of storing 7.6 cm (3 in.) of runoff when initially
installed. The Lamberton location in southwest Minnesota and a south-cen-
tral Texas location of Del Rio were used for these projections. The Del
Rio location [annual rainfall of 470 mm (18.5 in.)] was selected because it
represents a semiarid climate with no period of snowpack or frozen soil,
in contrast to the Lamberton location.

The results of the projections at both locations showed that both forms
of basin tillage had higher profile soil moisture levels than the conventional
tillage system. However, with the basins in place during the growing season
only, the benefit appears to be too small to justify the cost and effort
involved in installing and maintaining the basins. The benefit provided
by the year-round basin tillage system is more evident if the comparison
is made in terms of crop stress. To quantify the effect of reduced soil
moisture on the growing crop, "daily stress" was arbitrarily defined for
this study as 10 (1.0 - K_w). This assumes that crop stress is linearly
related to the factor K_w. Thus, a day with K_w = 0.90 produces one stress
unit.

Table 4 gives the average annual stress units for the three tillage
treatments and the percentage reduction in stress units for basin tillage.
At Lamberton, the average reduction was small (4%) for basins during the
growing season only but substantial (50%) for all-year basins. The average
reduction at the Texas location was also good (41%).

Table 4.   Average Effect of Basin Tillage on Soil Moisture Content Under Corn, as Indicated by Stress Units[1] at Lamberton and Del Rio.

Stress Units for:	Lamberton, MN 7-yr. average[2]	Del Rio 3-yr. average[3]
No basins (conventional)	387	802
Basins during growing season only	380	---
Basins all year	214	465
Percent Reduction[4] over Conventional Tillage for:		
Basins during growing season only	3.7	---
Basins all year	49.9	40.8

1/ See text for definition of stress units.
2/ Years are 1966, 1968, 1970, 1972, 1975, 1976, and 1979.
3/ Years are 1976, 1977, 1978, and 1979.
4/ Calculated as mean value of percent reduction for individual years.

DISCUSSION

The results of the projections (Table 4) indicate that basin tillage has considerable potential for reducing moisture stress but only if the basins are in place throughout the year. This leads to the conclusion that the storage and infiltration of off-season precipitation (snowmelt in Minnesota) provides most of the benefit from basin tillage.

Since little benefit interms of soil moisture was provided by basin tillage during the growing season, this suggests that the form and storage capacity of the basins are not as important during the growing season as during the off-season. This point is significant because it is difficult to maintain ridges at full height at all times, especially in the period following planting. With existing machinery, the ridges are partially broken down during planting and cannot be restored to full height until the corn has grown to a height of about 16 inches, roughly six weeks later.

Where soil erosion is a problem, basin tillage would undoubtedly reduce erosion by a high percentage providing an added benefit. In this respect, the period immediately following planting usually is critical.

The question of model validity and accuracy deserves further discussion. Figs. 3 and 4 show that the annual variation at Lamberton in soil-water content at a given depth is relative small, in most years less than 0.15 (vol./ vol.). Since the model predictions are often on the order of half this amount and sometimes higher, it is appropriate to ask whether the mean of the observed values would serve just as well as the model in predicting soil-water content. Table 3 (last column) shows that this is not the case since the mean square error of the model values is smaller than the base line root mean square error for each soil zone, which is based on observed soil-water contents and their seasonal averages.

The soil-water contents obtained from the projections with basin tillage are, of course, unverified. Their reliability can be judged only in terms of the nature of the model and its accuracy under the conditions in which it was tested (discussed earlier). Since the model is physically based, there is no reason to believe that the change in surface configuration would significantly affect the model parameters controlling ET and water movement with the soil.

The model described here was developed for a particular purpose. It might prove useful for other purposes, depending on the application. The model was developed and tested using data for one soil type, one crop and one location. Application elsewhere would require use of the appropriate input (precipitation and pan evaporation) and evaluation of the model parameters to suit local conditions. Soil parameters affecting infiltration and redistribuiton can be measured. The main difficulty would be in evaluating the parameter values needed for the ET submodel, but this applies to any ET model.

Comparison of the model results to those for other models has not been made. More accurate predictions are no doubt possible by using an ET submodel based on the energy balance approach, e.g. Saxton et al. (1974). However, this requires meteorological data that are less commonly available than evaporation pan data.

CONCLUSIONS

A soil-water model was developed and tested against observed soil moisture at a location in southwest Minnesota. The model is physically based and utilizes evaporation pan data as input. On the average, the model predicted soil moisture contents within ± 0.05 (vol./vol.) with corn as the crop. In addition to conventional tillage, two variations of basin tillage were used with the model to predict their effects on soil moisture levels under corn at two locations, the Minnesota location and one in south central Texas.

The main conclusions that can be drawn from the results of this study are:

(1) The model which was developed satisfactorily predicts soil moisture levels throughout the growing season for corn and provides a basis for comparing the effects of surface configuration on soil-water content.

(2) The model projections indicate that basin tillage with the basins in place during the growing season only does not increase profile soil moisture significantly.

(3) At both locations, the model results show that having the basins in place all year increases soil moisture levels and decreases crop stress significantly. It follows that most of the benefit is attributable to retention of off-season precipitation.

ACKNOWLEDGEMENTS

For the field data used in this study, the authors are indebted to the Southwest Agricultural Experiment Station at Lamberton, Minnesota, and to the Soil Science Department, both of the University of Minnesota. Special thanks are expressed to Dr. Wallace W. Nelson, Superintendent of the Station and to Dr. Donald G. Baker, Professor of Soil Science for their kind assistance.

The study was partially supported by funds from the Agricultural Experiment Station, University of Minnesota and assisted by a computer grant from the University of Minnesota Computer Center.

This paper has been approved for publication as Scientific Journal Series Paper No. 11, 587 of the Agricultural Experiment Station, University of Minnesota, St. Paul, Minnesota 55108.

REFERENCES

1. Allmaras, R. R., R. E. Burwell and R. F. Holt. 1967. Plow-layer porosity and surface roughness from tillage as affected by initial porosity and soil moisture at tillage time. Soil Science Society of America Proceedings, Vol. 31, No. 4, pp. 550-556.

2. Bilbro, J. D. and E. B. Hudspeth, Jr. 1977. Furrow diking to prevent runoff and increase yields of cotton. The Texas Agricultural Experiment Station Publication PR-3435, 3 p.

3. Brankensiek, D. I. 1979. Comments on empirical equations for some soil hydraulic properties by Roger B. Clapp and George M. Hornberger. Water Resources Research, Vol. 15, pp. 989-990.

4. Campbell, G. S. 1974. A simple method for determining unsaturated hydraulic conductivity from moisture retention data. Soil Science, Vol. 117, No. 6, pp. 311-314.

5. Clapp, R. B. and G. M. Hornberger. 1978. Empirical equations for some soil hydraulic properties. Water Resources Research, Vol. 14, pp. 601-604.

6. Denmead, O. T. and R. H. Shaw. 1962. Availability of soil-water to plants as affected by soil moisture content and meteorological conditions. Agronomy Journal, Vol. 54, pp. 385-390.

7. Denmead, O. T. and R. H. Shaw. 1959. Evapotranspiration in relation to the development of the corn crop. Agronomy Journal, Vol. 51, pp. 725-726.

8. Holmes, R. M. and G. W. Robertson. 1963. Application of the relationship between actual and potential evapotranspiration in dryland agriculture. Transactions, American Society of Agricultural Engineers, Vol. 6, pp. 65-67.

9. Hudspeth, Jr., E. B. 1978. Basin tillage for water conservation and maximum dryland cotton production. Presented at the 1978 Beltwide Cotton Production and Research Conferences, Dallas Texas, January 9-11, 1 p.

10. Idike, F. I. 1981. Modeling the effect of conservation practices on

soil moisture. Ph.D. thesis, (unpublished). Graduate School University of Minnesota, St. Paul, 176 p.

11. Kolstad, O. C. 1981. An evaluation of ridge forming tools for reduced tillage. M.S. thesis, (unpublished). Graduate School, University of Minnesota, St. Paul, 78 p.

12. Mein, R. G. and C. L. Larson. 1971. Modeling the infiltration component of the rainfall-runoff process. Bulletin 43, Water Resources Research Center, University of Minnesota, St. Paul, 72 p.

13. Mitchell, J. K. and B. A. Jones, Jr. 1976. Micro-relief surface depression storage: Analysis of models to describe the depth-storage function. American Water Resources Association Bulletin, Vol. 12, pp. 1205-1222.

14. Mitchell, J. K. and B. A. Jones, Jr. 1978. Micro-relief surface depression storage: Changes during rainfall events and their application to rainfall-runoff models. American Water Resources Association Bulletin, Vol. 14, pp. 777-802.

15. Moore, I. D., C. L. Larson, D. C. Slack, B. N. Wilson, F. I. Idike and M. C. Hirschi. 1981. Modeling infiltration: a measurable parameter approach. Journal of Agricultural Engineering Research (England), Vol. 26, pp. 21-32.

16. Moore, I. D. and C. L. Larson. 1979. Effects of drainage projects on surface runoff from small depressional watersheds in the north central region. Bulletin 99, Water Resources Research Center, University of Minnesota, 122 p.

17. Richardson, C. W. and J. T. Ritchie. 1973. Soil-water balance for small watershed. Transactions, American Society of Agricultural Engineers, Vol. 16, pp. 72-77.

18. Saxton, K. E., H. P. Johnson and R. H. Shaw. 1974. Modeling evapotranspiration and soil moisture. Transactions, American Society of Agricultural Engineers, Vol. 17, pp. 673-677.

19. Shanholtz, V. O. and J. H. Lillard. 1970. A soil-water model for two contrasting tillage systems. Bulletin 38, Water Resources Research Center, Virginia Polytechnic Institute and State University, 217 p.

20. Slack, D. C. 1980. Modeling infiltration under moving sprinkler irrigation systems. Transactions, American Society of Agricultural Engineers, Vol. 23, pp. 596-600.

21. Stegman, E. C., A. Bauer, J. C. Zubriski and J. Bauder. 1977. Crop curves for water balance irrigation scheduling in southeast North Dakota. North Dakota Research Report No. 66, Agricultural Experiment Station, North Dakota State University, 11 p.

22. Van Bavel, C. H. M. 1953. A drought criterion and its application in evaluating drought incidence and hazard. Agronomy Journal, Vol. 45, pp. 167-172.

23. Van Bavel, C. H. M. 1956. Estimating soil moisture condition and time for irrigation with the evapotranspiration method. U. S. Department of Agriculture, Agricultural Research Service No. 41-11, 16 p.

UNSATURATED FLOW AND EVAPOTRANSPIRATION MODELING AS A COMPONENT OF THE EUROPEAN HYDROLOGIC SYSTEM (SHE)

Karsten Hogh Jensen
Danish Hydraulic Institute
Agern Alle 5, DK-2970 Horsholm, Denmark

Institute of Hydrodynamics and Hydraulic Engineering
Technical University of Denmark, DK-2800, Lyngby, Denmark

Torkil Jonch Clausen
Danish Hydraulic Institute

ABSTRACT

The European Hydrologic System (Système Hydrologique Européen - acronym SHE) is a general deterministic and distributed hydrologic modelling system currently under joint development by the Institute of Hydrology (U.K.), SOGREAH (France) and Danish Hydraulic Institute (Denmark). The SHE is physically based in the sense that it is derived directly from equations of flow and mass conservation for the hydrologic processes it aims to represent, and it is distributed by describing the catchment on a rectangular grid system. A brief description of the general structure of the SHE is given, and the need for physically based distributed modelling is discussed.

The simulation of the flow in the unsaturated zone is a crucial part of the SHE, since all the other components of the system depend on boundary data from the unsaturated flow component, in which also most of the computational time will be spent. The unsaturated flow description is based on a finite difference approximation of Richards' equation in one dimension, assuming that the flow is predominantly vertical. This paper presents simulation results from the unsaturated flow component of SHE coupled with the interception/evapotranspiration component. The processes of interception, evapotranspiration, root extraction and vertical unsaturated flow are simulated, and the results are verified against field measurements of soil moisture content and tension from a location in southern Denmark.

INTRODUCTION

The SHE has been developed from the non-linear partial differential equations of flow for the processes of overland and channel flow, unsaturated and saturated subsurface flow, solved by finite differences techniques. The system is completed by descriptions of snow melt, interception and evapotranspiration. While considerable effort has been made to ensure that the solutions in each component are accurate and efficient, the structure of the system represents a compromise between the restrictions of computing and data requirements on the one hand, and the aim of representing the complexity of real catchments on the other.

Since it is not yet economically viable to develop a system fully three-dimensional in space which allows the required accuracy of discretization in both horizontal and vertical planes, the SHE has been rationally simplified by assuming that in the unsaturated zone vertical flow is, on most slopes, far more important than lateral flow. This results in a model structure in which independent one-dimensional unsaturated flow components of variable depths are used to link a two-dimensional surface flow component with a two-dimensional groundwater flow component. The horizontal plane is divided into rectangular grid squares, and an independent discretized unsaturated component is associated with each grid square. It is planned to use up to 2000 grid points in the horizontal and up to 30 in the vertical.

The model structure is illustrated in Figure 1.

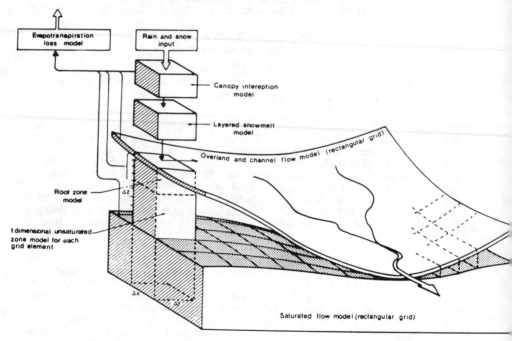

Fig. 1 Structure of the European Hydrologic System.

SHE is structured with a Frame, which coordinates and controls the operation of the individual components. This feature in combination with the modular form of structure of the SHE provides a considerable flexibility by allowing the components to assume different levels of complexity or be omitted in any given application, depending on the purpose of the application and the availability of data. Further, the individual components are structured as separate units which can be used independently, as for example the unsaturated flow component described later. Since SHE is intended as a general hydrologic modelling system, this flexibility is of vital importance.

THE NEED FOR DISTRIBUTED MODELS LIKE SHE

For many problems distributed and physically based hydrologic models like SHE have advantages over traditional hydrologic simulation models. Obviously, compared to simpler models, complex modelling systems like SHE are more demanding with respect to data requirements and computer time, and hence manpower and financial resources. Thus in applications such as extension of streamflow records in well gauged catchments, simple parametric models can provide hydrologic predictions of satisfactory accuracy at less expense. However, when one considers the prediction of the consequences of human interference in a catchment, this can only be fulfilled with distributed hydrologic models.

Most of the parameters in SHE have a direct physical interpretation, and their range can be established reasonably well on the basis of field- and laboratory investigations. Human activities in a watershed, as for example forest clearance or irrigation schemes, can be related directly to changes in physical catchment characteristics at certain locations. The parameters in SHE can be adjusted accordingly, and hence it will be possible to predict the effects of catchment changes prior to data becoming available. Further, by being able to establish reasonable ranges for most of the parameters on the basis of short term intensive field investigations, SHE can be used to generate at least approximate hydrologic predictions without the benefit of long simultaneous records of precipitation and streamflow for calibration. In fact, practically any hydrologic information can be utilized in the system.

Moreover, when the data base of a project is inadequate for accurate prediction, SHE can be used to indicate the range of possible outcomes of the project and hence the uncertainty of outcome, if the project is constructed on inadequate data base.

The capacity of SHE to account for spatial variations in meteorologic and hydrologic inputs represents an important advantage over traditional lumped catchment models. Particular examples are the movement of rainstorms over a catchment and localized river and groundwater abstractions and recharge.

INTRODUCTION TO COMPONENTS IN SHE

In the following, the components in SHE will briefly be described. Since the field test, which is described later, is a point simulation carried out with the interception/evapotranspiration and unsaturated flow components alone, these are discussed in more detail.

Overland-Channel Flow Component

The overland-channel flow component represents the surface runoff and river flow processes. They are modelled by finite difference solutions to the simplified Saint Venant equations for open channel flow (inertia terms neglected). In the modelling of overland flow the slope of the water surface is assumed to be parallel to the ground slope (kinematic wave assumption), but for channel flow, a water surface slope term is included in the mathematical formulation so that backwater effects can be modelled.

An equivalent roughness coefficient is used to characterize resistance to overland flow, while depth-dependent flow resistance functions are required for the river flow description.

Groundwater Flow Component

The groundwater flow component is, in the first version of the SHE model, restricted to a single layer, unconfined aquifer with direct links to any surface water bodies. The model is based on an alternating direction implicit finite difference solution to the non-linear Boussinesq equation. This formulation restricts application of the model at present, but it is envisaged that this component will be extended to cope with multilayered, confined/unconfined aquifer systems in the future.

Interception Component

The interception process is modelled in the SHE by using a variant of the Rutter model (Rutter et. al., 1975), which is essentially an accounting procedure for canopy storage. From the canopy storage, which can vary during the growing season, e.g. as a function of the leaf area index, the intercepted water may either evaporate directly or drain to the soil surface according to a prescribed function. Consideration of interception loss is particularly important when dealing with forest areas. For a grass vegetation, as in this field study, interception calculations are of minor importance.

Evapotranspiration Component

The evapotranspiration component determines the total evapotranspiration in each grid square of the catchment. The component interacts directly with the root zone, which is the upper part of the unsaturated zone component.

Transpiration

The Penman-Monteith equation (Monteith, 1965), constitutes the basis for the transpiration calculations:

$$E = f \cdot \frac{\Delta R_{NC} + \rho \cdot c_p \cdot \delta e / r_a}{\lambda (\Delta + \gamma (1 + r_s / r_a))} \tag{1}$$

E = transpiration under prevailing meteorological and physiological conditions (mm/time interval)

f = unit conversion factor

R_{NC} = net radiation absorbed by the plant (W/m^2)

ρ = air density (kg/m^3)

c_p = specific heat capacity (J/kg/^{o}C)

δe = vapour pressure deficit (mb)

r_a = aerodynamic resistance (s/m)

r_s = surface resistance (s/m)

λ = latent heat of water (J/kg)

γ = psychrometer constant (mb/^{o}C)

Δ = slope of saturation vapour pressure curve (mb/^{o}C)

Standard meteorological observations are used, measured at a specific height z above ground level.

The external or aerodynamic resistance between the crop surface and the air at height z can be calculated from a relationship, which assumes that effects of deviations from neutral atmospheric conditions are either compensating or negligible.

The surface resistance or canopy resistance r_s can be considered mostly as a physiological parameter, representing the resistance to vapour flow through the stomata. r_s is likely to be determined by a complex function of both past and present environmental variables, including soil moisture deficit in the root zone, light intensity and atmospheric vapour pressure deficit, and it will also vary with vegetation type.

When the evaporation from the leaves exceeds the capability of the soil-root system to supply water to the leaves, the stomata begin to close (increasing surface resistance r_s), and the transpiration is thereby reduced. By specifying r_s as a function of environmental variables, the actual transpiration rate can be predicted by the Penman-Monteith equation. This is one of the possible modes in the SHE evapotranspiration model. However, no well established functional relationship for the prediction of r_s can be found in the literature.

This leads to the other mode in SHE. The potential transpiration is calculated from Penman-Monteith's equation by setting r_s to some minimum value r_{sm}, which is supposed to exist when neither water nor light limits stomatal opening (Federer, 1975). Actual transpiration is then calculated according to a specified relationship between relative transpiration and soil moisture content in the root zone (equation 2). This is a relationship adapted from Kristensen and Jensen (1975).

$$ EAT/EPT = 1 - (\frac{\theta_{FC} - \theta}{\theta_{FC} - \theta_{WP}})^{\frac{C_3}{EPT}} \qquad (2) $$

EAT = actual transpiration rate

EPT = potential transpiration rate

θ = actual soil moisture content

θ_{FC} = soil moisture content at field capacity

θ_{WP} = soil moisture content at permanent wilting point

C_3 = parameter

A qualitative impression of this function is given in Figure 2. When θ is larger than θ_{FC}, EAT/EPT is set to 1.

The function (2) is applied at all computational points over the root zone depth. These "actual" values are subsequently multiplied by a root distribution or transpiration distribution function to account for the extraction pattern of the plant roots.

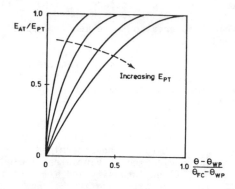

Fig. 2   Relative transpiration as a function of soil moisture content
and potential transpiration.

Soil evaporation

The potential soil evaporation, EPS, is calculated separately after the
formula suggested by Ritchie (1972), which assumes that the term in (1)
containing the vapour pressure deficit and aerodynamic resistance is negligible.

$$EPS = \frac{\Delta}{\lambda(\Delta + \gamma)} (R_{NS} - Q) \qquad (3)$$

where R_{NS} (W/m^2) is the net radiation at the soil surface and Q (W/m^2) is
the soil heat flux. Net radiation at the soil surface is calculated according to Beer's law:

$$R_{NS} = R_N \exp(- 0.4 \text{ LAI}) \qquad (4)$$

where R_N (w/m^2) is the net radiation measured at standard height and LAI is
the leaf area index. The remaining part of the net radiation is absorbed by the
plant and used for evaporation of intercepted water and/or transpiration.

When no vegetation is present, i.e. bare soil, equation (3) is not applicable. Instead the full Penman-Monteith equation is used for predicting the potential evaporation from the bare soil. The surface resistance
r_S is set to zero and an appropriate roughness length, characterizing the
soil surface, is selected.

With decreasing soil moisture content the soil evaporation is reduced in
a manner as illustrated in Figure 3. The constant level C_2 describes the
process of vapour diffusion between the moist soil atmosphere and the
generally drier atmosphere above the soil. Also this principle of reduction is adapted from Kristensen and Jensen (1975).

The evaporation zone is assumed only to contain the upper computational
point, and the actual evaporation is determined by the moisture content
in that point according to the function depicted in Figure 3.

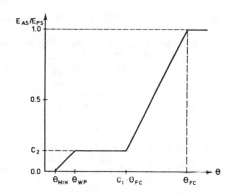

Figure 3 Relative soil evaporation versus soil moisture content
EAS = actual soil evaporation
EPS = potential soil evaporation

θ_{FC} = soil moisture content at field capacity

θ_{WP} = soil moisture content at permanent wilting point

θ_{MIN} = irreducible soil moisture content

C_1, C_2 = parameters

Interception

The potential interception loss rate EPI is calculated from (1) with $r_s = 0$ and otherwise the same data and parameters as described above. The interception evaporation requirement is met first, and the remaining evaporation capability is passed on for transpiration.

Unsaturated Flow

Unsaturated vertical flow is in SHE described by the capillary pressure version of Richard's equation.

$$C(\psi) \frac{\partial \psi}{\partial t} = \frac{\partial}{\partial z} \left(K(\psi) \frac{\partial \psi}{\partial z} \right) + \frac{\partial K(\psi)}{\partial z} - S(z) \qquad (5)$$

ψ = capillary pressure (the absolute value of capillary pressure will here be termed tension)
C = water capacity as a function of ψ, $C = d\theta/d\psi$

t = time

z = vertical coordinate, positive downwards

K = conductivity as a function of ψ

S = sink term

Hysteresis and effects from compressed air are not considered.

INPUT/OUTPUT DESCRIPTION OF THE SHE-COMPONENTS

The application of a distributed hydrologic model like SHE requires considerable inputs of parametric and exogenous data, including parameter values that change over time (e.g. crop parameters). Such data will not always be readily available and therefore options have been built into the system to allow components to degenerate to simpler modes of operation and thereby reducing the data requirements. It is stressed that parameter values are in principle measurable in the field and it is hoped that a general availability of models like SHE, which are able to utilize almost any hydrological information, will instigate more widespread measurements of the data required, if not on a routine basis, then at least as part of the application of the model to specific projects.

Parametric values and data input to the model can vary from grid square to grid square or from point to point in the vertical. However, in most cases the same input data and parametric values will be associated with an assembling of f.ex. grid squares. Both input data and parametric values are assumed to be valid over the entire area associated with a grid point. An option for variable grid spacing both in the horizontal and vertical direction is included in the system. Thus a refined grid may be introduced around rivers, pumping fields and other such areas, characterized by a significant variation in natural or man-influenced hydrologic processes.

In the following a description of the required input of data and parametric values for each component is given. The description is given according to the most complex form of the SHE system. The components interact with each other through a series of internal boundary conditions, either as flow or pressure conditions, which appear as a result of progress in time in the other components. No iteration between the components are performed but they are run in parallel. This means that time steps are taken which are based on "old" information.

Interception Component

Data input	1) rainfall rate
Model parameters	1) leaf are index
	2) drainage parameters in the Rutter model
Input to other components	1) net rainfall rate (unsaturated zone component or overland/channel flow component)

Evapotranspiration Component

Data input	1) meteorological data
Model parameters	1) aerodynamic resistance
	2) minimum surface resistance
	3) leaf area index
	4) parameters in relationships defining the ratio between actual and potential rates of transpiration and soil evaporation (defined in terms of soil moisture levels)
Input to other components	1) transpiration/soil evaporation rate

(unsaturated zone component)

2) evaporation from water surfaces
(overland/channel flow component)

Unsaturated Zone Component

Model parameters

1) soil moisture pressure - moisture content relationship for each soil layer

2) unsaturated hydraulic conductivity as a function of water content for each soil layer

3) distribution function of transpiration over root zone

Input to other components

1) infiltration (overland/channel flow component)

2) recharge/discharge (saturated zone component)

3) soil moisture levels in the root zone (evapotranspiration component)

Overland-Channel Flow Component

Data input

1) specified flows or water levels at the boundaries

2) man-controlled diversions and discharges

3) topography of overland flow plane and channels

Model parameters

1) manning roughness for overland flow

2) flow resistance function for river flow

3) coefficients of discharge for weir formulas

Input to other components

1) depth of overland flow (unsaturated flow component)

2) surface levels of rivers (saturated zone component, possibly unsaturated zone component if an unsaturated zone develops below the river)

Saturated Zone Component

Data input

1) topography of impervious bed

2) specified groundwater flows or potentials at the boundaries

3) aquifer management (pumping, artificial recharge etc.)

Model Parameters

1) porosities

2) saturated hydraulic conductivities

Input to other components

1) position of phreatic surface (unsaturated

243

zone component, possibly overland/channel flow component)

2) stream/aquifer interaction (overland/channel flow component)

NUMERICAL FORMULATION

In SHE all the governing partial differential equations of flow in different domains are solved by finite difference methods. In this context only the numerical approximation of the unsaturated flow equation will be presented.

The difference scheme, which is implicit and solved in an iterative procedure, reads:

$$\tilde{C}_j^{n+\frac{1}{2}} \frac{\psi_j^{n+1}-\psi_j^n}{\Delta t} = -\frac{1}{\Delta z}[(-\frac{\tilde{K}_{j+1}^{n+\frac{1}{2}}+\tilde{K}_j^{n+\frac{1}{2}}}{2} \cdot \frac{\psi_{j+1}^{n+1}-\psi_j^{n+1}}{\Delta z} + \frac{\tilde{K}_{j+1}^{n+\frac{1}{2}}+\tilde{K}_j^{n+\frac{1}{2}}}{2})$$

$$- (-\frac{\tilde{K}_j^{n+\frac{1}{2}}+\tilde{K}_{j-1}^{n+\frac{1}{2}}}{2} \cdot \frac{\psi_j^{n+1}-\psi_{j-1}^{n+1}}{\Delta z} + \frac{\tilde{K}_j^{n+\frac{1}{2}}+\tilde{K}_{j-1}^{n+\frac{1}{2}}}{2})] - \frac{S_j^{n+1}}{\Delta z} \qquad (6)$$

j = address in z

n = address in time t

Δz = distance increment

Δt = time increment

$\tilde{C}_j^{n+\frac{1}{2}}$, $\tilde{K}_j^{n+\frac{1}{2}}$ = capacity and conductivity parameters as explained below

Because of the non-linear coefficients C and K serious convergence and stability problems have been experienced. However, one method that seems to work quite satisfactorily with respect to these problems is to keep a running integration of the coefficients over the iterations carried out and average these values and the coefficients from the previous time step. This means that the coefficients will be almost time-centered. On the other hand the weight has to be put forward on the differential terms, for correct information transfer. The scheme is solved in a double sweep procedure. The evapotranspiration component is integrated with the unsaturated flow component in the root zone, where the estimated transpiration is withdrawn as root extraction from each computational point in the root zone at a rate S. Soil evaporation is treated as a flux component across the soil surface. The various water flow components in a single integrated evapotranspiration/interception and unsaturated flow model as described in the SHE system is illustrated in Figure 4.

The upper boundary condition for the numerical solution of Richards' equation is provided by the interception calculations in the form of an estimated flux across the soil surface (possibly none). The flux is the resultant of net precipitation (precipitation minus interception loss) and soil evaporation. If the incoming flux is greater than the infiltration capacity of the soil, surface runoff is generated. Surface storage

244

is not considered directly, only through the specification of the topography the overland flow plane and the equivalent roughness coefficients.
The lower boundary condition is defined by the level of the phreatic
surface.

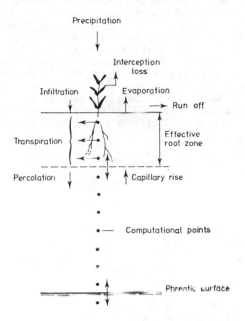

Fig. 4   Vertical flow components in the evapotranspiration/interception
and unsaturated flow models.

FIELD TEST

The model of the interception/evapotranspiration and unsaturated flow is
tested on data from an agricultural experimental station in the southern
part of Denmark, St. Jyndevad. The station is situated in an outwash
plain shaped during the latest glacial age, and the soil type can be
characterized as sandy.

A column simulation is carried out for the years 1974 and 1975. Distance
steps in the vertical are 10 cm adjacent to the soil surface and otherwise
20 cm. The distance to the phreatic surface varies between 2 and 4 metres.
The timestep for the numerical solution is 1 hour at the maximum. When
precipitation falls on an extremely dry soil, the timestep is reduced in
proportion to the intensity down to a minimum of 2-3 minutes.

Input data consist of meteorological data and precipitation data on an
hourly basis. The precipitation, which is recorded on an pluviograph, is
corrected for wind effects. (Allerup and Madsen, 1980). The level of the
phreatic surface is recorded at weekly intervals.

The field site is covered with short cut grass, and the model parameters
are adjusted accordingly. The depth of the root zone is assumed to be max.
70 cm and its variation over growing season is estimated from tensiometer

recordings. The variation in leaf area index is given a seasonal variation
with 5.0 as the maximum.

Two sets of retention curves are applied - one for the upper 0-50 cm and
another for the depths below. The retention curves are determined in the
laboratory as averages of several soil samples. The curves are modified
at small tensions, since the moisture content in field situations never
attains total porosity because of entrapped air (Fig. 5). The retention
curves are in the program represented by analytical relationships. Unsaturated conductivities are calculated from the retention curves following the method described by Kunze et.al. (1968). The calculated functions are adjusted by the use of a matching factor, defined as measured
to calculated conductivity at complete saturation. For computational
convenience the unsaturated conductivity functions are represented in
the simulation program by Averjanov's empirical formula (Averjanov,
1950).

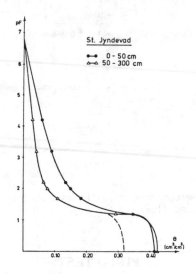

Figure 5   Retention curves.

Measurements of moisture content and capillary pressure (tension) on a
weekly basis in different depths are used as verification data.

The computer time required for a 2-year simulation run is a little less
than 2 CPU minutes on an IBM 3033 computer.

RESULTS AND DISCUSSION

A graphic representation of the input data, boundary conditions, verification data and simulation results are shown in Figures 6, 7, 8 and 9
on a daily basis for the period 1974-1975. Measurements are shown as
asterisks and simulation results are the lines.

In Fig. 6 various input data and calculated quantities are illustrated on
a daily basis. As can be seen, both leaf area index and root zone depth
are given a generalized seasonal variation with maximums of 5.0 and 70 cm

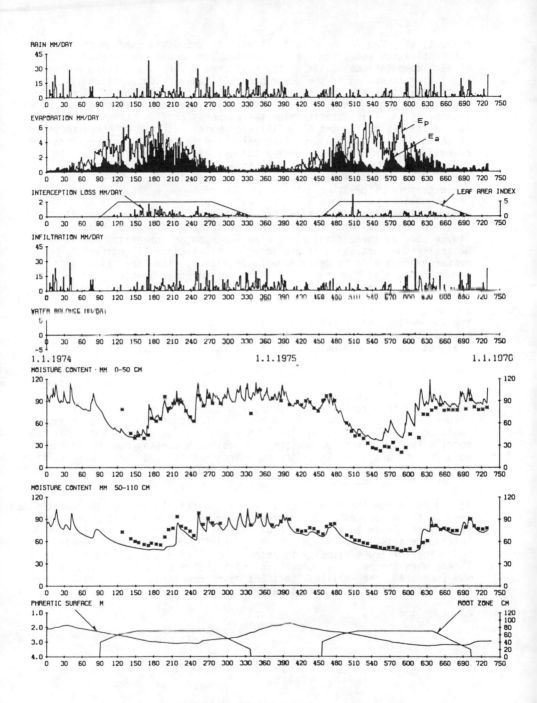

Fig. 6   Input data, boundary conditions and comparison of simulated and
         measured moisture contents integrated over the depths 0 to 50 cm
         and 50 to 100 cm. * measured ——— calculated.

respectively. The depth of root activity is the depth over which the evapotranspiration model and the unsaturated zone model are coupled, since
at each computational point in the root zone water is withdrawn as root
extraction according to the described relationship. Computationally the
root extractions are treated as sink terms in the solution of Richards'
equation. Integration of the sink terms over the root zone gives the total
transpiration. This quantity plus the evaporation flux across the soil surface describes the amount of infiltrated water returned to the atmosphere.
If the simulated and measured time variation in soil moisture content are
coinciding to an acceptable degree through the soil profile, then the
division of the infiltrated water into evapotranspiration and deeper
percolation as well as the transport of percolating water presumably
are described properly.

Fig. 7 illustrates a comparison between predicted and measured soil
moisture contents for various depths for the 2-year period, and in
Fig. 6 some of these results are also shown on an integrated basis.
Generally one must say that predicted moisture contents and field
measurements are in close agreement. Inevitably, some deviations will
occur, because of the complexities of the processes involved and the
simplifications made. For instance are deviations evident in the depth
of 40 cm. However, this is due to the restriction of two soil types in
the vertical. Obviously, at this level the soil properties lie between
the two defined ones, but it is believed that the model performance, as
regards the prediction of actual evapotranspiration and deep percolation,
will not improve considerably, if an additional soil type is introduced.

Fig. 8 is a comparison between predicted and measured soil water tensions.
It should be pointed out that tensiometers only are applicable for tensions
up to about 8 metres. When the measurements are 8 metres or above, this is
indicated with arrows in the figure, and correspondingly the simulated
values are not shown when above 8 metres. The accordance in Fig. 8 is not
quite as satisfactory as in Fig. 7, and this is despite the fact that the
model actually solves for tension, and moisture content can be regarded
as a derived quantity. Since the moisture content is of most concern, this
discrepancy is not serious, and it can be explained by a slight uncertainty
in the retention curves.

Fig. 9 shows how the input from infiltration travels through the unsaturated
zone to the water table. During the summer periods no infiltration water
reaches the lower layers because of the deficit in the root zone. Deep percolation starts in August-September and continues until February-March. In the
beginning the front velocity is rather slow (5-7 metres per month), but
later, when the moisture content and with that the hydraulic conductivity
has increased, the infiltration front moves downwards with a smaller phase
shift with depth (10-15 metres per month).

ACKNOWLEDGEMENT

The Danish agricultural experimental stations in St. Jyndevad and Højer
have provided the data material for this study, which is greatly appreciated.

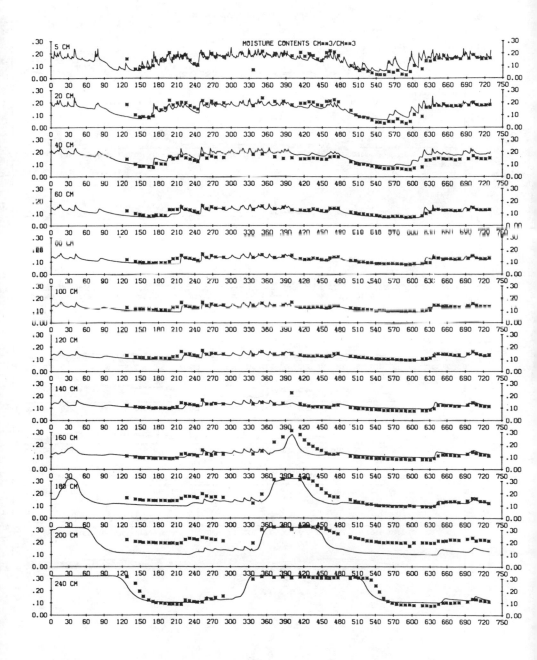

Fig. 7 Measured and calculated moisture contents.
    * measured   —— calculated.

249

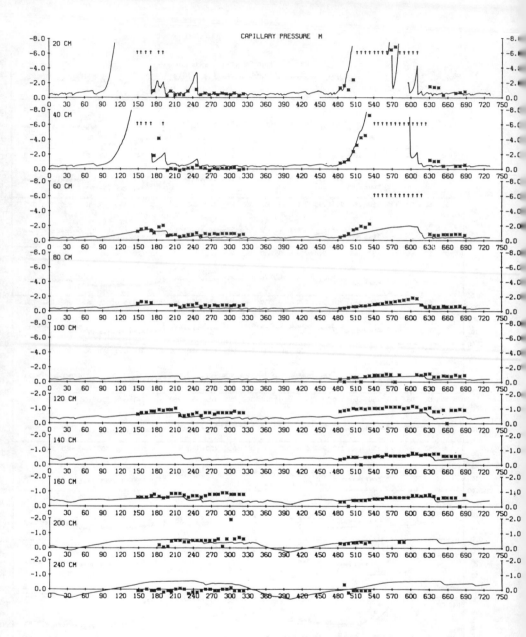

Fig. 8 Measured and calculated tensions.
* measured —— calculated.

250

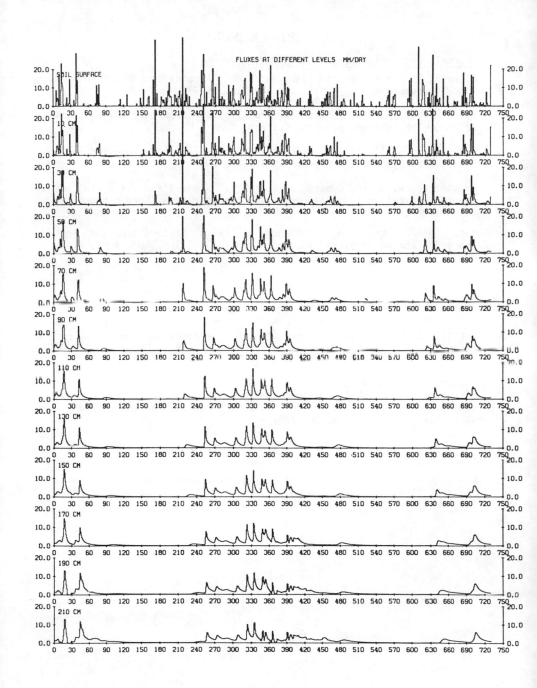

Figure 9 Calculated water fluxes at various levels.

251

REFERENCES

Journals and Periodicals

Allerup, P. and H. Madsen (1980). Accuracy of point precipitation measurements. Nordic Hydrology, vol. 11, no. 2, p. 57-70.

Averjanov, S. F. (1950). About permeability of subsurface soils in case of incomplete saturation. Eng. Collect. 7.

Federer, C. A. (1975). Evaportranspiration. Reviews of geophysics and space physics, vol. 13, no. 3, p. 442-445.

Kristensen, K. J. and Jensen, S. E. (1975). A model for estimating actual evapotranspiration from potential evapotranspiration. Nordic Hydrology, 6, p. 170-188.

Kunze, R. J., Uehara, G. and Graham, K. (1968). Factors important in the calculations of hydraulic condictivity. Soil Sci. Soc. Amer. Proc. 32, p. 760-765.

Molz, F. J. and Remson, I.(1970). Extraction term models of soil moisture use by transpiring plants. Water Resources Research, vol. 6, no. 5, p. 1346-1356.

Ritchie, J. T. (1972). Model for predicting evaporation from a row crop with incomplete cover, Water Resources Research, 8, p. 1204-1213.

Rutter, A. J., Morton, A. J. and Robins, P.C. (1975). A predictive model of rainfall interception in forests II. Generalisation of the model and comparison with observations in some coniferous and hardwood stands. J. Applied Ecology 12, p. 367-380.

Edited Books

Monteith, J. L. (1965). Evaporation and environment. in: G. E. Fogg (Editor), The state and movement of Water in living organism, p. 205-234. Acad. Press, New York, London.

Section 4
GEOMORPHOLOGY

GEOMORPHOLOGY OF UPPER PETERS CREEK CATCHMENT, PANOLA COUNTY, MISSISSIPPI: PART I, CONSTRAINTS ON WATER BUDGETS

E. H. Grissinger, Soil Scientist
USDA Sedimentation Laboratory
P. O. Box 1157
Oxford, Mississippi 38655

J. B. Murphey, Geologist
USDA Sedimentation Laboratory
P. O. Box 1157
Oxford, Mississippi 38655

R. L. Frederking, Assistant Professor
Department of Geology and Geological Engineering
University of Mississippi
University, Mississippi 38077

ABSTRACT

Near-surface materials in northern Mississippi have widely varying hydraulic characteristics. The hydraulic conductivity of some materials is rapid whereas other materials function as aquitards or aquicludes. We have identified three sets of aquitards or aquicludes in the study area. Each set has a characteristic distribution within the catchment and each set has the potential for inducing water redistribution within a catchment or, in some cases, between catchments. The significance of each of these impending materials ultimately depends upon (a) the individual material properties, (b) their distributions relative to catchment boundaries and (c) the continuity of pertinent surfaces. A thorough understanding of these three controls will aid in the evaluation of whether a specific catchment is or is not representative of a given area.

INTRODUCTION

The USDA Sedimentation Laboratory in cooperation with the Vicksburg District Corps of Engineers, is presently involved in a comprehensive study of stream channel stability. One premise of this investigation is that channels are not independent entities but must be studied as an integral part of the total geomorphic system. A logical extension of this premise is that within-channel flow characteristics are not independent variables; they are dependent variables which vary temporaly and spacially with interfluve conditions, including land management activities for sediment and water producing areas. Upland conservation activities which reduce the rate or magnitude of sediment and water production may be viable management tools for use in the design and management of stable channels. Efficient use of upland conservation practices will require knowledge of sediment and water routing through the drainage system and of runoff rates (including both sediment and water) from areas within the catchment. This report describes our initial effort to identify soil-profile and near-surface geologic conditions which may materially influence groundwater conditions, thereby influencing runoff rates or magnitudes.

This investigation is complementary to a detailed investigation of hydraulic properties of typical soils in the Goodwin Creek catchment. It has been implemented as a process-oriented geomorphic study to interface with engineering-oriented studies of open-channel hydraulics which are similarly process oriented. Additionally, a process-oriented approach was deemed necessary in order to adequately characterize pertinent controls. Such characterization is prerequisite to evaluating if a specific catchment is or is not representative of a given area.

BACKGROUND

The term geomorphology refers to various lines of inquiry concerning the Earth's landforms and landscapes. Higgins (1975) identified three lines of inquiry: (a) the physical description of the present surface, (b) the evolution of the surface through time and (c) the identification and quantification of processes responsible for forming and modifying the present land surface. Process geomorphology differs from the first two approaches both in technique and in application. The first two approaches are primarily descriptive and frequently tacitly assume a space-time continuum. Process geomorphology assumes no such continuum. It encompasses morphometry and, more significantly, adds a depth component to the system. Ruhe (1975) refers to this depth component as surficial geology. The depth is axiomatically established as that depth necessary to fully evaluate pertinent processes (hereafter referred to as process controls). Identification of process controls is essential in this approach and may be critical in establishing the degree to which elements in a given catchment are representative of catchments within a region. Clarke (1977) discusses in detail the difficulties involved in extrapolating data from study catchments to other catchments. Certainly, this is an area critical to model application and, at the same time, is an area that has received little attention. A long-term objective of this study is to evaluate the utility of process geomorphology for evaluating if a specific catchment is or is not representative of a larger area.

STUDY AREA

The study area includes Goodwin, Hotophia, Johnson and Long Creek catchments in the southeast quarter of Panola County in northern Mississippi. This area is within the Coastal Plain Physiographic Province with the western two-thirds in the Loess or Bluff Hills Subprovince and the eastern one-third in the North Central Hills Subprovince. Loess caps all interfluves but thins rapidly from west to east. Holocene alluvial deposits are present in all valleys. Drainage is westerly to the Mississippi alluvial valley via the Tallahatchie River for Hotophia Creek and via the Yocona River for Peters Creek and its tributaries; Johnson, Goodwin, and Long Creeks.

Field identification of geologic units in northern Mississippi in general, and in Panola County in particular, has been a major problem due to (a) the loess veneer, (b) the absence of diagnostic maker beds, (c) the similar nature of the numerous sand exposures which either directly underlie the loess cap or are exposed at the surface and (d) the lack of subsurface information. Extensive drilling in the study area has established several errors or insufficiencies in the presently mapped near-surface geology. Errors or insufficiencies pertinent to this discussion include (a) the lack of differentiation of the Holocene alluvial valley deposits, (b) the lack of differentiation of the loess deposits in the interfluves and (c) the stratigraphic irregularities involving units presently mapped as Citronelle, Tallahatta and Zilpha-Winona.

RESULTS AND DISCUSSION

Nine lithologically distinct materials identified in the study area influence subsurface water conditions. Three of these materials -- the Peoria, Roxana and Loveland loess units -- are distributed in the interfluve area and three -- the massive silt, meander-belt alluvium and postsettlement alluvium -- are fluvial valley-fill deposits. The remaining three materials underlie both interfluve and valley positions. Vestal (1956) identified these three units as the Citronelle formation of Pliocene age and the Tallahatta and Zilpha formations of Eocene age. The Winona formation was not positively identified by Vestal for Panola County. Materials of equivalent lithology have been observed in this study, however, and they typically interfinger with Zilpha-like materials. For this study area, the Zilpha and Winona materials are considered a facies complex due to this alternating occurrence. Grissinger et al., (1981) reported the initial results of a subsurface study and noted several stratigraphic irregularities in the relative positions of these formations in Panola County, Mississippi. They (Grissinger et al., 1981) used the unit names as proposed by Vestal (1956) but enclosed the unit names in quotes to disclaim any stratigraphic significance. This unit nomenclature has been used herein.

"Citronelle," "Tallahatta" and "Zilpha-Winona" Materials

We have examined formation exposures identified by Vestal (1956) and concur with his lithologic description. Characteristic properties are:
"Citronelle"--sand, sandstone, gravel, and clay.
  The sand is coarse to fine, cross-bedded to the southeast, and cemented in places. Gravel is common to sparse and occurs as stringers to thin to thick beds. Clay is present as lenses or is disseminated in the sand phase as a minor component.
"Tallahatta"--shale, clay, sand, silt, sandstone and siltstone.
  The sandy phase is composed of clean to argillaceous fine sand and is usually yellow to gray with some red to brown staining. Clay is present as matrix material, laminae, stringers or thin beds. The sandy phase is frequently micaceous and occasionally cemented. The fine-textured phase is shalelike to clayey, usually light colored but occasionally brown to red, micaceous, and has scattered thin seams of organic material. Outcrops of this member are frequently cemented.
"Zilpha"--clay, sandy silt, lignite, sandstone and siltstone.
  The fine sediments are shalelike, carbonaceous and brown to black when moist but dry to a gray color. They contain marcasite concretions and have a sulfide smell. They are layered and have laminae of micaceous silt to fine sand. The sands are fine, carbonaceous, gray to black, micaceous and have a sulfide smell.
"Winona"--sand, silt, clay, and claystone.
  This formation is slightly to very glauconitic, micaceous to very micaceous, carbonaceous, and has variable colors ranging from grayish-tan to greenish-brown to brownish-black. Clay is frequently present as thin stringers, laminae or beds. Outcrops oxidize rapidly to bright red to brown colors.
About 100 test holes were drilled in the study area (Fig. 1). The

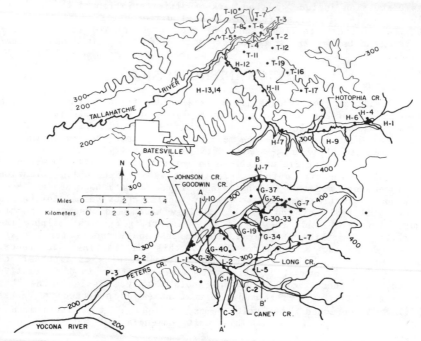

Figure 1. Location of exploratory holes in intensive study area, Panola
County, Mississippi.

typical sequence of lithologic units is shown in Fig. 2. "Tallahatta"
materials occur as scattered bodies overlying the gravels, or in some
locations within or below the gravels. The size and position of the
"Tallahatta" bodies appear to be random. Both the "Tallahatta" and
"Citronelle" lithologic units unconformably overlie the "Zilpha-Winona"
facies complex. Grissinger et al., (1981) discuss these units and their
distributions in greater detail. They present the argument that this
lithologic sequence resulted from processes controlled by post-Eocene
sea level fluctuations.

     Based upon field observation, typical "Zilpha-Winona" materials
were only slightly moist and "Zilpha-type" clayey materials were
effectively dry. The overlying gravel to sand materials were generally
saturated. The "Zilpha" materials apparently function as an aquiclude,
perching ground water and controlling subsurface hydrology.
Additionally, the paleosurface developed on the "Zilpha" has high relief
which is not conformable with present catchments (Figs. 2 to 4). These
two features, the relief on the "Zilpha" paleosurface and the relative
impermeability of the "Zilpha" materials, create the potential for
appreciable groundwater transfer between adjacent catchments.
Additional potential ground-water transfer results from the random
distribution of the fine-textured "Tallahatta" materials. These
materials are cohesive and function as aquicludes or aquitards depending
upon body thickness.

     At this time, instrumentation in the study area is not complete and
data are not sufficient to positively establish the significance of the
lithologic units with respect to ground-water movement. The units,
however, are thought to be responsible for the observed base flow
variability between many northern Mississippi streams. For instance,
Clear Creek and Hudson Creek in Lafayette County drain adjacent

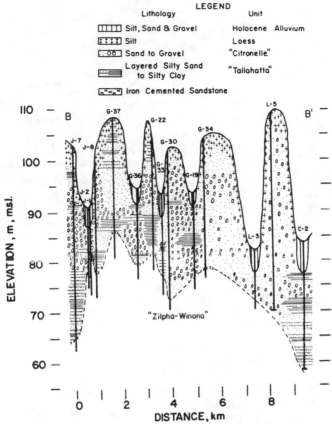

LEGEND

Lithology	Unit
▥ Silt, Sand & Gravel	Holocene Alluvium
▦ Silt	Loess
▨ Sand to Gravel	"Citronelle"
▤ Layered Silty Sand to Silty Clay	"Tallahatta"
▧ Iron Cemented Sandstone	

Figure 2. Cross section from hole J-7 to hole C-2 (B to B', Fig. 1).

catchments with similar soils, geology (as presently mapped) and land use. Clear Creek, with an area of 26.7 km² has an expected 7-day low flow (per two year interval) of 0.161 cms whereas Hudson Creek with an area of 24.2 km² has a comparable expected base flow of 0.006 cms. Water budgets based on surface catchment definition are thus tenuous and will be in error to the degree that the two aquicludes, the "Zilpha-Winona" and "Tallahatta" materials, influence subsurface water gain or loss for a study catchment.

Loess Materials

Three loess units cap the interfluves. These units generally thin from west to east. The upper loess unit, the Peoria, is relatively friable. It overlies the more dense, brittle Roxana loess. This contact is generally conformable in areas of low relief but unconformable in hillslope positions. The Loveland is a dense clayey loess. When present, it overlies the previously discussed "Tallahatta" or Citronelle" materials and unconformably underlies the Roxana loess. Buntley et al., (1977) describe the loess stratigraphy of west Tennessee, particularly as it is related to the distribution of fragipan horizons in the Memphis-Loring-Grenada soil sequence. They present stratigraphic, geomorphic and morphologic evidence that the fragipan horizons in these soils are a relic of pre-Peorian weathering of the

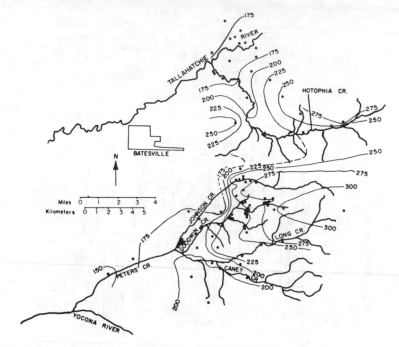

Figure 3. Surface of "Zilpha-Winona" complex, elevations in feet mean sea level.

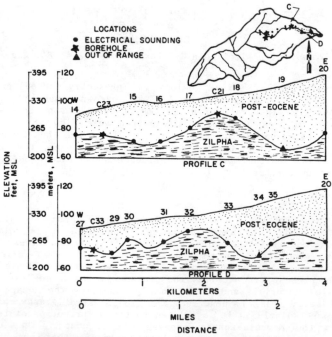

Figure 4. Surface of "Zilpha-Winona" complex under upper reaches of Goodwin Creek.

Roxana paleosurface. Axiomatically, the relative position of the fragipan horizon is thus not genetically related to the present landscape, as had been previously assumed. The general trends reported by Buntley et al., (1977) are that the Peoria loess thins and the (Roxana) fragipan horizon becomes closer to the surface (a) in a west to east direction and (b) from interfluve to hillslope position. Our observations in the Upper Peters Creek catchment are identical.

Römkens and Whisler (1980) have evaluated the hydraulic conductivity of comparable loess soils (Fig. 5). They reported "the fast equilibrium response of the soil profile above the fragipan and the

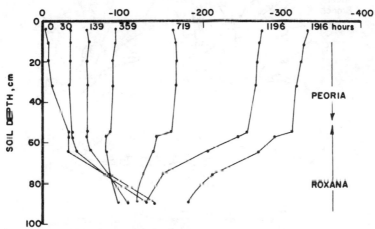

Figure 5. Hydraulic head changes with time for the Peoria and Roxana loess units.

very slow adjustment within the fragipan were indicative of appreciable differences in hydraulic characteristics between soil horizons." They observed that free water above the fragipan, that is above the Roxana, moved laterally rather than vertically. The Loveland loess unit is thought to function like the Roxana unit.

These results indicate the possible influence of loess stratigraphy on rainfall - runoff relations. Again, instrumentation is not complete and data are not sufficient to positively establish the significance of loess stratigraphy. Such definition will depend not only upon hydrologic characterization, such as that reported by Römkens and Whisler (1980), but more significantly upon the areal distribution and continuity of the paleosurface separating the Peoria and Roxana loess units. Definition of this surface and its continuity throughout the study area is cost prohibitive. Such studies will be attempted in restricted areas within catchments where rainfall-runoff relations indicate the paleosurface may be a significant control. This evaluation will require measurement of hydraulic conductivity in both vertical and horizontal dimensions for both materials (the materials above and below the paleosurface). We expect relatively low runoff rates in the western part of the study area where the fragipan horizon is covered by a maximum thickness of the more permeable Peoria loess and in the eastern part of the study area where the loess cap has minimum thickness and the surface soils are relatively sandy. Between these two areas of (expected) relatively-low runoff, the fragipan horizon has been observed to be close to the surface and runoff rates are expected to be materially greater.

Channel or gully incision into or through the fragipan horizon, that is the paleosurface, may significantly influence rainfall-runoff relations. This influence is, however, extremely complex and is probably site specific. Low-order streams respond rapidly to environmental conditions. An example is presented in Fig. 6 for Cypress Creek in Yalobusha County, Mississippi. Yalobusha County borders Panola

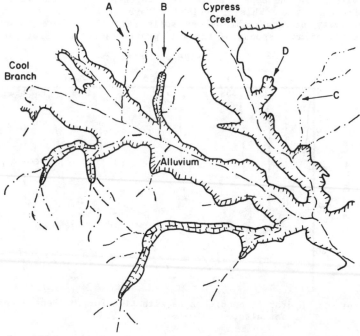

Figure 6. Alignment of tributary drainage with alluvial deposits, Cypress Creek (from Sheet 30, Huddleston et al., 1978).

County to the south. Tributary B has a well developed flood plain and is obviously older than tributaries A or C which have no flood plains. The confluence of Tributary C with Cypress Creek was evidently at point D sometime in the past. The present degree of tributary incision relative to the fragipan (Roxana) depth and the amount of alluvial fill related to the tributary age will determine the amount of transmission loss and hence the rainfall-runoff relations for each tributary drainage area. The most significant feature of the loess stratigraphy, however, is the congruence of the Roxana aquitard with an erosional unconformity (the unconformity between the Peoria and Roxana loess units). This paleosurface appears to be continuous in the hillslope position and may be a significant control of water redistribution within a catchment.

Valley-Fill Materials

Most Holocene valley-fill deposits in northern Mississippi are presently undifferentiated. Grissinger et al.,[1] however, have identified seven depositional units which reflect paleoclimatic process controls. Three of these units are pertinent to this discussion. These

[1] Grissinger, E. H., Murphey, J. B. and Little, W. C., Late-Quaternary valley-fill deposits in north-central Mississippi. In preparation.

three valley-fill units are, from youngest to oldest, the postsettlement alluvium, the meander-belt alluvium and the massive silt.

Postsettlement alluvium (PSA), produced in historic times largely by man's activities, caps almost all flood-plain surfaces. This material is frequently less than 1 m thick but may locally exceed a thickness of 4 m. The PSA has well preserved fluvial bedding features. It is unweathered with an Ap horizon directly overlying a C horizon. Iron diffusion halos have not been observed. This unit is identifiable in the field by the presence of man-made artifacts above a disconformity. It has been the subject of many reports, including those by Happ (1968, 1970) and Trimble (1974). It is highly permeable except when plow or traffic pans have been induced by cultural practices.

Meander-belt alluvial materials are typically vertical accretion overlying lateral accretion deposits with occasional oxbow deposits of layered fines. These two types of deposits have not been separated. Wood is scattered throughout this deposit but is usually not as well preserved as the older wood. This state of wood preservation is probably due to the greater permeability of this material relative to that of the massive silt. In all cases, bedding is readily observable. Weathering is less intense than that on the massive silt. The paleosol on the meander belt alluvium, noted as paleosol I, has an A_1 horizon which varies in thickness from more than 25 cm to only a few cm. In places, the profile is truncated and has no observable A horizon. Iron diffusion halos are usually present in the subsoil but are typically small. This unit has no A_2 horizon and no pronounced B_2 horizon. The thickness of the A_1 horizon and the absence of an A_2 horizon suggest that this paleosol developed under grass cover. It has no polygonal structure, is relatively fertile and is well drained.

The massive silt unit is a widespread predominantly fine-textured valley-fill deposit. This deposit is distributed throughout the study area and frequently exceeds 4 m in thickness. The deposit fines upward from a silty sand or sandy silt basal material to a silt, with no observable textural breaks except for occasional small relict channels. The weathering profile formed on the massive silt, noted as paleosol II, is distinctive. It possibly formed by ferrolysis-type weathering, as described by Brinkman (1970). This paleosol has a thick A_2 and a dense B_2 horizon, both unique in the study area to this paleosol. Gray is the dominant color in the upper part of the profile. Iron and manganese stains and concretions are present in the lower B. The B_2 horizon has a well developed polygonal structure with seams often wider than 2 cm. The massive silt is relatively infertile and restricts the vertical movement of water.

The general hydrologic features of this valley-fill sequence are similar to those of the loess units. Paleosol II materials adjust slowly whereas paleosol I materials adjust rapidly to changes in the soil water status (Fig. 7).[2/] The contact between these two paleosol materials is unconformable (Fig. 8); the contact is an erosional paleosurface more or less continuous under most of the present flood plain. Lateral water movement along this paleosurface is common. Although this flow component is probably insignificant in relation to total water budgets of catchments, it may be significant for base flow return times and for storm-flow versus base-flow relations. Channel incision through the lower paleosol, however, will result in transmission loss and may initiate transfer of water between catchments. Transfer would be influenced by "Tallahatta" bodies and the paleosurface on "Zilpha-Winona" materials. The flow component along this paleosurface on the massive silt, paleosol II, is important in channel stability relations and possibly in nutrient loss evaluations. Seepage forces along this paleosurface frequently induce bank instability

---

2/ Unpublished data, M. J. M. Römkens, USDA Sedimentation Lab, Oxford, MS.

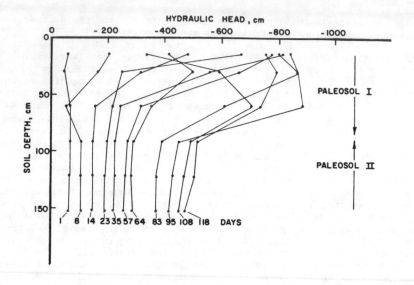

Figure 7. Hydraulic head changes with time for paleosols I and II.

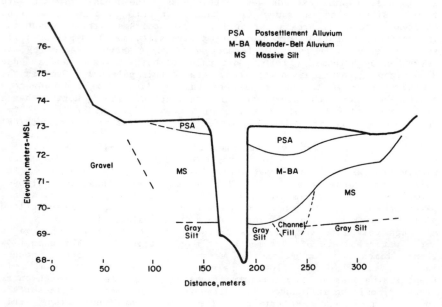

Figure 8. Distribution of valley-fill units in lower Goodwin Creek Valley.

problems in otherwise stable reaches. This seep originates by flow through the highly fertilized cultivated soils and is usually discharged to the channels during recession- or low-flow times. Such seep may be a significant consideration for water quality modeling.

SUMMARY

Three sets of lithologic constraints, i.e., the valley, interfluve and subsurface constraints, may affect water movement within and between catchments. Definition of these or other near-surface lithologic constraints is prerequisite to application of detailed hydrologic models concerned with complete water budgets or with base flow conditions. Base flow conditions are certainly of minor significance for flood or sediment routing models but are pertinent to water quality models, particularly those designed to interface with biological models, and to models of channel bed and/or bank stability.

The individual aquicludes or aquitards have three distinct distributions relative to present catchment definition. The paleosurface on the "Zilpha-Winona" materials is continuous whereas the "Tallahatta" bodies are discontinuous. Relief on each is independent of present-day topography. The paleosol aquitards are distributed within the valleys. In the interfluve area, the loessial aquitards vary with the loess thickness. The loess thins from west to east and the influence of the loessial aquitards varies inversely with loess depth. The nature of the contact between the valley and loess aquitards has not been defined at this time. Historic changes in the channel system have undoubtedly interacted with both the paleosol and loess aquitards, and such changes will probably continue. These changes include gullying and tributary incision and extension. Main channel thalweg lowering has progressed through the paleosol II weathered massive silt, into the "Citronelle" sands and gravels in many locations resulting in transmission loss. Although the significance of these changes cannot be measured, such changes have undoubtedly influenced the relative quantities and rates of subsurface and surface flow.

Dominant process controls have been tentatively identified for the three sets of lithologic constraints. The dominant process control of the valley-fill sequence was the late-Quaternary paleoclimate and base-level controls, that of the loess sequence was the paleoclimate conditions associated with glacial waning and that of the near-surface geology was post-Eocene sea-level controls. We hypothesize that each control has a unique distribution which must be defined before any catchment can be evaluated as a representative or nonrepresentative catchment. Inherently, this evaluation is prerequisite to efficient application of predictive models.

REFERENCES

Brinkman, R. 1970. Ferrolysis, a Hydromorphic Soil Forming Process. Geoderma, Vol. 3, pp. 199-206.

Buntley, G. J., Daniels, R. B., Gamble, E. E. and Brown, W. T. 1977. Fragipan Horizons in Soils of the Memphis-Loring-Grenada Sequence in West Tennessee. Soil Science Society of America Journal, Vol. 41, pp. 400-407.

Clarke, R. T. 1977. A review of research on methods for the extrapolation of data and scientific findings from representative and experimental basins. Technical Documents in Hydrology, 47 pp., United Nations Educational, Scientific and Cultural Organization, Paris, France.

Grissinger, E. H., Murphey, J. B. and Little, W. C. 1981. Problems with the Eocene Stratigraphy in Panola County, Northern Mississippi. Southeastern Geology (in press).

Happ, S. C. 1968. Valley sedimentation in north-central Mississippi. Proceedings of the Third Mississippi Water Resources Conference, 8 pp., Jackson, Mississippi.

Happ, S. C. 1970. North Tippah valley sedimentation survey. Research Report No. 415, 30 pp., United Stated Department of Agriculture, Agricultural Research Service, Southern Branch, Soil and Water Conservation Research Division.

Higgins, C. G. 1975. Theories of Landscape Development: a Perspective. in: W. N. Melhorn and R. C. Flemal (Editors), Theories of Landform Development, pp. 1-28, Publications in Geomorphology, State University of New York, Binghamton, New York.

Huddleston, J. S., Bowen, C. D. and Ford, J. G. 1978. Soil survey of Yalobusha County, Mississippi. Soil Conservation Service and Forest Service, United States Department of Agriculture in cooperation with Mississippi Agricultural and Forestry Experiment Station, 87 pp. plus maps, United States Government Printing Office, Washington, District of Columbia.

Römkens, M. J. M. and Whisler, F. D. 1980. In situ hydraulic conductivity determinations on a loessial soil. American Society of Agricultural Engineers Paper No. 80-2525, 15 pp., American Society of Agricultural Engineers, St. Joseph, Michigan.

Ruhe, R. V. 1975. Geomorphology, Geomorphic Processes and Surficial Geology. Houghton Mifflin Co., Boston, Massachusetts, 246 pp.

Trimble, S. W. 1974. Man-induced soil erosion on The Southern Piedmont 1700-1970. Special Report, 180 pp., Soil Conservation Society of America, Ankeny, Iowa.

Vestal, F. E. 1956. Panola County geology. Bulletin 81, 157 pp., Mississippi State Geological Survey, Jackson, Mississippi.

GEOMORPHOLOGY OF UPPER PETERS CREEK CATCHMENT, PANOLA COUNTY, MISSISSIPPI: PART II, WITHIN CHANNEL CHARACTERISTICS

E. H. Grissinger, Soil Scientist
USDA Sedimentation Laboratory
P. O. Box 1157
Oxford, Mississippi 38655

J. B. Murphey, Geologist
USDA Sedimentation Laboratory
P. O. Box 1157
Oxford, Mississippi 38655

R. L. Frederking, Asst. Prof.
Geology Department
University of Mississippi
University, Mississippi 38677

ABSTRACT

This geomorphic study of channel characteristics, including bed and bank stability, was organized along two complementary lines, one morphometric- and the other process-oriented. The morphometric investigation involved photogrammetric and field measurements of present channel widths, depths, sinuosities and channel slopes and the changes of these properties since 1937. Results indicate that the present channel morphometry is controlled by the nature and distribution of certain valley-fill and older deposits. The channels are thus not true alluvial channels.

The process-oriented investigation was organized to better describe the individual valley-fill or older deposits, to evaluate the significance of individual deposits with respect to channel characteristics and to identify dominant controls of formation and modification. Seven valley-fill deposits have been identified and each influences channel behavior and over-all characteristics. These deposits reflect paleoconditions. The areal distribution of these deposits is thus not related to current environmental conditions but to characteristic distributions of each pertinent paleocondition. A full definition of such controls is prerequisite to any evaluation of whether a catchment is or is not representative of a larger area. In turn, this type of evaluation is prerequisite to the optimal application of predictive models.

INTRODUCTION

Channels are an integral part of the total geomorphic surface and must be studied as one part of the total surface. Certainly, geomorphic properties which influence sediment and water production in the

267

interfluve area will influence channel characteristics such as drainage density, sinuosity, bed and bank stability and channel shape and size. On the other hand, channel characteristics influence sediment and water routing from the source area and thus influence interfluve degradation.

Although this approach has many advantages, it also has several limitations which restrict immediate implementation for solving today's environmental problems. One of these limitations involves the definition of relict features and of response times for processes affecting these features. In this context, the response time is a relative measure of the dependence of individual landforms upon present-day, historic or older controls. Such controls include climate, geology (including soils), base level and historic cultural activities. Geomorphic features or properties inherited from past controls, although pertinent to present-day processes, may result in an erroneous evaluation of whether a catchment is or is not representative of a given area. Apparent relations between present geomorphic properties may be erroneous predictors to the degree that such properties reflect varying controls which have varying response times and varying areal distributions. This paper summarizes our initial findings concerning late-Quaternary valley-fill sedimentation processes, controls and significance with respect to current channel conditions

STUDY AREA

The study area includes four catchments in the southeast quarter of Panola County in northern Mississippi; specifically Hotophia, Johnson, Goodwin and Long Creek catchments. This area lies within the Coastal Plain Physiographic Province with the western two-thirds in the Loess or Bluff Hills Subprovince and the eastern one-third in the North Central Hills Subprovince. Loess caps all interfluves but thins rapidly from west to east. Holocene alluvial deposits are present in all valleys but are presently undifferentiated. Drainage is westerly to the Mississippi alluvial valley via the Tallahatchie River for Hotophia Creek and via the Yocona River for Peters Creek, including Johnson, Goodwin, and Long Creek tributaries.

RESULTS AND DISCUSSION

As an initial phase of a comprehensive study of stream channel instability, we visually inspected many streams in northern Mississippi. Bed and bank materials were not uniform for any individual stream channel. We observed, however, that most materials could be grouped into one of several units and that individual units or sequences thereof affected channel characteristics. Each of these individual units appeared to be consistent throughout the study area and each possessed distinctive properties facilitating differentiation between units. Additionally, each unit occupied a consistent relative position in all catchments. These observations indicated that the alluvial deposits which are presently undifferentiated may include several definable units which individually affect present channel behavior. Results are presented in the following order:

a) valley-fill deposits, including the differentiation, properties and distributions of individual units,
b) paleoclimatic control of the valley-fill sequence,
c) channel stability, including relations between valley-fill units and failure processes,
d) channel morphometry, including relations between valley-fill units and channel width, depth, sinuosity and thalweg slope.

Valley-Fill Deposits[1]

Seven identifiable valley-fill units regularly crop out in channels of the study area. The units contain abundant wood or other organic detritus and 60 of these samples have been dated. A frequency histogram for outcrop samples of ages less than 13,000 ^{14}C years Before Present (yr BP) is presented in Fig. 1a. All ages were calculated using the Libby half-life of 5568 years. No correction has been made for variation in atmospheric ^{14}C concentration. These units, from oldest to youngest, are (a) consolidated sandstone, (b) bog-type deposit, (c) channel lag deposit, (d) massive silt, (e) channel fill, (f) meander-belt alluvium and (g) postsettlement alluvium.

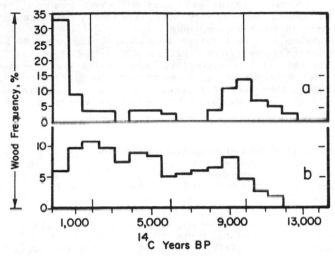

Figure 1. Age frequencies. (a) Frequency histogram of ^{14}C dates for 57 outcrop samples (source: this paper). (b) Frequency histogram for 815 ^{14}C dates selected from the journal Radiocarbon (source: Wendland and Bryson (1974).

Consolidated sandstone

Consolidated sandstones are present in many catchments. These sandstones are usually cross-bedded and frequently contain gravel. At three locations the sandstone contained wood which was all older than 40,000 ^{14}C yr BP. Outcrops are usually limited in size, rarely exceeding several tens of meters in horizontal distances. They are truncated and are typically disconformably overlain by younger bog-type and channel lag deposits which had a maximum age of 12,050±180 yr BP.

Bog-type and channel lag deposits

As used herein, bog-type sediments are fine-grained, organic-rich materials deposited from low-energy fluvial systems. Most of these deposits appear to have formed in either channel cutoffs or in separation zones downstream of point-bar deposits along the inner bank of a

---

1/ A more complete description of the valley-fill deposits is contained in "Late-Quaternary valley-fill deposits in northern Mississippi" by E. H. Grissinger, J. B. Murphey and W. C. Little. This manuscript is in review.

bendway. Channel lag materials are coarse-grained, frequently cross-bedded materials deposited from high-energy fluvial systems. Gravel is common where not limited by source availability. Both deposits contain abundant organics. In general, the organic debris in the lag deposit is relatively coarse, ranging up to 50 cm diameter logs. Bog-type organics include leaves, twigs, various nuts, and scattered stumps and logs similar in size to those found in the lag deposit. Typically the organics in both deposits show little evidence of abrasion or aerobic decomposition, indicating rapid burial. Heartwood, alburnum and bark are usually well-preserved and the cellular structure of woody tissue is intact. In addition, acorns with caps attached; complete leaves; and walnuts, butternuts, and hickory nuts, frequently complete with husks have been seen in the bog-type deposit.

All samples of the bog-type and channel lag deposits have ages defined by the frequency mode about 10,000 yr BP (Fig. 1a). The age span of this mode is interesting; it is generally synchronous with the time man first appeared in the lower Mississippi River valley (Saucier, 1974) and with the time of excessive Pleistocene generic extinction (Grayson, 1977). We interpret this period as transitional between a preceding period of valley erosion and a subsequent period of deposition of fine-grained materials.

Massive salt, channel fill and meander-belt alluvium

The massive silt and meander-belt alluvium developed in relatively fine-grained deposits, and both are buried beneath postsettlement (historic) alluvium. Each of these two units has a consistent set of depositional features and a distinctive, characteristic weathering profile. The properties of these two units have been produced by both depositional and weathering processes. We refer to the weathering profile in the massive silt as paleosol II and the profile in the meander-belt alluvium as paleosol I. This designation minimizes possible confusion involving soil classification units and associated weathering features. The soil classification units presently mapped are materially influenced by the overlying postsettlement alluvium.

The massive silt unit is a widespread, predominately fine-textured, valley-fill deposit. This deposit is distributed throughout the study area and frequently exceeds 4 m in thickness. The deposit fines upward, from a silty sand or sandy silt basal material to a silt, with no observable textural breaks except for occasional small relict channels. No large relict channels have been observed which would indicate channelized flow at this time of valley aggradation. Additionally no organics have been found in this unit and bedding is rare. Bedding has been observed only in the sandy basal material.

At most sites, the massive silt overlies bog-type and channel lag deposits which have an age of about 10,000 yr BP. At Johnson Creek, three samples at the contact between the massive silt and bog-type deposit had ages of about 8,700 yr BP. Relict entrenchment into or through the massive silt deposit is common. Five wood samples have been obtained at outcrops of such channel-fill deposits. These wood samples range in age from 4,050 to 6,120 yr BP and comprise the frequency mode at about 5,000 yr BP (Fig. 1a). The age of the massive silt is, therefore, older than 6,120 yr BP but younger than 8,700 yr BP.

Based upon the absences of large relict channels, bedding and textural breaks, we interpret the massive silt as a low-energy fluvial deposit, possibly associated with periodic inundation resulting from valley plugging. Such plugging has been described by Pflug (1969) for tributaries in eastern Brazil, but at a slightly earlier time than that for this deposit. Aeolian materials may have been an additional source for this massive silt deposit. The contact between the massive silt and channel lag deposits, where present, is gradational, indicating that the silt is only slightly younger than the underlying deposits. We interpret this massive silt deposit as the end member in the sequence of valley erosion → channel lag or bog-type → massive silt, this sequence representing a continuing decrease in energy resulting from decreasing pluvial activity and rising base level controls.

The paleosol II type weathering profile formed in the massive silt is distinctive. It possibly formed by ferrolysis-type weathering, as described by Brinkman (1970). This paleosol has a thick A_2 and a dense B_2 horizon, both unique to paleosol II. Gray is the dominant color in the upper part of the profile. Iron and manganese stains and concretions are present in the lower B. The B_2 horizon has a well developed polygonal structure, with seams often wider than 2 cm. The paleosol II unit is relatively infertile and restricts the vertical movement of water. Vegetative cover is rare on outcrops.

As noted previously, the channel fill deposit comprises the frequency mode at about 5,000 yr BP. Deposits are less extensive than those for either the massive silt or the meander-belt alluvium. Materials are typically coarser-textured than those of the massive silt but have the same gray color. These materials are usually highly weathered, probably due to the sandy texture. They have no polygonal structure and no well-developed B_2 horizon. Bedding is frequently difficult to discern.

A major entrenchment of streams into the massive silt began about 3000 yr BP. This time of entrenchment agrees with the paleoclimatic interpretation of Wendland and Bryson (1974) who reported a major botanic discontinuity at 2760 yr BP associated with increased rainfall. The distribution of ages in the youngest frequency mode (Fig. 1a) indicates that entrenchment in our study area was relatively minor until about 1600 yr BP. Activity increased gradually to a peak frequency about 200 to 400 yr BP. The entrenching streams apparently meandered across the flood plains, eroding the massive silt and channel fill materials and depositing the unit identified as meander-belt alluvium (Fig. 2). These materials are typically vertical accretion deposits overlying lateral accretion deposits with occasional oxbow deposits of

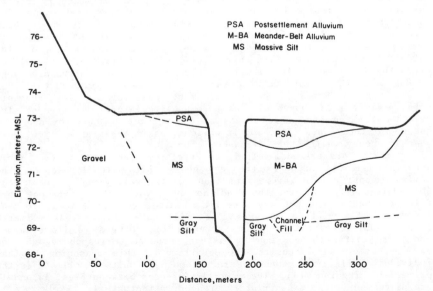

Figure 2. Distribution of valley-fill units in lower Goodwin Creek Valley.

layered fines. These two types of deposits have not been separated. Wood is scattered throughout this deposit but is usually not as well preserved as older wood. This state of wood preservation is probably due to the greater permeability of the meander-belt alluvium relative to that of the massive silt. In all cases, bedding is readily observable.

Paleosol I weathering on the meander-belt alluvium is less intense than that of paleosol II. Paleosol I has an A_1 which varies in thickness from more than 25 cm to only a few cm. In places, the profile is truncated with no observable A. Iron diffusion halos are usually present in the subsoil but are typically small. This unit has no A_2 horizon and no pronounced B_2 horizon, suggesting that it developed under grass cover. Paleosol I has no polygonal structure, is relatively fertile and is well drained.

Postsettlement alluvium

Postsettlment alluvium (PSA), produced in historic times largely by man's activities, caps almost all flood-plain surfaces. This material is frequently less than 1 m thick but may locally exceed a thickness of 4 m. The PSA has well preserved fluvial bedding features. It is unweathered with an Ap horizon directly overlying a C horizon. Iron diffusion halos have not been observed. Although this unit is too young to be identified by radiocarbon procedures, it is identifiable in the field by the presence of man-made artifacts above a disconformity. It has been the subject of many reports, including those by Happ (1968, 1970) and Trimble (1974).

Paleoclimatic Control

Wendland and Bryson (1974) used 815 dates published in the journal Radiocarbon "-- to identify times of large-scale hemispheric discontinuity --" based on geologic-botanic discontinuities. Their data base included ^{14}C dates which defined the age of discontinuities within peat beds, pollen profiles, glacial records and sea level stands. A parallel data base included 3,700 dates associated with 155 cultures to identify cultural discontinuities. The primary source for these dates was also Radiocarbon. Globally synchronous discontinuities occurred in both records and they argued that such results could only be produced if climate was the primary forcing function. Of the seven major geologic-botanic discontinuities which they identified by fitting partial collectives to a multimodal distribution, three are included in the age range of 850 to 2760, one at 5060 and three in the range from 8490 to 10,030 yr BP. Fig. 1b is a frequency histogram of their total geologic-botanic data, organized in 800-yr classes to emphasize large-scale trends. The relation between the three groups of discontinuities and the age frequency for their data (Fig. 1b) is obvious, as is the similarity of this age frequency with the frequency for our data (Fig. 1a). The distributions are trimodal with (generally) comparable modal ages. This apparent fit (a) supports the hypothesis that paleoclimate was the dominant control (the forcing function) of the valley-fill deposits and (b) additionally supports the Wendland and Bryson (1974) argument that climatic change, and not any specific climate, was the primary forcing function. In relation to our valley-fill data, large-scale climatic changes would logically produce corresponding changes in flow regime, and these latter changes would, in turn, produce corresponding changes in the valley-fill deposits. A corollary of this reasoning is that individual valley-fill deposits are chronologic units if climate is the primary forcing function. The major differences between Figs. 1a and 1b are discussed in the manuscript referenced in footnote 1.

Channel Stability

The late-Quaternary valley-fill units, together with present hydro-logic conditions, control present-day channel stability in this study area. Each of the units exhibits typical types of failure, depending upon their position in the channel bank and/or bed. The types of failure, in turn, depend upon both depositional and weathering properties of the units. Bank stability is influenced by postsettlement alluvium, meander-belt alluvium, the massive silt and the bog-type and channel lag deposits. Bed (thalweg) stability is influenced by the massive silt, bog-type and channel lag deposits and the consolidated sandstone. Although discussed separately, bank stability cannot be evaluated independent of bed stability; both must be considered for realistic solutions to the massive channel instability problems of the areas bordering the Lower Mississippi River Valley.

Postsettlement and meander-belt alluvium most frequently occur in an upper-bank position. These materials are well drained, relatively fertile and are usually well vegetated. Scour by high velocity flow is a minor consideration for stability due to the vegetative cover and the infrequency of exposure to such flow. Scour is proportionately more significant for these materials in a lower-bank position. The most frequent erosion problems result from gravity failure accentuated by tension crack development. These tension cracks are vertical and parallel to the channel bank. Their development is undoubtedly related to the minimal weathering, typical of paleosol I, and hence isotropic nature of these deposits.

Paleosol II developed in the massive silt has a distinctive polygonal structure which controls stability. Although individual blocks are resistant to channelized flow, the seam materials are only marginally stable. Erosion or weakening of the seam material reduces interped strength, resulting in gravity-induced block failure. We believe this polygonal structure is probably the result of desiccation due to the combined effects of the early- to mid-Holocene hydromorphic conditions and temperature maximum. As such, paleosol II is a relict of early-Holocene weathering. Although failure of both fine-textured units is gravity induced, removal of the slough material is undoubtedly controlled by weathering (break-up) of the slough blocks and by flow parameters. An additional influence of paleosol II on stability results from the low relative permeability of the B_2 horizon. This horizon is less permeable than the overlying materials and seep commonly occurs at the interface, further stressing stability of the overlying materials.

The bog-type and channel lag deposits underlie the massive silt. Both are unconsolidated materials of low cohesion and are easily eroded by channelized flow. Channel incision into either of these units invar-iably results in excessive channel widening due to excessively weak toe conditions.

Thalweg incision through the massive silt has occurred by headward migration of knickpoints. Two types of migration have been observed, the usual overfall-type failure and a more complex type of failure which is initiated by the development of chutes through the polygonal-structured paleosol II materials. Weak seam materials between individual blocks are winnowed by base flow, structurally isolating individual blocks which are easily displaced by high velocity flow. For Johnson Creek, the rate of knickpoint movement averaged 160 m/yr from 1940 through 1975 (Fig. 3, Ethridge, 1979). Thalweg elevations upstream of such knickpoints were relatively stable. Channel beds were cohesive and width-to-depth ratios were consistent (Fig. 4a[2/]). Exposure of the unconsolidated bog-type and channel lag deposits downstream of the knickpoint, however, resulted in channel widening and changed the flow regime. Downstream, the channels have sand to gravel beds with variable

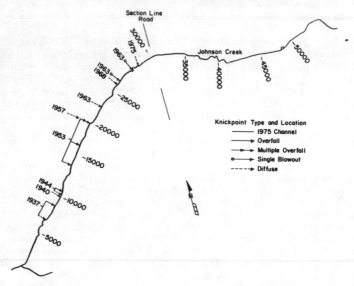

Figure 3. Knickpoint migration in Johnson Creek Valley (after Ethridge, 1979).

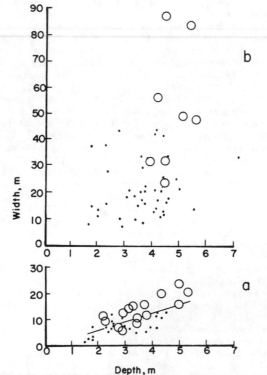

Figure 4. Width-to-depth relations for Johnson Creek upstream (a) and downstream (b) of the knickpoint from stereoscopic analysis of ASCS aerial photographs for the years 1937, 1940, 1944, 1953, 1957, 1963 and 1968 (identified by dots) and for the year 1975 (identified by circles) (Ethridge, 1979).

width-to-depth ratios (Fig. 4b$\underline{2/}$). Transport processes are dominant. Inherently, the bed stability of these sand-bed channel reaches is primarily dependent upon sediment supply to, and transport properties of, the hydraulic system. Consolidated sandstones, such as those which outcrop in Goodwin Creek channel, limit thalweg lowering; they function as local grade controls. Width-to-depth ratios for Goodwin Creek are similarly variable, in this case due primarily to variations in channel width (Fig. 5$\underline{2/}$). Excessive channel changes were determined for both channels for the period preceding the 1975 aerial photographs.

Channel Morphometry

Goodwin Creek channel was surveyed and flown for photographic record in 1977. Fig. 6 is the plan view of Goodwin Creek drawn from this 1977 photographic record. The main channel has been divided into 29 reaches (1 through 18 and 18-1 through 18-11) based upon the location of survey cross sections and upon "apparent" channel morphology. Most individual reaches include multiple cross sections. Average widths and depths for each reach have been calculated from the survey data and are listed in Table 1 along with standard deviations and maximum and minimum widths and depths.

Channel widths for Goodwin Creek are variable both between reaches and within reaches. The coefficient of variability (100 x standard deviation/average width) for individual reaches ranged from 10 to 72%. The average coefficient of variability for all reaches was 24%. Channel depth was less variable, having an average coefficient of variability of 7.5%. This degree of variability for individual measurements, particularly for channel width measurements, is accordant with the photogrammetric results presented in Fig. 5. Individual measurements of channel widths and depths showed no consistent relation. Similarly, for the 1977 survey data, average widths and depths per reach showed no consistent relation. Field reconnaissance, however, had established three process controls of channel morphometry for Goodwin Creek and the average values of channel width and depth have been organized into three groups based on these three controls (Fig. 7).

A knickpoint was present at the downstream end of reach 8 at this time (1977), separating the two downstream groups. Paleosol II materials were more or less continuous immediately upstream of this knickpoint but were absent in a downstream direction. Reaches 1 through 7 comprise the channel length characterized by an absence of stratigraphic controls of bed elevation. The correlation coefficient for this group (r = -.87) is significant but negative. Upstream of the knickpoint at reach 8 through reach 18 the channel width is independent of channel depth. Through this length of channel, the thalweg elevation has been stabilized by paleosol II materials, cemented gravel sills and numerous iron-cemented sandstone sills. Gravel bed material is common between outcrops of the bed controls and this gravel has probably lessened the failure rate for the controls, that is by lessening the probability of undercutting of the bed controls. Upstream of reach 18, the channel bed and bank materials are (stratigraphic) units considerably older than the Holocene valley-fill materials. These materials are lithologically different from bed and bank materials downstream of reach 18. For this headward-most length of channel, reaches 18-2 through 18-10, the correlation coefficient is significant

---

2/   Each data point in Figs. 4 and 5 is a measurement at a point location along the channel. Conventional photogrammetric methods were used for this interpretation of aerial photographs for the years 1937, 1940, 1944, 1953, 1957, 1963, 1968 and 1975.

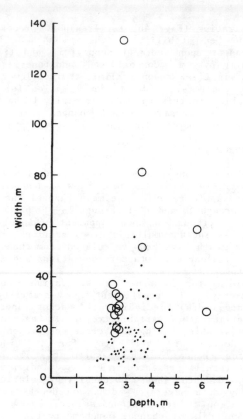

Figure 5. Width-to-depth relations for Goodwin Creek from stereoscopic analysis of ASCS aerial photographs for the years 1937, 1940, 1944, 1953, 1957, 1963 and 1968 (identified by dots) and for the year 1975 (identified by circles) (Ethridge, 1979).

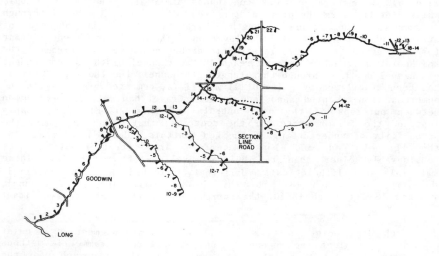

Figure 6. Goodwin Creek reaches.

276

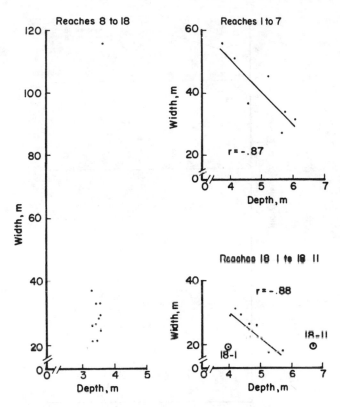

Figure 7. Average widths and depths for Goodwin Creek reaches by process control groups, from 1977 survey.

and negative (r = -.88). Reaches 18-1 and 18-11 were not included in this calculation; the former reach is transitional between the two groups and the latter reach is biased by a road-culvert control.

Iron-cemented sills crop out in several of these headward-most reaches but occur at greater depths below ground surface than similar outcrops downstream. Additionally, thalweg slopes are generally greater for this headward-most length of channel. These two features limit the influence of the iron-cemented sills as morphometric controls.

As noted previously, the channel width and depth are positively related for Johnson Creek upstream of the knickpoint (Fig. 4a). Channel width increases as depth increases. This relation is logical for channels that are enlarging in a uniform manner to some new regime. The two significant correlation coefficients for Goodwin Creek (Fig. 7), however, are both negative. Width-to-depth relations are not simple in this case but involve a more complex relation with channel sinuosity (Fig. 8). Width-to-depth ratios increase with increasing sinuosity for reaches 18-1 through 18-11. This sinuosity is associated with meanders that have minimum present-day lateral movement. For reaches 1 through 8, two relations are obvious. The width-to-depth ratio increases from reach 1 to reach 5 which is typified by an exceptionally large bendway (Fig. 6). A similar sequence is obvious for reaches 6 to 8. As previously noted, channel width was independent of channel depth for the middle reaches and the width-to-depth ratio for these reaches is independent of sinuosity.

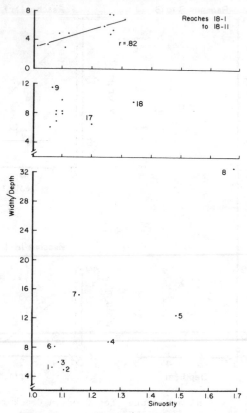

Figure 8.   Width/depth ratios versus sinuosities for Goodwin Creek reaches
by process control groups, from 1977 survey.

Sinuosity has no obvious relation with thalweg slope for Goodwin
Creek in general. As used in this discussion, the thalweg slope is the
difference in elevation from the upstream to the downstream end of the
reach divided by the thalweg length. The two large bendways, in reaches
5 and 8, however, apparently induce some local control of thalweg slope
(Fig. 9). These bendways are both atypical of channel conditions in the
lower reaches of Goodwin Creek. As such, they may impose excessive
resistance to flow primarily as they induce secondary circulation. This
effect of bendways and meanders in general has been discussed by Hickin
(1978) and by Ackers (1980). Hickin (1978) reported as much as an
8-fold increase in flow resistance for meanders relative to downstream
relatively straight reaches of the Squamish River, British Columbia.
Hickin (1978) and Bathurst et al., (1979) both discuss the significance
of secondary flow associated with bendways. For Goodwin Creek, the
thalweg slope of reach 4 is greater than that immediately upstream or
downstream. Reach 5, immediately upstream, contains one of the
exceptionally large bendways. This bendway has developed by erosion of
relatively weak channel-fill material in the bank. It has migrated 81
feet downstream in the last 3 years. A similar sequence is associated
with the bendway at reach 8 but is confounded by the knickpoint at the
lower end of this reach. The thalweg slope for reach 8, presented in
Fig. 9, has been reduced by an amount equivalent with this drop and is
identified as 8a. This adjustment results in the same sequence as

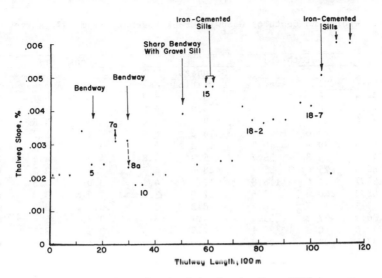

Figure 9. Thalweg profile for Goodwin Creek, from 1977 survey.

described for the bendway at reach 5. This bendway at reach 8 was
present on the 1937 photographic record but has meandered considerably
since that time.

We speculate that these large bendways induce excessive, atypical
flow resistance sufficient to disrupt the movement of coarse sediment
through the bendways. The coarse sediment is stored as point bar
materials at the bendways and as bed material immediately upstream.
This sediment accumulation controls the thalweg slope upstream of the
bendway. Downstream of the bendway, the flow is sediment-deficient
resulting in bed erosion and greater thalweg slopes. This condition is
probably aggravated by the relatively large proportion of secondary
flow. It is interesting to note that reaches upstream of the large
bendways are presently stable and the banks are typically well-
vegetated. This observation includes the reaches upstream of the
bendways at reaches 5, 8 and 18.

The iron-cemented sandstone and cemented gravel sills produce a
comparable change in thalweg slopes. Such sills are present in reaches
13, 15, 16, 18-8, 18-10 and 18-11 (Fig. 9). These sills are effectively
low-drop structures with overfalls, in some cases approaching 2-feet in
relief. Reaches containing sills, therefore, have excessive thalweg
slopes whereas reaches upstream of such sills have atypical low thalweg
slopes. These sills have no observable relation with sinuosity.

SUMMARY

Morphometric values for Goodwin Creek channel are highly variable.
Channel widths and depths at point locations are highly variable and
width-to-depth ratios are inconsistent. Average widths and depths per
reach are similarly inconsistent for Goodwin Creek channel in its
entirety but are coherent within each of three functional segments.
This functional segmentation of the channel results from the properties
and distributions of the valley-fill and older materials which
constitute present day bed and bank materials. Thalweg slope is
similarly controlled by certain of the valley-fill and older materials
which crop out at point locations along the channel thalweg.
Additionally, this slope is influenced by the several large bendways

Table 1. Width and depth of Goodwin Creek reaches, from 1977 survey.

Reach	Width (m)				Depth (m)			
	Max.	Min.	Average	Standard Deviation	Max.	Min.	Average	Standard Deviation
1	34.4	28.1	31.6	3.2	6.2	5.8	6.1	0.2
2	30.6	18.7	27.4	4.9	5.8	5.5	5.7	0.1
3	51.6	26.3	34.1	9.4	6.0	5.7	5.8	0.2
4	64.1	26.6	45.1	14.6	5.9	4.7	5.2	0.5
5	57.3	30.9	50.9	11.3	4.6	3.1	4.1	0.6
6	43.1	28.8	36.7	5.9	4.8	4.4	4.6	0.2
7	88.8	25.0	55.7	23.6	4.3	2.9	3.7	0.6
8	135.7	101.6	116.8	17.0	4.1	3.0	3.6	0.5
9	68.8	20.6	37.6	27.0	3.6	2.9	3.3	0.3
10	39.1	18.7	26.2	8.3	3.5	3.0	3.3	0.2
11	36.0	27.2	33.4	4.1	3.5	3.3	3.4	0.1
12	40.6	20.3	28.1	7.9	3.6	3.2	3.5	0.2
13	37.5	20.3	29.3	6.6	3.9	3.4	3.5	0.2
14	34.4	19.1	26.6	5.7	3.8	3.0	3.4	0.3
15	24.1	18.7	21.1	2.7	3.7	3.2	3.5	0.3
16	31.3	19.4	24.8	5.9	4.1	2.8	3.6	0.6
17	26.3	19.4	21.4	3.2	3.8	3.1	3.3	0.3
18			33.2				3.5	
18-1	24.4	14.3	19.3	4.4	4.3	3.7	4.0	0.3
18-2	43.1	20.9	29.7	10.0	5.1	3.9	4.4	0.4
18-3	45.3	21.2	29.6	9.2	4.5	2.8	4.0	0.7
18-4	35.4	24.4	31.4	5.2	4.4	4.0	4.2	0.2
18-5	27.4	15.2	22.5	3.9	4.8	4.6	4.7	0.1
18-6	27.4	16.8	22.2	4.0			4.6	
18-7	38.7	19.8	26.9	10.3	4.6	4.6	4.6	0.0
18-8	36.0	13.7	25.9	8.0	5.3	4.7	4.9	0.3
18-9	18.7	15.6	17.7	1.8	5.5	5.1	5.3	0.2
18-10	20.3	15.6	18.3	2.0	6.0	5.5	5.7	0.2
18-11	21.9	16.2	19.8	2.3	7.3	6.1	6.7	0.5

which have apparently evolved from the interaction of channel modifications with valley-fill controls.

We interpret the preceding morphometric properties to indicate that Goodwin Creek channel is not a true alluvial channel; channel dimensions, shapes, patterns and slopes have not adjusted freely to the present hydraulic regime. In this context, the present drainage system is immature. It is not stable. It will change through time and this change is the long-term corollary of the present, relatively short-term bank and/or bed instability problems. For such short-term instability problems, the mechanism of failure and probably the rate of failure is related to the properties and distributions of the valley-fill or older materials. More significantly for these short term problems, channel morphology is relatively independent of the present runoff regime. A logical continuation of this line of reasoning is that flow properties, such as total flow resistance, secondary flow conditions and the distribution of bed shear stresses, may be dependent to some degree upon channel morphology. The conclusion of this reasoning is that such flow properties are ultimately dependent to some degree upon the properties and distributions of pertinent valley-fill or older materials.

The properties and distributions of pertinent valley-fill or older materials are not related to current environmental conditions. Certainly, neither Johnson or Goodwin Creek channels are representative of the other, even though these two channels occur in adjacent catchments. The functional control or influences of each of the pertinent bed and/or bank materials, however, is constant for these catchments. These materials were deposited and modified by paleo-conditions, primarily climatic and base level controls. In essence, these are relict controls with each control having a characteristic distribution. Whether a catchment is or is not a representative catchment for a specific application will ultimately depend upon the distribution of these relict controls and hence the dominant paleoconditions. Morphometric studies are not sufficient for such evaluations of representative catchments or basins. Process-oriented geomorphic studies are required and such studies are prerequisite to optimum use of predictive models.

REFERENCES

Ackers, P. 1980. Meandering Channels and the Influence of Bed Material. in: Proceedings of the International Workshop on Engineering Problems in the Management of Gravel-Bed Rivers, 23 pp., John Wiley and Sons, Inc. (in press), New York.

Bathurst, J. C., Thorne, C. R. and Hey, R. D. 1979. Secondary Flow and Shear Stress at River Bends. Journal of the Hydraulics Division, American Society of Civil Engineers, Vol. 105 (HY10), pp. 1277-1295.

Brinkman, R. 1970. Ferrolysis, a Hydromorphic Soil Forming Process. Geoderma, Vol. 3, pp. 199-206.

Ethridge, L. T. 1979. Photogrammetric interpretation of stream channel morphology, Johnson and Goodwin Creeks, Panola County, Mississippi. Master of Engineering Science Project, 24 pp., Department of Geology, The University of Mississippi, University, Mississippi.

Grayson, D. K. 1977. Pleistocene Avifaunas and the Overkill Hypothesis. Science Vol. 195, pp. 691-693.

Happ, S. C. 1968. Valley sedimentation in north-central Mississippi. Proceedings of the Third Mississippi Water Resources Conference, 8 pp., Jackson, Mississippi.

Happ, S. C. 1970. North Tippah valley sedimentation survey. Research Report No. 415, 30 pp., United Stated Department of Agriculture, Agricultural Research Service, Southern Branch, Soil and Water Conservation Research Division.

Hickin, E. J. 1978. Mean Flow Structure in Meanders of the Squamish River, British Columbia. Canadian Journal of Earth Science, Vol. 15(11), pp. 1833-1849.

Pflug, R. 1969. Quaternary Lakes of Eastern Brazil. Photogrammetria, Vol. 24, pp. 29-35.

Saucier, R. T. 1974. Quaternary geology of the Lower Mississippi Valley. Research Series No. 6, 26 pp., Arkansas Archeological Survey, Fayetteville, Arkansas.

Trimble, S. W. 1974. Man-induced soil erosion on the Southern Piedmont 1700-1970. Special Report, 180 pp., Soil Conservation Society of America, Ankeny, Iowa.

Wendland, W. M. and Bryson, R. A. 1974. Dating Climatic Episodes of the Holocene. Quaternary Research, Vol. 4, pp. 9-24.

DYNAMICS OF ALLUVIAL CHANNELS—A PROCESS MODEL

Waite R. Osterkamp and P. E. Harrold, Hydrologists

U.S. Geological Survey

ABSTRACT

A process model of alluvial-channel dynamics is proposed that is based on the differences between delivery and discharge of fluvial sediment in the stream network. It is suggested that when discharge characteristics of a stream are reasonably steady and base-level changes have not been imposed, various channel changes occur on a continuing basis while a balance is maintained among the channel characteristics.

Periods of channel widening are short lived, occurring only during erosive flow. At all other times of discharge, stream channels are narrowing slowly or else are changing in a manner to promote narrowing. With the possible exception of periods after destructive floods, a complex balance exists among discharge characteristics, channel size and shape, delivery rate and size distribution of the sediment load, channel sediment, and channel gradient. In general, widening is accompanied by removal from storage of fine material and the transfer of coarse bank and flood-plain material to the channel bed. Results are an increase of the bed load, an increase in gradient, and a possible decrease in mean water velocity. During periods of channel narrowing opposite changes occur at much slower rates.

BACKGROUND AND TERMS

Alluvial channels exhibit a delicate balance between the discharge of water and sediment and the observable channel characteristics. Over the years, qualitative terms, such as graded stream and dynamic equilibrium, were employed to describe this balance and to suggest that any change in the characteristics of water or sediment discharge "will cause a displacement of the equilibrium in a direction that will tend to absorb the effects of the change" (Mackin, 1948, p. 471). Use of a general term such as displacement was necessary because several channel variables, including width, depth, gradient, and sediment properties, can reflect the change. In addition, the rates of change of these variables are highly variable. Although the complexity of fluvial systems generally has been acknowledged, many quantitative studies have related two selected variables while ignoring other variables that may be responsible for or responsive to this displacement of equilibrium. If other variables are held constant, these quantitative studies provide valid relations, but divert attention from the complex nature of fluvial systems. Additionally, attention is often placed on the quantitative relations without fully considering the processes operating in the system. Because the various fluvial processes occur at different rates, an understanding of the processes and their time scales seems essential to a quantitative description of alluvial channels.

The graded-stream concept, emphasizing displacement, provided the valuable function of pointing out the equilibrium that occurs in some stream channels. The concept was applied to channels that show a "long-term" balance between erosion and deposition (Mackin, 1948, p. 470). Based on earlier work by Gilbert (1880, p. 117-118), Leopold and Maddock (1953, p. 51) and Hack (1960, p. 85-86) extended the concept to degrading channels that nevertheless show a balanced condition. They proposed that all essentially balanced streams have a condition of quasi or dynamic equilibrium. It is proposed here that use of these terms should imply recognition that, during a short interval of geo-morphic time, channel conditions can change markedly, although mean delivery rates from upland areas and variability of water and sediment discharge remain measurably unchanged (that is, a nearly balanced condition has prevailed). Consideration of fluvial processes can suggest why a stream channel, measurably in grade at two different times, might exhibit significantly different channel characteristics.

The purpose of this discussion is to propose a generalized scheme or model representing the manner in which alluvial perennial-stream channels change through time. In this paper, an alluvial channel is defined as one fully bounded at most times by sediment derived from pre-vailing conditions of the water-sediment discharge, regardless of the particle-size distributions. In most respects the model is not original but combines a number of well known and accepted ideas. Although the model is based on and partially supported by a variety of previous studies, the principal support is obtained from hundreds of hydrologic and geomorphic data sets collected from streamflow gage sites, mostly in the western half of the conterminous United States. Although some parts of the model are inferences, others are based on statistical relations among width, dis-charge, channel-sediment, and gradient data collected from the various gage sites. Previous models of alluvial-channel dynamics have, like this one, suggested a tendency for decreasing gradient by means of increasing sinu-osity (Dury, 1969; Keller, 1972; Langbein and Leopold, 1966; and Tinkler, 1970). This paper differs from previous discussions by emphasizing the effect of fluvial sediment on channel form. Because changes through time are discussed, brief consideration is given to the amount of time required for various channel changes to occur and how those changes may be con-trolled by sediment supply. A final purpose is to offer suggestions concerning how terms such as quasi equilibrium, which are essentially time independent, relate to a model of channel dynamics that is concerned with changes through years to decades.

CHANNEL DYNAMICS

A basic premise of this paper, that the characteristics of an alluvial stream channel tend to adjust to the water (Q) and sediment (Q_s) discharges conveyed through the channel, is identical to the implications of grade or quasi equilibrium. In a given time interval, a graded condition for a channel reach requires that no net erosion or deposition occurs. Thus, the shear stress, T_o, along the wetted perimeter of the channel must be distributed to cause no net erosion or deposition. This requirement demonstrates that:

$$T_o = f(Q, Q_s) \tag{1}$$

and that any change in Q or Q_s produces a change of T_o. Furthermore, the shear stresses on the bed and banks must be sufficient to move the coarse sediment sizes supplied to the stream from upland areas as well as to maintain transport of suspended sediment without causing

bank erosion. Because no natural stream has constant discharges of
water and sediment, relation 1 represents a time integrated value of T_o
resulting from the entire ranges of Q and Q_s.

For a balanced condition, the measurable expressions of the shear-
stress distribution are the variables of channel geometry and sediment
forming the perimeter. Hence:

$$[W, D, G, n, d_{50}, SC, ...] = f (Q, Q_s) \qquad (2)$$

where W, D, G, n, d_{50}, and SC, respectively, are channel width, mean depth,
gradient, roughness, median particle size, and silt-clay content of channel
material. Relation 2 is not limited to these variables, but those included
appear to be most prominent. For example, hydraulic variables such as
velocity distribution and turbulence could be included, but they also would
cause redundancy. Further discussions in this paper assume that no
changes in the discharge characteristics of water and sediment occur
through geomorphic time periods, and that no base-level changes occur that
would cause aggradation or degradation.

Storage of Sediment

The concept of dynamic equilibrium was used by Gilbert (1880) to
explain the land forms of an entire drainage basin, not just the stream
channel. In relation 2, therefore, the independent variables Q and Q_s
refer to water and sediment delivered to the stream network from upland
areas—surfaces higher than flood-plain levels. Departures from relations among water discharge, sediment characteristics, and channel
geometry later were treated stochastically, without providing a more
detailed analysis of the processes causing the deviations. The present
analysis restricts attention to the drainage network in order that a
portion of the deviations from the geometry-discharge relations, which
are necessarily treated as random in large-scale models, can be considered.

If a graded stream reach, or one in dynamic equilibrium, had a balance through time between water and sediment received from upland areas
and water and sediment discharged at the lower end, empirical relations
could accurately predict the channel characteristics at a site. Variability of discharge complicates the problem, however, with the result
that the amount and sizes of sediment transported through time and
leaving the reach may be significantly different from the sediment
delivered from the uplands. Instead, Q_s of relation 2, as used here,
is the mean sediment discharge through time at a site or short channel
reach and is determined by the availability of sediment from both upland
and flood-plain sources. Owing to variation of water discharge through
time, availability and sizes of sediment from flood-plain (stored)
sources can vary greatly. For the purposes of this paper, delivery of
sediment from upland sources is assumed generally constant, but the
fraction and sizes of the sediment that are stored as channel and flood-
plain deposits are assumed variable. It is in part this variability of
storage rates that accounts for the randomness inherent in the large-
scale dynamic-equilibrium model. Likewise, it is these changes or
processes that are basic to this restricted model of fluvial dynamics.

Elements of the Model

The following observations describe the manner in which the characteristics of alluvial stream channels change through time. It is
emphasized that the changes can occur without implying that a general,

relatively long-term fluvial balance has been upset to cause the change. Referring to relation 2, it is suggested that as one of the dependent variables changes, such as width, accompanying changes in the other dependent variables also occur but that mean and peak discharges of water and sediment may remain constant. For this discussion, channel width is defined at the lower limit of permanent vegetation--the active-channel level.

1. Periods of channel widening are short-lived, occurring in large part during erosive flow and perhaps facilitated by destruction of riparian vegetation or other channel-stabilizing properties (fig. 1). The widening might result from a single flood, as has that of numerous streams throughout the United States, or it might occur in stages, as at the Cimarron River in southwestern Kansas (Schumm and Lichty, 1963) and the Gila River in Safford Valley, Arizona (Burkham, 1972). Accompanying the rapid widening of a channel is reduction of mean depth, channel straightening and increase of gradient, generally an increase of median particle size of bed material, and an increase in the bed-load discharge of the stream after the flood (fig. 1).

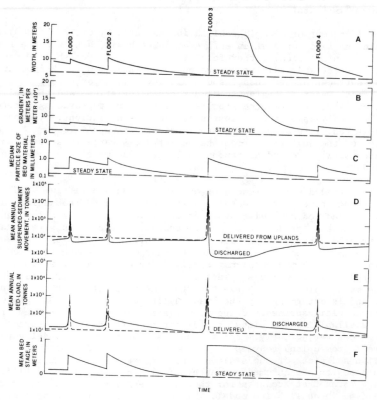

Figure 1. Graphs showing changes with time of channel variables.

Solid lines are hypothetical changes of width, gradient, particle size, suspended and tractive-sediment discharge, and stage; dashed lines represent sediment delivery; and broken lines represent steady-state conditions.

2.   At all other times of discharge, alluvial channels are
narrowing slowly or else are changing in a manner to promote
narrowing. Other channel properties change in a manner consistent with the narrowing. Relations among the various channel
properties and the rates of change of those properties are
dependent primarily on the delivery characteristics of water
and sediment to the alluvial channel and the prevailing conditions of sediment storage in the flood-plain area.

Implicit in this generalization of narrowing is a nearly
continual net storage of a part of the total sediment load. If
the sediment delivered to a stream has an abundance of fine
sizes, the rate of narrowing after erosive flow may be relatively
fast. Bank accretion at relatively stable channel sections
develops by the addition to the channel sides of fine-grained,
suspended sediment.   Support is provided by field observations,
particle-size data (Osterkamp and Wiseman, 1980), and petro-
graphic data of bank-material samples. For example, samples
collected for an M.S. thesis (C. F. Glazzard, personal commun.,
1980) show that bank material of channels draining basins with
active coal mines of eastern Tennessee generally contains about
twice the amount of coal found in bank material of geographically
similar but unmined basins. Banks are defined here as the
vertical to subvertical surfaces forming the sides of the active-
channel section.

When available sediment from upland areas is deficient
in fine sizes, an extended period of time may be required
for storage of the cohesive fine sizes after a channel-widening
discharge. For example, much of the Sand Hills area in Nebraska
has soils with little silt and clay; streams of the area tend to
be braided. If a normally well-shaped stable channel is braided
owing to a highly destructive flood, the braided channel pattern
may persist for many years until sufficient fines can be restored
in the channel alluvium to cause narrowing and formation of well-
defined channel banks. Exceptions to the generalization include
stream channels with very stable discharges (a lack of erosive
flow) and highly ephemeral stream channels, which are unable to
heal or narrow after floods because discharge of water and
sediment occurs infrequently.

3.   In general, the rapid widening of an alluvial channel is
accompanied by the removal from storage (as cohesive alluvium)
of fine-grained material and the transfer of sand and coarser
sizes from storage in the banks and flood-plain deposits to the
channel bed. The changes in available sediment result in an
increase of the bed load, a change in the medial particle size
of bed material consistent with the increased availability of
coarse sediment, an increase in channel gradient (as a decrease of
sinuosity), and a possible decrease of mean water velocity. During the longer periods in which erosive floods or other channel-
widening events do not occur, the channel narrows, but at a progressively decreasing rate, thus storing sediment as bed and bank
material. Owing to the smaller surface area of the channel and
consequent change in shear-stress distribution, the tractive movement
of sand and coarser sediment decreases, thereby decreasing median-
particle size. Other characteristics also change, these include
decreases of channel roughness and channel gradient (as an increase
in sinuosity), and a possible increase of mean velocity of flow.

Normal processes of fluvial sorting, and those of erosive

floods in particular, tend to enrich a channel bed in medium-grained sand and coarser sediment sizes. Likewise, stable al-luvial banks are normally enriched in fine sizes (diameter less than 0.35 millimeter) and are deficient in sand sizes between 0.35 and 1.3 millimeters (Osterkamp, 1980). Channel narrowing by accretion of fine sediment sizes can occur at most discharge rates; net loss of the cohesive fine sizes occurs only during flood discharges when the shear stresses on the banks exceed the critical shear stress for the bank sediment.

The widening and other alteration of a channel by an erosive flood discharge distributes the shear stresses more uniformly across the channel bed than in the period before the flood (see data by Chow, 1959, p. 1969; Kartha and Leutheusser, 1970; and Prasad and Alonso, 1976). A relatively uniform shear-stress distribution and an increased channel gradient favor the tractive transport of the relatively coarse material that normally forms a channel bed after an erosive flood (Gessler, 1971, p. 83). This relation has been demonstrated in laboratory studies by Khan (1971), who showed an exponential increase in sand dis-charge with width-depth ratio for streams of constant water discharge (fig. 2).

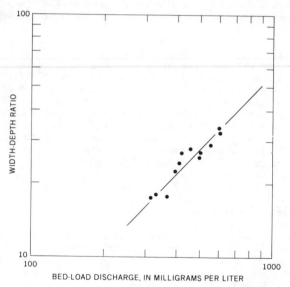

BED-LOAD DISCHARGE, IN MILLIGRAMS PER LITER

Figure 2.   Relation between width-depth ratio and bed load in straight laboratory channels of constant discharge (modified from Khan, 1971). Data points are indicated by dots.

4.   A steady state can be represented conceptually, however, by conditions in which the channel size, shape, sediment character-istics (roughness), and gradient are constant through time, and the transport rates of all particle sizes of sediment are equal to the delivery rates of these sizes to the channel. Steady-state conditions for natural fluvial systems probably never occur owing to discharge variability.

Assuming nearly continuous additions to storage of part of the fine sediment delivered from upland areas, the suspended load discharged by a stream is less than that supplied by upland erosion. Only if a channel narrows to a condition in which storage of sediment, as channel or flood-plain material, no longer occurs can the system be regarded as having a steady state. This condition has been approximated by laboratory (flume) studies, regime canals (Leopold and Maddock, 1953, p. 43-44), and some spring-effluent channels, but generally is not found in relatively large streams unless the discharge is nearly constant owing to regulation.

DISCUSSION

Numerous papers have given bivariant relations between discharge characteristics and variables of channel geometry, but only a small number of these provide relations as a function of the sediment characteristics of a stream channel (for example, Lane, 1957; Schumm, 1960a; 1960b; 1968; Henderson, 1961; Osterkamp, 1977; 1978; 1980). A summary conclusion of these studies is that for an alluvial channel of specified water- and sediment-discharge characteristics, the geometry of the channel can be related directly to the material forming the bed and banks. Thus, if the discharge characteristics remain constant but the channel-material properties change, changes in the geometry of the channel can be expected.

Changes Through Time

The schematic graphs of figure 1 suggest changes through time of several channel variables, but are based on various power-function relations developed from data collected at several hundred streamflow gage sites of the western United States. Typical of the power functions are those relating mean discharges to widths of channels having similar sediment properties (Osterkamp, 1977; 1980) and those relating mean discharges to gradients of channels having similar sediment properties (Osterkamp, 1978).

The results suggest that when general conditions of water and sediment discharge are known, a variety of relatively stable channel conditions is possible, each possibility being associated with the recent flood history of the stream (figure 1). Any two of three variables may determine the character of a stream reach. These are: 1) channel materials, 2) recent flood history, and 3) channel geometry. Thus, if two of these variables are known, the other can be assumed.

The graphs of figure 1 are hypothetical and represent changes that might occur in a channel with a mean discharge of $1.0 \text{ m}^3/\text{s}$ (cubic meters per second) and a sediment load predominantly of clay, silt, and fine-sand sizes. Similar graphs can be drawn for streams of other water- and sediment-discharge characteristics (Osterkamp, 1980). Changes of channel width (measured from the lower limit of perennial vegetation in the channel section) and gradient are among the easiest variables to measure (fig. 1A and B). Both are increased by moderately erosive floods (floods 1 and 2), but the increase in gradient may be negligible if the channel is merely widened without significant change of channel position. Narrowing and reduction of gradient, by increased meandering, follow the floods. A large flood (flood 3), that destroys the channel pattern as well as extensive flood-plain area, produces a wide, braided, and relatively straight channel. Significant reductions of channel width and gradient after the destructive flood cannot occur until sufficient amounts of fine sediment sizes, represented as d_{50}, the median-particle size of bed material (figure 1C), have been restored as channel and flood-plain material.

Storage of fine, cohesive sediment sizes permits formation of stable alluvial banks that can resist relatively high shear stresses. After storage of fines, the change from a braided to meandering and vegetated channel pattern may be relatively rapid (within a few years). Flood 4 (fig. 1) represents an interruption of the channel adjustments after flood 3.

Included in figure 1 (D and E) are graphs of inferred changes through time of suspended sediment and tractive, or bed-load, movement. The solid lines represent the sediment discharged past a channel site, whereas the dashed lines suggest the relative amounts of sediment delivered from upland sources to the stream network above the channel site. Differences between the solid and dashed lines, of course, represent the amounts of sediment being stored or taken from storage.

The shapes of these curves assume constant conditions of mean runoff and variability of water discharge except for the four floods depicted. Natural channels, of course, show sufficient annual variation of discharges that hydrologic data would rarely, if ever, well define the changes through time that are illustrated. Therefore, the two graphs, particularly the curves representing the amounts of sediment delivered from upland areas relative to sediment discharge at a channel site, are necessarily speculative. The generalized shapes of these curves, however, are supported indirectly by documented changes of other channel variables, such as width and gradient, by relations such as that between width-depth ratio and bed load (fig. 2), and known processes of flood-plain construction and destruction. For example, a significant part of the sand sizes delivered during most floods is stored as overbank deposits, thereby suggesting greater amounts delivered than discharged at these times (fig. 1E).

Additional indirect support for graphs D and E (fig. 1) is provided by typical changes with time of channel-bed stage during and after an erosive flood (fig. 1F). Numerous examples (and exceptions) can be provided to support figure 1F. Included here are generalized stage-time relations for five alluvial channels of Kansas immediately before and after historic regional flooding in the summer of 1951 (fig. 3). When the flood plain of a narrow channel with stable banks is destroyed by an erosive flood, the new wide and possibly braided channel assumes a stage higher than the previous bed level owing to the redistribution of flood-plain deposits and possibly to an influx of sediment from upland areas (fig. 1F). After the flood, a steepened channel gradient promotes tractive sediment movement in excess of the amounts of sand and coarser sizes delivered to the stream network. Thus, a net loss of the material forming the bed occurs, and the loss is reflected by channel degradation (fig. 3). Concurrent with the net loss of coarse sizes from storage is net storage of fine sizes (fig. 1D), which is necessary for reconstruction of stable alluvial banks and a flood plain (Schumm and Lichty, 1963). The total delivery and discharge of sediment during a time interval may be equal, therefore, but the size composition of channel material changes during the period. Because alluvial channels appear to lose coarse sediment during periods of normal discharge rates, it is inferred that mass balance is maintained by an excess delivery of these sizes during floods (fig. 1E).

The broken lines of figure 1A, B, C, and F suggest steady-state conditions, the hypothetical lower limits for channel width, gradient, bed-material size, and stage. Although most channels probably never reach a steady state as defined by figure 1, a time immediately before flood 3 illustrates this ideal. Not only is it assumed that minimum

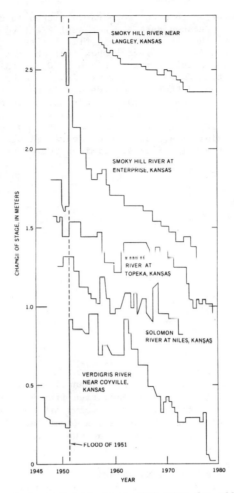

Figure 3.   Changes of channel-bed stage with time for alluvial channels
of Kansas (compiled from published hydrologic records of
the U.S. Geological Survey).

values of width, gradient, and particle size must be associated with a
steady state, it is inferred also that a mass balance must exist between
delivery and discharge of all particle sizes. Hence, the time preceding
flood 3 shows equal amounts of delivery and discharge for both suspended
and tractive sediment (fig. 1D and E).

Owing to a scarcity of pre-1950 records, few published data for
specific streams that directly support the suggested changes of figure 1
are available. Among the best documentation, however, is that of Burkham
(1972) for the Gila River in Safford Valley, Arizona. Figure 4, which
is modified and expanded directly from Burkham's work, is a representation of channel-width changes of the Gila River between 1870 and 1970.
The Gila River serves as a convenient example for the present discussion
because the stream has a substantial and well-graded sediment supply and
is susceptible to extensive channel erosion owing to its flashy nature
and relatively fragile vegetation. Pronounced channel changes for the

Gila River are inferred to occur more rapidly than is typical for most alluvial channels. Because data are not sufficient to define fully the curves of figure 4, the changes through time are necessarily in part interpretive. The data do, however, show that channel widths of the Gila River have been highly variable since 1890 and that the variation can be related to the timing of major floods. If other width data were available, it is assumed that the curves of figure 4 could be redrawn to show the effects of medium and small floods not shown now. Although not presented here, gradient data (Burkham, 1972) agree well with the trends proposed in figure 1B; stage data seem inadequate to support figure 1F, and indicate only minimal aggradation or degradation. Available channel-sediment data for the Gila River are limited to recent years.

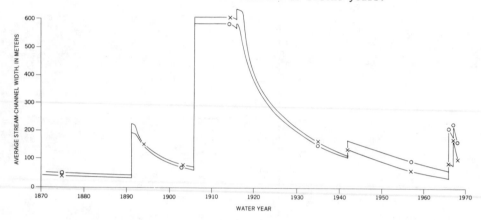

Figure 4. Graph showing average stream-channel widths, Gila River in Safford Valley, Arizona (modified from Burkham, 1972, pl. 3). Circles and crosses, respectively, are data points for channel reaches adjacent to Thatcher, Arizona, and Fort Thomas, Arizona. Vertical portions of the curves are times of rapid channel widening during major floods.

Although it is an oversimplification to relate only the occurrence of floods to channel changes of the Gila River, a substantial correspondence seems evident. Additional support is provided by similar, though less detailed, channel-geometry information for the Cimarron River in southwestern Kansas (Smith, 1940; Schumm and Lichty, 1963). Sediment-size data as well as photographic evidence for the Cimarron River are inferred to be in general agreement with figure 1C, D, and E.

The graphs of figure 1 pertaining to sediment (graph C in particular) and figure 4 are consistent with the fining-upwards alluvial sequences that have been recognized by various geomorphologists and sedimentologists. In addition, the model presented here provides a partial explanation for the cyclicity of these sequences in the stratigraphic record (Allen, 1965).

Time Scales

The recoveries indicated in figures 1, 2, and 4 are measured in time intervals of years to decades. The rapidity with which channel changes can occur is a function of the mean rate at which a stream can accomplish the work required to affect the change. Hence, a much greater rate of work is accomplished during periods of flooding than during equal periods of more normal streamflow. Returning to relations 1 and

2, the shear stresses, excess or erosional shear stresses, and changes in the channel characteristics are greatest during times of peak discharges.

The rates at which different channel changes proceed during periods of normal streamflow differ due to the differing amounts of work, or sediment movement, required for the various changes. For example, channel narrowing requires the redistribution or accretion of a small amount of channel sediment relative to that needed for a proportional reduction of gradient. Except for times of major flooding or change of channel pattern (during and after flood 3), therefore, the rates of gradient change in figure 1 are depicted as being smaller than those for width. For this reason, when the adjustment of a channel is upset, changes in width are normally apparent before changes in gradient. For example, recent degradation of the Kansas River channel near Bonner Springs, Kansas, seems to be mostly the result of sand and gravel extraction for construction. No evidence of channel degradation is apparent in the channel upstream from the Bonner Springs reach, but recent channel widening has been measured.

Sediment transport and copping increases of most perennial streams are adequate to produce significant narrowing at the active-channel reference level within several years. If widths are measured at the bankfull (or flood-plain) level, however, which is normally higher and inundated much less frequently than the active-channel level, a change in width requires the reworking of large amounts of channel sediment relative to the ability of the stream to accomplish work. Thus, width adjustments at the bankfull level may continue through many decades or centuries in response to a change in discharge characteristics. Because gradient change through increased meandering also necessitates the redistribution of channel and flood-plain material below the bankfull level, relatively large reductions of channel gradient require similar time periods as do changes of bankfull width. It is inferred that median particle size of bed material and gradient are largely interdependent. Thus, the two variables probably change at similar rates. Other measures of channel material, such as the silt-clay content of the banks, are related to active-channel width and may change significantly within a few years.

SUMMARY

Implicit in much of this discussion is the suggestion that the terms graded stream, dynamic equilibrium, and quasi equilibrium serve the important function of drawing attention to the balance that exists in most alluvial stream channels. Assuming that the variables of a large majority of channels maintain a balance, however, the terms have scant utility for distinguishing differences among channels. If measurement techniques were adequate, perhaps differences among graded channels could be described in terms of the imbalance between delivery and discharge of specified particle-size ranges of fluvial sediment (fig. 1D and E). At present, of course, such measurement abilities are not available on a practical level, but an expression of the imbalance could suggest the deviation from a condition of steady state that a channel exhibits.

A steady-state condition for an alluvial channel is defined here as the condition of no net aggradation or degradation (during a short span of geomorphic time) for any of the particle sizes delivered to or discharged from the stream network. This condition is an ideal state which is probably not reached by most streams. As indicated by figure 1, however, it provides a basis for comparing the relative adjustment of a channel toward that ideal. The hypothetical channel of figure 1 is in a balanced or graded condition at all times except during floods and a period after flood 3. The degree of adjustment within that balanced condition, however, shows continual change. It is proposed, therefore, that the term adjustment be reserved in this context when describing fluvial systems.

REFERENCES

Allen, J. R. L. 1965. Fining-Upwards Cycles in Alluvial Successions. Geological Journal, Vol. 4, pp. 229-246.

Burkham, D. E. 1972. Channel Changes of the Gila River in Safford Valley, Arizona, 1846-1970. U.S. Geological Survey Professional Paper 655-G, 24 pp.

Chow, V. T. 1959. Open-Channel Hydraulics, McGraw-Hill Book Co., New York, p. 169.

Dury, G. H. 1969. Relation of Morphometry to Run-Off Frequency. in: R. I. Chorley (Editor), Water, Earth, and Man, pp. 419-430, London Methuen.

Gessler, Johannes 1971. Aggradation and Degradation. in: H. W. Schen (Editor), River Mechanics, pp. 8-1 to 8-24, H. W. Schen, Fort Collins, Colorado.

Gilbert, G. K. 1880. Report on the Geology of the Henry Mountains, 2d edition. U. S. Geographical and Geological Survey of the Rocky Mountain Region, 170 pp., U.S. Government Printing Office, Washington, D.C.

Hack, J. T. 1960. Interpretation of Erosional Topography in Humid Temperate Regions. American Journal of Science, Bradley Volume, Vol. 258-A, pp. 80-97.

Henderson, F. M. 1961. Stability of Alluvial Channels. American Society of Civil Engineers Proceedings, Journal Hydraulics Division, Vol. 87, HY6, pp. 109-138.

Kartha, V. C., and Leutheusser, H. J. 1970. Distribution of Tractive Force in Open Channels. American Society of Civil Engineers Proceedings, Journal Hydraulics Division, Vol. 96, HY7, pp. 1469-1483.

Keller, E. A. 1972. Development of Alluvial Stream Channels: A Five-Stage Model. Geological Society of America Bulletin, Vol. 83, pp. 1531-1536

Khan, H. R. 1971. Laboratory Study of Alluvial River Morphology. Unpublished Ph.D. dissertation, Colorado State University, Fort Collins, Colorado, 189 pp.

Lane, E. W. 1957. A Study of the Shape of Channels Formed by Natural Streams Flowing in Erodible Material. Sediment Series No. 9, 106 pp., U.S. Army Engineer Division, Missouri River, M.R.D.

Langbein, W. B., and Leopold, L. B. 1966. River Meanders--Theory of Minimum Variance. U.S. Geological Survey Professional Paper 422-H, 15 pp.

Leopold, L. B., and Maddock, Thomas, Jr. 1953. The Hydraulic Geometry of Stream Channels and Some Physiographic Implications. U.S. Geological Survey Professional Paper 252, 57 pp.

Mackin, J. H. 1948. Concept of the Graded River. Geological Society of America Bulletin, Vol. 59, pp. 463-511.

Osterkamp, W. R. 1977. Effect of Channel Sediment on Width-Discharge Relations, with Emphasis on Streams in Kansas. Kansas Water Resources Board Bulletin 21, 25 pp.

Osterkamp, W. R. 1978. Gradient, Discharge, and Particle-Size Relations of Alluvial Channels in Kansas, with Observations on Braiding. American Journal of Science, Vol. 278, pp. 1253-1268.

Osterkamp, W. R. 1980. Sediment-Morphology Relations of Alluvial Channels. Proceedings, American Society of Civil Enginners Symposium on Watershed Management, Vol. 1, pp. 188-199, Boise, Idaho.

Osterkamp, W. R., and Wiseman, A. G. 1980. Particle-Size Analyses of Bed and Bank Material from Channels of the Missouri River Basin. U.S. Geological Survey Hydrologic Data Open-File Report 80-429, 31 pp.

Prasad, S. N., and Alonso, C. V. 1976. Integral-Equation Analysis of Flows over Eroding Beds. Proceedings, American Society of Civil Engineers Symposium on Inland Waterways for Navigation, Flood

Control, and Water Diversions, Vol. 1, pp. 760-772, Colorado State University, Fort Collins, Colorado.

Schumm, S. A. 1960a. The Effect of Sediment Type on the Shape and Stratification of Some Modern Fluvial Deposits. American Journal of Science, Vol. 258, pp. 177-184.

Schumm, S. A. 1960b. The Shape of Alluvial Channels in Relation to Sediment Type. U. S. Geological Survey Professional Paper 352-B, pp. 17-31.

Schumm, S. A. 1968. River Adjustment to Altered Hydrologic Regimen, Murrumbidgee River, Australia. U.S. Geological Survey Professional Paper 598, 65 pp.

Schumm, S. A., and Lichty, R. W. 1963. Channel Widening and Flood-Plain Construction along Cimarron River in Southwestern Kansas. U.S. Geological Survey Professional Paper 352-D, pp. 71-88.

Smith, H. T. U. 1940. Notes on Historic Changes in Stream Courses of Western Kansas, with a Plea for Additional Data. Kansas Academy of Science Transactions, Vol. 43, pp. 299-300.

Tinkler, K. J. 1970. Pools, Riffles, and Meanders, Geological Society of America Bulletin, Vol. 81, pp. 547-552.

Section 5
RUNOFF

A STUDY ON MATHEMATICAL
RAINFALL - RUNOFF MODELING

Hua Shi Qian. Professor, Senior Engineer

Wen Kang, Li Die Juan, Kan Gui Sheng
Jim Guan Sheng and Li Qi. Engineers

Nanjing Hydrological Research Institute, Min.
of Water Conservancy, Nanjing, China

ABSTRACT

It is necessary to distinguish between the mechanism and the model in studying the rainfall-runoff problem. For example, in computing runoff so far as a point is concerned, the hydraulic method would be adopted to describe the flow movement both in non-saturated and saturated soil with the partial differential equations developed by solving the continuity equation and Darcy's Law simultaneously. Several well known infiltration formulae of Child, Philip and Horton, etc. could be obtained for some special initial and boundary conditions. Consequently, we may obtain the law of point infiltration when the parameters of these formulae have been determined. When the rate of water supply or rainfall is greater than that of infiltration f, i.e. i>f, then the surface runoff would occur; the condition of i>f is necessary for the formation of surface flow, it would be called the mechanism.

However, the basin is much more complex than a point. It is very difficult to tackle the problem of watershed runoff production and its flow concentration based on hydraulic method entirely, but some characterization must be adopted to establish the conceptual model. For example, the watershed runoff production modeling would be developed by making use of the point infiltration rate for simulating the average basin infiltration rate and by considering the stochastic areal distribution of the point infiltration rate simultaneously. This model with its simplicity and clear logical inference is suitable for the ruoff computation either in humid or in arid regions. According to the preliminary texts in several basins of China, we found this model to exhibit successful results.

For flow concentration, a lot of discussion centered on the IUH and its relation to the general linear differential equation. The initial value problem of solving the general differential equation was pointed out and settled. We have derived the formulae for converting the Nash IUH as well as the Muskingum successive routing IUH into UH.

The non-linearity of flow concentration attracted our attention too. Early in the fifties, Prof. Hua put forward a diagrammatic decomposition method to compute the total runoff without the effect of storage in stream nets and derived the formula of IUH exhibiting steep rising segment and slow falling segment in regions where subsurface flow was prominent.

POINT INFILTRATION FORMULAE

The point infiltration formula under the condition that rainfall rate being always greater than infiltration capacity would be developed from hydraulics. The partial differential equation, or so called Richards equation, for describing the flow movement within un-saturated soil may be used,

$$\frac{\partial \theta}{\partial t} = \frac{\partial}{\partial z}\left[D(\theta)\frac{\partial \theta}{\partial z}\right] + \frac{\partial}{\partial z}K(\theta) \tag{1}$$

where θ = moisture content, $D(\theta)$ = hydraulic diffusivity, and $K(\theta)$ = hydraulic conductivity.

The solution of equation (1) is far from easy. Therefore a number of authors have suggested some empirical relationships between the unsaturated hydraulic conductivity K or the hydraulic diffusivity (D) and the moisture content, and then obtained several infiltration equations. For example, if

$$\frac{\partial \theta}{\partial t} = \frac{\partial^2 \theta}{\partial z^2} \tag{2}$$

$$\theta(z,o) = \theta_o; \quad \theta(o,t) = \theta_5$$

The solution of equation (2) for infiltration f is,

$$f = (D/\P)^{1/2} t^{-1/2} (\theta_5 - \theta_o) + K_o \tag{3}$$

where θ_o and θ_5 denote respectively the initial moisture content and the saturated moisture content on the surface soil . K_o denotes the initial infiltration which is equal to the hydraulic conductivity corresponding to the initial moisture content θ_o. Equation (3) is called the Child equation.

When

$$\frac{\partial \theta}{\partial t} = \frac{\partial^2 \theta}{\partial z^2} + k\frac{\partial \theta}{\partial z} \tag{4}$$

$$\theta(z,o) = \theta_o; \quad \theta(o,t) = \theta_5$$

The solution of equation (4) would be approximated by

$$f = (D/\P)^{1/2} t^{-1/2} (\theta_5 - \theta_o) + \frac{k_1 + k_2}{2} \tag{5}$$

where K_2 denotes the hydraulic conductivity corresponding to the saturated moisture content θ_5, and k_1 that of the initial moisture content θ_o.

When

300

$$\frac{\partial \theta}{\partial t} = \frac{\partial}{\partial z}\left(D \frac{\partial \theta}{\partial z}\right) + \frac{\partial k}{\partial z}$$

$$\hspace{10cm} (6)$$

$$\theta(z,o) = \theta_o \ ; \quad \theta(o,t) = \theta_s$$

then one obtains the Philip infiltration equation

$$f = S/2 \ t^{-1/2} + A$$

$$\hspace{10cm} (7)$$

where S and A are parameters.

It is clear from the infiltration equations that the infiltration capacity is proportional to $t^{\frac{1}{2}}$. It is known that the Horton equation can be derived under certain boundary conditions by using Richard equation (Eagleson, 1970).

From an analysis of infiltration data on three experimental basins in semi-arid, semi-humid and humid regions of China, we found the Philip-Type and Horton-Type infiltration equation can be used. Table 1 gives an example.

Table 1.--Example of point infiltration formula.

Regions	Name of Stations	Area (km^2)	Formula (mm/min)
semi-arid	Chabagou	187	$f=5.75 \ t^{-\frac{1}{2}}$
semi- humid	Qiyi	44	$f=0.77 \ t^{-\frac{1}{2}}$
humid	GuanZhow	30	$f=0.2+1.25 \ \exp(-0.025t)$

WATERSHED RUNOFF MODELING

Point Infiltration Capacity

Owing to variation of geography, soil type, moisture content,

and vegetative cover within a basin it is impossible to make use of the point infiltration curve. Suppose that there exists an average infiltration curve (Horton-Type or Philip-Type) within a basin. The average infiltration rate f_t in any time interval consists of an infinity of random point infiltration rates. Thus there exists a stochastic areal distribution of point infiltration rates within the basin.

Distribution within the entire basin. This kind of distribution supposes that increasing of moisture content leads to a decreasing rate of every point infiltration within a basin; also this will be true for the basin average infiltration rate. Since the infiltration capacity changes with time the areal distribution of point infiltration capacity is also time dependent and will be varied with time interval as shown in Fig. 1. In the figure α denotes the percent of area with an infiltration capacity equal to or less than the indicated value, f'_m denotes the maximum value of the infiltration capacity of basin. The area between the curve

301

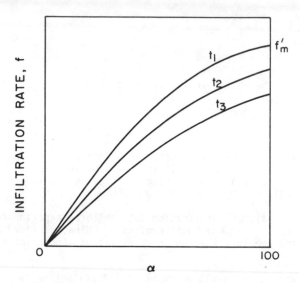

Fig. 1   Distribution Curve within the entire basin. α represents
the percent of area with an infiltration capacity equal to
or less than the indicated value.

and the abscissa in Fig. 1 represents the average infiltration capacity
of the entire basin $\bar{f}_{\Delta t}$ in any time interval Δt.

Distribution within partial basin. This distribution supposes that as
the moisture content increases, a part of basin would be losing its infiltration capacity so that the successive average infiltration capacities
can only be distributed on the remaining basin area as shown in Fig. 2.

How to determine the kind of distribution as well as its mathematical
model has not yet been settled theoretically. A linear distribution within the entire area was used in Stanford model (Crawford and Linsley, 1966).
A distribution within partial area was used in Popov storage model (Popov,
1980; Zhao, et al., 1980).

WATERSHED RUNOFF MODEL

Two principle mechanisms were set up in the model: First, the Philip-
Type and Horton-Type infiltration formulae were adopted for modeling the
average area infiltration capacity of a basin. Second, the parabola of
n-th order would be adopted for the duration curve, i.e.

$$\alpha = i - (1 - f/f'_m)^n \tag{8}$$

When the Horton-Type formula is used, then

$$\bar{f} = f_0 \exp(-kt) \tag{9}$$

The amount of infiltration from 0 to t yields

$$F = \int_0^t \bar{f}\, dt = f_0/k\, [1 - \exp(-kt)] \tag{10}$$

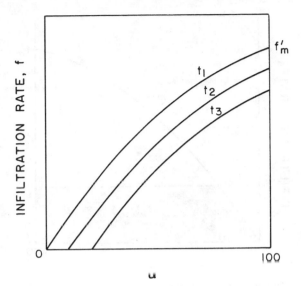

Fig. 2   Distribution Curve within partial basin. α represents the percent of area with an infiltration capacity equal to or less than the indicated value.

or

$$\bar{f} = f_o - kF \tag{11}$$

Since F being the soil moisture content before time t or so called antecedent precipitation p_a, then we have

$$\bar{f} = f_o - kP_a \tag{12}$$

Because of the runoff generation being a time process, the computation for runoff yield must be carried out in time interval Δt, the average areal infiltration amount in that time interval would be

$$\bar{f}_{\Delta t} = \int_t^{t+\Delta t} \bar{f}\ dt = \left[\frac{1 - \exp(-k\Delta t)}{k}\right](f_o - kP_a) \tag{13}$$

From equation (8) we obtain

$$\bar{f}_{\Delta t} = \int_0^{f'_m}(1 - \alpha)\ df = \int_o^{f'_m}(1 - f/f'_m)^n\ df = \frac{1}{1+n}\ f'_m \tag{14}$$

For distribution within the entire area subject to the rainfall of time interval Δt equal to and less than the maximum point infiltration capacity of that time interval, f'_m, as shown in Fig. 3, the actual infiltration amount within the time interval would be

$$\bar{f}'_{\Delta t} = \int_o^{\bar{p}}(1 - \alpha)\ df$$

$$= \left[\frac{1 - \exp(-k\Delta t)}{k}\right](f_o - kP_a)\left[1 - \left[1 - \frac{\bar{p}}{(1+n)\frac{1-\exp(-k\Delta t)}{k})(f_o - kP_a)}\right]^{\frac{1}{1+n}}\right]$$

303

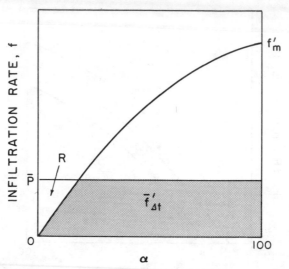

INFILTRATION RATE, f

f'_m

R

$\bar{P}$

$\bar{f}'_{\Delta t}$

O

100

α

Fig. 3   Diagram of runoff production. α represents the percent of
area with an infiltration capacity equal to or less than
the indicated value.

(15)

The runoff yield of the time interval is

$$R = \bar{P} - \bar{f}'_{\Delta t}$$

$$= \bar{P} - \left[\frac{1 - \exp(-k\Delta t)}{k}\right](f_o - kP_a)\left[1 - \left[1 - \frac{\bar{P}}{(1+n)\ \dfrac{1 - \exp(-k\Delta t)}{k}\ (f_o - kP_a)}\right]^{1+n}\right]$$

(16)

where $\bar{P}$ = rainfall of time interal Δt (m,m); f_o, k = parameters of Horton
equation determined by optimization, and n = parameter of distribution
curve determined by optimization.

For distribution within partial area, one obtains

$$R = \bar{P} - \left[\frac{1 - \exp(-k\Delta t)}{k}\right](f_o - kP_a)\left[1 - \left[1 - \frac{\bar{P}}{(1+n)\ \dfrac{1 - \exp(-k\Delta t)}{k}\ f_o - S'_o}\right]^{1+n}\right]$$

(17)

where

$$S'_o = (1 + n)\left[\frac{1 - \exp(-k\Delta t)}{k}\right]f_o\left[1 - (1 - kP_a/f_o)^{\frac{1}{1+n}}\right]$$ (18)

S'_o being the ordinate corresponding to p_a.

When k = 1, $\Delta t \geq 4$, equation (17) reduces to Xinanjiang Model (Zhao,
et al., 1980)

$$R = \bar{P} - (f_o - kP_a)\left[1 - (1 - \frac{\bar{P}_\cdot}{(1+n)\ f_o - S_o'})^{1+n}\right] \qquad (19)$$

By adopting the symbols used in Xinanjiang Model (Zhao, et al., 1980)

$$W_m = f_o,\ W_o = P_a,\ W_m' = (1+n)f_o,\ s_o' = a$$

we obtained the same equation.

It is clear that Xinanjiang Model is only the special case of our general model mentioned above. In the same manner, we obtained another runoff yield formula based on Philip curve.

For distribution within the entire area, using the Philip equation,

$$f = A/2\ t^{-\frac{1}{2}} \qquad (20)$$

$$F = A\ t^{\frac{1}{2}} \qquad (21)$$

The runoff yield of the time interval is

$$R = \bar{P} - A\left[(t + \Delta t)^{\frac{1}{2}} - t^{\frac{1}{2}}\right]\left\{1 - \left[1 - \frac{\bar{P}}{(1+n)\ A\left((t+\Delta t)^{\frac{1}{2}} - t^{\frac{1}{2}}\right)}\right]^{1+n}\right\} \qquad (22)$$

The proposed general runoff yield model has been tested on several basins with different hydrological behaviour in China. This model is suitable for arid regions as well as humid regions. The model consists of an infiltration curve- the very important infiltration process and then computes the runoff yield. This model is easier to use and to expand by means of its construction.

THE INITIAL VALUE PROBLEM

It is well-known that by assuming the catchment as a dynamic system we can obtain a general differential equation linking the input and output,

$$Q(t) = \frac{-b_m D^{m+1} - b_{m-1} \ldots\ldots -b_o D + 1}{a_n D^{n+1} + a_{n-1}\ D^n \ldots +a_o D + 1}I(t)\ =\ L[D]I(t) \qquad (23)$$

where $L[D]$ denotes the differential operator system, $D = d/dt$, $I(t)$ and $Q(t)$ denote input and output respectively.

As $I(t) \to \delta(t)$ (impulse function) we obtain

$$U(t) = L[D]\ \delta(t) \qquad (24)$$

where $U(t)$ means impulse response or IUH.

For convenience, $L[D]$ may be written as

$$L[D] = \frac{1}{\prod_{j=1}^{n}\ (1+k_j D)^n} \qquad (25)$$

then

$$\prod_{j=1}^{n} (1+k_j D)^n U(t) = \delta(t) \qquad (26)$$

We found that in using laplase Transform for solving the equation mentioned above we can never assume zero for the initial condition, i.e. $U^{(n-1)}(t) = 0$. To our knowledge, the IUH can't be obtained by this initial condition. For example, when n=1, the IUH is

$$U(t) = 1/k \exp(-t/k) \qquad (27)$$

and $\quad U(0) = 1/k \qquad (28)$

This result is opposite to the initial condition. According to our study, to solve equation (26) should be reduced to the following initial value problem:

$$\prod_{j=1}^{n} (1 + k_j D) U(t) = 0$$

$$U(0) = U'(0) = \ldots = U^{(n-2)}(0) = 0 \qquad (29)$$

$$U^{(n-1)}(0) = a$$

where a denotes the degree of jump. It is easier to prove by mathematical induction that the degree of jump would be $1/k^n$ (supposing k_j being the same). The general solution would be

$$U(t) = \exp(-t/k) \sum_{j=0}^{n} \alpha_j \, t^{j-1} \qquad (30)$$

On the basis of the initial condition involved in equation (29) we obtain

$$\alpha_1 = \alpha_2 = \ldots = \alpha_{n-1} = 0, \quad \alpha_n = \frac{1}{k^n \, \overline{\varGamma(n)}}$$

then $U(t) = \dfrac{1}{k \overline{\varGamma}(n)} \, (t/k)^{n-1} \, \exp(-t/k) \qquad (31)$

This is the well known Nash model.

THREE PARAMETER ROUTING MODEL

It can be shown that various well known linear models are obtained by selecting the parameters in equation (23) (Chow, 1975). In equation (23) let the numerator be 1, the denominator be $(1+a_o D)$, and $a_o=k$. We may get a linear reservoir model with its impulse response as

$$U(t) = 1/k \exp(-t/k) \qquad (32)$$

let the numerator be 1 and the denominator be $(1+kD)^n$. Then the Nash Model would be obtained,

$$U(t) = \frac{1}{k\Gamma(n)} (t/k)^{n-1} \exp(-t/k)$$

let the numerator be

$$\left[\sum_{\lambda=1}^{\infty} (-1)^{\lambda-1} (TD)^{\lambda}/\lambda! + 1 \right]$$

and the denominator be $(1+kD)^n$, the impulse response of Lag-Route model can easily be obtained,

$$U(t) = \frac{1}{k\Gamma(n)} \left(\frac{t-\tau}{k}\right)^{n-1} \exp\left[-\frac{(t-\tau)}{k}\right] \tag{33}$$

In the same way, let

$$I[n] = \left[\frac{A}{1+k_1 D} - Ax\right]^n$$

$$-A^n \left[\frac{1}{(1+k_1 D)^n} - \frac{nx}{(1+k_1 D)^{n-1}} + \frac{\frac{n(n-1)}{2} x^2}{(1+k_1 D)^{n-2}} + \dots + (-1)^n x^n\right]$$

where $A = \frac{1}{1-x}$, $k_1 = k(1-x)$, the impulse response or IUH for the Muskingum Successive Routing Model is obtained,

$$U(t) = \frac{1}{(1-x)^n} \left\{ \frac{1}{k^n(1-x)^n \Gamma(n)} t^{n-1} \exp\left[-\frac{t}{k(1-x)}\right] + \right.$$

$$+ \sum_{j=1}^{n} (-1)^j C_j^n x^i \frac{t^{n-j-1}}{k^{n-j}(1-x)^{n-j}\Gamma(n-j)} \exp\left[-\frac{t}{k(1-x)}\right] \right\}$$

$$\left. + (-1)^n \frac{x^n}{(1-x)^n} \delta(t) \right. \tag{34}$$

Because of the negative part of the impulse response, when equation (34) was used to route an input through a channel, we are bound to obtain the negative output. Therefore, we made some refinements to it.

Let

$$L[D] = \left[\frac{B}{1+k_2 D} + B(x-1)\right]^n$$

$$= B^n \left\{\frac{1}{(1+k_2 D)^n} + \frac{n(x-1)}{(1+k_2 D)^{n-1}} + \frac{\frac{n(n-1)}{2!}(x-1)^2}{(1+k_2 D)^{n-2}} + \ldots + (x-1)^n\right\}$$

where

B = 1/x, k_2 = kx. The corresponding impulse response is

$$U(t) = 1/x^n \left[\frac{t^{n-1}}{k^n x^n \, \Gamma(n)} \exp(-t/kx) + \sum_{i-1}^{n-1} C_i^n (x-1)^i\right.$$

$$\left.\frac{t^{n-i-1}}{k^{n-i} x^{n-i} \, \Gamma(n-i)} \exp(-t/kx) + (x-1)^n \delta(t)\right], (x \geq 1) \qquad (35)$$

Equations (34) - (35) constitute the three parameter routing model.

CONVERSION OF IUH TO UH

In hydrology we usually use the S-curve to convert the IUH to the UH. We found the equation for converting the Nash Model to the UH by means of Duhamel integration. When n is an integral value,

$$U(\Delta t, t) = \begin{cases} 1 - \exp(-m) \sum_{\lambda=0}^{n-1} m^\lambda/\lambda! , & (m < m_x) \\ \\ \exp(-M) \sum_{\lambda=0}^{n-1} M^\lambda/\lambda! - \exp(-m) \sum_{\lambda=0}^{n-1} m^\lambda/\lambda! , & (m \geq m_k) \end{cases} \qquad (36)$$

where $U(\Delta t, t)$ denotes UH, m = t/k, M = $m - m_k$, m_k = $\Delta t/k$, Δt = time interval.

Similarly the equation for converting the three parameter successive routing model, equation (34), to the UH was obtained,

when $t \leq \Delta t$, then

$$U(\Delta t, t) = \frac{1}{(1-x)^n} \left\{1 - \exp\left[-\frac{m}{(1-x)}\right] \sum_{\lambda=0}^{n-1} \frac{(\frac{m}{1-x})^\lambda}{\lambda!}\right.$$

$$\left. + \sum_{j=1}^{n} (-1)^j C_j^n x^j [1 - \exp(-\frac{m}{1-x}) \sum_{\lambda=0}^{n-j-1} \frac{(\frac{m}{1-x})^\lambda}{\lambda!}]\right.$$

$$+ (-1)^n t^n \Bigg\} \tag{37}$$

When $t > \Delta t$, then

$$U(\Delta t, t) = \frac{1}{(1-x)^n} \Bigg\{ \exp(-\frac{M}{1-x}) \sum_{\lambda=0}^{n-1} \frac{(\frac{M}{1-x})^\lambda}{\lambda!}$$

$$- \exp(-\frac{m}{1-x}) \sum_{\lambda=0}^{n-1} \frac{(\frac{m}{1-x})^\lambda}{\lambda!}$$

$$+ \sum_{j=1}^{n} (-1)^j C_j^n x^j [\exp(-\frac{M}{1-x}) \sum_{\lambda=0}^{n-j-1} \frac{(\frac{M}{1-x})^\lambda}{\lambda!}$$

$$- \exp(-\frac{m}{1-x}) \sum_{\lambda=0}^{n-j-1} \frac{(\frac{m}{1-x})^\lambda}{\lambda!}] \Bigg\} \tag{38}$$

Solution of the initial value problem in using laplace transform for solving the equation and obtaining the IUH might be useful to the IUH theory. The proposed three parameter routing model would be a contribution to flood routing work, for it is easier to use.

NONLINEAR FLOW CONCENTRATION MODEL

There has been a number of hydrologists tackling the nonlinear problem of flow concentration by means of nonlinear reservoir model. In order to consider the effect of the infiltration during the formation of runoff and the nonlinearity in computing the overland flow, early in 1959, Professor Hua (Hua, 1959) put forward the concept about dividing the runoff yield area and made use of the nonlinear storage equation derived from the total input to substitute for dynamic equation. Conceptually the two methods of Professor Hua and Dr. Horton are somewhat alike, but there is an essential difference between them.

First, there is a distinguishable concept of runoff yield area in the Hua method. Second, the storage equation was established on the basis of the total runoff and the overland depth rather than the section discharge at the basin outlet so as to get rid of the effect of storage in stream nets.

Suppose the overland flow per unit width q' is proportional to the total runoff in stream nets, or $q = bq$, where b is the average flow width varing with rainfall intensity, and q the total runoff in stream nets. According to the hydraulic method, the following equation can be obtained,

$$S = k'q^{1/m} \tag{39}$$

309

where S is the storage on overland, $1/m$ the exponential depending on the type of flow with the value of 0.5 in general, and k' the exponential coefficient. The relation of $S \propto q$ can be derived as follows.

It is well knwon that the total runoff after the cessation of rainfall is supplied by the storage of overland, a part of which turns into total runoff and the other part infiltrates into the soil. Thus

$$S_c = \Sigma q_c + \Sigma f_c \qquad (40)$$

where S_c denotes the overland storage after the termination of rainfall, q_c denotes the accumulated total runoff, and f_c denotes the infiltration amount during the period of overland flow concentration.

At the end of rainfall, $\partial S/\partial t$ may be zero in a short time interval ∂t and the infiltration rate at that instant would be

$$f_c = p_c - q_c \qquad (41)$$

where f_c is infiltration rate, p_c the rainfall intensity and q_c the total runoff at the instant near the end of rainfall. At this instant, the maximum value of overland storage would be reached, since then it would decrease gradually and the overland flow width and depth become more and more small. Owing to the variation of infiltration area, the infiltration rate may be considered proportional to the total runoff. Therefore,

$$S_c = \Sigma q_c (1 + \frac{\Sigma f_c}{\Sigma q_c}) - \Sigma q_c (1 + f_c/q_c) = (1 + k)\Sigma q_c \qquad (42)$$

Since the values of k, q_c and f_c were known, S_c would be derived. In the same way, the values of Σq_c and the corresponding values of S_c at every instant t can then be obtained by equation (42). Therefore, the relationship of $S \propto q$ may be developed and the infiltration process $f \sim t$ would be obtained easily by solving the dynamic equation $S \propto q$ and the water balance equation simultaneously. To our knowledge, $S \propto q$ and $f \sim t$ reflect the characteristics of horizontal and vertical flow movements during the stage of nonlinear overland flow concentration. By making use of these two important tools the total runoff for storm event can be obtained.

Toward the end of fifties we began to use the mathematical models of the unit graph. Based on the flood hydrograph with steep rising segment and slow falling segment in the southern regions of China, a mathematical model given by the product of the parabola with an arbitrary power and the sine curve was proposed in 1957 by Professor Hua (Hua, 1957),

$$U(t) = k \sin(\pi/\tau_m - t)(\tau_m - t)^n \qquad (43)$$

Our experience, gained on more than 20 rivers, suggests that it is sufficient to adopt the first three terms of the expansion in series,

$$U(t) = k \tau_m^n [1 - n\, t/\tau_m + \frac{n(n-1)}{2}(t/\tau_m)^2] \sin \pi/\tau_m - t \qquad (44)$$

where $U(t)$ is the IUH, τ_m the maximum flow concentration time of the basin and k and n are the parameters to be determined by optimum seeking method or by diagrammatic decomposition method the latter is shown as follows:

By substituting equation (44) for discrete form of convolution equation, we obtain

$$Q(t) = k \tau_m^n \sum_1^{\tau_m} q \sin(\pi/\tau_m - t) - kn\, \tau_m^{n-1} \sum_1^{\tau_m} q \sin(\pi/\tau_m - t)t$$

$$+ \frac{n(n-1)}{2} k \tau_m^{n-2} \sum_1^{\tau_m} \sin(\pi/\tau_m - t)t^2 \qquad (45)$$

Let

$$k \tau_m^n = A \qquad\qquad \sum_1^{\tau_m} q \sin(\pi/\tau_m - t) = \Sigma y_1$$

$$nk \tau_m^{n-1} = B \qquad\qquad \sum_1^{\tau_m} q \sin(\pi/\tau_m - t)t = \Sigma y_2$$

$$\frac{n(n-1)}{2} k \tau_m^{n-2} = C \qquad\qquad \sum_1^{\tau_m} q \sin(\pi/\tau_m - t)t^2 = \Sigma y_3$$

then

$$Q(t) = A \,\Sigma y_1 - B\, \Sigma y_2 + C\, \Sigma y_3$$

$$= f_1(Q)\, \Sigma y_1 - f_2(Q)\, \Sigma y_2 + f_3(Q)\, \Sigma y_3 \qquad (46)$$

where q is input, and A, B, C are the functions of time t. By establishing the graphical correlation for $Q = f(\Sigma y_1, \Sigma y_2, \Sigma y_3)$, the values of $f_1(Q)$, $f_2(Q)$ and $f_3(Q)$ can be obtained and then the parameters n and k depending on Q are determined.

$$\begin{cases} n = \tau_m\, f_2(Q)/f_1(Q) \\[2em] k = \dfrac{f_1(Q)}{\tau_m\, f_2(Q)/f_1(Q)} \end{cases} \qquad (47)$$

311

The proposed model by Hua (1957) has been successfully used in several basins with steep rising segment and slow falling segement of flood hydrograph.

REFERENCES

Crawford, N. H. and Linsley, R. K. 1966. Digital Simulation in Hydrology: Stanford Watershed Model IV: Tech. Rept. No. 39, Stanford University, Stanford, California.

Eagleson, P. S. 1970. Dynamic Hydrology. McGraw Hill Book Co., New York.

Hua, S. Q. 1957. On the total runoff and flow concentration curve as well as the proposal about new flow concentration curve. Journal of Hydraulic Engineering, No. 4, Beijing.

Hua, S. Q. 1959. The new method for estimating the effect of mass water conservancy and soil conservation on the flood runoff. Journal of Hydraulic Engineering, No. 5, Beijing.

Popov, E. G. 1980. Forecasts of seasonal inflow to reservoirs as one of the components of water control system. Proceedings of the Oxford Xymposium, April, IAHS Publ. No. 129.

Chow, V. T. 1975. Hydrologic Modeling. Selected Works in Water Resources, IWRA, March 1975.

Zhao, R. J., Zuang, Y., Fang, L. R., Lin, X. R., and Zhanquan, S. 1980. The Xinangjiang Model. Proceedings of the Oxford Symposium, April 1980, IAHS Publ. No. 129.

DETERMINATION OF FLOOD HYDROGRAPH FOR LARGE WATERSHEDS

Kedar N. Mutreja
Executive Engineer
Irrigation Design Organization
Roorkee, INDIA

ABSTRACT

A methodology is developed to compute the flood hydrograph due to a given rainfall occurring on a large watershed. The study uses the simple concept of the unit hydrograph. The watershed is divided into small subareas. The unit hydrograph of each subarea is derived, with outlet at project site, by available hydrologic records supplemented by synthetic unit hydrograph computations. These are converted to flood hydrographs with the respective effective rainfalls of subareas and then combined together to give a single flood hydrograph for the whole watershed.

The methodology developed has been successfully applied to predict the time and magnitude of the flood at project site for the given four hourly rainfall of six raingauge stations of the watershed during monsoon seasons of the construction period.

The advantage of the approach adopted for computing the routed flood hydrograph at project site for any subarea presented here is that it does not need a complicated procedure of subarea flood hydrograph to be routed down the stream channel to the project site. Moreover it considers the irregular distribution of rainfall in the watershed in a simple way.

INTRODUCTION

There appears to be a growing awareness in the developing world of the importance of timely exploitation/utilization of available water resources to meet the demand of power hungry industries. Consequently major projects involving large watersheds are being investigated and planned. The planning of such projects needs not only the estimation of peak flows by statistical analysis of runoff data, but also the complete flood hydrograph.

Various mathematical models are available for the study but no model gives results accurate enough for large watersheds where the response of different watershed subareas is different and, therefore, to work out the composite time response of the watershed makes the model all the more complex. Moreover, the data required for such models are so enormous that they are rarely available in the developing countries, besides the problem of computer facility.

The choice, thus, is to make use of the basic principle of the unit hydrograph to compute the flood hydrograph with given rainfall pattern. However, for large watersheds, the major floods in the upper-portions of the total basin, in some cases, get considerably dissipated before they reach the project site and the contributing watershed is usually sub divided into subareas. Thus a single unit hydrograph for the whole watershed with single weighted average rainfall by, say, Thiesson Polygon would not give the correct flood hydrograph.

The correct approach for the analysis of large watersheds is to divide the total watershed into subareas. Flood hydrograph may then be determined for each subarea, which may be routed down the stream channel to the project site before they are combined to give the flood hydrograph for the whole watershed.

It is important to note that the routing process besides being complicated is subject to considerable error unless based on accuarate data of the channel characteristics such as cross sections, slope and roughness coefficient. Therefore, the problem should be carefully examined and the division of total watershed into subareas should be done only when it is indicated that omission of the step would cause incidence of greater error than would occur with the undivided basin (C.W.C., 1972, p. 86).

The fact of the matter is that the routing of the unit hydrographs of the subareas and their computation due to non-availability of necessary data in the real world makes the problem all the more complex. Consequently large watersheds are invariably analysed in practice on the principle of small watersheds while computing the hydrograph with the known rainfall.

PROBLEM SOLUTION

Stepwise procedure to tackle the problem of computing the flood hydrograph with the rainfall is as follows:

(1) Collection and examination of data.

(2) Analysis of observed flood hydrographs and their sub-division into baseflow and direct runoff hydrograph.

(3) Analysis of rainfall data related to observed flood hydrographs to work out the runoff coefficient for intense storms.

(4) Derivation of the unit hydrographs and selection of the unit hydrograph for flood computations as an average of these and then finding its unit time.

(5) Derive Synthetic unit hydrograph for the whole watershed as given by Snyder (1938) with the unit duration of step (4).

(6) Compare the unit hydrograph derived under steps (4) and (5) and work out the coefficients C_t and C_p of Snyder's formulae as follows:

$$t_p = C_t(LL)^{0.3} \qquad (1)$$

$$q_p = C_p(A_d/t_p) \qquad (2)$$

$$t_r = t_p/5.5 \tag{3}$$

$$t_p' = t_p + (t_r' - t_r)/4 \tag{4}$$

$$T = 3(1+t_p/24) \tag{5}$$

where L = distance of the farthest point in the watershed from the outlet point in kms; $\overline{L}$ = distance of C.G. of the watershed area from the
outlet point in kms; A_d = watershed area in sq.kms; t_p = lag time from
mid point of effective rainfall t_r to peak of a unit hydrograph in hours;
t_r = standard duration of effective rainfall in hours; q_p = peak discharge of unit hydrograph for standard duration t_r in cumec; T = base length
of the unit hydrograph for standard duration t_r in days; t_r = duration
of effective rainfall in hours; and t_p = lag time from mid point of effective rainfall duration t_r to peak of a unit hydrograph in hours.

(7) Divide the whole watershed into small subareas by Thiessen
Polygons. However, they may be modified a bit such that they conform
to the watershed boundaries.

(8) Work out the synthetic unit hydrograph for each of the small
subareas with outlet point at project site by a simple approach of subtracting from a unit hydrograph of a larger area with desired subarea
as its part and with outlet at project site, a unit hydrograph of the
remaining area other than the subarea so as to give the net unit hydrograph of subarea with outlet at project site.

(9) Work out the flood hydrographs with outlet at project site
for those subareas which experience rainfall.

(10) Superimpose flood hydrographs of step (9) and find out a
composite flood hydrograph.

(11) Add baseflow to the composite flood hydrograph to get the
complete flood hydrograph at the project site.

Basic Data

Continuous discharge data of the project site for at least 4 to 5
rainy seasons with adequate concurrent rainfall data should be available.
For complete shape of the flood hydrograph it is better to have a self-
recording gauge at site.

Rainfall data are required for all rainfall stations in and near the
project basin, including hourly rainfall data from available self-recording
rain gauge stations to define adequately the areal and time distribution
of rainfall.

Analysis of Observed Flood Hydrographs

The first step in the analysis is to separate the base-flow from the
direct runoff. Separation of base flow is generally done in any arbitrary
manner. However, the errors made in the separation of baseflow are small
and unimportant as long as one method is used consistently especially
with regard to the analysis of floods. Thus, a simple procedure of drawing a constant rainfall intensity line for each hydrograph such that rainfall exceeding this intensity equals surface runoff.

To derive the unit hydrograph for the isolated observed flood hydrograph, measure the volume of the hydrograph and get the depth of rainfall excess by dividing hydrograph and get the depth of rainfall excess by dividing the same by watershed area. On dividing the flood hydrographs by their respective depth of rainfall excess gives the unit hydrograph. A number of such unit hydrographs are developed from the observed hydrographs for a project site. An average unit hydrograph conforming to the general shape of the unit hydrographs passing through the average peak ordinate and having a volume of 1 cm. is derived for design purposes. In doingso, extra weight may be given to those unit hydrographs which have been derived from reliable data, more uniform storm patterns and higher floods.

Subarea Unit Hydrograph with Outlet at Project Site

As already discussed, the subarea is generally the area controlled by a rain gauge station. Since runoff record only due to rainfall at one rain gauge station is rarely available in practice, the alternative is to derive synthetic unit hydrograph as given by Snyder (1938). To use this its constants C_t and C_p may be determined for the project basin with the help of design unit hydrograph as below.

(a) Measure L and L from the map. The centroid of the basin is located by cutting out a stiff pattern of the basin and intersecting plumb lines drawn for different rotations of the pattern.

Knowing t_r' and t_p' as the rainfall duration and time lag of design unit hydrograph, t_p can be found from Eqs. (3) and (4). The value of t_p is then substituted in Eq. (1) to yield the value of C_t. Similarly knowing q_p of design unit hydrographs Eq. (2) gives the value of C_p.

Having determined the coefficients C_t and C_p the synthetic unit hydrograph for the subarea B (Fig. 1) can be computed with outlet at d. The next problem is to modify this so as to give hydrograph at outlet h, which

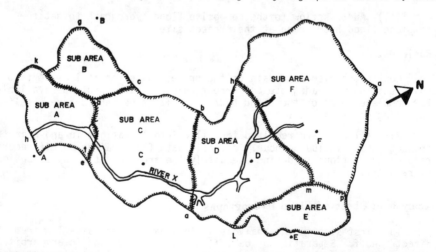

SCALE :– 1 : 2,50,000

Fig. 1. Sub-division of Project Watershed.

is the project site. Two alternatives are available for this purpose.

(i) Route the unit hydrograph with outlet at d through the stream channel by Muskingum's Method to get the unit hydrograph of the subarea with outlet at project site h.

(ii) To derive a synthetic unit hydrograph of subarea B, derive synthetic unit hydrograph of subareas (A+B) with outlet at project site h and then subtract from it a synthetic unit hydrograph of subarea A with outlet at project site h to give a synthetic unit hydrograph of subarea B with outlet at project site, h.

Initially the first alternative was tried, but it could not be applied as it required observed data of inflow and outflow at small intervals of time at points d and h. Moreover the problem of local ungauged inflow between d and h created a lot of complications. The second method, however, has been successfully tried.

The practical applicability of the method lies in the fact that it does not require any observed data at d and h, etc. It may also be mentioned that if no data for the basin is available then the coefficients C_t and C_p could be computed for the adjoining basin and the same value used for this ungauged basin.

Thus the unit hydrographs for all the subareas can be derived and then converted to flood hydrograph by their respective ranfalls before they are superimposed to give the composite hydrograph.

APPLICATION OF THE METHOD

The method outlined for computing the flood hydrograph for large basins by subdividing the area has been tried to tackle a problem of a project.

Project is a multi purpose one with 412 foot high earth and rock-fill dam. Salient features of the dam are:

1. Watershed at dam site   = 3100 sq.km.

2. Mean annual rainfall   = 1552 m.m.

3. Mean annual runoff at dam site   = 2683 M m^3.

4. Design flood (1 in 750 years).   = 9374 m^3/sec.

The watershed shown in Fig. 1 has six wireless raingauge stations, namely, A B C D E F, from where four hourly rainfall of monsoon period was being intimated at h which is the project site.

Case Study

The problem is to compute the runoff hydrograph at the project site from a particular storm of 1969 with the rainfall intimated from these six wireless stations of the catchment.

In 1969 the maximum observed flood was of the order of 6600 cumecs and was peculiar in the sense that it was due to rainfall up to station D and there was no rainfall in the remaining catchment. This flood hydrograph has been computed by the methodology developed herein and compared with the actual observed hydrograph.

The solution to the problem is of great importance during the construction period as the machines are pouring dam material day and night and the decision to evacuate the project site is to be taken during monsoon period based on the computed runoff from four hourly ranifall of the watershed.

The problem proposed to be tackled does arise at all such projects. However, the solution to this is peculiar in the sense that unlike the usual studies of determining flood wherein combinations of different factors are considered to maximise the flood the present study envisages the computations of flood hydrograph at project site for a particular storm with a possibility of partial area experiencing a rainfall and that too with durations different from time of concentration of watershed and hence the need to consider the subareas contributions to the runoff.

DATA AVAILABLE

The maximum flood discharge of 6600 cumecs was observed at project site h on 5-8-1969. Rainfall data of different stations and runoff data at station h is given in Tables 1 to 5.

TABLE 1   Rainfall at Station A

Date	Time duration of Rainfall in hrs.						Total rainfall (mms)
	00 to 04	04 to 08	08 to 12	12 to 16	16 to 20	20 to 24	
1.8.69	12.00	-	-	3.00	-	-	15.00
2.8.69		63.00	34.00	14.00	-	-	111.00
3.8.69	-	-	-	-	-	-	-
4.8.69	-	-	-	-	-	-	-
5.8.69	-	4.00	16.00	7.00	-	-	27.00
6.8.69	30.00	4.50	-	-	-	-	34.50

TABLE 2   Rainfall at station B

Date	Time duration of Rainfall in hrs.						Total rainfall (mms)
	00 to 04	04 to 08	08 to 12	12 to 16	16 to 20	20 to 24	
1.8.69	-	-	-	4.00	-	-	4.00
2.8.69	16.40	38.00	9.30	-	10.00	-	73.70
3.8.69	-	-	4.00	-	-	-	4.00
5.8.69	-	27.00	70.20	15.00	3.30	18.00	133.50
6.8.69	23.00	25.00	-	-	-	-	48.00

318

TABLE 3 Rainfall at Station C

| Date | Time Duration of Rainfall in hrs. | | | | | | Total rain- fall (mms) |
	00 to 04	04 to 08	08 to 12	12 to 16	16 to 20	20 to 24	
1.8.69	-	-	-	-	21.00	-	21.00
2.8.69	4.00	76.00	74.00	6.00	-	-	160.00
3.8.69	-	-	-	-	-	-	-
4.8.69	-	-	-	-	-	-	-
5.8.69	30.00	32.00	12.00	-	5.00	-	79.00
6.8.69	8.00	-	-	-	-	-	8.00

TABLE 4 Rainfall at Station D

| Date | Time Duration of Rainfall in hrs. | | | | | | Total rain- fall (mms) |
	00 to 04	04 to 08	08 to 12	12 to 16	16 to 20	20 to 24	
1.8.69	-	-	-	-	-	-	-
2.8.69	6.2	7.4	17.6	11.00	12.00	0.8	55.00
3.8.69	1.0	5.0	-	-	-	-	6.00
4.8.69	-	-	-	-	-	-	-
5.8.69	19.2	68.2	20.00	5.6	19.4	8.0	140.40

TABLE 5 Observed runoff at Station h

Date	Time	Discharge (cumec)
1.8.69	0.00	104.80
	4.00 AM	102.00
	8.00 AM	102.00
	12.00 PM	93.50
	4.00 PM	107.50
	8.00 PM	104.80
2.8.69	12.00 AM	107.60
	4.00 AM	93.50
	8.00 AM	1545.00
	10.30 AM	2830.00
	13.00 PM	2675.00
	4.00 PM	1129.00
	8.00 PM	788.00
3.8.69	12.00 AM	713.00
	4.00 AM	702.00
	8.00 AM	533.00
	12.00 PM	447.00
	4.00 PM	399.00
	8.00 PM	343.00
4.8.69	12.00 AM	314.00
	4.00 AM	274.50
	8.00 AM	243.50
	12.00 PM	223.50
	4.00 PM	204.00
	8.00 PM	190.00

Table 5 continued.

5.8.69	12.00 AM	178.50
	4.00 AM	190.00
	8.00 AM	456.00
	12.00 PM	4440.00
	1.50 PM	6600.00
	4.00 PM	3625.00
	8.00 PM	1779.00
6.8.69	12.00 AM	1050.00
	4.00 AM	850.00
	8.00 AM	675.00
	12.00 PM	600.00
	4.00 PM	475.00
	8.00 PM	400.00
7.8.69	12.00 AM	350.00

Sub Areas of Catchment

An area over which a raingauge is effective has been found out by drawing Thiessen Polygons. The boundaries of the area controlled by each station polygon have been slightly adjusted such that they correspond to the water divide lines. This is done to convert the rainfall effective boundaries into drainage basin boundaries. The subareas controlled by each station polygon are shown in Fig. 1 and given in Table 6.

TABLE 6   Area Controlled by Different Raingauge Stations

Sl.No.	Stations	Area Controlled by the Station (sq.km.)
1.	A	323.50
2.	B	315.50
3.	C	861.00
4.	E	310.00
5.	D	710.00
6.	F	580.00

Design Unit Hydrograph

Flood hydrographs for three normal storms were available. They have been analysed and the design unit hydrograph with 24-hour duration was obtained by averaging the three unit hydrographs of observed normal storms and is given in Figure 2.

Subarea Unit Hydrograph for 12-hour Duration

The first step is to find out the values of the coefficients C_t and C_p in Snyder's equation. For this design the unit hydrograph for the whole watershed of Fig. 2. is compared with the synthetic unit hydrograph of 24-hour duration.

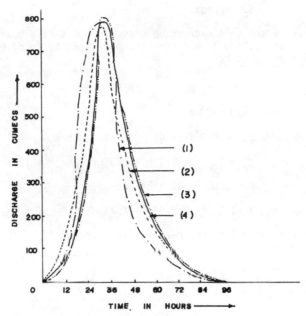

Fig. 2. Derivation of Design Unit Hydrograph:
(1) Unit Hydrograph of 4-10 August 1980; (2) Unit Hydrograph
of 26-31 July 1962; (3) Unit Hydrograph of 13-18 September
1957; and (4) Design Unit Hydrograph of One Day Duration.

$$L \quad = \quad 157 \text{ Km. (measured from map)}$$

$$\bar{L} \quad = \quad 77 \text{ Km. (measured from map)}$$

$$t_p \quad = \quad C_t(L\bar{L})^{0.3}$$

$$= \quad C_t(157 \times 77)^{0.3}$$

$$t_p' \quad = \quad t_p + (24 - t_r)/4 \text{ since } t_r' = 24 \text{ hrs.}$$

$$21 \quad = \quad t_p + (24 - t_p/5.5)/4 \text{ as } t_p' = 21 \text{ hr. from Fig. 2}$$

$$t_p \quad = \quad 15.7 \text{ hrs.}$$

Thus,

$$15.7 \quad = \quad C_t(157 \times 77)^{0.3}$$

$$C_t \quad = \quad 0.9 \text{ Approx.}$$

$$q_p \quad = \quad 782 \text{ cumec (Fig. 2.)}$$

$$782 \quad = \quad C_p \times A_d/t_p = C_p \times 3100/21 \text{ as } A_d/ = 3100 \text{ sq.km.}$$

C_p =   5.0 Approx.

To derive the unit hydrograph of subarea C a card board is cut in the shape of abcghea_(Fig. 1) and its centre of gravity found out to give the values of L and L as below:

L =   76.8 kms. and $\bar{L}$ = 57.6 kms.

A_d =   1500 sq. km.

t_p = $0.9 \ (76.8 \times 57.6)^{0.3}$ = 11.20 hrs.

t_p' = $t_p + (t_r' - t_r)/4$ = 13.69 hrs as t_r' = 12 hrs.

q_p = $5 \times 1500/13.69$ = 548 cumec.

T = $3 + (3/24 \times 13.69)$ = 4.71 days.

The unit hydrograph is drawn in Fig. 3.

To compute the unit hydrograph of area (A+B), the card board is cut in the shape of edcgkhe and the above process in repeated to get synthetic unit hydrograph of this area as drawn in Fig. 3.

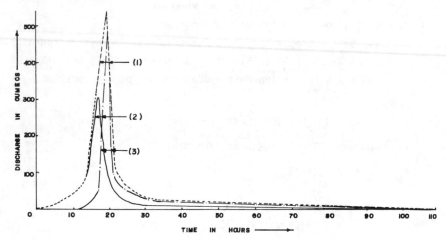

Fig. 3.   Derivation of Unit Hydrograph for Stations C:

         (1)   Unit Hydrograph for Station (A+B+C) Polygon;

         (2)   Unit Hydrograph for Station (A+B) Polygon; and

         (3)   Unit Hydrograph for Station (C) Polygon for 12 Hours Duration.

To get the unit hydrograph of station C subarea abcdea for 12-hr. duration with outlet at h, the project site, the unit hydrograph of area edcgkhe is subtracted from the unit hydrograph of area abcghea. The ordinates of this unit hydrograph so obtained are given in Table 7. Similarly the unit hydrograph ordinates have been calculated for other subareas.

TABLE 7   Unit Graph for Station C Polygon

Sl.	Unit Graph of Station (A+B+C) Polygon (i.e. area abcghea)	Unit Graph of Stations (A+B) Polygon (area edcgkhe)	Unit Graph of Station C Polygon (abcdea)
00	0.00	0.00	0.00
04	7.50	7.50	0.00
08	25.00	25.00	0.00
12	60.00	60.00	0.00
15-22	220.00	190.00	30.00
16	290.00	260.00	30.00
16-22	340.00	308.00	32.00
19-40	548.00	62.00	486.00
20	340.00	58.00	282.00
24	63.00	23.00	40.00
28	44.00	8.00	36.00
32	31.00	4.50	26.50
36	24.00	4.25	191.75
40	19.00	4.00	15.00
44	18.00	3.75	14.25
48	17.00	3.50	13.50
52	16.00	3.25	12.75
56	15.00	3.00	12.00
60	14.00	2.75	11.25
64	13.00	2.50	10.50
68	12.00	2.25	9.75
72	11.00	2.00	9.00
76	10.00	1.75	8.25
80	9.00	1.50	7.50
84	8.00	1.25	6.75
88	7.00	1.00	6.25
92	6.00	0.75	5.25
96	5.00	0.50	4.50
100	4.00	0.25	3.75
104	3.00	0.00	3.00
108	1.50	0.00	1.50
112	0.00	0.00	0.00

Runoff Coefficient

Rainfall and runoff relationship has been studied for some natural events. Based on elaborate studies, a 75% runoff factor for intense storms has been calculated and the same is adopted over here.

Estimation of Flood Hydrograph

Rainfall excess has been worked out by multiplying the total rainfall with runoff factor. The flood hydrographs of different subareas have been worked out by multiplying unit graph ordinates with respective rainfall excesses. These flood hydrographs have then been superimposed to give the composite flood hydrograph. Assuming baseflow of 200 cumecs (as separated from original hydrographs), the calculations for complete hydrograph are given in Table 8 and plotted in Fig. 4, along with the observed discharge hydrograph.

Discussion of Results

The following major differences have been noticed between the observed and the computed flood hydrographs:

(1) The computed flood is occuring about 7 hours after the observed flood.

Date	Time	Flood Ordinates at Stations				Total Hydrograph ordinates (3+4+5+6+ Base flow) (Cumecs)
		A	B	C	D	
1	2	3	4	5	6	7
1.8.69	0.00	0.00	0.00			200.00
	4.00 AM	6.75	135.00			341.75
	8.00 AM	22.50	32.40			254.90
	12.00 AM	54.00	16.20			270.20
	4.00 PM	110.70	50.15			350.85
	8.00 PM	25.40	12.95			238.35
2.8.69	12.00 AM	18.00	7.40			225.40
	4.00 AM	26.80	3.60	0.00		230.40
	8.00 AM	61.60	2.50	47.10		311.20
	12.00 PM	185.40	2.35	442.74		830.50
	4.00 PM	448.08	718.73	409.3	0.00	1776.11
	7.40 PM	1000.00	2000.00	5613.30	133.38	7146.70
	8.00 PM	831.32	174.00	3313.62	725.40	5244.34
3.8.69	12.00 AM	125.50	87.90	503.60	203.58	1120.58
	4.00 AM	153.75	29.20	460.30	46.80	890.05
	8.00 AM	41.13	125.32	456.52	579.88	1402.85
	12.00 PM	22.27	39.09	258.49	180.60	700.44
	4.00 PM	18.03	24.85	210.64	59.00	512.52
	8.00 PM	17.05	14.90	196.51	183.09	611.55
4.8.69	12.00 AM	15.98	11.69	183.64	79.79	491.10
	4.00 AM	14.78	10.79	71.67	46.69	443.93
	8.00 AM	13.50	10.00	159.49	41.03	424.02
	12.00 PM	12.35	9.21	151.30	39.69	397.96
	4.00 PM	11.25	8.40	141.13	37.18	383.13
	8.00 PM	9.89	7.45	130.90	34.89	369.68
5.8.69	12.00 AM	8.80	6.60	120.78	33.50	369.68
	4.00 AM	7.65	5.85	110.59	31.21	355.50
	8.00 AM	6.60	5.25	100.81	30.10	342.76
	12.00 PM	16.65	4.60	90.25	26.87	338.37
	4.00 PM	45.85	3.85	249.95	24.56	524.23
	8.00 PM	106.45	1096.65	1634.99	1518.69	5556.78
	8.22 PM	110.00	1735.00	388.50	3800.00	6233.50
	8.44 PM	120.00	1604.00	360.75	5152.00	7436.75
6.8.69	12.00 AM	198.80	264.90	281.71	720.36	1864.57
	4.00 AM	91.95	540.90	248.27	178.75	1259.87
	8.00 AM	18.85	140.27	195.18	877.97	1432.27
	12.00 PM	6.10	66.57	233.29	316.60	822.56
	4.00 PM	4.30	32.33	99.73	140.77	477.21
	8.00 PM	4.11	22.13	93.75	112.36	432.35
7.8.69	12.00 AM	4.85	20.68	85.43	107.22	418.18
	4.00 AM	3.58	19.50	78.06	97.14	398.28

(2)   There are two peaks for every flood while observed flood has only one peak.

(3)   Computed peaks are higher than the observed peaks.

As far as the first point is concerned, it may be due to (a) wrongly reported time of start of rainfall and (b) higher value of C_t adopted in synthetic unit hydrograph.

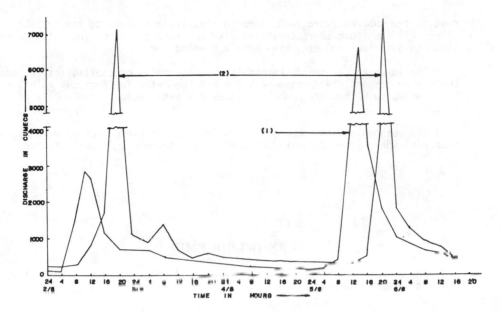

Fig. 4.   Flood Hydrograph of 2-6 August 1969; (1) Observed Hydrograph;
and (2) Computed Hydrograph.

    In fact, no importance was being attached to the correct reporting of
time of start of rainfall. Just taking time of start of rainfall four hours,
this way or that way will shift the computed flood time by four hours either
way. Similarly reducing the value of C_t the basin lag t_p is reduced, reducing
thereby the time difference between the computed and the observed peaks.
However, reduction in the value of C_t can be taken when the methodology is
applied to some other flood events and its value so modified that it satisfies
the majority of floods.

    Single flood peak can be observed only when the rainfall starts from
higher regions of watershed and comes towards lower regions with such a time
of start that the time of concentration at station h outlet is at one time
from each subareas. Since such a condition can hardly be satisfied, therefore in almost all the cases flood peak is always preceded by small peak.
The same phenomena occurs in computations also, but has not been found in
the observed flood. It appears that one peak is missed from observation
because of small time difference between the two peaks.

    Regarding higher peak of computed flood hydrograph, the reason is
the higher value of C_p adopted in the analysis. This can also be suitably
modified like that of C_t to suit the majority of floods. It is also possible
that the observed flood discharge may be on lower side due to its measurement being made by gauge discharge relationship which is likely to change
from time to time due to change in control.

CONCLUSIONS

    The methodology developed to analyse the storms in large basins is simple
and applicable in practice. It does not need the complicated analysis of

routing the subarea hydrographs through the stream channel to the project site. All the storm characteristics like the type of storm, intensity, duration and distribution, etc. have been accounted for.

The advantage of the technique is that the value of coefficients C_t and C_p, the former responsible for the peak lag and the later for the peak value, can be adjusted such that the computed floods are satisfactory in majority of the cases.

The methodology to compute flood hydrograph is useful for developing countries where storm data are abundant while the runoff is quite scarce.

ACKNOWLEDGEMENT

The author is very thankful to Mr. R. K. Jindal, Assistant Engineer, for his valuable help in making the necessary computations.

REFERENCES

Central water Comission (CWC), 1972. Estimation of Design Flood Recommended Procedures. No. 258, September, Central Water Comission, Ministry of Irrigation, New Delhi, India.

Snyder, F. F., 1938. Synthesis of unit hydrographs. Trans. Amer. Geophy. Union 19: 447-454.

MODELING OF STREAMFLOW ROUTING FOR FLOOD EMBANKMENT PLANNING

Subhash Chander
Professor in Civil Engineering
Indian Institute of Technology
New Delhi - 110016, India

P.N. Kapoor
Asstt. Professor in Civil Engineering
Indian Institute of Technology
New Delhi - 110016, India

S.K. Spolia
Asstt. Professor in Civil Engineering
Indian Institute of Technology
New Delhi - 110016, India

P. Natarajan
Professor in Civil Engineering
Indian Institute of Technology
New Delhi - 110016, India

ABSTRACT

A cascade model of linear reservoirs has been used to model the propagation of a flood wave in a river reach. Each reservoir of the cascade is identified with a pre-embanked sub-reach of the river in terms of the flood-plain storage for its inflow hydrograph. Subsequent placement of embankments would modify the storage in the sub-reach by an amount designated as the "cutout storage". A relationship is established for determining the modified storage delay time constant representing the post-embanked situation. Further, it has been possible to evaluate the increase in stage in the post-embanked situation by identifying the storage pertinent to this increase in stage in the model. Effective use of this method has been demonstrated in evaluating the modifications to peak discharge and stage at Delhi for the flood of September, 1978 in the River Yamuna.

INTRODUCTION

Often a flood-plain manager is concerned with protecting the flood-plain from extensive flooding; and can frequently achieve this by raising flood banks thus denying some of the flood-plain storage otherwise available to the passing flood wave. Construction of flood banks may protect land along the embanked reach but has an adverse effect on the downstream flooding. For arriving at better decisions, the flood-plain manager needs to predict the modifications in the flood peaks

327

further downstream in case of a historic or synthetic flood
likely to occur in future. Several routing models are
available based on Saint Venant's equations. To incorporate
detailed channel geometry and flood-plain features in such
models adds to the complexities of the problem. To superimpose
alternate levee plans on such models for evaluating their
effect on the downstream flow conditions makes the problem
intractable. Conceptual models with simpler formulations and
limited parameters can conveniently aid the decision-making if
their parameters can be identified with the significant physical
phenomenon representative of both the pre-embanked and post-
embanked river systems.

MODELING OF THE PROPAGATION OF THE DESIGN FLOOD WAVE

Models based on approximate solutions of Saint Venant's
equations are available for routing streamflow in the river
channel. Any system of flood protection by embankments has to
be designed and evaluated for a given flood. In order to
evaluate the overall effect of embankments on the passage of
the design flood, one is initially required to formulate a model
which adequately represents the propagation of this flood in the
pre-embanked condition. Modeling of the propagation of such
flood waves is a function of both the physical features of the
flood-plain and the available data. In most alluvial flood-plains
the area likely to be submerged is vast. Also, the overbank flow
is obstructed by cross-drainage structures or other features on
the flood-plain. The formulation of a physical model incorporating all such flood-plain details is quite a complex problem and
besides being time-consuming, does not ensure reliability of
results. However, formulations based on conceptual modeling
employing linear reservoirs can advantageously be used for
modeling the propagation of the flood wave.

The limiting processes in which a flood wave can
propagate are: propagation as a purely translatory wave in
channels or as a dispersive wave through a reservoir. Most flood
waves in rivers propagate by both of these processes. It is
known that:

(i)   if the number of reservoirs in a cascade model approaches
        infinity for a fixed lag (Kulandaiswamy, 1964), the
        input suffer pure translation; and

(ii)   in case of a single reservoir in the model, the input
        peak is attenuated and the outflow peak falls on the
        recession limb of the outflow hydrograph.

On the other hand, when the flood wave propagates as in a
channel, the input is, of course, again attenuated but the
outflow peak occurs well after the time at which discharge of
the same magnitude occurs on the recession limb of the inflow
hydrograph. This last aspect can be modeled by an appropriate
choice of the number (n) of linear reservoirs and their
respective storage constants in a cascade model.

The input and output in case of a cascade model of n
equal linear reservoirs are related by the equation

$$Q(t) = \frac{1}{(1+KD)^n} I(t) \qquad (1)$$

in which I(t) is the inflow into the reach, Q(t) is the outflow
from the reach, n is the number of linear reservoirs in the
cascade, D is the differential operator with respect to time
and K is the storage delay time constant $\approx \frac{T_L}{n}$ where T_L is
the time lag between the centroids of the inflow and outflow
hydrographs. The number of reservoirs, n, in the cascade can
be determined either by solving Eq.1 or by an iterative
procedure so as to get the peak and the shape of the routed
hydrographs to closely match. For every reservoir in the cascade
the storage-outflow relationship can then be expressed
(with n = 1)

$$S(t) = K.Q(t) \tag{2}$$

in which S(t) is the storage in the reservoir at time t; taken
with the hydrologic equation of

$$I(t) - Q(t) = \frac{d\,S(t)}{dt} \tag{3}$$

one gets

$$Q(t) = (1 - a)\,\overline{I}_t + a\,Q_{t-1} \tag{4}$$

in which $a = \exp(-\Delta t/K)$ wherein Δt is the routing interval and
$\overline{I}_t = (I_t + I_{t-1})/2$. If the river reach under study is represented
by a cascade of reservoirs, the calculations by Eq.(4) are
carried out successively from reach to reach with the outflow
at the end of any reach as the inflow for the subsequent reach.

MODELING OF THE EFFECT OF EMBANKMENTS

Construction of embankments for flood protection denies
a certain portion of the flood-plain storage otherwise available
to the propagation of flood waves. If the extent of storage
behind the embankments (i.e., farther away from the stream
course) below the pre-embanked maximum flood level (Fig.1) is

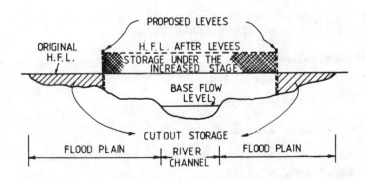

FIG.1. DEFINITION SKETCH OF CUTOUT STORAGE

defined as the cutout storage, then this storage is not
available to the spilling from future flood waves. The effect
of the cutout storage on the outflows is discussed in the
following paragraphs.

Consider a prismoidal reservoir (Fig.2) with a linear
spillway at one end. Let its side walls be moved inwards
causing reduction in the storage available in the reservoir
by an amount ΔS. Let the storage delay time constant for
the modified reservoir change to K' from its original value
of K. Assume a constant inflow I into the reservoir for a
time interval T. The outflows Q_T and Q_T' in the original and
constricted position of the sidewalls, respectively, would be
given by

$$Q_T = I\{1 - \exp(-T/K)\} \tag{5}$$

and

$$Q_T' = I\{1 - \exp(-T/K')\} \tag{6}$$

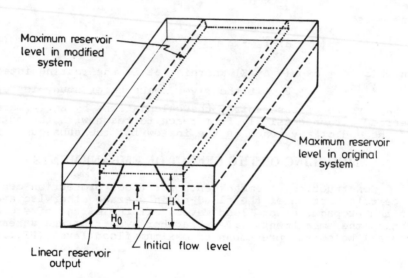

FIG.2.   CONCEPTUAL REPRESENTATION OF THE ORIGINAL AND
         MODIFIED FLOOD PLAIN SYSTEM

The storage volumes S_T and S_T' in the two situations at the end
of the interval (Fig.3) are

$$S_T = \text{Area ABCDA} = KQ_T \tag{7}$$

$$S_T' = \text{Area ABCFDA} = K'Q'_T \tag{8}$$

In the modified system, an outflow of magnitude Q_T will occur
at a time $t' < T$, given by

$$t' = \frac{K'}{K} \; T. \tag{9}$$

330

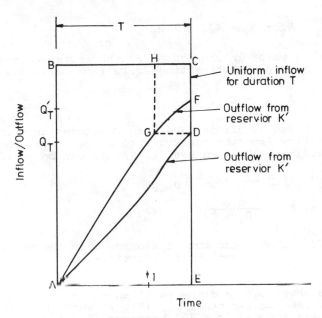

FIG.3. EFFECT OF MODIFICATION IN THE STORAGE DELAY TIME
CONSTANT ON THE OUTFLOWS FROM A LINEAR RESERVOIR
FOR UNIFORM INFLOW

The volume in storage, $S_{t'}$, in the modified reservoir at this
instant will be

$$S_{t'} = \text{Area ABHGA} = K'Q_T \qquad\qquad (10)$$

Eq.7 gives the storage in the original system at time T
when the discharge is Q_T; and Eq.10 gives the storage in the
modified system for the same discharge which occurs at time t'.
The difference in these two storage volumes equals the cutout
storage ΔS in the system which, in turn, equals the sum of
excess outflow volume (ΔV_1 which has occurred from the modified
system (AGFDA), and of the volume (ΔV_2) stored under the increased
stage in the modified system (GHCFG). This cutout storage can
be written as

$$\Delta S = \text{Area AGHCDA} = S_T - S_{t'}$$

or, $$\Delta S = (K - K')\, Q_T. \qquad\qquad (11)$$

For any cutout storage being considered the modified storage
constant K' can now be computed by re-arranging Eq.11 as

$$K' = K - \frac{S}{Q_T} \quad . \qquad\qquad (12)$$

The volume of water (ΔV_1) flown out as excess outflow from the
subreach in the duration T is

$$\Delta V_1 = S_T - S_T' = KQ_T - K'Q_T' \tag{13}$$

The volume of water in storage under the increase in stage (ΔV_2) is obtained by subtracting Eq.13 from Eq.11 giving

$$\Delta V_2 = K'(Q_T' - Q_T). \tag{14}$$

The above equations have been developed assuming the initial outflow from the reservoir to be zero. In case the initial outflow is Q_o then Eqs. 11 and 12 can be modified as

$$\Delta S = (K - K') (Q_T - Q_o) \tag{15}$$

$$K' = K - \frac{S}{(Q_T - Q_o)} \tag{16}$$

Eq.12 relates the cutout storage with the modified storage constant K', and Eq.13 identifies the volume which is responsible for increase in stage.

The above Eqs. are now verified on an idealised situation. Consider a rectangular reservoir (Fig.2) whose water-spread area has been reduced from 202 hectares (500 acres) to 162 hectares (400 acres) by moving the sidewalls of the reservoir inwards. Considering that this reservoir is fitted with a spillway whose storage discharge relationship is

$$Q(m^3/sec) = 20 H(m) \tag{17}$$

in which H is the head above the spillway crest. The storage constants corresponding to the original and modified reservoirs are calculated to be K = 28.06 hours and K' = 22.5 hours respectively. The inflow hydrograph given in Table 1 is routed through each of the two reservoirs. The ordinates of the inflow hydrograph and the corresponding outflow hydrographs are listed in Table 1. The storage volumes according to Eqs.12 and 14 are compared with those estimated from the physical system and are given in Table 2. The comparison verifies the Eqns. 12 and 14.

APPLICATION TO THE RIVER YAMUNA: A CASE STUDY

The upper catchment of the River Yamuna (above Tajewala) (Fig.4) is fan-shaped and hilly. The middle reach of the catchment (from Tajewala upto Delhi and beyond) is narrow and elongated with flat countryside. There is no tributary joining the main river between Kalanaur and Delhi. As for any rainfall in the Kalanaur - Delhi reach, the run-off contribution of this is perceptibly passed down Delhi long before arrival of the flood wave from above.

The river also serves as a boundary between the States of Haryana and Uttar Pradesh. The land in this reach is very fertile and both States have constructed embankments along the river for protecting their agricultural land and developmental activity on the flood-plains from inundation and damage from

332

TABLE 1

INFLOW AND OUTFLOW HYDROGRAPHS FOR LINEAR RESERVOIRS OF
202 and 162 HECTARES
(SPILLWAY RELATIONSHIP Q $(m^3/sec) = 20$ H(m))

Time Hours (m^3/s)	I_t (m^3/s)	Q_t (m^3/s)	Q'_t (m^3/s)	$I_t - Q_t$ (m^3/s)	$I_t - Q'_t$ (m^3/s)	Remarks
(1)	(2)	(3)	(4)	(5)	(6)	(7)
0	28.34	28.34	28.34	0	0	H_o = 1.417m
3	34.01	28.63	28.70	5.38	5.31	
6	45.34	29.75	30.07	15.59	15.27	
9	59.51	32.05	32.86	27.46	26.65	
12	74.53	35.80	37.13	38.73	37.40	
15	83.60	40.09	42.37	43.51	41.23	
18	86.43	44.58	47.70	41.85	38.73	
21	85.02	48.76	52.46	36.26	32.56	
24	73.68	51.86	55.82	21.83	17.86*	
27	65.18	53.64	57.52	11.54	7.66*	H=2.724m
30	58.50	54.48	57.96	4.02	0.54*	H'=2.898m
33	50.00	54.45	57.66	–	–	
			SUM	246.16	223.21	

NOTES: 1. Q_t and Q'_t are outflows from 202 and 162 hectares
respectively.

2. Volume of outflow = (246.16–223.21)×3×3600 = 24.78 ha.m.

3. Volume in storage under the increased stage = (17.86+7.66+
0.54)×3×3600 = 28.14 ha.m.

4. Cutout storage as estimated from the inflow and the two
outflow hydrographs = 24.78 + 28.14 = 52.92 ha.m.

TABLE 2

CUTOUT STORAGE VOLUMES AND VOLUMES OF WATER UNDER INCREASED STAGE

Estimated from physical system		Estimated using inflow and outflow hydrographs	
Discharge m^3/sec	Depth over spillway crest (m)	Storage description	Storage amount ha.m
29.34	1.417 ... (H_o)	Cutout storage	
54.48	2.724 ... (H)		52.92*
57.96	2.898 ... (H')		
Cutout storage=(A-A') (H-H_o)		Storage under the increased stage	
= 40(2.724-1.417)			28.14*
= 52.28 ha.m			
Increase in stage = H'-H = .174m		Increase in stage = $\frac{28.14}{102}$ = 0.1737m.	

*For these figures see Notes below Table 1.

FIG. 4. THE CATCHMENT MAP OF RIVER YAMUNA

recurring flooding. Every time an embankment experiences a failure it is invariably replaced with a higher embankment with much stronger sections.

During the unprecedented flood of September, 1978, the embankments between Kalanaur and Delhi were largely washed out. Their failure made available to the river most of the pre-embanked flood-plain storage and this reduced the flood havoc further downstream considerably. Subsequently, the planners are, however, again keen to build a set of embankments which will protect the flood-plains in this reach against inundation by any future flood of the September 1978 magnitude. Accordingly, an action policy needs to be investigated for the heights and spacing of the embankments to be rebuilt consistent with the acceptable peak discharge and stage at Delhi.

Flood hydrographs at Kalanaur, Mawi and Delhi for the September 1978 flood were used in this study. A cascade of six linear reservoirs (Fig.5) was found to be appropriate to model the propagation of this flood from Kalanaur to Delhi. The four error parameters as in Eqs. 18 to 21 (Flood Studies Report, 1975) were used to judge the adequacy of the model through comparison of the observed hydrographs at Mawi and Delhi with the routed hydrographs obtained from the model.

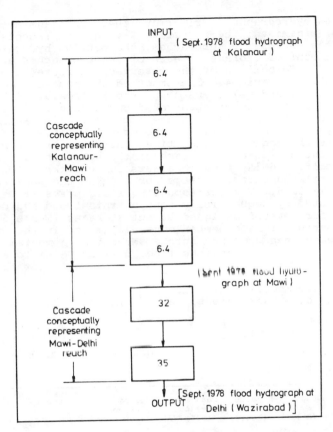

INPUT

(Sept.1978 flood hydrograph at Kalanaur)

6.4

Cascade conceptually representing Kalanaur-Mawi reach

6.4

6.4

6.4

(Sept. 1978 flood hydro-graph at Mawi)

32

Cascade conceptually representing Mawi-Delhi reach

35

OUTPUT [Sept.1978 flood hydrograph at Delhi (Wazirabad)]

FIG.5. A CASCADE OF SIX LINEAR RESERVOIRS REPRESENTING KALANAUR-DELHI REACH OF THE RIVER YAMUNA FOR THE SEPTEMBER,1978, FLOOD

$$\text{Percentage error in attenuation} = \frac{Q_{po} - Q_{pr}}{Q_{po}} \times 100 \tag{18}$$

$$\text{Percentage error in time to peak} = \frac{t_{po} - t_{pr}}{t_{po}} \times 100 \tag{19}$$

$$\text{Percentage mean deviation} = \frac{\bar{Q}_{to} - \bar{Q}_{tr}}{Q_{to}} \times 100 \tag{20}$$

$$\text{Percentage standard deviation} = \left\{ \sum_{1}^{n} \frac{(Q_{tp} - Q_{tr})^2}{n \times Q_{to}} \right\} = 100 \tag{21}$$

in which Q_p, t_p, Q_t and $\bar{Q}_t$ denote, respectively, the peak discharge of the hydrograph, time to peak, discharge intensity at time t and the mean discharge under the hydrograph; and the further suffixes o and r, refer, respectively, to the observed and routed hydrographs. Primarily, the percentage attenuation error was minimised; the resulting K-values representative of each of the successive subreaches, the outflow hydrograph peaks and the amount of storage in each of the subreaches are given in Table 3. Comparison of the observed and routed hydrographs is available in Table 4. The observed and routed flood hydrographs at Mawi and Delhi are displayed in Figs.6 and 7 respectively. The significantly higher observed flows during Sept. 1- 4 are attributed to the rainfall in the catchment below Kalanaur on these days. Having identified the model, Eq.16 was adopted to compute the modified delay time constants corresponding to assigned cutout storages in each of the subreaches. Sample values of the modified delay time constants for embanked flow together with the related cutout storage for each of the sub-reaches are given in Table 5. The corresponding flood peaks at Delhi have also been computed by sequentially routing this flood hydrograph at Kalanaur for several assigned cutout storages (Table 5). The several permutations of embankment alignments in the six sub-reaches have been combined into a sequential nomogram by Chander et al (1979) for relating the flood peak at Delhi to any combinations of the cutout storages in the individual subreaches.

The same model has further been employed to illustrate the effect of embankments spaced continuously 3km apart in the

TABLE 3

K-VALUES FOR THE SEPTEMBER 1978 FLOOD-FLOW (KALANAUR-DELHI REACH)

Reach	Length (km)	K-value (hours)	Peak outflow ($1000 m^3/s$)	System storage (1000 ha.m)	Remarks
(1)	(2)	(3)	(4)	(5)	(6)
KM1	22.5	6.4	23.876	52.15	Kalanaur (Designated 'K')
KM2	22.5	6.4	20.323	43.966	
KM3	22.5	6.4	18.368	39.459	
KM4	22.5	6.4	17.010	36.332	
					Mawi (Designated 'M')
MD1	43	32	10.684	108.789	
MDL	53	35	7.862	83.423	
					Delhi (Wazirabad) (Designated 'D')

NOTES:
1. One hectare-meter = 8.1 acre-feet

2. Subreaches between Kalanaur and Mawi are designated as KM_1, KM_2 ....., and those between Mawi and Delhi as MD_1^1, MD_2^2.

3. Total storage in the four reservoirs upto Mawi = 171.907x 10^3ha.m. and in six reservoirs upto Delhi = 364.119x103 ha.m.

4. Initial flow, Q_o = 1241 m^3/s all along the reach.

5. Storage figures given in col.3 are worked out as $K(Q_p - Q_o)$. For reach KM_1 it is 6.4(23.876-1.241)x3600 = 52,150ha.m

TABLE 4

COMPARISON OF THE OBSERVED AND ROUTED HYDROGRAPHS
AT MAWI AND DELHI (WAZIRABAD)

S.No.	Description	Comparison between the observed and routed hydrographs	
		at Mawi	at Delhi(Wazirabad)
1.	% attenuation error	0.00	-0.15
2.	% time to peak error	3.33	-4.76
3.	% mean deviation error	17.1	5.43
4.	% std.deviation error	3.68	2.39

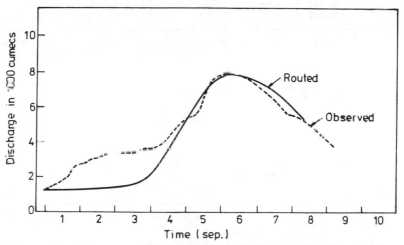

FIG.6. COMPARISON OF THE OBSERVED AND ROUTED FLOOD HYDROGRAPHS
AT DELHI (WAZIRABAD) SEPTEMBER,1978 FLOOD

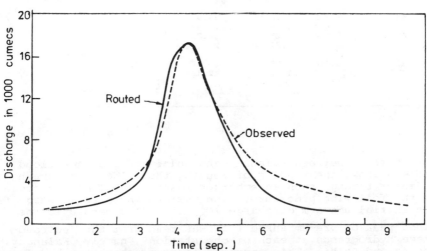

FIG.7. COMPARISON OF THE OBSERVED AND ROUTED FLOOD
HYDROGRAPHS AT MAWI (SEPTEMBER,1978 FLOOD)

TABLE 5

K-VALUES REPRESENTING THE EMBANKED FLOOD-PLAIN SYSTEM
WITH THE FOLLOWING CUTOUT STORAGE THROUGH KALANAUR-MAWI-
DELHI REACH AND THE ANTICIPATED PEAK - FLOWS AT DELHI

	Cutout storage in acre feet/hectare meter					
Reach	40,000 (4940)	80,000 (4880)	120,000 (14820)	160,000 (19760)	200,000 (24700)	240,000 (29640)
KM1	5.794	5.188	4.582	3.976	3.370	2.764
KM2	5.681	4.962	4.244	3.525	2.806	2.087
KM3	5.599	4.798	3.997	3.197	2.396	1.595
KM4	5.530	4.660	3.791	2.921	2.051	1.181
MD1	30.548	29.095	27.643	26.190	24.738	23.286
MD2	32.928	30.857	28.785	26.714	24.642	22.570
Peak Discharge at Delhi	8216	8604	9021	9470	9970	10511 m^3/s

NOTE: Figures in ( ) are cutout storages in ha.m rounded of to 10 ha.m.

TABLE 6

OUTFLOW PEAK AND AVERAGE STAGE INCREMENT WITH EMBANKMENTS
3KM APART IN KALANAUR-DELHI REACH (FLOOD: SEPTEMBER,1978)

Subreach	Cutout storage (ha.m)	K' (Hours)	Modified Flood Peak ($1000m^3$/s)	Stage Increment (m)	Remarks
(1)	(2)	(3)	(4)	(5)	(6)
KM1	39.275	1.83	29.903	0.644	
KM2	31.614	2.08	27.491	0.845	
KM3	27.580	2.23	26.714	1.098	
KM4	24.859	2.34	25.606	1.150	
					-- Mawi
MD1	73.886	12.79	17.97	2.103	
MD2	37.318	21.85	12.408	2.443	
					-- Delhi (Wazirabad)

entire reach between Kalanaur and Delhi for the same flood.
Table 6 shows the results of routing this flood hydrograph at
Kalanaur through such an embanked reach. The Table is organised
by the following procedure. The maximum increase in stage from
the initial uniform discharge (Q_o) of 1241 cumecs upto the

flood peak of pre-embanked flow was read from the recorded
stage hydrographs at each of the stations, namely, Kalanaur,Mawi,
Kutana and Delhi. This increase in stage above the baseflow was
approximated for the mean depth to estimate the volume of water

within the planned embankments.* The deficit between the pre-embanked storage (Table 3) and the above-said volume was taken as the estimate of the respective cutout storage in each of the sub-reaches. Eq.16 would then give the respective K' (modified storage constant values). These K' values were used for outing the flood sequentially through the subreaches using Eq.4. Part of the cutout storage which would cause increase in the embanked stage condition was then determined through Eq.14. The average increase in stage through the subreach between the embankments was then computed.

CONCLUSIONS

A cascade model of linear reservoirs has been success-fully used to model the propagation of a flood wave in a river reach. It has been possible to identify storage in each of the component reservoirs of the cascade in terms of the flood-plain storage, thereby identifying also the representative sub-reaches. Modified storage delay time constants for the post-embanked system are then determinable for any pre-assigned cutout storage. The results are directly adoptable for evaluating the effect of any levee plans on downstream flow conditions. The method is particularly useful in situations where only limited data are available as in the case study.

ACKNOWLEDGEMENT

The authors are grateful to the Central Water Commission, Government of India, for having sponsored this study.

---

*This procedure was adopted because of non-availability of contour maps of the region.

LIST OF SYMBOLS

A	Water spread area in pre-embanked sub-reach
A'	Water spread area in post-embanked sub-reach
H	Head over spillway in pre-embanked sub-reach
H'	Head over spillway in post-embanked sub-reach
H_o	Head over spillway corresponding to initial flow in the sub-reach
$I(t), I_t$	Inflow hydrograph ordinates
$\bar{I}_t$	Average inflow
K	Storage constant representing pre-embanked sub-reach
K'	Storage constant representing post-embanked sub-reach
n	Number of reservoirs in the cascade
$Q(t), Q_t$	Outflow hydrograph ordinates
Q_o	Initial outflow from the sub-reach
Q_{po}	Observed hydrograph peak
Q_{pr}	Routed hydrograph peak
$S(t), S_t$	Storage in the reservoir at time t in the pre-embanked river system
$S_{t'}$	Storage in the reservoir at time t in the post-embanke river system
ΔS	Cutout storage
T	Time interval
T_L	Lag time between inflow and outflow hydrographs
t_{po}	Time to peak of the observed hydrograph
t_{pr}	Time to peak of the routed hydrograph
Δt	Routing time interval
ΔV_1	Volume flown out under excess outflow
ΔV_2	Volume in storage under increase in stage

REFERENCES

1.   Chander,S., P.N.Kapoor, S.K.Spolia, and P.Natarajan,1979, Simulation of the 1978 Flood in River Yamuna, Consultancy Report submitted to the Central Water Commission, Government of India; Civil Engineering Department, Indian Institute of Technology (Delhi),New Delhi.

2.   Flood Studies Report, Vol.III, Natural Environmental Research Council, London, 1975.

3.   Kulandaiswamy, V.C.,1964. A Basic Study of the Rainfall-Excess-Surface Runoff Relationship in a Basin System, Ph.D. Thesis, University of Illinois, Urbana, Illinois.

WATER QUALITY

WATER QUALITY MODELING IN RELATION
TO WATERSHED HYDROLOGY

Anthony S. Donigian, Jr., P.E.
Senior Engineer
Anderson-Nichols & Co., Inc.

ABSTRACT

The past decade has witnessed a dramatic increase in the
development, testing, and application of mathematical models
of the water quality characteristics of natural water
systems. Modeling efforts during this period have
emphasized the integration of hydrology and water quality,
with chemical and biological processes super-imposed on the
relevant transport components of hydrologic models. This
joining of hydrologic and water quality models is reviewed
in order to assess the current state-of-the-art of modeling
runoff, receiving water, and groundwater quality of natural
water systems. Current environmental problems, involving
the fate, transport, and effects of chemicals in the
environment, are discussed in terms of needed research and
capabilities of integrated hydrology-water quality modeling
techniques for analysis and evaluation.

INTRODUCTION

The past decade has witnessed a dramatic increase in the
development, testing and application of mathematical
modeling for analysis of water resources problems. This
increase has been especially notable in the areas of water
quality and nonpoint pollution modeling. With the reduction
in pollutant discharges from cities and industries through
treatment and recycling, the importance of nonpoint sources
(NPS) of pollution and its effect on water quality has been
recognized. Moreover, we now fully realize that current
water quality problems associated with nonpoint pollution
result from the superposition of man's activities on the
hydrologic components of nature. Consequently, model
development in the 70's has emphasized the integration of
hydrology and water quality, with chemical and biological
processes super-imposed on the relevant transport components
of hydrologic models.

This paper attempts to review the integration of water
quality and hydrologic modeling in the 70's, to assess the
current state-of-the-art of modeling the water quality of
natural water systems, and to project current and future
environmental problems that will require integrated

hydrology-water quality modeling techniques for analysis and evaluation. Primary emphasis is devoted to the linked-process, water quality models of runoff, receiving waters, and groundwater. These models attempt to represent the important physical, chemical, and biological processes and their linkage to hydrologic and hydrodynamic transport components. Water quality processes in natural water systems are briefly discussed as a foundation for reviewing the current state-of-the-art of modeling water quality. The general structure and components of water quality models are reviewed, with emphasis on particular selected models, in order to define current capabilities and future needs for modeling sediment, nutrients, pesticides, toxic chemicals, and biological processes.

The concept of "watershed management" embodies the integration of hydrology and water quality concerns and requires the use of joint modeling techniques. Environmental problems and concerns of the 80's are discussed with watershed management providing the encompassing philosophy for analysis of best management practices (BMPs), environmental exposure and risk assessment, hazardous waste assessment and control, and multi-media chemical fate and transport. Contemporary problems within these topical categories, such as pesticides, toxic chemicals, acid rain, eutrophication, etc. will require greater analytical capabilities and sophistication in future integrated hydrologic/water quality modeling techniques.

WATERSHED HYDROLOGIC PROCESSES AFFECTING WATER QUALITY

The concern for the effects of NPS pollution has provided the realization of the importance of watershed hydrologic processes on water quality conditions. Whereas municipal and industrial point sources of pollution are discharged directly to receiving waters, nonpoint sources must rely on the vagaries of the hydrologic cycle to provide the means of movement from the land. The pollution of our lakes and streams from nonpoint sources is the direct result of the interaction between land use and the hydrologic cycle. In effect, land use activities are the basic sources of nonpoint pollutants, while the hydrologic cycle provides the transport mechanisms to move the pollutants to the streams or groundwater. Thus, NPS pollution is a "source-transport" problem that does not exist without both components. Climate and soil characteristics of a watershed combine in the hydrologic cycle to determine surface runoff and subsurface flow components. These components in turn interact and transport pollutants generated from land use activities to the water environment where, in conjunction with point sources, the total water quality impact is exerted. Moreover, NPS pollutants undergo chemical changes on their journey from the land to the stream. Table 1 lists the sources, transport mechanisms, chemical interactions, and transformations that collectively determine the extent and severity of water quality problems in any specific region. How these components combine to produce pollution problems is shown schematically in Figure 1.

TABLE 1. COMPONENTS OF WATER QUALITY PROBLEMS AND MODELS

POLLUTANT SOURCES

    ACCUMULATION/DEPOSITION
    APPLICATION/DISPOSAL
    SOIL STORAGE
    PRECIPITATION
    SURFACE MODIFICATION (e.g., TILLAGE,
     CONSTRUCTION, MINING)

TRANSPORT MECHANISMS

    SURFACE RUNOFF
    INFILTRATION/PERCOLATION
    INTERFLOW
    GROUNDWATER FLOW
    EROSION
    STREAM TRANSPORT BY FLOW AND SEDIMENT

CHEMICAL INTERACTIONS

    ADSORPTION/DESORPTION PROCESS
    DISSOLUTION OF SOLUBLE POLLUTANTS
    POLLUTANT ATTACHMENT TO SEDIMENT

TRANSFORMATIONS

    BIOLOGICAL/CHEMICAL REACTIONS
    VOLATILIZATION
    DEGRADATION

FIGURE 1 COMPONENTS OF WATER QUALITY PROBLEMS AND POLLUTION

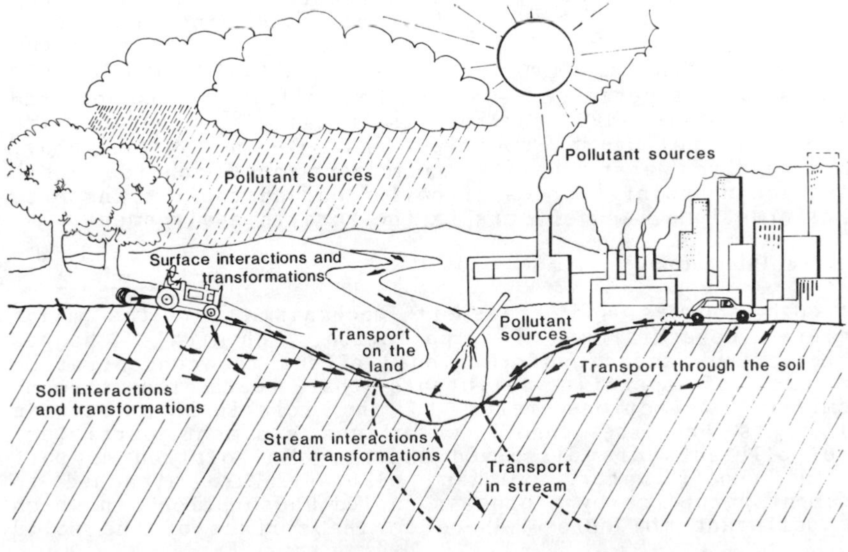

Pollutant Sources.

As we all know, the sources of water pollution are numerous. In urban areas, pollutants deposit and accumulate on the land surface as a result of human and industrial activities. In addition to direct disposal of wastes both on land and in receiving waters, applications of agricultural chemicals and their resulting runoff are other sources of water contamination. The land itself can be a source of nutrients, chemicals, and sediment which is both a pollutant and a means of transport for other pollutants. The disclosure of wide-spread 'acid rain' conditions and its effects, indicate that distant sources of air pollution can be significant local sources of water pollutants through precipitation. Disruption of the land surface, through tillage, construction, mining, etc. makes certain pollutants more readily available for transport to receiving waters. All these pollutant sources can combine in varying proportions within a complex watershed system to produce a variety of water quality problems.

Pollutant Transport Mechanisms

The transport mechanisms that move nonpoint pollutants are universal irrespective of the land use or whether the land surface is pervious or impervious. Surface runoff is the prime mover of pollutants contributed directly from the land surface. The vertical movement of infiltrating and percolating water is the vehicle for moving soluble pollutants from the surface through the soil to the groundwater. Subsurface flow components, such as interflow and groundwater flow may then transport these pollutants to a lake or stream. Although the erosion process contributes sediment as a pollutant, it is also a tranpsort mechanism for those pollutants that are attached and move with the eroded sediment. Once the pollutant contributions from surface runoff, erosion, and subsurface flow reach a stream channel, transport in the stream occurs. Since a stream generally receives both point and nonpoint pollutant contributions, the origin of a pollutant once it is in the stream is often indeterminate. Water quality impacts and effects are experienced after the pollutant reaches the stream. While NPS control is usually concerned with preventing pollutant contributions to a water body, comprehensive water quality management must consider both point and nonpoint sources of pollution and their transport, transformations, and effects in the aquatic environment.

Chemical Interactions

Although sources and transport mechanisms are the major factors determining NPS pollution, chemical changes (interactions and transformations) affect the concentration and mass of specific pollutants that reach the stream. Chemical interactions refer to the relationship between pollutants and the runoff water and sediment particles. Some pollutants are dissolved and move in solution as part of the runoff water. Other pollutants are attached to sediment particles by a process called "adsorption" in which the pollutant molecules adhere to the surface of the solid sediment particles. In addition, some pollutants will be carried in suspension by the moving water. The specific mode of transport is the result of the adsorption/desorption process whereby a pollutant either adsorbs onto a sediment particle or desorbs from sediment back into solution. This process is important because it determines whether a pollutant can move in solution through the soil to contaminate groundwater, or if it will be suspended or attached to sediment and thus can only move over the land surface. Most pollutants can move by either mode but one means will generally predominate. For pollutants that move only in solution, the runoff water gradually dissolves and washes off the available material or it moves with the percolating water. Pollutants associated with sediment must await the transport of the resident sediment particles.

Pollutant Transformations

As opposed to the chemical interactions that determine the solution or sediment phases of a pollutant, transformations actually change the form or composition of a pollutant. A wide range of biological and chemical reactions affect pollutants on the land surface and in the soil profile. Nutrients in particular undergo many transformations that determine the pollution potential of specific forms. For example, oxidation of ammonia to nitrate and the reduction of nitrate to nitrogen gas may occur. Since ammonia is transported both in solution and on sediment, nitrate moves only in solution, and nitrogen gas is effectively removed from the soil systems, the specific form that nitrogen takes determines the severity of nitrogen as a NPS pollutant. Most pesticides and ammonia can change from a solid or liquid to a gaseous form which is of no concern as a water pollutant. Degradation processes can transform substances from polluting to nonpolluting forms, such as for pesticides and other toxic substances. However, in some cases the degradation products of a transformation are more toxic than the original substance.

In summary, the extent and characteristics of the pollutant sources, the magnitude of the NPS transport mechanisms, and the nature and impact of chemical interactions and chemical and biological transformations jointly determine the severity of water quality problems in a watershed.

CURRENT STATUS OF WATER QUALITY MODELING

The decade of the 1970's produced accelerated growth in the development, testing, application, and resulting sophistication of mathematical water quality models for runoff, receiving waters, and groundwater systems. The environmental consciousness of the times, coupled with resulting federal regulations, such as the Federal Water Pollution Control Act Amendments of 1972 (PL 92-500), the Clean Water Act of 1977, the Toxic Substances Control Act, etc., forced the recognition and development of integrated hydrologic-water quality analysis techniques (models). Whereas, prior to 1970 (approximately) the primary focus of water quality models was steady-state BOD/DO problems (with occasional consideration of nitrogenous oxygen demand) from municipal discharges, the current (1980) frontier includes dynamic continuous simulation of pesticides and toxic substances from various sources; sediment transport and sediment-contaminant interactions; particle-size erosion, pesticide and nutrient models; ecologic system models with multiple trophic levels; and two or three dimensional groundwater flow and contaminant models. Obviously, this growth in sophistication of water quality modeling was primarily possible due to the earlier development of hydrologic and hydraulic modeling techniques providing the pollutant transport components, and the continued advancement of these methods during the 1970's. This symposum clearly provides the most relevant state-of-the-art assessment of rainfall-runoff modeling; consequently I will concentrate on the water quality aspects of models, and defer to other speakers and authors for the hydrologic and hydraulic components.

Water Quality Runoff Models

The water quality runoff models have, perhaps, undergone the most dramatic increase in development of all water quality models, since for all practical purposes this class of models was born in the late 60's and early 70's. Negev (1967) superimposed equations for sheet erosion, gully erosion, and instream sediment transport on the relevant hydrologic flow components calculated by the Stanford Watershed Model IV (Crawford and Linsley, 1966). Huff (1967) advanced the concept one step further by adding vegetal interception, ion exchange processes, and migration through the soil profile in order to simulate the movement of radioactive aerosols. Huff, to my knowledge, introduced the term 'hydrologic transport model' (Figure 2) emphasizing

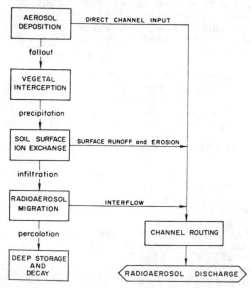

FIGURE 2.   SCHEMATIC OF THE HYDROLOGIC TRANSPORT MODEL
(Huff, 1967)

the importance of the transport mechanisms provided by the watershed hydrology. Although earlier methods, such as the Universal Soil Loss Equation (USLE) (Wischmeier and Smith, 1965), allowed estimation of average annual or seasonal sediment loss, Huff's work appears to be the first significant attempt to 'piggy-back' sediment erosion and chemical processes onto a detailed hydrologic model to develop an integrated hydrologic/water quality runoff modeling system.

In 1969, Meyer and Wischmeier (1969) presented the mathematical formulations for detailed simulation of erosion components including detachment and transport by both rainfall and runoff. Clearly, all these investigators drew upon earlier research into individual hydrologic, erosion, and chemical/biological soil processes as the foundation for their integrated modeling efforts. In essence, the 1965 to 1970 time period initiated and set the stage for the extensive model development work, at various levels of complexity, that occurred in the next decade. Obviously, a review of all the resulting models is beyond the scope of this paper. However, I would like to highlight selected modeling efforts to demonstrate the scope and complexity of the work.

In 1971, the U.S. EPA Environmental Research Laboratory in Athens, Georgia, launched an extensive, and continuing, program to develop and test mathematical models of agricultural runoff (Figure 3). The goal of the program was to develop tools for evaluating the effects of proposed agricultural management practices on sediment, pesticide, and nutrient runoff. The tools would also be useful in assessing the fate and transport of current and newly-proposed pesticidal compounds. Comprehensive field data collection efforts were sponsored in cooperation with various universities (e.g., Baker et al, 1979) and federal agencies (e.g., Smith et al, 1978) to establish the data base necessary for model testing. The field programs included continuous monitoring of meteorologic conditions and runoff from small field sites, with sampling for sediment, pesticide, and nutrients in the runoff. Soil cores were taken and analyzed, crop canopy development was measured, and agronomic (tillage and chemical application) practices were monitored. The programs were multi-year efforts producing complete data for model testing.

The initial product of the EPA program was the Pesticide Transport and Runoff (PTR) model (Crawford and Donigian, 1973) which was similar in structure to Huff's radioaerosol model but with special emphasis on pesticide processes. It used a version of the Stanford Watershed Model as the hydrologic submodel and superimposed submodels for sediment detachment and erosion, pesticide adsorption/desorption and vertical movement, and pesticide degradation. With further model development and testing, the PTR model was succeeded by the Agricultural Runoff Management (ARM) model (Donigian and Crawford, 1976a) which included additional capabilities for snowmelt and plant nutrient (nitrogen and phosphorus) simulation in the soil and runoff. The ARM model has undergone additional testing (Donigian et al, 1977) and

FIGURE 3. ATHENS ENVIRONMENTAL RESEARCH LABORATORY MODEL DEVELOPMENT, TESTING, AND DATA COLLECTION PROGRAMS (Barnwell and Johanson, 1981)

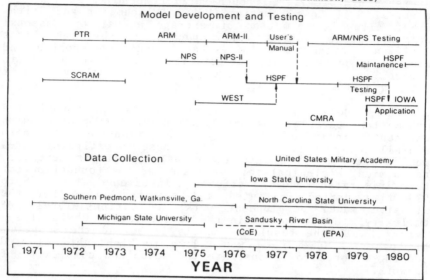

Model acronyms and references:
PTR: Pesticide Transport and Runoff Model (Crawford and Donigian,1973)
ARM/ARM-II: Agricultural Runoff Management Model (Donigian and Crawford,1976a;
                                                    Donigian et al.,1977)
NPS/NPS-II: Nonpoint Source Model (Donigian and Crawford, 1976b)
SCRAM: Simulation of Contaminant Reactions and Movement (Adams and Kurisu,1976)
WEST: Watershed Erosion and Sediment Transport Model (Leytham and Johanson,1979)
CMRA: Chemical Migration and Risk Assessment Method (Onishi et al.,1979)
HSPF: Hydrologic Simulation Program-Fortran (Johanson et al.,1980;1981)

documentation (Donigian and Davis, 1978) with continued testing and applications in different geographic areas (L.A. Mulkey, personal communication, 1980). The EPA-Athens program has produced a variety of models at different levels of complexity (e.g., Adams and Kurisu, 1976; Donigian and Crawford, 1976b; Zison et al, 1977; Onishi et al, 1979), the most recent of which is the Hydrologic Simulation Program - FORTRAN (HSPF) (Johanson et al, 1980; 1981) which includes both runoff and receiving water quality simulation capabilities. HSPF includes and expands upon the ARM model, and incorporates the sediment transport, pesticide decay, sediment-contaminant partitioning, and risk assessment procedures developed by Onishi et al (1979). HSPF will be discussed further in the 'receiving water quality models' section below.

In 1972, the Agricultural Research Service of the U.S. Department of Agriculture initiated a research effort to develop an agricultural chemical transport model. The resulting model, named ACTMO (Frere et al, 1975), was published in June 1975. It uses the USDAHL-74 model (Holtan et al, 1975) for the hydrologic component, and includes an erosion submodel based on the Williams modification (Williams, 1972) of the USLE, chemical partitioning (between water and sediment) and vertical movement, chemical degradation, and a separate option for nitrate simulation.

Other than the hydrologic submodel, ACTMO has undergone limited testing in a few watersheds.

More recently, the Science and Education Administration of USDA in 1976 assembled an impressive group of well-known agricultural scientists to begin the development of a modeling system for analyzing agricultural management systems. In May 1980, the product of this effort was published as the CREAMS model (Knisel W. (ed), 1980), a field-scale model for Chemicals, Runoff, and Erosion from Agricultural Management Systems. CREAMS represents the same surface processes as the ARM model (except snow processes) and includes simulation of small channels, grassed waterways, sedimentation basins, and impoundment terraces; CREAMS does not consider subsurface processes. The hydrologic submodel allows either detailed continuous simulation with short time interval rainfall or daily simulation using the SCS Curve Number approach. The sediment submodel computes total storm event erosion by size fractions (including aggregates) and allows deposition and scour in the channel module. CREAMS was initially tested during its development on selected test sites across the country.

Since 1972 the Oak Ridge National Laboratory (ORNL), under the sponsorship of the National Science Foundation, has been developing models under the collective title of the Unified Transport Model (UTM) to describe the atmospheric and hydrologic fate and transport of heavy metals (Munro et al, 1976). The UTM actually consists of the Atmospheric Transport Model (Mills and Reeves, 1973), the Wisconsin Hydrologic Transport Model (Patterson et al, 1974) which is a modification of the Stanford Watershed Model, and a suite of models to simulate the atmosphere-soil-plant water system, plant growth, soil chemical exchange of heavy metals, and solute uptake by vegetation. The models and submodels can be linked in various configurations to allow flexibility for particular applications. For example, the hydrologic, plant growth, and atmosphere-soil-plant models have been structured as the Terrestrial Ecology and Hydrology Model (TEHM) (Huff et al, 1977) to simulate heavy metal distribution and accumulation in the soil and plants, and study the effects of heavy metals and sulfate on plant growth and litter decomposition. To my knowledge, the UTM has been subjected to limited testing and applications in a few test watersheds.

The above models clearly represent the 'Cadillacs' of water quality runoff models in their degree of representation of the physical, chemical, and biological processes. These models were developed particularly for agricultural, forested, and rural lands although the basic processes are universal with only changes in their relative importance for different land areas. Close examination of these models shows that they differ primarily by their methods of computing hydrologic flow components and sediment erosion. Chemical partitioning is usually by a linear or Freundlich isotherm, and chemical decay and transformations are performed exclusively by first-order kinetics, with rate adjustments for environmental conditions. This again emphasizes the importance of the early development of hydrologic models and the role they have played as the foundation for the structure for runoff water quality models developed in the 70's.

In urban areas runoff water quality modeling has followed a different historical development. Urban stormwater modeling has focused almost exclusively on the impervious land surfaces as the primary source (some models assume the sole source) of runoff and pollutants. Thus, urban models are much simpler than the non-urban models discussed above since simulating soil processes is not a major concern. As a corollary, the initial urban models were single-event models; it wasn't until the mid 70's that the need for continuous simulation was recognized as being relevant for both urban and non-urban areas.

In 1971, the publication of the EPA Storm Water Management Model (SWMM) (Metcalf and Eddy, Inc. et al, 1971) ushered in the era of modeling urban stormwater quantity and quality. The initial version of SWMM was a single-event model for urban stormwater runoff and combined sewer overflow phenomena (Huber et al, 1975). SWMM is a comprehensive model that simulates runoff quantity and quality from pervious and impervious areas, dry weather flow and quality, flow and quality routing in sewers, sedimentation and scour, receiving water quality, storage and treatment options, and performs cost calculations. The runoff water quality is simulated with input solids (dust and dirt) accumulation rates and empirical washoff functions that assume the pollutant concentration is a linear function of the solids concentration. SWMM continues to be supported by the EPA (U.S. EPA, 1980a) and has undergone numerous modifications and enhancements by various agencies, universities, and private consultants. The November 1977 Release of SWMM included capabilities/modifications for continuous simulation and urban snowmelt/removal (U.S. EPA, 1980); SWMM Version III was scheduled for release in late 1980.

In September 1973, the Hydrologic Engineering Center (HEC) of the U. S. Army Corps of Engineers released the Storage, Treatment, Overflow, Runoff Model (STORM). Actually, portions of STORM were developed earlier by Water Resources Engineers Inc. for the EPA and the City of San Francisco, CA. STORM is a continuous runoff water quality model designed primarily for urban areas and for use with many years of hourly precipitation data. The initial version simulated rainfall runoff (using runoff coefficients and the rational formula), pollutant accumulation and washoff, treatment rates, storage, and overflow from the storage/treatment system (U.S. Army Corps of Engineers, 1975). Subsequent improvements have included additional runoff computation options (unit hydrograph and SCS Curve Number method), quantity and quality of dry weather flow, land surface erosion (using USLE) and snowmelt (U.S. Army Corps of Engineers, 1977). Pollutant accumulation and washoff algorithms are essentially identical to those in SWMM.

STORM, and its forerunners, appears to be the first major

urban model based on continuous simulation. As evidenced by
the agricultural runoff models discussed above, the need for
continuous simulation was recognized much earlier in the
agricultural realm than in the urban realm. A 1976 review
of models of urban systems showed that only two out of 18
models included continuous simulation of runoff water
quality (Brandstetter, 1976). Thus, the majority of urban
runoff models developed prior to 1975 were single-event
models. The growing emphasis on nonpoint pollution and
realization of the critical importance of antecedent
conditions on runoff quantity and quality continued to
expand until, by the middle to end of the decade, continuous
simulation became the accepted methodology for planning
decisions in urban areas. In 1976 the EPA published a
simplified version of SWMM that performed continuous
simulation on daily or hourly time steps (Lager et al,
1976). As noted above, the 1977 version of the original
SWMM included a continuous simulation capability. A
companion receiving water model designed for continuous
operation has been developed (Medina, 1979).

Although many more urban runoff models exist, STORM and
various versions of SWMM are the most commonly used models;
this was confirmed in a survey of planning and public works
agencies on the use of modeling in urban water planning
(Donigian and Linsley, 1978). Essentially all urban models
use the same formulations for solids accumulation and
empirical washoff equations that assume a first-order
washoff of the accumulated solids, with the rate related to
runoff volume or intensity; pollutant washoff is then
calculated as a function of the solids washoff (Donigian and
Crawford, 1976b). Unfortunately, these basic equations have
not been improved upon since their incorporation into the
original SWMM in 1971. The emphasis on fundamental
processes (physical, chemical and biological) that has been
apparent in the development and improvement of agricultural
runoff models over the last decade has been lacking in the
development of urban runoff models (Thomann and Barnwell,
1980, pp. 71-77). However, there are indications that the
Nationwide Urban Runoff Program (U.S. EPA, 1978) may collect
the type of data needed to begin to correct this deficiency.

The above discussions concentrated on selected models and
modeling efforts that occurred during the 1970's; obviously
the discussion is not comprehensive due to the scope of this
paper. A recent, uniform and comprehensive runoff model
review was performed by Huber and Heaney (1980), and I have
reproduced their summary (with addition of the CREAMS model)
in Table 2. Huber and Heaney reviewed 73 models that are
known to have been applied to actual problems, and selected
the 14 models (15 including CREAMS) in Table 2 considered to
be 'operational'. Their definition of operational was based
on the following criteria:

    (1). It must have been successfully applied to, and
         verified for, at least one application and can be
         used to model another different, but roughly
         similar water body (watershed) without extensive
         internal modification.

TABLE 2. CAPABILITIES OF OPERATIONAL WATER QUALITY RUNOFF MODELS (Huber and Heaney, 1980)

PROBLEM CHARACTERISTICS

Column groups and characteristics (left to right):

- **Applicable Land Area:** Urban; Agriculture; Forests; Wetlands
- **Applicable Land Area (temporal):** Single Storm Events; Continuous Simulation; Annual or Seasonal Average
- **Temporal Properties:** Single Catchment; Multiple Catchments
- **Spatial Prop.:** Surface/Total Hydrograph Generation; Subsurface Processes
- **Hydrology:** Snowmelt; Dry-weather / Base Flow; Flow Routing in Channels/Pipes
- **Hydraulics:** Backwater, Surcharging, Pressure Flow; Flow Controls and Diversions; Storage / Reservoir Routing; Surface Generation; Routing in Channels/Pipes
- **Quality:** Sediment / Erosion; Scour/Deposition in Channels/Pipes; Parameter Interaction; Soil/Sediment–Parameter Interaction; Routing through Storages
- **Processes:** Treatment in Storages; Treatment/Removal in Storages; Organics / BOD / COD; Nitrogen Species; Phosphorus; Suspended Solids; Coliforms
- **Residuals:** Pesticides; Arbitrary or other Conservative; Arbitrary or other Non-Conservative; Economic Analysis

MODEL NAME:

- HYDROSCIENCE
- MRI
- SWMM-Level 1
- FPARRB
- Simpl. SWMM
- ACTMO
- ARM / HSP } HSPF
- NPS
- QQS
- STORM
- AGRUN
- CAREDAS
- SWMM
- CREAMS

(2) Sufficient written documentation must be available
     about the model to enable a user to apply the
     model in a location other than that for which it
     has been applied.

Table 2. and the referenced work by Huber and Heaney provide
an excellent summary of the major runoff water quality
models and their basic capabilities.

The runoff model embodies the inherent concept of
superposition of water quality physical, chemical, and
biological processes onto the soil moisture and flow
components (surface runoff, percolation/infiltration,
subsurface flow) of the hydrologic model. Thus, the
accuracy of current runoff water quality models depends
equally on how well the hydrologic and water quality
processes are modeled. For pervious areas (and possibly
porous-type pavements in urban areas), the distribution
between surface and subsurface (interflow and groundwater)
flow components is critical; surface runoff transports
primarily sediment and sediment-associated pollutants,
whereas subsurface flow will not transport sediment (and
adsorbed pollutants) but it will include soluble pollutants
originating from the land surface or subsurface. Moreover,
proposed control practices or best management practices
(BMPs) may change the division between surface and
subsurface flow with resulting effects on the pollutants
transported by each component. The ability of a hydrologic
model to represent this division is critical to the accurate
modeling of existing conditions and proposed BMPs.

Receiving Water Quality Models

Receiving water quality models have a significantly longer
history than runoff models, but they have experienced the
same type of explosive growth in sophistication and
capabilities during the last decade. These models simulate
water quality conditions, at various degrees of complexity,
in streams, rivers, lakes, reservoirs, estuaries, and
offshore waters resulting from pollutant loadings (often
provided by runoff models) and the transport, interactions,
and transformations within the waterbody. This concurrent
growth of both runoff and receiving water quality models was
not coincidental, since the ability to predict receiving
water quality is highly dependent on the ability of the
runoff models to predict pollutant loadings. This linkage
of runoff and receiving water models further emphasizes the
important foundation provided by watershed hydrologic
modeling, including runoff and hydrodynamics of receiving
waters, for water quality modeling. Analogous to runoff
models, receiving water quality models superimpose physical,
chemical, and biological mechanisms and interactions onto
the hydrodynamic transport components of the waterbody.

According to Thomann (Thomann and Barnwell, 1980, pp. 37-
61), water quality modeling has a 50-year history dating
from the Ohio River dissolved oxygen (DO) studies in the
1920's. Table 3 is a synthesis of the information presented
by Thomann on the progression of water quality models from
1925 through to the present. Up to about 1965, these models

TABLE 3. PROGRESS OF RECEIVING WATER QUALITY MODELING,
1925-1980 (after Thomann and Barnwell, 1980)

Time Period	System Representation	Water Quality Constituents/ Processes
1925 - 1965	Linear system(s), One-dimensional rivers and estuaries	BOD, DO
1965 - 1970	Multiple linear systems, One to two-dimensional water bodies	BOD, DO, nitrogenous oxygen demand
1970 - 1975	Non-linear interactive systems, One and two-dimensional water bodies	BOD, DO Nutrient cycle effects Phytoplankton/zoo-plankton dynamics
1975 - 1980	Multiple linear and non-linear interactive systems, One, two, and three dimensional water bodies	Added capabilities: Sediment transport Sediment-chemical processes Toxic chemical processes Multiple trophic levels and biological effects

consisted of linear relationships of biochemical oxygen
demand (BOD) and DO to evaluate the effects of municipal
seweage discharge on DO in primarily one-dimensional steady-
state representations of streams and estuaries.
Occasionally, benthal demand on the DO and the effects of
photosynthesis and respiration would be considered.

From 1965 to 1970, the models expanded to consider
nitrogeneous oxygen demand, with modeling of the nitrogen
cycle, in one and two dimensional system representations.
In the period from about 1970 to 1975, model capabilities
were augmented to include nutrient cycles (nitrogen and
phosphorus) within non-linear interactive systems for
biological simulation of zooplankton and phytoplankton
dynamics; multi-layer lake ecologic and eutrophication
models (Chen and Orlob, 1972) were also emerging during this
period. From about 1975 to the present, the degree of
realism has increased to include the interactive dymmanics
of many non-linear natural systems. Additional capabilities
include sediment transport, sediment/chemical partitioning
(Onishi et al, 1979) carbon cycle dynamics, multiple trophic
levels (including multiple fish species) (Park et al, 1974;
Park, 1978), and toxic substances transport and aquatic
impact. Many of these areas require additional research and
improvement before wide-spread acceptance of these
capabilities is established. However, the state-of-the-art
of receiving water quality modeling has advanced

sufficiently so that simulation of the above topical areas, which was unthinkable prior to 1970, is within our grasp.

Obviously, not all current receiving water quality models include (or need to include) all the capabilities noted above, and a review of the current population of models is beyond the scope of this paper. A recent review by Hinson and Basta (1979) appears to be the most recent and comprehensive available. Table 4 from their report lists and summarizes the capabilities of 27 'operational' receiving water quality models; the definition of 'operational' is the same as used by Huber and Heaney (1980), discussed above. To this table I have added the HSPF model (Johanson et al, 1981) because of its recent publication and comprehensive scope. Table 4 and the report by Hinson and Basta (1979) provide an excellent 'snapshot' of current water quality models and the state-of-the-art for receiving waters. Also, at a recent conference a group of modeling professionals assembled to evaluate and establish the veracity or validity of water quality modeling in various topical areas (Thomann and Barnwell, 1980). From those sources, the following list of capabilities was compiled in an order indicating increasing uncertainty or

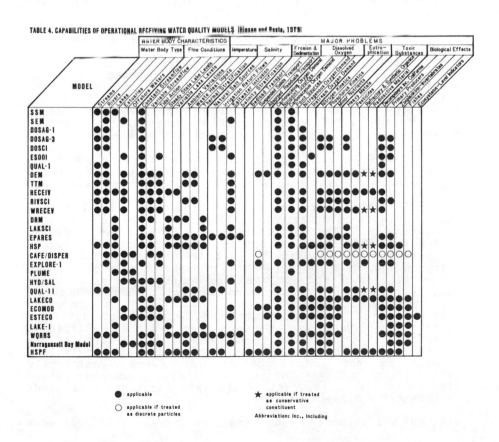

TABLE 4. CAPABILITIES OF OPERATIONAL RECEIVING WATER QUALITY MODELS (Hinson and Basta, 1979)

357

decreasing confidence in the capabilities of current models:

> Hydrodynamic transport
> Dissolved Oxygen/temperature
> Salinity/conservatives
> Nutrients/eutrophication
> Sediment transport
> Toxic/hazardous substances fate and transport
> Biological/ecosystem effects

Clearly transport modeling is the critical foundation for any water quality model, and well-established techniques exist for most water systems. Methods include hydrologic routing (e.g., Muskingum), kinematic wave, and full equations (continuity and momentum) approaches. The kinematic wave approach is commonly used and adequate for many fresh-water systems. For stream segments with flat slopes or possible backwater effects, such as in estuaries, tidal streams, or complex urban drainage systems, full equation methods may be required. However, estuarine transport modeling is one area that requires greater study and model verification. In general, the required sophistication of the transport component should be determined by the sensitivity of the specific water quality processes and model to transport processes (Thomann and Barnwell, 1979, pp. 78-90).

The kinetics of DO and temperature processes have been sufficiently studied and analyzed to be the most reliable and accurate area of water quality modeling. Salinity and other conservative constituents are highly dependent on the accuracy of the specified loadings to the receiving water quality model, either from measured data or runoff models. Thus with reliable loadings and transport simulation, problems associated with irrigation return flows, certain heavy metals, and other conservatives can be analyzed.

Next in order of increasing uncertainty are the so-called eutrophication models that simulate nutrient cycle dynamics, and sometimes biomass, in water bodies. Nitrogen and phosphorus are the primary nutrients of concern and, as shown in Table 4, the majority of the operational models include their simulation. The carbon cycle is also simulated in selected models as a basis for modeling carbonate-based alkalinity and pH. Nutrient transformations in streams are relatively well-defined, especially for nitrogen and, to a lesser degree, phosphorus; the models are technically sound and are used for decision-making in water quality management (Thomann and Barnwell, 1980). Much of the current uncertainty is due to the lack of relevant information on rate constants and the lack of data for calibration. Phosphorus simulation is especially difficult since much of it is sediment-bound, and thus limited by the accuracy of sediment transport modeling, and analytical methods do not often measure the forms functionally available for uptake. Nutrient modeling in receiving waters also suffers from the lack of accurate input nutrient loads from runoff and precipitation.

Sediment transport and modeling of toxic/hazardous

substances are closely associated because sediment is a transporting medium for many toxic chemicals. Although sediment transport formulas have existed for more than 40 years, it remains one of the most difficult areas of water quality modeling. The previously mentioned workshop on water quality modeling concluded the following:

> Modeling of sediment transport (suspended and settleable) suffers from a lack of quantification which results from inability to specify correctly settling velocities, critical erosion velocities, and a general lack of understanding of the physics and mechanisms involved in sediment transport processes

> (Thomann and Barnwell, 1980, pg. 81)

Table 4 shows that most of the operational water quality models are lacking erosion and sediment transport capabilities. Such models do exist, e.g., Fields (1976) and Onishi et al (1979) which also contain sediment-chemical interactions, but testing and applications have been limited. The primary weakness is in the modeling of the cohesive sediments (silt and clay) which are also the most reactive in terms of adsorbing organics and chemicals; bedload (sand) transport is somewhat better defined and less calibration dependent.

The capabilities for modeling toxic substances in Table 4 are somewhat misleading, since most of the models that claim to represent toxic substances do not include sediment simulation. The apparent underlying assumption must be that the heavy metals, pesticides, organics, etc. are represented strictly as solubles and/or conservatives - a tenuous assumption for many compounds and water bodies. This demonstrates the need to review carefully model assumptions and algorithms before accepting the stated capabilities.

In addition to sorption processes with cohesive sediments, toxic chemicals can undergo a variety of decay mechanisms and transformations in receiving waters. Although laboratory research to define individual transformations is not new, a comprehensive approach to enumerate the range and types of possible toxic chemical reactions and their linkage to hydrodynamic and sediment transport models is a recent endeavor. In 1977, the Stanford Research Institute (Smith et al, 1977) integrated the work of other researchers and identified photolysis, hydrolysis, oxidation, volatilization, and biodegradation as the primary mechanisms determining the fate of chemicals in aquatic environments. This work provided the basis for exposure assessment methods (U.S. EPA, 1980b) and pesticide fate and transport modeling with HSPF (Johanson et al, 1981) and by Onishi et al (1979). Obviously much additional work is needed particularly in the identification and evaluation of relevant rate constants (Mill et al, 1980) and the effects of environmental conditions.

Although the ecologic and biological effects models are at the bottom of the list in terms of confidence and reliability, many would argue (and rightfully so) that they should be ranked closer to the nutrient models since they are the next logical step in the aquatic food chain. The ecologic models invariably include nutrient simulation as the food source for the lower organisms (algae, phytoplankton, benthic algae), and then consider higher organisms (zooplankton, invertebrates) up to and including multiple species and/or life stages of fish. The simulation of nutrients and lower organisms has generally been tested and applied to a much greater degree than the processes related to the higher organisms. Due to their comprehensive scope, scarcity of data to calibrate and test many portions of these models is a common problem. They have been applied almost exclusively to lakes and estuaries for analysis and evaluation of the aquatic ecosystem and effects of management and control practices.

Although the population of receiving water quality models covers a range of capabilities, the process formulations for many models are remarkably similar. Zison et al (1978) have reviewed the physical (excluding sediment transport), biological, and chemical formulations for a variety of models, and presented the data for the rates and constants used in various studies. Although many exceptions exist, generally first-order kinetics are used for most chemical transformations; equilibrium partitioning, or a first-order approach to equilibrium, for sediment-chemical interactions; first-order or Michaelis-Menton kinetics for metabolic processes (respiration, growth, death, etc.); and second-order kinetics for predation since this is a function of both the predator and prey populations. The work by Zison et al (1978), Hinson and Basta (1979), and specific model documentation should be consulted for details.

Sediment transport formulations for sand and cohesive sediments are quite variable including both empirical and theoretical approaches; the interested reader should consult the documentation on the specific model of interest, or general references such as Vanoni (1975) or Graf (1971).

Groundwater Quality Models

Although this author cannot claim to be well-versed on the topic of groundwater modeling, no discussion of water quality modeling in relation to watershed hydrology would be complete without mention of groundwater quality modeling. Consequently, this section is drawn primarily from a review by Bachmat et al (1978) of numerical groundwater models and their use in water resource management; Table 5 provides an overview of the characteristics of the models surveyed.

Groundwater modeling is a relatively new field that was not extensively pursued until about 1965. For the first ten years, the primary emphasis was on groundwater flow modeling since, like the runoff and receiving water quality models, the transport mechanisms had to be established before water quality could be considered. Beginning about 1973 to 1975, groundwater quality models have been emerging in the literature. Much progress and advancement has been made in the past decade, but gaps and deficiencies do exist. In general, groundwater quality modeling appears to be 5 to 10 years behind the state-of-the-art of surface water modeling. This is not an indictment of the field, but simply an indication of the complexity of conceptualizing and modeling a natural system that cannot be seen and is extremely difficult to monitor. In fact, groundwater and runoff water quality models share similar deficiencies in representing contaminant movement by subsurface flow components.

As shown in Table 5, Bachmat et al (1978) classified the models reviewed in terms of capabilities, i.e. quantity (supply problems), quality (contamination problems), and environmental impact (subsidence), and the categories of prediction, management, parameter identification/evaluation, and data manipulation. The prediction models, which comprised 80% of the 250 models reviewed, are the deterministic simulation models of interest here. The mass transport category includes groundwater quality models, representing only 16% of the total. The quality models, which usually include a flow submodel, are divided about equally between those that consider only conservative constituents and those that allow reactions. Generally the reactions assume equilibrium or linear first-order kinetics, such as for adsorption and radioactive decay. According to Bachmat et al (1978) only two of the 39 groundwater quality models reviewed included biochemical transformations of nitrogen which are important in waste disposal and reclamation problems. Also, ecologic and biologic aspects of groundwater have not been addressed.

Of the entire population of 250 models, only 30 were judged to be 'usable', with the relevant criteria being adequate documentation (including a user's manual), availability, and prior application to a field situation. Only two of the water quality models were judged to be usable.

With regard to areas for needed model development, the following processes were listed:

-flow in media of secondary porosity (e.g., fractures and karstic formations)

-flow for immiscible fluids (e.g., oil in water, for hazardous waste problems)

-fully integrated surface, unsaturated, and saturated flow models

-contaminant transport with chemical and biological reactions

-ecological aspects and processes

Clearly, model development is hindered in some areas by a lack of understanding of the relevant phenomena. Kinetics of chemical, physical, and biological processes; pollutant transport through the unsaturated zone; and the effects of scale and heterogeneity on transport phenomena are particular areas for further research.

TABLE 5. OVERVIEW OF GROUNDWATER MODELS SURVEYED BY BACHMAT ET AL. (1978)

Model Category	Number of Reports	Groundwater					Groundwater & Surface Water				
		Quantity		Quality		Environmental Impact	Quantity		Quality		Environmental Impact
		Lumped	Distributed	Lumped	Distributed		Lumped	Distributed	Lumped	Distributed	
Prediction											
Flow											
Water	127	2	119**b		2*e			4*			
Water & other fluids	11		5*		6e						
Mass transport											
Conservative	20			2a	17**f				1*a		
Nonconservative	19			2*	17**d						
Heat transport	9				9c						
Deformation	8					8					
Others	68g										
Management	29	10*			3a		7	6*	2	1	
Identification	16	2	14*								
Data manipulation	5										
Total	250										

* One "usable" model in group.
** Two or more "usable" models in group.
a Treats waste disposal and reclamation.
b Sixteen models treat coupled saturated-unsaturated systems.
c Treats thermal problems 2 of which are geothermal.
d Two models treat biochemical reactions.
e Treats interface.
f Two models treat interface.
g Includes 1 frost propagation, 1 coupled heat-mass transport, 2 coupled subsidence and heat transport models and 1 general purpose code.

It should be noted that the work by Bachmat et al concentrated exclusively on published groundwater models of regional and areawide scope, down to the level of a single well, and that the work was performed in the 1975-1977 period. Models published and progress made since that time have concentrated on the above areas but the general conclusions remain valid at this time. (J. Keeley, personal communications, 1981). Also, considerable work related to microscale transport and reactions of contaminants (e.g., leaching from waste piles and/or landfills) in the unsaturated zone, such as work by Davidson et al (1978), Rao et al (1976), and Enfield and Carsel (1981) was not considered. These models include chemical and biochemical transformations to a much greater degree of sophistication than the regional groundwater quality models reviewed by Bachmat. They are representative of research that has concentrated on the small scale (soil column) vertical transport and interactions of chemicals in the unsaturated zone; these models will likely be the foundation for future improvements in watershed or regional groundwater models.

WATER QUALITY MODELING IN THE 80'S

Just as the 1970's was the decade for the emergence of water quality and general environmental concerns, the 1980's will examine the problem, and hopefully some solutions, of toxic chemicals in the environment; their sources, fate, transport, and effects will be considered in water quality modeling efforts during the next decade. With the problems of point source pollution from industries and municipalities reasonably under control, nonpoint and groundwater pollution by toxics and conventional pollutants have been identified as key water quality issues in the 1980's (Burmaster, 1980). In addition, the phenomena and impacts of acid rain and energy development (e.g., oil shale, strip mining, synthetic fuels), the products of our industrialized society, will bring often conflicting economic and environmental concerns into the public eye. In this section I will try to examine what role water quality models, and especially integrated hydrologic-water quality models, will play and the demands that will be placed on these models by the water quality issues of the 80's.

Table 6 lists four major topics or problem areas that will require or utilize water quality models during the next decade, and associated needed capabilities not fully satisfied by current models. The topics listed - watershed management and BMP evaluation, environmental exposure and risk assessment, hazardous waste assessments and control, multi-media chemical fate and transport - include obvious overlaps in many areas; the common thread is the need to better quantify the sources, fate, transport, and effects of chemicals in the air, soil, and water environment. Advances in one topical area will benefit other areas. However, each of the topics is directed toward specific problems and analytical requirements.

Watershed Management and BMP Evaluation

Watershed management and Best Management Practice (BMP)

TABLE 6. TOPICS AND NEEDS FOR WATER QUALITY
MODELING IN THE 80'S

TOPIC	NEEDED WATER QUALITY MODELING CAPABILITIES
Watershed management and BMP evalulation	Integrated runoff/receiving water/groundwater models Sediment and pollutant delivery from field to stream Representation of BMP effects Solute transport by subsurface flow Erosion modeling by particle size Soil nutrient processes
Environmental exposure and risk assessment	Soil persistence and decay mechanisms of chemicals Land surface chemical behavior Instream chemical transformations Sediment transport and sediment-contaminant interactions Aquatic (ecologic) impact of toxicants
Hazardous waste assessment and control	Contaminant transport and reactions through/in the unsaturated zone Groundwater pollutant transport Effects of disposal practices on transport and reaction parameters
Multi-media chemical fate and transport	Integrated atmospheric, soil, and water models Wet and dry pollutant deposition and accumulation Soil and stream processes for pH and alkalinity

evaluation is listed first because, in many ways, it
encompasses the basic concerns of all the topics. The term
'watershed management' refers to a combination of
structural, non-structural, and other control practices
implemented on a watershed to achieve a specific purpose, be
it flood reduction, water supply increase, water quality
improvement, or multiple purposes. The term was earlier
used by the agricultural community for effective management
of land resources to maintain productivity and by flood
control specialists for use of nonstructural techniques. In
the water quality realm, watershed management includes the
comprehensive management of the land and water resources to

minimize the detrimental effects of all pollutant sources (point and nonpoint) on the watershed ecosystem (U.S. EPA, 1979a, pg. 36). This requires a fundamental understanding of the dynamics of the physical, chemical, and biological interactions within the watershed.

The concept of 'best management practices' is closely related to watershed management since it is commonly defined as follows:

> A practice or combination of practices that is determined to be the most effective, practicable (including technological, economic, and institutional considerations) means of preventing or reducing the amount of pollution generated by nonpoint sources to a level compatible with water quality goals.

With reductions in point sources, the two terms - watershed management and BMPs - are essentially equivalent with BMPs providing the means of achieving effective watershed management and relevant water quality goals. Thus, watershed management, in a systems context, includes consideration of toxic and hazardous materials, employs environmental exposure and risk assessment as a basis for evaluating current conditions and effects of BMPs, and involves evaluation of multi-media processes to the extent that pollutant deposition and input by precipitation affects the soil and aquatic environment.

The types of water quality models needed in the future for complete watershed management and accurate assessment of proposed BMPs will be comprehensive, integrated runoff/receiving water/groundwater modeling systems. Such systems will encompass the full range of simulation capabilities needed for integrated surface and groundwater quality studies, and will be closely tied, as are current models, to the hydrologic and hydrodynamic transport components. To what extent such integrated systems will be achieved during the 1980's is uncertain. Currently, the HSPF model appears to be the most comprehensive, with both simple and complex runoff modules, relatively sophisticated receiving water simulation, but a simple storage/outflow approach to groundwater. Also, various runoff models such as the STORM, SWMM, ARM, and NPS models have been linked to receiving water models such as the WQRRS, RECEIV, EXPLORE-I, and QUAL-II models to perform comprehensive water quality assessments. Such linkages, generally affected by data file transfers from the runoff to the receiving water model, have become more frequent in recent years indicating a need for such integrated models. Linkages of receiving water and groundwater models are still relatively rare, partially due to the traditional division between surface and groundwater modeling.

Additional specific problems and associated improvements in water quality modeling capabilities needed in the future include:

Sediment and pollutant delivery from field to stream.
It is well known that sediment erosion from small

field-size areas is generally not equivalent to the sediment load measured at the watershed outlet, due to intermediate deposition/scour and transport and instream processes. Baker et al (1979) have shown that the same problem occurs for pesticides and nutrients in agricultural watersheds; presumably urban watersheds experience the same phenomena possibly to a lesser degree. The implication is that runoff water quality models for small areas may not accurately represent the pollutant load reaching the receiving water. If BMPs are evaluated on their ability to achieve instream pollutant concentration and water quality goals, intermediate transport between field and stream must be better understood and modeled in the future.

Representation of BMP effects. The accuracy of models in evaluating the overall effects of proposed BMPs depends significantly on how well the model user can estimate changes in model parameters to represent the way a practice affects particular processes being modeled. The data base available for estimating parameter changes is usually fragmented, non-existent, or only indirectly related to the conditions being modeled. Except for street cleaning practices in urban areas for which some data has been collected (Pitt, 1979) and some collection system controls or treatment (Field et al, 1977), parameter estimates are usually based on a combination of judgment and extrapolation of data from other areas.

The work by the Northern Virginia Planning District Commission (1979) and Hartigan et al (1981) in estimating parameter changes for urban BMPs is demonstrative of the best that can be done at the current time. For agricultural BMPs, the supporting documentation of the CREAMS model (discussed earlier) and a study by Donigian et al (1980) for HSPF model parameters draws upon the extensive body of process-oriented research by the agricultural research community to provide some guidance for quantifying parameter adjustments for selected BMPs.

However, what is needed in the 1980's is post-implementation data collection efforts on watersheds with adequate baseline (pre-implementation) data to calibrate and verify the ability of models to represent BMPs. The Nationwide Urban Runoff Program presents some hope that relevant data of this type will be collected in urban watersheds. In addition, a number of opportunities exist for agricultural watersheds with extensive historic data (e.g., Baker et al, 1979) where post-BMP implementation data collection would be extremely valuable. Even in a period of budget cutting and dwindling financial resources, it would be wise to devote a fraction of the millions of dollars spent on implementation to improve the data and analytical tools for predicting BMP effects. Until this is done models will continue to be used in BMP analyses, with parameter changes based on available data, indirect information, and judgment, primarily because there are no alternatives. For BMP analyses in many situations, models are 'the only game in town'.

Solute transport by subsurface flow. Modeling the transport of reactive solutes, such as nutrients, pesticides, and other toxics, is a difficult area that is not well represented, especially subsurface transport, by currently available models. In fact, ACTMO and the family of models based on the Stanford approach (i.e., ARM, HSP, HSPF, UTM) are the only operational runoff models that attempt to evaluate subsurface pollutant contributions to receiving waters. A major part of the problem is the inability to measure, and then model, the subsurface flow component that exists between surface runoff and base flow (often called 'interflow'). Agricultural practices designed to reduce surface runoff and erosion often increase percolation and subsurface flow providing an increased opportunity for subsurface loss of soluble pollutants (Haith and Loehr, 1979). Better understanding and methods of representing subsurface solute movement are needed for more reliable evaluation of BMP effects.

Erosion modeling by particle size. Since instream sediment transport and chemical partitioning are highly dependent on sediment particle size, modeling the erosion process by separate size fractions should provide an opportunity to better represent chemical partitioning and both sediment and pollutant loading to receiving water quality models. Recent advances in this area have been made: Simons, Li, and Stevens (1975) developed a distributed (grid-type) parameter model with particle-size erosion algorithms, and Foster et al (1980) developed the erosion submodel of CREAMS including event-based simulation of both aggregates and primary particles (sand, silt, clay) of sediment. For urban areas, Ellis and Sutherland (1979) and Alley et al (1980) have proposed a more deterministic approach to urban water quality runoff modeling with specific consideration of particle size fractions in the accumulation and washoff/transport algorithms. As noted previously, such considerations have been sorely lacking in urban models. Although more testing and parameter evaluation is needed, erosion and chemical runoff models of the mid-80's will likely include erosion by size fractions as an accepted capability.

Soil nutrient processes. Nutrient runoff is a concern primarily from agricultural croplands, but it can be a problem from both rural and urban areas, leading to accelerated eutrophication of water bodies. Modeling soil nutrient processes, especially for nitrogen and phosphorus, on agricultural and rural lands is quite advanced; the primary transformations are generally assumed to follow first-order kinetics. Although particular mechanisms may need further study, such as nitrogen fixation by legumes, nutrient leaching from crop residues, plant root contributions to soil nutrients, etc., the primary problem is the evaluation of model parameters (e.g., crop uptake, partition coefficients, mineralization rates, nitrification/denitrification rates) for local conditions. A better data base is needed from which model parameters for base conditions and proposed BMPs can be evaluated. Also, the process-oriented approach of the agricultural runoff models should be considered for the pervious fractions of urban areas to replace the simple washoff equations currently used.

Environmental Exposure and Risk Assessment

With regard to water quality, exposure and risk assessment pertain to the process of evaluating the incidence or exposure of aquatic organisms to toxic chemicals, and the associated risk of possible lethal or chronic effects. Pesticides and other toxic chemicals that are actively released into the environment are the primary targets. Exposure assessment relies upon the integrated runoff/receiving water modeling techniques discussed above; the time series of chemical concentrations provided by the model(s) is statistically analyzed to assess the frequency (and resulting risk) of concentration levels lethal for a particular organism. Such procedures will be commonly used in the 1980's by both industry and government, under the mandate of regulations for pesticides and toxic chemicals (e.g., FIFRA, TSCA), to evaluate the possible environmental risk associated with both current and proposed chemical compounds.

Falco et al (1978) demonstrated the overall procedures by linking the ARM model with simplified stream routing/decay equations to assess expected environmental concentrations of a newly proposed pesticide in projected use areas of the south. As part of the RPAR (Rebuttable Presumption Against Registration) process, Mulkey and Hedden (1979) linked the ARM model with a sediment/pesticide transport model to evaluate the frequency and duration of toxaphene concentrations in the Mississippi Delta region, under recommended application rates and conventional agronomic practices of the region.

Onishi et al (1979) advanced the methodology further by linking the ARM model with a detailed sediment transport model including sediment-contaminant interactions and pesticide decay processes. They analyzed the simulated pesticide concentration time series using lethality-duration information and maximum acceptable toxicant concentration (MATC) values (Figure 4) to assess the 'percent of time' that lethal (acute), potentially lethal (chronic), and sub-lethal (below MATC) conditions exist. The capabilities developed by Onishi et al have been incorporated into the HSPF model.

These studies show how comprehensive modeling studies and exposure/risk assessment methods can be used to evaluate the potential lethal effects for old and new compounds. Furthermore, with appropriate parameter changes the above lethality analyses can be used as a basis for evaluating proposed BMPs.

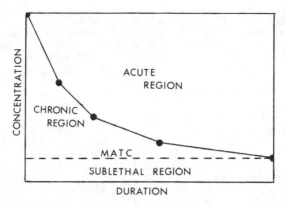

FIGURE 4.   LETHALITY ANALYSIS OF CHEMICAL
CONCENTRATION DATA

Since exposure assessment is based on comprehensive water quality modeling, the deficiencies and needed improvements in capabilities discussed above are also relevant here. Other needs specifically related to modeling pesticides and other toxicants include the following:

Soil persistence and decay mechanisms of chemicals.

Since pesticide losses in surface runoff are generally a maximum of 1 to 5% of the application amount (Wauchope, 1978), pesticide loss by volatilization, microbial degradation, photochemical degradation and other mechanisms account for the remaining 95 to 99% (excluding application losses).   Thus, accurate modeling of chemical attenuation mechanisms is critical to determining amounts in the soil available to runoff when a storm occurs.   Since most currently used pesticides have half lives in the range of a few weeks to a few months, the critical time period for pesticide runoff is usually one to two months following application.   The time between application and the first major storm event is usually a primary factor determining the amount of pesticide lost during the event.

Current models of pesticide decay in soils are generally based on lumped first-order kinetics without corrections for environmental conditions. Decay rates will vary greatly depending on soil moisture, organic matter, pH, soil characteristics, etc.   Moreover, location in the soil profile will affect persistence since surface - applied pesticides will undergo volatilization, photodegradation, and other surface processes; whereas incorporated pesticides and those leached below the surface will often decay primarily by microbial action.   Each process obviously proceeds at different rates.

More accurate prediction of chemical persistence in the soil including effects of natural environmental conditions is needed to improve the accuracy of chemical runoff models. Although the above discussion emphasizes pesticides, other chemicals reaching the land will demonstrate the same general behavior with specific mechanisms varying in importance.

Land surface chemical behavior. In addition to decay mechanisms, the general behavior of chemicals in the near surface zone (down to about 1 cm) is another weakness of current models. A common procedure is to assume a uniform 'mixing' zone of one centimeter or less. Adsorption/desorption processes, under widely varying moisture and soil conditions at the surface, complicate the extrapolation of laboratory data to field conditions which are extremely difficult (and costly) to measure. Leaching from the surface zone generally removes the chemical as a possible component of surface runoff, but increases the likelihood of subsurface transport and contamination. More research and model development is needed to improve the representation of chemical behavior in the surface zone.

Instream chemical transformations. As noted in the section on receiving water models, the Stanford Research Institute identified the primary mechanisms determining the fate of chemicals in aquatic (freshwater) environments. This work allowed the development of chemical fate models including the specific processes of photolysis, hydrolysis, oxidation, volatilization, and biodegradation. Although more work is needed in the process representation, the primary weakness at this time is the needed data base to evaluate the pertinent rate constants for specific chemicals of concern, and the effects of environmental conditions on these rate constants.

A related area that is not considered by most models is the formation of toxic daughter products from the parent material (e.g., DDE and DDD from DDT). The capability to track daughter products produced by various mechanisms has been included in improvements to the HSPF model, but the inability to identify and evaluate the relevant rate constants remains.

Sediment transport and sediment - contaminant interactions. As discussed above, the primary weakness in sediment transport modeling is representation of the cohesive (silt and clay) sediments which are the most reactive, and thus critical to modeling sediment - contaminant interactions. Studies on shear stress, settling velocities, and scour/deposition processes in natural freshwater systems are needed.

The sediment - contaminant interface must be better defined and the choice of equilibrium vs. kinetic partitioning formulations should receive special emphasis in the future. Sediment and contaminant processes in stream and lake beds are not well

understood or modeled. Data by Schnoor (1981) indicate
that concentrations of the chlorinated hydrocarbon
dieldrin are still at significant levels in the Iowa
River up to 5 years after it was banned by the EPA.
Release from bottom sediments was a partial cause for
the continuing concentrations. This indicates the
importance of the bed as a continuing source of
contamination, and the need for further study.

Aquatic (ecologic) impact of toxicants.   Few of the
biologic and ecologic receiving water models in Table 4
consider the effects of toxicants on aquatic
organisms. Recent work by Leung (1978) and Schnoor
(1980) indicate the progress that has been made in the
biologic realism of ecologic models and the
possibilities of representing bioaccumulation of
chemicals in aquatic organisms. With recent and future
advances in toxics simulation (fate and transport), the
integration of ecologic and toxics models would seem to
be a natural extension needed for reliable aquatic
exposure and risk assessment studies This would allow
the analysis of both direct toxic effects on organisms,
and indirect effects, such as loss of a predator's food
source.

Hazardous Waste Assessment and Control

With the discovery of problems like Love Canal and the
Valley of the Drums, the assessment and control of hazardous
wastes will be an enduring activity that will receive major
attention and support in the 1980's. Just as the TSCA and
FIFRA were enacted to regulate the manufacture and use of
chemicals, the Resource Conservation and Recovery Act (RCRA)
of 1976 was developed to provide control of hazardous wastes
resulting from our chemical-based society, develop improved
practices in solid and hazardous waste disposal, and
establish resource conservation as the preferred waste
management approach (U.S. EPA, 1978, p. 5). The role of
water quality models in this management and regulatory
process is not clear, primarily because the relevant
procedures and guidelines are still evolving. However, it
is clear that groundwater contamination is the most common
problem associated with hazardous waste disposal
practices. Currently, landfill and surface impoundments are
the most common methods of waste disposal, and engineered
landfilling procedures will remain in common use for the
foreseeable future until recovery, reuse, and treatment
methods are better developed (U.S. EPA, 1980c, p. 14).

Although surface water contamination by runoff from waste
impoundments and disposal sites is a significant concern,
and one that will be analyzed with runoff water quality
models, groundwater contamination is the primary focus of
hazardous waste assessment and control. The use of both
unsaturated and saturated groundwater quality models will
likely be a significant part of the problem assessment and
analyses. In addition to the model development areas
discussed previously, groundwater quality models for
assessment of hazardous waste problems will need improvement
in the following areas:

371

Contaminant transport and reactions through/in the unsaturated zone.
The movement of contaminants from surface disposal sites and impoundments will require better modeling capability of their transport and behavior in the unsaturated zone between the land surface and the groundwater table. Work by Davidson et al (1976, 1980) and Rao and Davidson (1979) on chemical movement and behavior for pesticide disposal is exemplary of the type of modeling studies needed for other hazardous wastes. The specific chemical and microbial adsorption and decay processes, and relevant parameters under field conditions, require better definition for the enormous variety of chemicals that may comprise a hazardous waste. Also, interactions among chemicals and possible impacts on the vertical and lateral flow transport cannot be overlooked. A major limitation may be characterization of the waste in terms of the movement, reactions, and interactions of the constituent chemicals. Models describing the movement of contaminants from waste sites will need to be integrated with watershed models for comprehensive water quality assessment and watershed management as discussed above.

Groundwater pollutant transport.
Once contaminant input to the groundwater table is determined, movement and reactions within the saturated groundwater zone presents another difficult problem. Many of the same deficiencies of the current unsaturated zone models are also relevant to the groundwater models; advances in modeling the kinetics of chemical and biological processes will benefit both areas. In fact, except for direct pollutant input to groundwater through injection wells or waste impoundments, coupling of unsaturated and saturated zone water quality models will be required in the 1980's for adequate analysis of hazardous waste problems. Further integration with comprehensive watershed hydrology and water quality models is inevitable; such models may become operational in the coming decade.

Effects of disposal practices on contaminant transport and reactions.
Just as BMPs will affect transport and reaction processes, proposed disposal practices will impact the transport and behavior of the wastes being disposed or contained. For example, disruption of the soil profile for landfill sites will change the infiltration and pollutant transport characteristics of the area. Analogous problems exist for strip mining activities and associated acid mine drainage and waste disposal. If modeling is used for assessment of proposed land-based hazardous waste control practices, accurate representation of the complete impacts of the practices will be needed.

Multi-Media Chemical Fate and Transport

The primary focus of multi-media chemical modeling is the integration of air-borne pollutants into the land and water phases of current water quality models. Although urban runoff models include pollutant accumulation functions for dry deposition of chemicals, the CREAMS model is the only one in Table 2 that specifically allows input of chemicals (in this case, nutrients) with precipitation. Essentially all other runoff models assume that precipitation is comprised of 'pure' rain and snow. To assess the linkage among air, land, and water pollution problems, a linkage dramatically demonstrated by the "Acid Rain" problem (U.S. EPA, 1979b; 1980d), multi-media chemical modeling systems of the 80's must include the air medium as a source of contamination.

Multi-media chemical models will be used to analyze a wide variety of problems, such as the effects of industrial air emissions on local or regional water quality, drift and aquatic impact of aerially applied chemicals (e.g., pesticides) on non-target areas, relative environmental effects of incineration versus land disposal or stream discharge, and the acid precipitation problem noted above. Some needed improvements in current modeling capabilities to address these problems are as follows.

Integrated atmospheric, soil, and water models. The Unified Transport Model (UTM) (Munro et al, 1976) discussed previously is the only model known to this author that integrates the various media to assess the effects of fugitive air emissions on water quality. It effects of mining and smelting operations on the movement of heavy metals through a forested watershed. Recently, the EPA has sponsored the linkage of current air transport/deposition, runoff and stream water quality models for multi-media chemical screening assessments. Greater efforts of this type are needed, ranging from the simple inclusion of the rainfall input of pollutants, nutrients, and other chemicals to the actual air transport and transformations of chemical stack emissions, and their washout and deposition onto land and water surfaces.

Wet and dry pollutant deposition and accumulation. As part of the integrated multi-media models, the pollutant deposition problems require special attention. Both dry deposition and wet deposition by rain and snow of chemicals and pollutants must be added to future water quality models. In the immediate future, this will likely be affected by more reliable accumulation/deposition functions and estimates of chemical concentrations in precipitation. Ultimately a comprehensive systems approach to the air, land, and water environments will be followed to track pollutant transport and transformations within and among the various environments.

Soil and stream processes for pH and alkalinity. The primary compounds that cause acid precipitation are oxides of sulfur and nitrogen that combine with water in the atmosphere, and return to the earth in the form of sulfuric and nitric acids. These substances can be harmful to a wide range of plants, agricultural crops, and aquatic organisms, and they can leach from the soil heavy metals such as aluminum and manganese that can have direct toxic effects. To model the effects of acid precipitation, advances in modeling both soil and stream processes for pH and alkalinity are needed. Although a few stream models include these processes, alkalinity is usually based on the carbonate system whereas the sulfate input with acid rain must also be considered. None of the current operational runoff models consider the relevant soil proceses. Model development work specific to the acid rain problem is currently underway (e.g., Chen et al, 1978) and will be an important topic in the coming decade. Integration of these models with the air transport modeling of acid rain components will provide a means of assessing the possible effects of continued or increased fossil fuel combustion processes, a major source of the acid rain problem (U.S. EPA, 1979a).

CLOSURE

In the decade of the 1970's water quality modeling has come of age. It has grown from the simple, steady-state dissolved oxygen models of the 60's to include complex runoff and receiving water quality models representing physical, chemical and biological processes; analogous models are currently being developed and tested for the complex topic of groundwater quality. Realization of the critical importance of watershed hydrology and the resulting superposition of water quality processes onto hydrologic and hydrodynamic transport components has been a primary vehicle for this growth. Water quality models are now used as important tools in decision-making for comprehensive watershed and water quality management programs.

Inspite of, and partly because of, this accelerated growth in water quality modeling, the next decade presents even greater challenges and demands on our analytical ingenuity as we begin to deal with the pervasive problem of chemicals in the environment. The concepts of watershed management, which is essentially a comprehensive systems approach to land and water systems, will require continued improvements in integrated hydrologic and water quality modeling techniques to assess the fate, transport, and effects of pesticides, toxic chemicals, hazardous wastes, etc. The medium of air transport and dispersion of pollutants will require multi-media modeling efforts to analyze the insidious impact of acid rain and other air pollutants on natural land and water systems. Clearly, not all these problems will be solved in the 1980's, and probably new ones will be discovered; water quality modeling will play a major role in developing possible solutions and providing a better understanding of the environment we are trying to protect.

REFERENCES

Adams, R.T. and Kurisu, F.M. 1976. Simulation of pesticide Movement on Small Agricultural Watersheds. U. S. Environmental Protection Agency, Athens, GA. EPA-600/3-76-066. 324 pp.

Alley, W.M., Ellis, F.W., and Sutherland, R.C. 1980. Toward a More Deterministic Urban Runoff-Quality Model. Presented at the International Symposium on Urban Storm Runoff, Lexington, Kentucky. July 18-31. pp. 171-182.

Bachmat, Y., Andrews, B., Holtz, D. and Sebastian, S. 1978. Utilization of Numerical Groundwater Models for Water Resource Management. R.S.K. Environmental Research Laboratory, U. S. Environmental Protection Agency, Ada, OK. EPA-600/8-78-012. 186 pp.

Baker, J.L., Johnson, H.P., Borcherding, M.A., and Payne, W. R. 1979. Nutrient and Pesticide Movement from Field to Stream: A Field Study. in: R.C. Loehr, D. A. Haith, M. F. Walter, and C. S. Martin (editors), Best Management Practices for Agriculture and Silviculture. Ann Arbor Science, Ann Arbor, MI. pp 213-245.

Barnwell, T. O., Jr. and Johanson, R.C. 1981. HSPF: A Comprehensive Package For Simulation of Watershed Hydrology and Water Quality. In: Nonpoint Pollution Control: Tools and Techniques for the Future. Interstate Commission on the Potomac River Basin. Rockville, MD. pp. 135-153.

Brandstetter, A. 1976. Assessment of Mathematical Models for Storm and Combined Sewer Management. Municipal Environmental Research Laboratory. U. S. Environmental Protection Agency. Cincinnati, OH. EPA-600/1-76-175a. 510 pp.

Burmaster, D. 1980. New Thrust in the 1980's for U.S. Water Quality Programs. Civil Engineering Magazine. September 1980 issue. pp. 78-81.

Chen, C. and Orlob, G. 1972. Ecologic Simulation for Aquatic Environments. Water Resources Engineers, Inc., Walnut Creek, CA. Prepared for Office of Water Resources Research, Washington, D.C. 156 pp.

Chen, C.W., Gherini, S., and Goldstein, R.A. 1978. Modeling the Lake Acidification Process. Chapter 5 in: M.G. Wood (editor), Ecological Effects of Acid Precipitation. Electric Power Research Institute, Palo Alto, CA. EA79-6-LD.

Crawford, N.H. and Donigian, A.S., Jr. 1973. Pesticide Transport and Runoff Model for Agricultural Lands. U.S. Environmental Protection Agency, Athens, GA. EPA-660/2-74-013. 211p.

Crawford, N.H. and Linsley, R.K. 1966. Digital Simulation in Hydrology: Stanford Watershed Model IV. Stanford University, Stanford, CA. TR No. 39. 210 pp.

Davidson, J.M., Graetz, D.A., Rao, P.S.C., and Selim, H.M. 1978. Simulation of Nitrogen Movement, Transformation, and Uptake in Plant Root Zone: U.S. EPA, Ada, OK. EPA-600/3-78-029. 105 pp.

Davidson, J.M., Ou, Li-Tse, and Rao, P.S.C. 1976. Behavior of High Pesticide Concentrations in Soil Water Systems. Proc. of Hazardous Wastes Research Symp., Tuscon, Arizona, EPA-600/9-76-015. p. 206-212.

Davidson, J.M., Rao, P.S.C., Ou, Li-Tse, Wheeler, W.B., and Rothwell, D.F. 1980. Adsorption, Movement, and Biological Degradation of Large Concentrations of Selected Pesticides in Soils. U.S. EPA Ecological Res. Series. EPA-600/1-80-124. 110 pp.

Donigian, A.S., Jr. and Crawford, N.H. 1976a. Modeling Pesticides and Nutrients on Agricultural Lands. U.S. Environmental Protection Agency, Athens, GA. EPA-600/3-76-043. 263 pp.

Donigian, A.S., Jr. and Crawford, N.H. 1976b. Modeling Nonpoint Pollution from the Land Surface. U.S. Environmental Protection Agency, Athens, GA. EPA-600/3-76-083. 292 pp.

Donigian, A.S., Jr. Beyerlein, D.C., Davis, H.H., Jr., and Crawford, N.H. 1977. Agricultural Runoff Management (ARM) Model - Version II: Testing and Refinement. U.S. Environmental Protection Agency, Athens, GA. EPA-600/3-77-098. 310 pp.

Donigian, A.S., Jr. and Davis, H.H., Jr. 1978. User's Manual for Agricultural Runoff Management (ARM) Model. U. S. Environmental Protection Agency, Athens, GA. EPA-600/3-78-080. 173 pp.

Donigian, A.S., Jr. and Linsley, R.K. 1978. Planning and Modeling in Urban Water Management. Office of Water Research and Technology, U. S. Dept. of the Interior, Final Report on Contract No. 14-34-0001-6222, 158 pp.

Donigian, A.S., Jr., Baker, J.L., Haith, D.A., and Walter, M.F., 1980. HSPF Parameter Adjustments to Evaluate the Effects of Agricultural Best Management Practices. Draft Report on EPA Contract No. 68-03-2895. U. S. Environmental Protection Agency, Athens, GA. 102 pp.

Ellis, F. W. and Sutherland, R. C. 1979. An Approach to Urban Pollutant Washoff Modeling. Presented at the International Symposium on Urban Storm Runoff, Lexington, Kentucky. July 23-26. pp. 325-340.

Enfield, C. G. and Carsel, R. F. 1981. Mathematical Prediction of Toxicant Transport Through Soil. U. S. Environmental Protection Agency, Ada, OK. 18 pp.

Falco, J.W.,. Mulkey, L.A., Hedden, K.F., Smith, C.N., Barnwell, T.O., Dean, J.D., Lipcsei, R.E., and Smith, M.C. 1978. Estimated Degradation and Transport of Dimilin in Selected Rivers of the Southern U.S. U. S. Environmental Protection Agency, Athens, GA

Field, R., Tafuri, A. N. and Masters, H.E. 1977. Urban Runoff Pollution Control Technology Overview. Municipal Environmental Research Laboratory. U.S. Environmental Protection Agency, Cincinnati, Ohio. EPA-600/2-77-047. 91pp.

Fields, D. E. 1976. CHNSED: Simulation of Sediment and Trace Contaminant Transport with Sediment/Contaminant Interaction. Oak Ridge National Laboratory, Oak Ridge, Tennessee. ORNL/NSF/EATC-19. 203 pp.

Foster, G. R., Lane, L.J., Nowlin, J.D., Laflen, J.M., and

Young, R.A. 1980. A Model to Estimate Sediment from Field-Sized Areas. In: W. G. Knisel (editor), CREAMS: A Field Scale Model for Chemicals, Runoff, and Erosion From Agricultural Management Systems. U.S.D.A. Conservation Research Report No. 26. pp. 36-64.

Frere, M.H., Onstad, C.A. and Holtan, H.N. 1975. ACTMO: An Agricultural Chemical Transport Model. 54 pp. Agricultural Research Service, U.S. Department of Agriculture. Hyattsville, Maryland. ARS-H-3. 54 pp.

Haith, D.A. and Loehr, R.C., Ed. 1979. Effectiveness of Soil and Water Conservation Practices for Pollution Control. Environmental Research Laboratory, U. S. Environmental Protection Agency, Athens, GA. EPA-600/3-79-106. 480 pp.

Hartigan, J.P., Biggers, D.J., Bonuccelli, H.A., and Wentink, B.E. 1981. Cost-Effectiveness Factors For Urban Best Management Practices. In: Nonpoint Pollution Control: Tools and Techniques For The Future, Interstate Commission on the Potomac River Basin, Rockville, M.D. pp. 199-212.

Hinson, M.O. Jr. and Basta, D.J. 1979. Analyzing Surface Receiving Water Bodies. Chapter 7. In: D.J. Basta and B.T. Bower (editors), Analysis For Regional Residuals - Environmental Quality Management: Analyzing Natural Systems. Draft Report. Resources for the Future, Washington, D.C.

Holtan, H.N.; Stiltner, G.I.; Henson, W.H., and Lopez, N.C. 11975. USDAHL-74 Revised model of Watershed Hydrology. Agricultural Research Service, U.S. Department of Agriculture, Washington, D.C. Technical Bulletin No. 1518.

Huber, W.C.; Heaney, J.P.; Medina, M.A.; Peltz, W.A.; Sheikh, H.; and Smith, G.F. 1975. Storm Water Management Model User's Manual, Version II. National Environmental Research Center. U. S. Environmental Protection Agency, Cincinnati, Ohio. EPA-670/1-75-017. 367 pp.

Huff, D.D. 1967. Simulation of the Hydrologic Transport of Radioactive Aerosols. Ph.D. Dissertation. Stanford University, Stanford, California. 206 pp.

Huff, D.D.; Luxmoore, R.J.; Mankin, J.B.: and Begovich, C.L. 1977. TEHM: A Terrestrial Ecosystem Hydrology Model. Oak Ridge National Laboratory, Oak Ridge, Tennessee. ORNL/NSF/EATC-17. 152 pp.

Johanson, R.C., Imhoff, J.C., and Davis, H.H. 1980. User's Manual for the Hydrologic Simulation Program - Fortran (HSPF). U. S. Environmental Protection Agency, Athens, GA. EPA-600/9-80-015. 684 pp.

Johanson, R.C., Imhoff, J.C., Davis, H.H., Kittle, J.L., and

Donigian, A.S., Jr. 1981. User's Manual for the Hydrologic Simulation Program - Fortran (HSPF): Release 7.0. U. S. Environmental Protection Agency, Athens, GA 745 pp.

Knisel, W., Ed. 1980. CREAMS: A Field Scale Model for Chemicals, Runoff, and Erosion From Agricultural Management Systems. U.S. Department of Agriculture. Conservation Research Report No. 26. 640 pp.

Lager, J.A.; Didriksson, T. and Otte, G.B. 1976. Development and Application of a Simplified Stormwater Management Model. 139 pp. Municipal Environmental Research Laboratory, U.S. Environmental Protection Agency, Cincinnati, Ohio. EPA-600/2-76-218.

Leytham, K.M. and Johanson, R.C. 1979. Watershed Erosion and Sediment Transport Model. U. S. Environmental Protection Agency, Athens, GA EPA-600/3-79-028. 373 pp.

Loung, D. 1978. Modeling the Bioaccumulation of Pesticides in Fish. Center for Ecological Modeling. Rensselaer Polytechnic Institute, Troy, New York. Report No. 5, 18 pp.

Loehr, R.C.; Haith, D.A.; Walter, M.F., and Martin, C.S. (Editors) 1979. Best Management Practices for Agriculture and Silviculture. Ann Arbor Science Publishers Inc., Ann Arbor, Michigan.

Medina, M.A. 1979. Level IIIL: Receiving Water Quality Modeling for Urban Stormwater Management. Municipal Environmental Research Laboratory, U.S. Environmental Protection Agency, Cincinnati, Ohio. EPA-600/2-79-100. 217 pp.

Metcalf & Eddy, Inc., University of Florida, and Water Resources Engineers, Inc. 1971. Storm Water Management Model. Water Quality Office. Environmental Protection Agency. Washington, D.C. 11024 DOC. 4 Volumes.

Meyer, L.D. and Wischmeier, W.H. 1969. Mathematical Simulation of the Process of Soil Erosion by Water. Transactions of American Society of Agricultural Engineers. Vol. 12, pp. 754-758, 762.

Mill, T., Mabey, W.R., Bomberger, D.C., Chou, T.W., Hendrey, D.G., and Smith, J.H. 1980. Laboratory Protocols for Evaluating the Fate of Organic Chemicals in Air and Water. Stanford Research Institute, Menlo Park, CA. Prepared for Environmental Research Laboratory, U.S. Environmental Protection Agency, Athens, GA. 329 pp.

Mulkey, L.A. and Hedden, K.F. 1979. Assessment of Toxaphene Exposure Levels in the Yazoo River Resulting From Basin-Wide Application of Toxaphene to Cotton and Soybeans. U. S. Environmental Protection Agency, Athens, GA.

Munro, J.K., Luxmoore, R.J., Begovich, C.L., Dixon, K.R., Watson, A.P., Patterson, M.R. and Jackson, D.R. 1976. Application of the Unified Transport Model to the Movement of Pb, Cd, Zn, Cu, and S through the Crooked Creek Watershed. Oak Ridge National Laboratory, Oak Ridge, Tennessee. ORNL/NSF/EATC-28. 92 pp.

Northern Virginia Planning District Commission. 1979. Guidebook for Screening Urban Nonpoint Pollution Management Strategies, Falls Church, Virginia. 122 pp.

Onishi, Y., Brown, S.M., Olsen, A.R., Parkhurst, M.A., Wise, S.E., and Walters, W.H. 1979. Methodology for Overland and Instream Migration and Risk Assessment of Pesticides. Battelle, Pacific Northwest Laboratories, Richland, WA. Prepared for U. S. Environmental Protection Agency, Athens, GA.

Park, R. 1978. A Model for Simulating Lake Ecosystems. Center for Ecological Modeling, Rensselaer Polytechnic Institute, Troy, New York. Report No. 3, 19 pp.

Park, R.A., et al (25 authors). 1974. A Generalized Model for Simulating Lake Ecosystems. Simulation Journal. pp. 33-56.

Patterson, M.R., Munro, J.K., Fields, D.E., Ellison, R.D., Brooks, A.A., and Huff, D.D. 1974. A User's Manual for the Fortran IV Version of the Wisconsin Hydrologic Transport Model. Oak Ridge National Laboratory. Oak Ridge, TN. ORNL-NSF-EATC-7. 252 pp.

Pitt, R. 1979. Demonstration of Nonpoint Pollution Abatement Through Improved Street Cleaning Practices. Municipal Environmental Research Laboratory. U.S. Environmental Protection Agency, Cincinnati, Ohio. EPA-600/2-79-161. 290 pp.

Rao, P.S.C., Davidson, J.M. and Hammond, L.C. 1976. Estimation of non-reactive and reactive solute front locations in soils. Proc. of Hazardous Wastes Research Symp., Tuscon, Arizona, EPA-600/9-76-015. p. 235-242.

Rao, P.S.C., and Davidson, J.M. 1979. Adsorption and Movement of Selected Pesticides at High Concentrations in Soils. Water Research. Vol. 13. pp. 375-380.

Schnoor, J.L. 1980. Field Validation of Water Quality Criteria for Hydrophobic Pollutants. Presented at the Fifth ASTM Symposium on Aquatic Toxicology, October 7-8, Philadephia, PA 25 pp.

Schnoor, J.L. 1981. Fate and Transport of Dieldrin in Coralville Reservoir: Residues in Fish and Water Following a Pesticide Ban. Science. Feb. 20.

Simons, D.B., Li, R.M., and Stevens, M.A. 1975. Development of Models for Predicting Water and Sediment Routing and Yield from Storms on Small Watersheds.

Prepared for U.S.D.A. Forest Service, Flagstaff, AR. 130 pp.

Smith, C.N., Leonard, R.A., Langdale, G.W., and Bailey, G.W. 1978. Transport of Agricultural Chemicals From Small Upland Piedmont Watersheds. U. S. Environmental Protection Agency, Athens, GA and U.S.D.A. Watkinsville, GA. EPA-600/3-78-056. 386 pp.

Smith, J.H., Mabey, W.R., Bohonos, N., Holt, B.R., Lee, S.S., Chou, T.W., Bomberger, D.C., and Mill, T. 1977. Environmental Pathways of Selected Chemicals in Freshwater Systems, Part I: Background and Experimental Procedures. Stanford Research Institute, Menlo Park, CA. Prepared for Environmental Research Laboratory. U.S. Environmental Protection Agency, Athens, GA. 81 pp.

Thomann, R.V. and Barnwell, T.O. 1980. Workshop on Verification of Water Quality Models. Environmental Research Laboratory, U. S. Environmental Protection Agency, Athens, GA. EPA-600/9-80-016. 274 pp.

U. S. Army Corps of Engineers. 1975. Urban Storm Water Runoff: "STORM". Hydrologic Engineering Center, Davis, CA 104 pp.

U. S. Army Corps. of Engineers. 1977. Storage, Treatment, Overflow, Runoff Model, "STORM": User's Manual. Hydrologic Engineering Center, Davis, CA. 170 pp.

U. S. Environmental Protection Agency. 1978. 1978-1983 Work Plan for the Nationwide Urban Runoff Program. Water Planning Division, Washington, D.C.

U. S. Environmental Protection Agency. 1979a. Research Outlook 1979. Office of Research and Development, Washington, D.C. EPA-600/9-79-005. 140 pp.

U. S. Environmental Protection Agency. 1979b. Research Summary: Acid Rain. Office of Research and Development, Washington, D.C. EPA-600/8-79-028. 23 pp.

U. S. Environmental Protection Agency. 1980a. Center for Water Quality Modeling Newsletter, September 19th issue. Athens, GA.

U. S. Environmental Protection Agency. 1980b. EXAMS: An Exposure Analysis Modeling System. Environmental Research Laboratory, Athens, GA.

U. S. Environmental Protection Agency. 1980c. Research Summary: Controlling Hazardous Wastes. Office of Research and Development, Washington, D.C. EPA-600/8-80-017. 25 pp.

U. S. Environmental Protection Agency. 1980d. Acid Rain. Office of Research and Development, Washington, D.C. EPA-600/9-79-036. 36 pp.

Vanoni, V.A. (editor) 1975. Sedimentation Engineering. Prepared by the ASCE Task Committee for the Manual on Sedimentation of the Hydraulics Division, American Society of Civil Engineers. New York, NY.

Wauchope, R.D. 1978. The Pesticide Content of Surface Water Draining from Agricultural Fields - A Review. Journal of Environmental Quality, Vol. 7, pp. 459-472.

Williams, J.R. 1972. Sediment Yield Prediction with Universal Equation Using Runoff Energy Factor. Presented at Interagency Sediment Yield Conference. U.S.D.A. Sedimentation Laboratory, Oxford, MS. November 28-30.

Wischmeier, W.H. and Smith, D.D. 1965. Predicting Rainfall Erosion Losses from Cropland East of the Rocky Mountains. U.S. Department of Agriculture. Agricultural Handbook No. 282. 47 pp.

Zison, S.W., Haven, K.F., and Mills, W.B. 1977. Water Quality Assessment: A Screening Method for Nondesignated 208 Areas. Environmental Research Laboratory, U.S. Environmental Protection Agency, Athens, GA. EPA-600/9-77-023. 549 pp.

Zison, S.W., Mills, W.B., Deimer, D., and Chen, C. 1978. Rates, Constants, and Kinetics Formulations in Surface Water Quality Modeling. Environmental Research Laboratory, U. S. Environmental Protection Agency, Athens, GA. EPA-600/3-78-105. 335 pp.

A UNIFIED APPROACH TO THE MODELING OF TRANSIENT STORAGE, TREATMENT AND TRANSPORT OF URBAN POINT AND NONPOINT WATER POLLUTANTS

Miguel A. Medina, Jr., Associate Professor
Department of Civil Engineering
Duke University
Durham, N.C. 27706

and

Jennifer Buzun, Environmental Engineer
Division of Environmental Management
N.C. Dept. of Natural Resources
& Community Development
P.O. Box 27687
Raleigh, N.C. 27611

ABSTRACT

A rational evaluation of urban water pollution control strategies must include an assessment of the pollutant removal efficiency of control measures for both point and nonpoint sources, and their combined impacts on receiving water quality. A unified approach is presented to apply the one-dimensional, transient conservation of mass equation throughout the urban environment to represent storage, treatment and transport systems. The optimal operation of storage/treatment facilities for wet weather and dry weather flows is examined in terms of their ultimate effect on downstream water quality and established standards. Continuous simulation allows the derivation of water quality frequency curves to aid decision-makers in the screening of alternatives.

Pollutant transport within each phase of the hydrologic cycle, and through the various components of the physical system, is governed by the principle of conservation of mass. Deterministic mathematical models were derived from this unifying concept to represent the movement, decay, storage, and treatment of stormwater runoff pollutants and dry weather wastewater flows through the urban environment and the receiving body of water. The general one-dimensional, transient conservation of mass equation may be simplified for application to the various pathways by retention of only the dominant terms in each instance. The detention time and decay coefficient are key parameters in establishing the final form of the governing differential equation.

The transient response of storage/treatment systems to variable forcing functions of flow and concentration was determined for completely mixed systems of constant and variable volumes and for one-dimensional advective systems with and without dispersion, and results were compared. Frequency analyses were performed on input and output concentrations and mass rates for a single event and for all wet weather events during the year of record simulated. The response of typical activated sludge and trickling filter systems (while accounting for storage effects) is also represented by models derived from the one-dimensional, transient advective-dispersive equation. A dual-purpose, wet weather and dry-weather, storage/treatment facility is represented by parallel and series operation of trickling filter models.

The unifying concept of continuity is extended to determine the receiving water response to waste inputs from (1) wet weather urban sources, (2) dry weather urban sources, and (3) upstream sources. The results are presented for a stream reach in terms of dissolved oxygen cumulative frequency curves. Interpretations can be based, conveniently, on established stream standards for the study area. The sensitivity of receiving water quality to control system parameters (detention time, dispersion characteristics) is of particular interest. Thus, urban water pollution control schemes can be evaluated in terms of their combined and separate impacts on receiving water quality and their respective costs.

Application of this methodology to Des Moines, Iowa and the Des Moines River indicated that receiving water response is sensitive to the length of the detention time in the wet-weather storage/treatment device, during periods of urban runoff. The choice of mathematical model for wet-weather flow storage/treatment systems may lead to different outflow pollutographs but in all cases, adequate detention reduces the peak and variability of the input pollutographs. Treatment is effected by natural decay processes while in storage.

INTRODUCTION

The complex interrelationships and interactions among urban stormwater runoff, man-made storage/treatment systems and receiving water response encompass a large number of variables--which can be grouped broadly as those studied from:

(1) a natural science perspective in order to better describe chemical, biological and physical phenomena;

(2) an engineering perspective to achieve useful purposes, such as deterministic and/or statistical prediction of pollutant transport/storage/treatment or the design of conveyance and containment systems; and

(3) a management and planning perspective to achieve a degree of control over the state of the system within the context of a rational decision-making process.

Scientific studies are based on the premise that all natural phenomena and processes are interrelated and interactions are governed by certain laws, thus involving numerous subsystems. The basic engineering problem, for example, is finding analytical relationships between the variables characterizing the inflow and outflow processes and parameters defining the state of the system. In principle, these analytic relationships are provided by solution of the complete equations of energy, mass, momentum and state. However, this is often impractical (if not impossible) because of: inadequate knowledge of chemical-biological-physical behavior, unknown spatial and temporal variabilities of system parameters, and numerical approximations among others. Whereas state variables (density, volume, temperature, etc.) define the condition of system components, decision variables act to modify the state. Storage and treatment may modify the concentration and mass rate of pollutants in an accelerated manner to prevent damaging shock loadings from entering receiving bodies of water. The degree of control defines an aspect of the management problem.

Urban hydrology, including water quality processes, is a combination of concepts and parameters that pertain to scientific, engineering, and management points of view. The essence of a rational approach to water quality control in the urban environment is the development of a conceptual model, based on scientific principles, which has the predictive capabilities required by the decision-making process. The high cost of pollution control facilities in terms of energy utilization, land requirements, engineering manpower and long-term financial burden obligates the planning agency to select an optimum strategy for area-wide wastewater management. Such a process must focus on a systematic procedure that identifies and defines (1) the cause/effect relation-ships in the environment, (2) the efficiency of control alternatives, and (3) the benefits to be derived from implementation of these controls.

The advantages of continuous simulation for hydrologic studies have been clearly presented and contrasted to the pitfalls of the single-event (design storm) concept (e.g., Linsley and Crawford, 1974). It is becoming more widely accepted that water quality planning should also be based on long-term characterization of the rainfall-runoff process and assessment of the cost-effectiveness of alternative control measures (e.g., Heaney et al., 1977; Donigian and Linsley, 1979; Medina et al., March 1981). Continuous simulation accounts for: (1) land use changes in the modeled area and variable pollutant accumulation and washoff rates, (2) precipitation and runoff characteristics and antecedent conditions for the entire time period, (3) design characteristics of storage/treatment systems, and (4) receiving water flow and quality characteristics. Worst-case conditions of receiving water quality are usually arbitrarily defined (e.g., 7-day, 10-year low flow) in conventional waste allocation studies. It is well-known that high-frequency storm events can generate the greatest pollutant loadings to a stream (Vilaret and Pyne, 1971; Heaney et al., 1977). Continuous simulation produces results which can be interpreted for a wide range of water quality standards rather than a fixed standard (e.g., 5 mg/l of DO). Sherwani (1971) studied low-flow data from 37 gauging stations in North Carolina with at least 25 years of record. He noted that, in setting stream and effluent water quality standards, use of the 7-day, 10-year low flow (7Q10) ignored the marked variation in behavior between different streams and that multiple standards specifying the requirements for different durations may be necessary.

METHODOLOGY

A Unified View of the Physical System

The principle of conservation of mass may be applied universally throughout the urban and natural environments to describe the transport of pollutants. Figure 1 represents a generalized component of the physical system, which may characterize (1) a pipe segment of the sewer system, (2) a storage/treatment device, or (3) a reach of the receiving

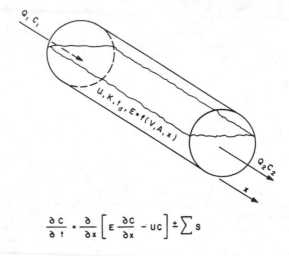

$$\frac{\partial C}{\partial t} = \frac{\partial}{\partial x}\left[E\frac{\partial C}{\partial x} - UC\right] \pm \sum S$$

Fig. 1. A Generalized Component of the Physical System

body of water. Essentially, all of these subsystems may be approximated by the one-dimensional version of the convective-dispersion equation,

$$\frac{\partial C}{\partial t} = \frac{\partial}{\partial x}\,[E\frac{\partial C}{\partial x} - UC] \pm \Sigma S_i \tag{1}$$

where C = concentration of water quality parameter (pollutant), M/L^3,

t = time, T,

$-E\frac{\partial C}{\partial x}$ = mass flux due to longitudinal dispersion along the flow axis, the x direction, M/L^2T,

UC = mass flux due to advection by the fluid containing the mass of pollutant, M/L^2T,

S_i = sources or sinks of the substance C, M/L^3T,

i = 1, 2, . . ., n,

n = number of sources or sinks,

U = flow velocity, L/T, and

E = longitudinal dispersion coefficient, L^2/T.

The source/sink term accounts for biochemical processes (e.g., decay, photosynthesis, algal respiration), boundary losses such as stream benthic deposits, and boundary gains (e.g., reaeration, and point or distributed waste discharges). Assuming that the longitudinal dispersion coefficient and the advective velocity are constant along the flow axis,

Eq. 1 may be expanded for the generalized component in Figure 1 to

$$\frac{\partial C}{\partial t} = E \frac{\partial^2 C}{\partial x^2} - U \frac{\partial C}{\partial x} - KC + \frac{q_1}{A} (c_1 - C) \tag{2}$$

where K = first-order reaction rate coefficient, $1/T$,

q_1 = influent fluid flow rate per unit width, L^2/T,

c_1 = concentration of water quality parameter in the inflow, M/L^3, and

A = wetted cross-sectional area in the component, L^2.

In Eq. 2 the influent fluid flow rate per unit width, influent concentration, and the wetted cross-sectional area may all be variable functions of time.

Solutions to differential equations derived from Eq. 1, and governing the behavior or urban stormwater storage/treatment systems, conventional dry-weather flow facilities, and a receiving body of water are presented in the following sections. All of the systems described by the generalized component in Figure 1 are essentially characterized by

(1) a flow velocity U,

(2) a reaction rate coefficient K,

(3) a residence time t_D, and

(4) a longitudinal dispersion coefficient E,

all of which are a function of the system geometry (volume V, cross-sectional area A, length x).

Urban Storage/Treatment Models

Well-mixed constant volume

For completely mixed storage/treatment systems ($\partial C/\partial x = 0$), Eq. 1 reduces to

$$\frac{dC}{dt} = \Sigma S_i$$

which may be rewritten, for the sources and sinks of the generalized component in Figure 1, as the ordinary differential equation

$$V \frac{dC}{dt} = Q_1 c_1 - Q_2 C - KC V$$

where Q_1 = influent fluid flow rate, L^3/T,

Q_2 = outflow rate, L^3/T, and

V = volume of the fluid mass in the component, L^3.

If the storage/treatment facility is modeled as a well-mixed <u>constant volume</u> tank, the system may be illustrated as shown in Figure 2. The volumetric rates of flow into and out of the tank are equal and

SYSTEM SCHEMATIC

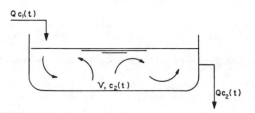

MASS BALANCE

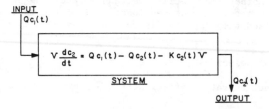

Fig. 2. Well-Mixed Constant Volume Model

constant ($Q_1 = Q_2 = Q$). However, the concentration forcing function and response are continuous functions of time. The pollutant mass in the tank is assumed to undergo a first-order reaction. From a mass balance of the dissolved material across the tank, the governing equation has been derived to be (Rich, 1973):

$$V \frac{dc_2}{dt} = Q\, c_1(t) - Q\, c_2(t) - K\, c_2(t)\, V \qquad (3)$$

where   V = volume of tank, L^3,

$c_2(t)$ = concentration of material in the tank and outflow, as a continuous function of time, M/L^3,

Q = volumetric flow into and out of tank, L^3/T,

$c_1(t)$ = concentration forcing function of material in inflow, M/L^3,

K = first-order decay rate constant, $1/T$, and

t = time, T.

Eq. 3 is the mathematical expression that describes the response of the system, $c_2(t)$, to a time-varying forcing function of concentration, $c_1(t)$. Since this model is a linear, first-order ordinary differential equation with constant coefficients, analytic solutions are not difficult to obtain. When the storage facility is subjected to time-varying forcing functions of both concentration and flow, $Q(t)$, the governing equation is given by

$$V \frac{dc_2}{dt} = Q(t) \left[c_1(t) - c_2(t) \right] - K c_2(t) V$$

and its general solution, obtained by the method of integrating factors, is

$$c_2(t) = \exp\left[-\int_0^t a(t)dt \right] \int_0^t \frac{Q(t)\, c_1(t)}{V} \exp\left[\int_0^t a(t)dt \right] dt$$

$$+ c_2(t=0) \exp\left[-\int_0^t a(t)dt \right] \tag{4}$$

where

$$\exp\left[\int_0^t a(t)dt \right] = \text{integrating factor,}$$

$$\text{and} \quad a(t) = \frac{Q(t)}{V} + K.$$

The complete integration of Eq. 4 is dependent on the form of the two forcing functions, $a(t)$ and $c_1(t)$. For computational convenience, the forcing function may be treated as a step function input. Thus, initial conditions are assumed to be defined at time t_i by $c_2(t)$, the solution at the end of the previous time step. This allows representation of Eq. 4 in discrete form as (Medina et al., August 1981):

$$c_2(t_i + \Delta t) = \exp\left[-\int_0^{\Delta t} \left(\frac{Q_i}{V} + K \right) dt \right] \cdot$$

$$\int_0^{\Delta t} \exp\left[-\left(\frac{Q_i}{V} + K \right) t \right] dt + c_2(t_i) \exp\left[-\int_0^{\Delta t} \left(\frac{Q_i}{V} + K \right) dt \right]$$

which integrates to

$$c_2(t_i + \Delta t) = \exp\left[-\left(\frac{Q_i}{V} + K\right)\Delta t\right] \cdot \qquad (5)$$

$$\left\{\frac{Q_i c_1}{Q_i + KV}\left[\exp\left(\frac{Q_i \Delta t}{V} + K\Delta t\right) - 1\right] + c_2(t_i)\right\}$$

where 　　Q_i = value of input flow rate from t_i to $(t_i + \Delta t)$, L^3/T,

　　　　c_i = value of input concentration from t_i to $(t_i + \Delta t)$, M/L^3,

　　　　Δt = length of time interval, say, 1 hour, and

　　　　t_i = the beginning of the time interval for which the system response is being evaluated, T.

Finally, the time-averaged concentration for each time step is (Medina et al., August 1981):

$$\overline{c}_2\left(t_i + \frac{\Delta t}{2}\right) = \frac{Q_i c_i}{Q_i + KV} - \left[\frac{Q_i c_i}{Q_i + KV} - c_2(t_i)\right] \cdot$$

$$\frac{1}{\Delta t}\left\{1 - \exp\left[-\left(\frac{Q_i}{V} + K\right)\Delta t\right]\right\}\frac{V}{Q_i + KV}$$

where 　$c_2(t_i)$ = concentration value obtained by Eq. 5 for the previous time step, M/L^3, and

$\overline{c}_2\left(t_i + \frac{\Delta t}{2}\right)$ = time-averaged concentration for each time step, M/L^3.

Advective-dispersive variable volume

Nonideal mixing is a more likely phenomenon where the transport of a pollutant mass in a fluid medium is described by advection and dispersion. Since the detention time in a constant volume system varies according to the influent fluid flow rate, the flow-through velocity is also variable. Most analytical solutions to the convective dispersion equation are for a constant advective velocity (Harleman, 1971). Thus, the model of nonideal mixing is developed for a storage/treatment system represented by a variable volume facility (i.e., constant velocity). The dispersive model may also be applied to a river or an estuary.

In Figure 3, a plane source perpendicular to the flow axis continuously injects a mass of pollutant into a variable volume storage/treatment system. A uniform longitudinal velocity advects the pollutant mass in the direction of flow. The governing equation may be derived from the general expression for the one-dimensional convective dispersion equation, Eq. 1. For a constant longitudinal dispersion coefficient and first-order decay (sink term), it may be rewritten as:

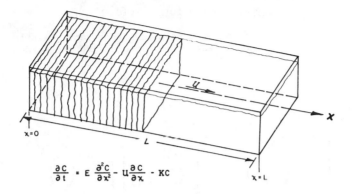

Fig. 3. Advection and Dispersion in the Variable Volume Model

$$\frac{\partial C}{\partial t} = E \frac{\partial^2 C}{\partial x^2} - U \frac{\partial C}{\partial x} - KC \qquad (6)$$

where C = concentration of pollutant in the fluid medium, M/L^3,

$\frac{\partial C}{\partial t}$ = temporal variation in the pollutant concentration, $M/L^3 T$,

E = longitudinal dispersion coefficient, constant along the flow axis, L^2/T,

U = longitudinal velocity in the storage/treatment system, L/T,

$\frac{\partial C}{\partial x}$ = spatial variation in the pollutant concentration, M/L^4, and

K = first-order decay rate of pollutant in the fluid medium, $1/T$.

Eq. 6 is a linear partial differential equation, parabolic, of the second order, and is formally equivalent to the one-dimensional heat equation (Berg and McGregor, 1966). The fundamental mathematical techniques to solve the Fickian equations of heat conduction are given rigorous treatment by Carslaw and Jaeger (1959).

A solution is presented by Harleman (1971) for the continuously discharging plane source by a time integration of the instantaneous plane source solution for a constant time rate of injection. Thus,

$$c_2(x, t-\tau) = \int_0^{t_1} \frac{q''(\tau)}{\rho\sqrt{4 \, E \cdot (t-\tau)}} \, \exp\left[- \frac{[x - U \cdot (t-\tau)]^2}{4E \cdot (t-\tau)} + K \cdot (t-\tau) \right] d\tau \qquad (7)$$

where $q''(\tau)$ = variable time of mass injection per unit area at the plane source, M/L^2T, and

ρ = density of the receiving fluid, M/L^3.

If the time rate of injection per unit area is constant, then the solution to Eq. 7 for any time t greater than or equal to the duration of injection t_1, is

$$c_2(x, t-t_1) = \frac{q''}{2\rho\Omega} \exp \left[\frac{xU}{2E}\right] \cdot$$

$$\left[\left\{erf\left(\frac{x+\Omega t}{\sqrt{4Et}}\right) - erf\left[\frac{x+\Omega \cdot (t-t_1)}{\sqrt{4E \cdot (t-t_1)}}\right]\right\} \exp\left(\frac{x\Omega}{2E}\right) \right. \qquad (8)$$

$$\left. - \left\{erf\left(\frac{x-\Omega t}{\sqrt{4Et}}\right) - erf\left[\frac{x-\Omega \cdot (t-t_1)}{\sqrt{4E \cdot (t-t_1)}}\right]\right\} \exp\left(-\frac{x\Omega}{2E}\right)\right]$$

where $c_2(x, t-t_1)$ = pollutant concentration at distance x along the flow axis and time $t \geq t_1$, mass of pollutant/mass of fluid, dimensionless,

$\Omega = \sqrt{U^2 + 4KE}$, has the dimensions of velocity, L/T, and

t_1 = duration of injection, T.

When the time t is equal to the duration of injection t_1, Eq. 8 reduces to

$$c_2(x, t-t_1) = \frac{q''}{2\rho\Omega} \exp \left(\frac{xU}{2E}\right)$$

$$\left[\left\{erf\left(\frac{x+\Omega t}{\sqrt{4Et}}\right) - 1\right\} \exp\left(\frac{x\Omega}{2E}\right) \right. \qquad (9)$$

$$\left. - \left\{erf\left(\frac{x-\Omega t}{4Et}\right) - 1\right\} \exp -\left(\frac{x\Omega}{2E}\right)\right]$$

for x > 0.

Eqs. 8 and 9 are the response of the system to one continuous injection for a duration equal to t_1. It is of interest to adapt these solutions to the case where a succession of injections, representing different mass loadings lagged in time, occurs. First, it should be noted that (Medina et al., August 1981):

$$q'' = \frac{\rho Q_1 c_1}{A}$$

where Q_1 = influent fluid flow rate, L^3/T,

c_1 = concentration of pollutant in the inflow, dimensionless, and

A = cross-sectional area of the plane source, equal to the width of the storage/treatment system times the depth of flow in the tank, L^2.

Also, since the system has a variable volume and a constant velocity,

$$U = \frac{Q_1(t)}{A(t)} = \frac{L}{t_D} = \text{constant}$$

where U = longitudinal velocity, L/T,

$Q_1(t)$ = fluctuating influent fluid flow rate, L^3/T,

$A(t)$ = fluctuating cross-sectional area due to a variable depth, L^2,

L = length of the storage/treatment basin, units of L, and

t_D = dispersive variable volume system detention time, T.

Therefore, it is convenient to define an input function to the system (Medina et al., August 1981):

$$I(\tau) = \frac{U \; c_1(\tau)}{2\Omega}$$

where $I(\tau)$ = system input function, dimensionless, and

$c_1(\tau)$ = concentration forcing function, dimensionless.

From Eq. 8, a system response function may also be defined (Medina et al., August 1981):

$$f(t-\tau) = \exp\left(\frac{xU}{2E}\right)\left[\left\{\text{erf}\left(\frac{x+\Omega t}{\sqrt{4Et}}\right) - \text{erf}\left[\frac{x+\Omega\cdot(t-\tau)}{\sqrt{4E\cdot(t-\tau)}}\right]\right\}\exp\left(\frac{x\Omega}{2E}\right) - \right. \tag{10}$$

$$\left\{ \mathrm{erf}\ \left(\frac{x-\Omega t}{4Et}\right)\ -\ \mathrm{erf}\ \left[\frac{4-\Omega\cdot(t-\tau)}{4E\cdot(t-\tau)}\right]\right\} \exp\ \left(-\ \frac{x\Omega}{2E}\right)\Biggr]$$

where $f(t-\tau)$ = system step response, dimensionless.

The essence of linearity is the principle of superposition. The dispersive variable volume storage/treatment system is subjected to a step function input of pollutant concentration, $I(\tau_i)$. Individual step responses to each injection are characterized by $f(t-\tau_i)$. It should be noted that the effects of the various injections are appropriately lagged in time. Thus, the general form of the discrete convolution equation is given by (Medina et al., August 1981):

$$C(t)\ =\ \sum_{i=1}^{n} I(\tau_i)f(t-\tau_i)$$

where $C(t)$ = system concentration distribution response to step function inputs, dimensionless,

$i = 1, 2, 3, \ldots, n,$

n = total number of inputs, and

$\tau_i = i\Delta t$, dimensions of time, T.

For computational convenience the step function input is again defined in terms of the discrete forcing function and constant terms, as follows:

$$I(\tau_i)\ =\ \frac{U\ c_1(\tau_i)}{2\Omega}$$

where $c_1(\tau_i)$ = discrete concentration forcing function, dimensionless.

Since the system response is of interest only at the storage/treatment unit's outfall, then Eq. 10 is evaluated for $x = L$.

The mathematical models of ideal and nonideal mixing presented above describe the transient response of such systems to highly variable forcing functions of flow and concentration. These models are well suited for representing stormwater detention facilities, but are also useful in approximating the response of unit processes of conventional dry-weather flow (DWF) wastewater treatment facilities. Storage and treatment are, in fact, inseparable. Wet-weather flow (WWF) facilities provide storage primarily but there is treatment by organic decay, settling, and sometimes disinfection. Wastewater treatment plants, whether biological or physical/chemical, are also characterized by a hydraulic residence time which provides some degree of equalization of flows, reducing mean effluent concentrations and concentration variability.

Primary clarifier and trickling filter

The primary clarifier is represented by the advective-dispersive variable volume model for all DWF treatment plant configurations. Its system response function is given by Eq. 10. The primary process in the trickling filter is the mass transfer of organic material to the bacterial film through intermittent dosing over the filter bed. The simplest geometrical model is a plane film of liquid in contact with a plane bacterial film (Petrie, 1978). Each dosing results in a uniform concentration (c_0) being applied to a zone at the top of the filter bed. Assuming no decay, a quiescent liquid film (no advection), and for flow normal to the plane of the filter bed, Eq. 6 becomes

$$\frac{\partial C}{\partial t} = D_m \frac{\partial^2 C}{\partial z^2} \tag{11}$$

where D_m = molecular diffusion coefficient, L^2/T, and

Z = distance from the bacterial film, units of L.

One initial and two boundary conditions are required for solution, as follows

initial condition:

$C = c_0$ @ $t = 0$, $0 < z < d$

boundary conditions:

$\frac{\partial C}{\partial Z} = 0$ @ $z = d$, $t > 0$

$C = 0$ @ $z = 0$, $t > 0$

in which d is the thickness of the film of liquid. The first boundary condition implies no mass transfer through the free surface, and the second that the organic material is consumed by the micro-organisms as soon as it reaches the surface of the bacterial film. By the method of separation of variables, the solution is (Petrie, 1978):

$$c(z,t) = \frac{4c_0}{\pi} \sum_{n=0}^{\infty} \frac{1}{(2n+1)} \sin \left[\frac{(2n+1)\pi z}{2d} \right] \cdot$$

$$\exp \left[- \frac{(2n+1)^2 \pi^2 D_m t}{4d^2} \right]$$

The mean concentration at the end of the dosing cycle, c_m, is given by

$$c_m = \frac{1}{d} \int_0^d c(x,\tau) dz$$

$$= c_0 \sum_{n=0}^{\infty} \frac{8}{(2n+1)^2 \pi^2} \exp \left[- \frac{(2n+1)^2 \pi^2 D_m \tau}{4d^2} \right]$$

395

where τ = time cycle for dosing, T,

   n = number of terms in the series needed, a function of the value of $(D_m\tau/4d^2)$.

Only about twelve terms are needed, even at the lower values of the dimensionless ratio. A solution to Eq. 11 can also be obtained with Laplace transform methods.

Primary clarifier and activated sludge

This configuration is represented by two advective-dispersive variable volume systems in series. The primary clarifier functions essentially as an equalization basin. Novotny and Stein (1976) reported a longitudinal dispersion coefficient of 68 ft^2/hr (6.3 m^2/hr) in a relatively well-functioning equalization basin at a batch organic chemical wastewater treatment plant. Murphy and Timpany (1967) reported an average axial dispersion coefficient of 5,730 ft^2/hr (532 m^2/hr) in an activated sludge aeration tank subjected to constant airflow from sparger-type air diffusers inducing spiral mixing.

Dual-purpose wet-weather and dry-weather flow facility

The concept of dual use is maximum utilization of wet-weather facilities during dry periods and maximum utilization of dry-weather facilities during storm flows--which is clearly not achieved with separate storage/treatment trains. The system is represented by a primary clarifier and two trickling filters, and operates in series during dry-weather flow. During series operation, a percentage of the flow from the primary clarifier bypasses the first filter and feeds into the second filter in order to maintain an effective biological film. This flow is weighted with the effluent from the first filter to obtain the concentration of organic material flowing into the second filter. During wet-weather flows, the filters operate in parallel. This configuration is triggered as soon as the design dry-weather flow is exceeded.

Receiving Body of Water

A non-tidal receiving stream can be adequately represented by a one-dimensional, advective-dispersive system with a constant, uniform cross-sectional area, constant velocity and longitudinal dispersion--over each stream segment (Δx) and time step (Δt). The governing equations for the biochemical oxygen demand (BOD)-dissolved oxygen (DO) coupled reaction are derived from Eq. 1 for continuously-discharging plane sources in the y-z plane:

$$\frac{\partial L}{\partial t} + U \frac{\partial L}{\partial x} = E_x \frac{\partial^2 L}{\partial x^2} - K_{13}L \tag{12}$$

$$\frac{\partial D}{\partial t} + U \frac{\partial D}{\partial x} = E_x \frac{\partial^2 D}{\partial x^2} + K_{13}L - K_2 D \tag{13}$$

where   L = BOD concentration,

K_{13} = biochemical oxidation rate and sedimentation rate coefficient for carbonaceous BOD,

D = DO deficit $\equiv C_s - C$,

C_s = saturation concentration of DO at stream temperature,

C = concentration of DO in stream, and

K_2 = reaeration rate coefficient.

It can be shown by using common transformation techniques (e.g., Bennett, 1971) that Eq. 12 is formally equivalent to the classical heat equation, for which many formal solutions have been derived for various boundary conditions (Carslaw and Jaeger, 1959). For an instantaneous plane source at the origin, x = o, the solution to Eq. 12 is

$$L = \frac{m''}{\rho\sqrt{4\pi E_x t}} \; \exp - \left[\frac{(x - Ut)^2}{4E_x t} + K_{13} \right] \tag{14}$$

where ρ is the fluid medium density and m'' the instantaneous mass of BOD per unit area. Substituting the instantaneous plane source solution for BOD for the "L" term in the DO deficit equation and applying the formal solution to the heat equation (Carslaw and Jaeger, 1959) the following expression is obtained for an instantaneous plane source:

$$D = \frac{1}{\rho\sqrt{4\pi E_x t}} \; \exp - \left[\frac{(x - Ut)^2}{4E_x t} \right] \left\{ n'' e^{-K_2 t} + \right. \tag{15}$$

$$\left. \frac{m'' K_{13}}{K_2 - K_{13}} \left(e^{-K_{13} t} - e^{-K_2 t} \right) \right\}$$

where n'' = instantaneous mass of DO deficit/unit area, and the DO deficit source is assumed to be at the origin.

For continuous (long-term) simulation of receiving water quality, solutions for continuous plane sources of BOD and dissolved oxygen deficit are required. Solutions for continuously discharging sources can be built up by summing the effect of closely spaced (in time) instantaneous injections (Harleman, 1971). Figure 4 demonstrates this concept. Thus, the solution for a continuous plane source can be obtained by a time integration of the instantaneous plane source.

To achieve this time integration of the instantaneous plane source solution for the BOD governing equation, Eq. 14 can be written as (Medina and Buzun, 1981):

$$dL = \frac{dm''}{\rho\sqrt{4\pi E_x (t -\tau)}} \; \exp - \left\{ \frac{[x - U(t - \tau)]^2}{4E_x (t - \tau)} + K_{13}(t - \tau) \right\} \tag{16}$$

397

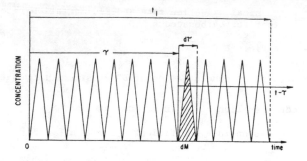

t   measured   from   start   of   injection   series

t_1 — duration   of   injection   series

dT   — incremental   amount   of   time   for   each   instantaneous   injection

dM   — incremental   amount   of   mass   injected   during   each   instantaneous
injection

Fig. 4.   Series of Instantaneous Injections Forming a Continuous
Injection Source (after Harleman, 1971)

Defining

$$q'' = dm''/d\tau,$$

and assuming a constant injection over a time period from $\tau = 0$ to $\tau = t_1$, Eq. 16 can be integrated with respect to time and yields:

$$
L = \frac{q'' e^{\frac{xU}{2E_x}}}{2\rho\Omega} \left[\left\{ \mathrm{erf}\left(\frac{x + \Omega t}{\sqrt{4E_x t}}\right) - \mathrm{erf}\left[\frac{x + \Omega(t - t_1)}{\sqrt{4E_x(t - t_1)}}\right] \right\} \exp\left(\frac{x\Omega}{2E_x}\right) \right.
$$

$$
\left. - \left\{ \mathrm{erf}\left(\frac{x - \Omega t}{4E_x t}\right) - \mathrm{erf}\left[\frac{x - \Omega(t - t_1)}{\sqrt{4E_x(t - t_1)}}\right] \right\} \exp\left(\frac{-x\Omega}{2E_x}\right) \right] \quad (17)
$$

where $\qquad \Omega = \sqrt{U^2 + 4K_{13}E_x}$.

The same technique can be used to obtain an analytic solution for a continuous plane source of DO deficit.   After writing Eq. 15 as

$$
dD = \frac{1}{\rho\sqrt{4\pi E_x(t - \tau)}} \exp -\left(\frac{[x - U(t - \tau)]^2}{4E_x(t - \tau)}\right) \left\{ dn'' e^{-K_2(t - \tau)} + \right.
$$

$$
\left. \frac{dm'' K_{13}}{K_2 - K_{13}} e^{-K_{13}(t - \tau)} - \frac{dm'' K_{13}}{K_2 - K_{13}} e^{-K_2(t - \tau)} \right\} d\tau, \quad (18)
$$

398

to reflect the conception of a continuous source as a series of instantaneous sources, then the assumption is made that p'' represents the time rate of mass injection of DO deficit per unit area at the plane source, or

$$p'' = dn''/d\tau.$$

For a constant injection over a period from $\tau = 0$ to $\tau = t_1$, Eq. 18 can be integrated with respect to time after several transformations to yield (Medina and Buzun, 1981):

$$D = \left[p'' - \frac{K_{13}q''}{K_2 - K_{13}} \right] \left[\frac{e^{\frac{xU}{2E_x}}}{2\rho\Omega_2} \right] \left[\left\{ \mathrm{erf}\left(\frac{x + \Omega_2 t}{\sqrt{4E_x t}} \right) - \mathrm{erf}\left[\frac{x + \Omega_2(t - t_1)}{\sqrt{4E_x(t - t_1)}} \right] \right\} \exp\left(\frac{x\Omega_2}{2E_x} \right) \right.$$

$$\left. - \left\{ \mathrm{erf}\left(\frac{x - \Omega_2 t}{\sqrt{4E_x t}} \right) - \mathrm{erf}\left(\frac{x - \Omega_2(t - t_1)}{\sqrt{4E_x(t - t_1)}} \right) \right\} \exp\left(\frac{-x\Omega_2}{2E_x} \right) \right] \qquad (19)$$

$$+ \frac{K_1 q'' e^{\frac{xU}{2E_x}}}{2(K_2 - K_{13})\rho\Omega} \left[\left\{ \mathrm{erf}\left(\frac{x + \Omega t}{\sqrt{4E_x t}} \right) - \mathrm{erf}\left[\frac{x + \Omega(t - t_1)}{\sqrt{4E_x(t - t_1)}} \right] \right\} \exp\left(\frac{x\Omega}{2E_x} \right) \right.$$

$$\left. - \left\{ \mathrm{erf}\left(\frac{x - \Omega t}{\sqrt{4E_x t}} \right) - \mathrm{erf}\left[\frac{x - \Omega(t - t_1)}{\sqrt{4E_x(t - t_1)}} \right] \right\} \exp\left(\frac{-x\Omega}{2E_x} \right) \right]$$

where $\qquad \Omega_2 = \sqrt{U^2 + 4K_2 E_x}$

which is the continuous plane source solution for DO deficit. The advantages of applying Eqs. 17 and 19 over the use of finite difference numerical techniques are that the latter often exhibit stability and numerical dispersion problems.

APPLICATION TO STUDY AREA

The city of Des Moines, Iowa and the Des Moines River were selected primarily because of data availability from a combined sewer overflow abatement plan (Davis and Borchardt, 1974). The center of the metropolitan area is near the confluence of the Des Moines River and the Raccoon River, as shown in Figure 5. The metropolitan area covers approximately 49,000 acres (19,830 ha) of land which has gently rolling terrain, and had an estimated population of 288,000 at the time of the sampling program, conducted from March 1968 to October 1969. The mean annual precipitation is 31.27 inches (795 mm), the average value for the State of Iowa. Most of the area, 45,000 acres (18,211 ha), is

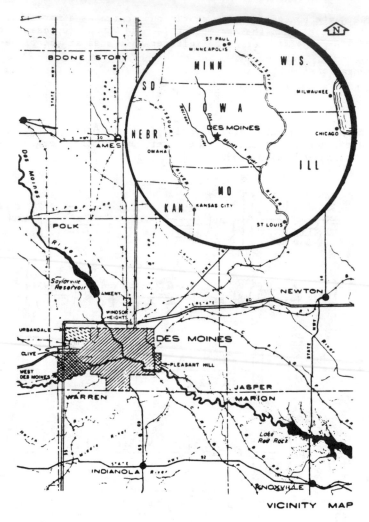

Fig. 5. Des Moines, Iowa Area (Davis and Borchardt, 1974)

served by separate sewers, while 4,000 acres (1619 ha) are served by combined sewers. Other considerations were that Des Moines is somewhat typical of many urban centers throughout the country:

(1) it has a medium-sized population;

(2) its domestic and industrial dry-weather flows receive secondary treatment;

(3) its wastewaters are discharged into a non-tidal receiving stream; and

(4) the urban area receives a mean annual precipitation equal to the national average.

The urban runoff time series and BOD loadings (BOD5) were generated by an hourly, continuous hydrologic and water quality model--STORM (Hydrologic Engineering Center, 1976)--from the 1968 rainfall time series. Calibration of the model developed in this study was preceded by calibration of the urban runoff BOD5 loading rates for Des Moines, Iowa as computed by STORM. The dust and dirt surface loading factors were adjusted to obtain an annual average flow-weighted BOD5 concentration of 53 mg/l for urban stormwater runoff. The above concentration was the average value determined by the field monitoring program in the separate sewer system. The developed mathematical model simulates the mixing of stormwater runoff and sanitary sewage in the combined sewer system. The <u>annual</u> <u>average</u> flow-weighted BOD$_5$ concentration of combined sewer overflows was computed to be 75 mg/l, including the effects of first flush. The average value measured in the combined sewer system was reported to be 72 mg/l. Model parameters were adjusted to obtain a fit between the <u>time-averaged</u> computed profile and the measured dissolved oxygen profile. These curves are shown in Figure 6 for the year 1968, and correspond to a point 5.6 mi (9.0 km) downstream from the confluence of the Raccoon and Des Moines Rivers.

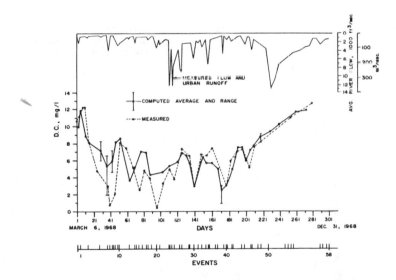

Fig. 6. Measured and Time-Averaged Computed Dissolved Oxygen
Profiles at 5.6 mi. (9.0 km) Downstream from
Confluence of Raccoon and Des Moines Rivers

During the sampling period (1968 to 1969), the average daily plant flow was 35.3 million gallons per day, and the average daily BOD was 95,800 pounds. The plant was a high-rate trickling filter facility which achieved secondary treatment, with average rate of BOD removal about 85 percent. This treatment plant discharges to the Des Moines River. Urban runoff collected in the separate sewers, and the combined sewer overflow are also discharged to the Des Moines River. There was no treatment of nonpoint sources. During the first half of 1968, the secondary treatment units of the plant were under construction. At times, raw sewage was bypassed to the river; and other times, primary effluent was bypassed. The records are incomplete (obtained directly from the wastewater treatment plant supervisor). Thus, the fit of predicted values is not as good as from late June to December. However, for the latter period the standard error of the estimate between time-averaged computed values and observed values is 1.08 mg/l, an excellent fit for continuous simulation. The urban treatment scheme was represented by a primary clarifier (advective-dispersive variable volume model) and a trickling filter (dispersive plane film model), with an overall BOD removal of about 85 percent--and, of course, no wet-weather flow control.

It is interesting to examine the potential improvement in receiving water quality if urban runoff control had been provided. Figure 7 illustrates cumulative minimum dissolved oxygen frequency curves for an urban treatment scheme represented by the DWF models described above, and a hyopthetical urban runoff storage/treatment system (well-mixed constant volume model). Appreciable water quality improvement in terms of DO is predicted at the higher detention times of the WWF facility, particularly in the range from 4 to 6 mg/l--which includes the minimum allowable stream standards for most states. Figure 8 depicts frequency histograms of consecutive hours of violation of a 5.0 mg/l DO standard for urban storage/treatment schemes resulting in various percent BOD removals. At the highest level of WWF control, very few extended violations of the DO standard occur (9 occurrences of at least 16 consecutive hours of violation). For no WWF control, and at least a secondary treatment level of control of DWF, the same standard is violated (for at least 16 consecutive hours) during 27 occasions.

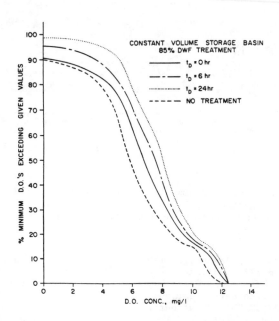

Fig. 7.   Minimum D.O. Frequency Curves in Des Moines River, for
          Varied Detention in Stormwater Storage/Treatment
          System and Secondary Treatment of Dry Weather
          Flows

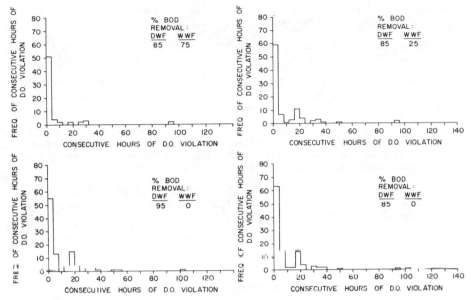

Fig. 8. Frequency Histograms of Consecutive Hours of Violation of 5.0 mg/l Dissolved Oxygen Standard for DWF Secondary Treatment Technology Minimum

CONCLUSIONS

A unified approach to the modeling of transient storage, treatment and transport of urban point and nonpoint water pollutants has been presented. Solutions to versions of the one-dimensional conservation of mass equation for several urban storage/treatment systems and a non-tidal receiving body of water are applied within the framework of a continuous simulation hydrologic and water quality model. The results of the application are site-specific; however, the utility of evaluating urban water pollution control schemes in terms of their impact on receiving water quality--on the basis of probability of occurrence-- has been demonstrated.

ACKNOWLEDGEMENTS

This investigation was conducted in partial fulfillment of the objectives of National Science Foundation Research Initiation Grant No. CME-7805531 A01. Mr. Donald Greeley, a Graduate Research Assistant at Duke University, contributed by programming segments of the dry-weather flow models.

REFERENCES

Bennett, J. P. 1971. Convolution Approach to the Solution for the Dissolved Oxygen Balance Equation in a Stream. Water Resources Research, Vol. 7, No. 3, pp. 580-590.

Berg, P. W. and McGregor, J. L. 1966. Elementary Partial Differential Equations, Holden-Day, San Francisco.

Carslaw, H. S. and J. C. Jaeger. 1959. Conduction of Heat in Solids. Oxford University Press, London, England.

Davis, P. L. and F. Borchardt. 1974. Combined Sewer Overflow Abatement
    Plan. EPA-R-73-170, Washington, D.C.

Donigian, A. S. and R. K. Linsley. 1979. Continuous Simulation for
    Water Quality Planning. Water Resources Bulletin 15(1):1-16.

Harleman, D. R. F. 1971. Transport Processes in Water Quality Control.
    Department of Civil Engineering, Massachusetts Institute of Tech-
    nology, Cambridge, Massachusetts.

Heaney, J. P.; Huber, W. C.; Medina, M. A., Jr.; Murphy, N. P.; Nix, S. J.;
    and Hasan, S. M. 1977. Nationwide Evaluation of Combined Sewer
    Overflows and Urban Stormwater Discharges, Volume II: Cost Assess-
    ment and Impacts. EPA-600/2-77-064, Cincinnati, Ohio.

Hydrologic Engineering Center. 1976. Storage, Treatment, Overflow,
    Runoff Model: STORM. 723-S8-L7520, U.S. Army Corps of Engineers,
    Davis, California.

Linsley, R. K., and N. Crawford. 1974. Continuous Simulation Models
    in Hydrology. Geophysical Research Letters, 1(1):59-62.

Medina, M. A., Jr. 1979. Level III: Receiving Water Quality Modeling
    for Urban Stormwater Management: EPA-600/2-79-100, Cincinnati, Ohio.

Medina, M. A., Jr.; Huber, W. C.; Heaney, J. P.; and R. Field.
    March, 1981. A River Quality Model for Urban Stormwater Impacts.
    ASCE Journal Water Resources Planning and Management 107(1):263-280.

Medina, M. A., Jr.; Huber, W. C.; and J. P. Heaney. August 1981.
    Modeling Stormwater Storage/Treatment Transients-Theory. Journal
    of the Environmental Engineering Division, ASCE, Vol. 107, No. EE4.

Medina, M. A., Jr., and Buzun. August 1981. Continuous Simulation
    of Receiving Water Quality Transients. Water Resources Bulletin,
    Vol. 17, No. 4.

Murphy, K. L., and P. L. Timpany. 1967. Design and Analysis of
    Mixing for an Aeration Tank. Journal of the Sanitary Engineering
    Division, ASCE, Vol. 93, No. SA5, pp. 1-14.

Novotny, V., and R. M. Stein. 1976. Equalization of Time Variable
    Waste Loads. Journal of the Environmental Engineering Division,
    ASCE, Vol. 102, No. EE3, pp. 613-625.

Petrie, C. J. S. 1978. The Modelling of Engineering Systems - Math-
    ematical and Computational Techniques. in: A. James (Editor),
    Mathematical Models in Water Pollution Control, pp. 39-65,
    John Wiley & Sons, Ltd., New York.

Rich, Linvil G. 1973. Environmental Systems Engineering, McGraw-Hill
    Book Co.

Sherwani, J. K. 1977. Effect of Low-Flow Hydrologic Regimes on Water
    Quality Management. Water Resources Research Institute of
    the University of North Carolina, No. 26, Raleigh, N.C.

Vilaret, M. and R. D. G. Pyne. 1971. Storm and Combined Sewer Pollution
    Sources and Abatement. EPA 11024 ELB 01/71, Black, Crow & Eidsness,
    Inc., Atlanta, Georgia.

A KINEMATIC MODEL FOR POLLUTANT CONCENTRATIONS DURING THE RISING HYDROGRAPH

Steven G. Buchberger
Engineer, Dravo Engineers and Constructors
Denver, Colorado 80202 U.S.A.

Thomas G. Sanders, Ph.D.
Assistant Professor Civil Engineering
Colorado State University, Fort Collins, Colorado 80523 U.S.A.

ABSTRACT

The partial differential equation which describes the one-dimensional transport of soluble particles in overland flow is developed. With kinematic wave theory, the transport equation is applied to conservative nonpoint source pollutants moving in rainfall runoff over an impervious plane. The kinematic formulation shows that temporal and spatial changes in pollutant concentrations result from several processes which characterize the two regions of overland flow. In particular, within the region of steady nonuniform rainfall runoff, pollutant concentrations are influenced by convection, diffusion and dilution; within the region of unsteady uniform rainfall runoff, dilution is the dominant influence on pollutant concentrations.

Considering only the dilution process, the kinematic transport equation reduces to a deterministic model which predicts continuously the concentration of nonpoint source pollutants during the rising hydrograph. For this case, the kinematic model expresses pollutant concentrations as a function of three measurable parameters: (1) the initial pollutant mass per unit area, (2) the intensity of the rainfall and (3) the time since the start of the rainfall. The predicted concentration distribution displays a descending exponential shape with concentrations inversely related to time.

Theoretical predictions of the kinematic transport equation are compared against experimental data collected during 26 laboratory runs with an indoor rainfall simulator. Continuous measurements of pollutant concentration and runoff discharge were made during all runs. Rhodamine WT, a soluble fluorescent dye was used to simulate the pollutant. Variables investigated include: pollutant mass, bed slope, rainfall intensity and rainfall duration.

Experimental results show that predicted pollutant concentrations are generally within 10% of observed concentrations. All observed concentration distributions exhibit the characteristic decay shape documented in many field studies and predicted by the model. These results demonstrate that the kinematic wave approximation, ordinarily used to synthesize the hydrograph of overland flow, also can be used to predict the concentration distribution of conservative soluble nonpoint source pollutants transported in rainfall runoff during the rising hydrograph.

INTRODUCTION

Environmental degradation attributable to the ubiquitous nonpoint source pollutant is evident in both urban and rural watersheds. The identification and control of these pollutants are key elements of the strategy to maintain and improve the quality of the nation's water resources. The U.S. Environmental Protection Agency (1977) defines a nonpoint source pollutant as:

> A pollutant which enters a water body from diffuse origins on the watershed and does not result from discernable, confined or discrete conveyances.

Pollutants from nonpoint sources exhibit other characteristics as well. While most point source pollutants originate from waste products, many nonpoint source contaminants (e.g., nutrients, pesticides, deicing salts) result from chemicals deliberately applied for beneficial purposes (Crawford and Donigian, 1973). Soluble, rather than insoluble, contaminants move greater distances in the drainage system and hence are more difficult to control (Stewart et al, 1975). In most cases, overland flow rather than subsurface flow is the primary transport medium of nonpoint source pollution. Consequently, the rate and amount of pollution removal are linked closely to the rate and volume of surface runoff. Field investigations of nonpoint source discharges often document the "first-flush" phenomenon in which disproportionately high pollutant concentrations occur during early stages of runoff (Sartor and Boyd, 1972; Shaheen, 1975; Asmussen et al, 1977).

Numerous mathematical models have been developed to estimate loadings from urban and rural catchments. A partial list of the modeling techniques includes: multiple regression (Colston, 1974), mass balance (Wulkowicz and Saleem, 1974), exponential decay (Roesner et al, 1974; Roffman et al, 1975), continuous simulation (Donigian and Crawford, 1976), and linear regression (Ponce and Hawkins, 1978). Most of these models include provisions which reflect the influence of either precipitation or runoff.

Of particular interest is a recent modeling approach which postulates that the bulk transport of conservative soluble pollutants is analogous to the convective movement of water particles in overland flow (Brazil, 1976). With principles of kinematic wave theory, this premise leads to a simple deterministic model which predicts the travel time of soluble point source and line source pollutants moving in unsteady uniform rainfall runoff. Experiments using tracer injections into simulated rainfall runoff show excellent agreement between predicted and observed travel times (Brazil, Sanders and Woolhiser, 1979).

The implications of this research suggest that fundamental principles of overland flow may also be used to estimate the discharge of soluble nonpoint source pollutants. Following the framework established by the work with point and line sources, this paper presents the theoretical derivation and experimental verification of a kinematic model for predicting the concentration of soluble conservative nonpoint source pollutants transported in rainfall runoff.

KINEMATIC WAVE THEORY

In a treatment of floodwave movement, Lighthill and Whitham (1955) introduced the notion of kinematic waves which propagate only in the downstream direction. Early applications of this theory to rainfall runoff were made by Henderson and Wooding (1964) and Woolhiser (1969).

The kinematic wave equations are derived from the governing equations for gradually varied unsteady flow in a wide open channel. For an impervious catchment with steady uniform lateral inflow, as shown in Figure 1, the continuity equation is (Woolhiser, 1975)

$$\frac{\partial h}{\partial t} + \frac{\partial q}{\partial x} = i \qquad (1)$$

and the simplified momentum equation is

$$s_o = s_f \qquad (2)$$

where
q = discharge per unit width,
h = depth of flow,
x = longitudinal distance,
t = time,
i = rainfall intensity,
s_o = bed slope, and
s_f = friction slope.

The kinematic wave relationship, a parametric expression of the simplified momentum equation, is given by

$$q = \alpha h^m \qquad (3)$$

where
α = flow resistance parameter, and
m = flow regime parameter.

For turbulent flow the parameter m has a value close to 1.5 while for laminar flow its value is close to 3.0. Woolhiser and Liggett (1967) demonstrate that kinematic wave theory is an adequate approximation for most instances of overland flow.

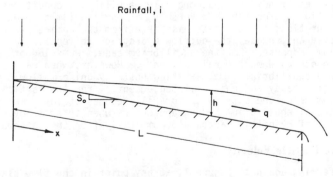

Figure 1.--Definition sketch of overland flow on an impervious plane.

Characteristic Wave Velocity

A distinct feature of kinematic waves is that they possess only one velocity at each point. The characteristic wave velocity is the slope of the depth-discharge curve or from equation 3

$$u_w = \alpha m h^{m-1} \tag{4}$$

Along the characteristic, the depth of flow is

$$h = it \tag{5}$$

An alternate formulation of the characteristic wave velocity is

$$u_w = \left(\frac{dx}{dt}\right)_w \tag{6}$$

Combining equations 4, 5 and 6 yields

$$\left(\frac{dx}{dt}\right)_w = \alpha m (it)^{m-1} \tag{7}$$

Integration of equation 7 provides an expression for characteristic wave travel times. Of particular interest is the time required for a wave to travel the entire length of the plane. Often referred to as time to equilibrium, t_e, it is deduced as

$$t_e = \left(\frac{L}{\alpha(i)^{m-1}}\right)^{1/m} \tag{8}$$

in which L is the length of the catchment.

Prior to equilibrium, as the first--or limiting--characteristic wave from the upstream boundary traverses the flow plan, it separates the flow into two regions. Illustrated in Figure 2, these regions are (1) steady nonuniform flow upstream from the characteristic and (2) unsteady uniform flow downstream from it.

Classification of the discharge hydrograph depends on the duration of rainfall and the time to equilibrium. If rainfall duration exceeds equilibrium time, the entire flow plan eventually contributes to steady-state runoff and, accordingly, equilibrium conditions apply to the outflow hydrograph. Conversely, if equilibrium time exceeds rainfall duration, steady-state discharge from the entire place is not achieved and, hence, the outflow hydrograph reflects a partial equilibrium response. For the equilibrium case, recession of the hydrograph occurs when the rainfall ends. However, when rainfall ends before equilibrium time, existing unsteady uniform flow conditions become steady and uniform. Consequently, for the partial equilibrium case, the outflow temporarily becomes steady. Recession of the partial equilibrium hydrograph occurs when this steady uniform region of runoff is discharged completely from the catchment.

Convective Particle Velocity

Referring again to Figure 1, at any point in the flow plane the mean velocity is

$$u = q/h \tag{9}$$

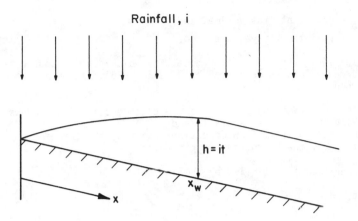

Rainfall, i

$h = it$

x_w

x

For $x < x_w$:
Steady Nonuniform Flow

For $x > x_w$:
Unsteady Uniform Flow

x_w = Location of Characteristic from Upstream Boundary
at Time = t

Figure 2.--Flow conditions during the rising hydrograph.

which, from equation 3, becomes

$$u = \alpha h^{m-1} \tag{10}$$

Comparing equations 10 and 4 reveals

$$u_w = mu \tag{11}$$

and demonstrates that at any point on the catchment the characteristic wave velocity is m times faster than the mean convective velocity.

The convective velocity of a soluble particle is defined by equation 12.

$$u_p = \left(\frac{dx}{dt}\right)_p \tag{12}$$

Assuming that the mechanism of transport for a soluble pollutant particle is the same as that for a water particle, it follows that their velocities must be equal. On this premise, the velocity of a soluble particle transported in overland flow is

$$\left(\frac{dx}{dt}\right)_p = \alpha(it)^{m-1} \tag{13}$$

which is obtained after substituting equations 12 and 5 into equation 10. Integrating and rearranging this expression gives

$$t_p = \left(\frac{mx_p}{\alpha(i)^{m-1}}\right)^{1/m} \tag{14}$$

409

where t_p = pollutant travel time, and

x_p = pollutant travel distance.

Equation 14 is the travel time model for soluble point source and line source pollutants developed by Brazil (1976). It is applicable when:

(1) The pollutant injection onto the catchment is simultaneous with the start of rainfall, and
(2) Unsteady uniform flow conditions exist on the catchment, that is, the hydrograph is rising.

The second condition is satisfied by the inequality

$$t_p \leq t_e \tag{15}$$

or after substituting equations 8 and 14 and simplifying

$$x_p \leq L/m \tag{16}$$

Equation 16 specifies the maximum pollutant travel distance for which the travel time model is applicable. It will be shown that an analogous interpretation of equation 16 applies to the kinematic nonpoint source model.

POLLUTANT TRANSPORT EQUATION

The concentration of a conservative pollutant transported in runoff depends on the volume of flow and the amount of pollutant. In a closed system with steady uniform conditions, concentration is defined as the mass of solute per volume of solution or

$$c = M/V \tag{17}$$

where c = concentration,
M = mass of solute, and
V = volume of solution.

For unsteady nonuniform conditions, equation 17 must be modified to reflect spatial and temporal variations in pollutant concentrations which occur as the pollutant mass is transported in overland flow.

Two simultaneous processes, diffusion and convection, are assumed responsible for the transport of soluble pollutants in overland flow. Consider first the diffusion process. Spatial variation in pollutant concentrations imply a concentration gradient which, from Fick's first law of diffusion, causes a mass flux given by

$$F_d = -D \frac{\partial c}{\partial x} \tag{18}$$

where F_d = mass flux due to diffusion, and
D = dispersion coefficient.

In equation 18 D is ordinarily defined as the diffusion coefficient. However, for this paper, D is designated as the dispersion coefficient in order to emphasize the combined effects of diffusion and turbulent mixing in rainfall runoff. The negative sign appears because diffusion occurs in the direction of decreasing rather than of increasing concentration.

The mass flux resulting from pollutant convection is due to the bulk movement of soluble pollutant particles in overland flow. The convective mass flux is given by

$$F_c = uc \tag{19}$$

The fundamental differential equation which describes the one-dimensional concentration of soluble conservative pollutants in overland flow is derived from equations 17, 18 and 19. Consider a control volume of unit width for gradually varied unsteady flow without lateral inflow as shown in Figure 3.

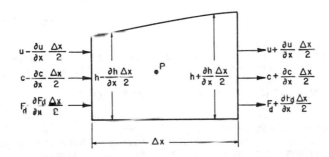

Figure 3.--Control volume for one-dimensional flow.

At a point P in the center of the control volume, the depth of flow is given by h, the flow velocity by u, the pollutant concentration by c, and the diffusive mass flux by F_d. The principle of mass conservation requires that the rate of change of mass within the control volume equals the net rate of mass flow from the control volume, or

$$\frac{\partial M}{\partial t} = F'_d + F'_c \tag{20}$$

where
$\frac{\partial M}{\partial t}$ = rate of change of mass in the control volume,

F'_d = net rate of mass flow due to diffusion, and

F'_c = net rate of mass flow due to convection.

Applying a difference scheme to the terms shown in Figure 3 and neglecting higher order terms yields

$$c \frac{\partial h}{\partial t} + h \frac{\partial c}{\partial t} = -F_d \frac{\partial h}{\partial x} - h \frac{\partial F_d}{\partial x} - q \frac{\partial c}{\partial x} - c \frac{\partial q}{\partial x} \tag{21}$$

Substituting equation 18 and rearranging gives

$$\frac{\partial c}{\partial t} = D \frac{\partial^2 c}{\partial x^2} + \frac{\partial c}{\partial x} \left(\frac{D}{h} \frac{\partial h}{\partial x} - u \right) - \frac{c}{h} \left(\frac{\partial q}{\partial x} + \frac{\partial h}{\partial t} \right) \qquad (22)$$

The above expression represents the one-dimensional equation for diffusive and convective transport of soluble conservative pollutants in gradually varied overland flow (Buchberger, 1979).

The continuity equation for the flow conditions shown in Figure 3 is (Liggett, 1975)

$$\frac{\partial q}{\partial x} + \frac{\partial h}{\partial t} = 0 \qquad (23)$$

Substituting equation 23 into 22 gives

$$\frac{\partial c}{\partial t} = D \frac{\partial^2 c}{\partial x^2} + \frac{\partial c}{\partial x} \left(\frac{D}{h} \frac{\partial h}{\partial x} - u \right) \qquad (24)$$

Equations 22 and 24 represent intermediate steps in the solution of the problem under consideration--the transport of conservative soluble nonpoint source pollutants in rainfall runoff. The effect of rainfall runoff on the transport and the concentration of soluble pollutants is incorporated with principles of kinematic wave theory.

KINEMATIC NONPOINT SOURCE MODEL

When combined with notions from kinematic wave theory, the mass transport relation (equation 22) provides expressions for the concentration distribution of soluble pollutants moving in both the steady and unsteady regions of rainfall runoff. The ensuing derivations incorporate these assumptions:

(1) The catchment is impervious,
(2) the rainfall is steady and uniform,
(3) the pollutant is conservative and soluble,
(4) prior to the start of rainfall, the pollutant is distributed uniformly over the entire catchment, and
(5) the pollutant is mixed instantaneously and completely in the runoff.

Steady Nonuniform Overland Flow

These flow conditions exist on the catchment upstream from the limiting characteristic wave. In this region the continuity equation becomes

$$\frac{\partial q}{\partial x} = i \qquad (25)$$

Substituting equation 25 into equation 22 and simplifying yields

$$\frac{\partial c}{\partial t} = D \frac{\partial^2 c}{\partial x^2} + \frac{\partial c}{\partial x} \left(\frac{D}{mx} - u \right) - \frac{ci}{h} \qquad (26)$$

This expression describes the concentration distribution of soluble pollutants in steady nonuniform rainfall runoff. Solution of equation 26 is not attempted in this paper. However, included is a brief discussion of the transport of nonpoint source pollutants in this region of flow.

Examination of each term in equation 26 reveals that pollutant concentrations are influenced by three factors: (1) dilution, (2) convection, and (3) diffusion. Because the depth of runoff increases in the direction of flow, dilution must occur as the conservative pollutant mass is transported downstream into progressively deeper flow. The effect of convection is best illustrated by considering the movement of two pollutant particles at different distances from the catchment outfall. Because the downstream particle is moving faster than the upstream particle, it follows that nonuniform convection causes the pollutant mass to spread longitudinally while moving down the catchment. The third process, diffusion, must occur because pollutant particles moving from the upstream boundary of the catchment are followed by uncontaminated runoff. This situation guarantees a concentration gradient which, then, causes a diffusive mass flux in the upstream direction.

Unsteady Uniform Flow

These conditions exist on the catchment downstream from the limiting characteristic wave. Here the continuity equation simplifies to

$$\frac{\partial h}{\partial t} = i \tag{27}$$

Substituting equation 27 into 22 yields

$$\frac{\partial c}{\partial t} = D \frac{\partial^2 c}{\partial x^2} - u \frac{\partial c}{\partial x} - \frac{ci}{h} \tag{28}$$

In this region of rainfall runoff, the depth of flow and, hence, the convective particle velocity are everywhere the same. Because the pollutant is distributed uniformly over the entire catchment prior to rainfall, it is reasonable to conclude that the concentration of the soluble nonpoint source is, likewise, everywhere the same, or

$$\frac{\partial c}{\partial x} = 0 \tag{29}$$

Applying this result to equation 28 gives

$$\frac{\partial c}{\partial t} = - \frac{ci}{h} \tag{30}$$

which upon substitution of equation 5 and use of the total derivative yields

$$\frac{dc}{c} = - \frac{dt}{t} \tag{31}$$

After integration, equation 31 simplifies to

$$c = b/t \tag{32}$$

where b is a constant of integration. The value of b is readily deduced. Let M equal the total mass of the pollutant on the catchment prior to rainfall. Then the mass per unit area is given by

$$w = M/A \tag{33}$$

where w = pollutant mass per unit area,
                M = total initial pollutant mass, and
                A = area of the catchment.

The fundamental premise of the kinematic travel time model asserts that the velocity of a soluble pollutant particle is equal to the velocity of a water particle at the same location. Extending this notion to nonpoint source pollutants requires that everywhere in unsteady uniform flow all pollutant particles are transported at the local convective velocity. Hence, at any lateral transect in this region of flow, the pollutant mass discharge per unit width is equal to the product of the convective particle velocity and the pollutant mass per unit area. Further, this rate is given by the product of the pollutant concentration and the runoff per unit width, or

$$uw = cq \tag{34}$$

Substituting equations 34, 9 and 5 into equation 32 and simplifying gives

$$b = w/i \tag{35}$$

and, hence, equation 32 becomes

$$c_p = w/it \tag{36}$$

where c_p = predicted pollutant concentration,
             t = time since the start of rainfall, and
             w and i as previously defined.

Equation 36 is a deterministic water quality model which describes the concentration of soluble conservative nonpoint source pollutants moving in unsteady uniform rainfall runoff. An alternative form of this kinematic nonpoint source model is

$$c_p = w/h \tag{37}$$

Considering the complex nature of the rainfall runoff nonpoint source phenomenon, a noteworthy feature of equations 36 and 37 is their simplicity. Equation 36 contains parameters which are readily measured and as such is the most convenient form of the model. Equation 37 provides a clear physical interpretation of the kinematic nonpoint source model. Predicted pollutant concentrations are inversely related to the depth of flow. Accordingly, the modeling approach is a dilution scheme which can be visualized as follows: the overland discharge of soluble pollutants from a nonpoint source is simulated by the accelerating convective transport of a pollutant mass which becomes progressively dilute as it moves downstream into regions of deepening flow.

Predicted pollutant concentrations at the catchment outfall for both the equilibrium and the partial equilibrium hydrographs are illustrated in Figure 4. In both instances, predicted concentrations vary inversely with time during the rising hydrograph. For the equilibrium case, predicted concentrations are not shown after steady state is reached because the nonpoint source model is not applicable to steady nonuniform flow conditions. The dashed line which appears is included to reflect changes in pollutant concentrations which result from

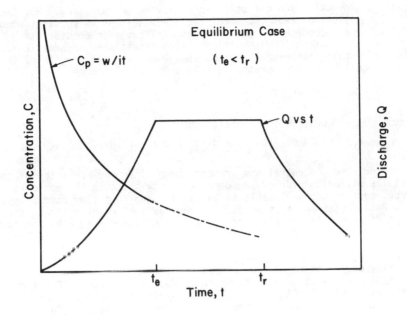

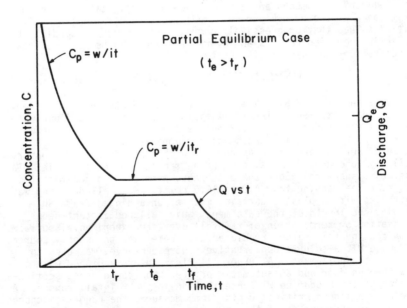

Figure 4.--Hydrograph of overland flow and concentration of
nonpoint source pollutants predicted by kinematic
wave theory.

dilution, convection and diffusion in steady nonuniform flow. For the partial equilibrium case, the predicted concentration is constant when the discharge is constant. This response follows from consideration of the steady uniform flow conditions which result when rainfall ceases before time to equilibrium. For the steady uniform flow region of the partial equilibrium hydrograph, equations 36 and 37 become

$$c_p = w/it_r = w/h_r \tag{38}$$

where t_r = duration of rainfall, and

h_r = depth of flow when discharge is steady.

Predicted concentrations are not shown after the recession of the partial equilibrium hydrograph, denoted by t_f because the nonpoint source model is not applicable to unsteady nonuniform flow conditions.

Recall that equation 16 defines the maximum travel distance for the point source travel time model. This limiting distance has similar significance for the kinematic nonpoint source model. In particular, the ratio L/m delineates that portion of the catchment from which the discharge of nonpoint source pollutants is described by equation 36. Similarly, the term $1/m$ represents the maximum percentage of pollutant mass which is removed prior to equilibrium time. For turbulent flow ($m = 1.5$), 67 percent of the nonpoint source pollutant reaches the outfall prior to equilibrium time; for laminar flow ($m = 3.0$), 33 percent is removed prior to equilibrium.

Note that the kinematic nonpoint source model simulates two well documented characteristics of nonpoint source pollution, namely, the "first-flush" phenomenon and a dependence on precipitation. In the first case, as shown in Figure 4, predicted concentrations exhibit high initial values that progressively decrease with time. Secondly, the prediction equation reveals an inverse relationship between pollutant concentration and rainfall intensity.

EXPERIMENTAL FACILITY

The indoor rainfall facility at Colorado State University served as the model catchment. The facility is constructed on a flume with inside dimensions of 18.3 by 1.2 meters and slope adjustable between 0 and 3 1/3 percent. The catchment portion extended over a 12.5 by 1.2 meter surface area and was covered with impervious butyl rubber. Capillary tube drop formers, protruding from the base of 40 Plexiglas reservoirs suspended above the catchment, were used to generate rainfall at rates ranging from 4.0 to 10.0 cm/hr. Rainfall duration was controlled with a plastic curtain, mounted on a hinged frame and spanning the length of the catchment, which allowed instantaneous application or termination of rainfall over the experimental surface. Deionized and temperature regulated city water was the supply source. Rhodamine WT, a soluble conservative fluorescent dye, was used to simulate the pollutant. A Turner 111 Fluorometer, equipped with a flow-through door and an automatic printer, monitored cross-sectional samples of the Rhodamine WT concentration in the rainfall runoff. A specially designed critical depth flume measured the runoff discharge. The general arrangement of the sampling equipment and the experimental catchment is shown in Figures 5 and 6. Oehler (1979) provides additional details about the operation of the experimental facility.

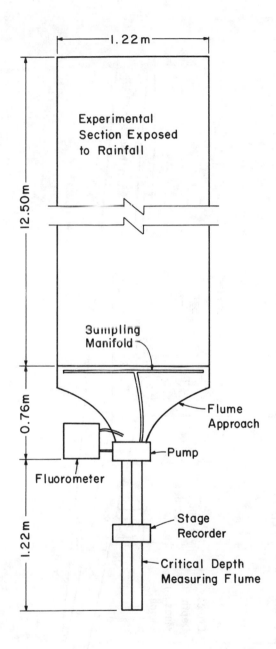

Figure 5.--Plan view of the indoor rainfall facility.

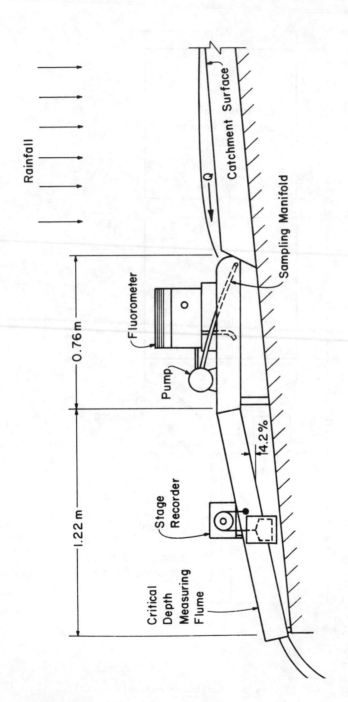

Figure 6.--Arrangement of sampling equipment at the outfall of the catchment.

EXPERIMENTAL PROCEDURE

Preparation for each nonpoint source experiment involved uniform application of a known amount of Rhodamine WT dye to the dry catchment surface. Following this surface treatment, steady uniform rainfall was applied instantaneously to the catchment. Continuous measurements of dye concentrations and runoff discharge were made at the outfall of the catchment. A stop watch was used to determine the time lag introduced by the sampling manifold.

The variables investigated were rainfall intensity, rainfall duration, bed slope and pollutant mass. A total of 26 experimental runs were performed at bed slopes of 1 1/3, 2 1/3, and 3 1/3 percent. Rainfall intensities ranged from 4.0 to 10.0 cm/hr, while Rhodamine WT mass applications ranged from 1.75 to 6.00 mg. For partial equilibrium runs the rainfall duration was 90 seconds. Equilibrium runs lasted 6 minutes.

RESULTS OF NONPOINT SOURCE EXPERIMENTS

The observed and predicted concentrations in parts per billion (ppb) for six nonpoint source experiments--four equilibrium runs and two partial equilibrium runs--are shown in Figure 7. Visual inspection indicates reasonable agreement between theoretical and measured concentrations for the rising hydrograph of both cases. As expected, during the steady portion of the equilibrium hydrograph, pollutant concentrations continue to decrease with time. In contrast, during the steady portion of the partial equilibrium hydrograph, pollutant concentrations tend to level off--a pattern which is consistent with the theory of the kinematic nonpoint source model.

To provide a quantitative evaluation of the nonpoint source model, two indices were developed to measure goodness-of-fit. The first index, known as the root mean square of the error is

$$\varepsilon_1 = \left(\frac{1}{n} \sum_{i=1}^{n} (c_{o_i} - c_{p_i})^2 \right)^{1/2} \tag{39}$$

where c_{o_i} = observed concentration at time t_i, and

c_{p_i} = predicted concentration at time t_i.

Pollutant concentrations measured during the early stages of each experiment were susceptible to large deviations between observed and predicted values because the rate of change of the pollutant concentration is greatest during this time. Due to the presence of the squared term in equation 39, a single large deviation can significantly bias ε_1. To reduce this bias, a second index denoted as the mean absolute error was formulated

$$\varepsilon_2 = \frac{1}{n} \sum_{i=1}^{n} | c_{o_i} - c_{p_i} | \tag{40}$$

For each nonpoint source run, ε_1 and ε_2 were computed from n pairs of observed and predicted concentrations selected at evenly spaced time intervals prior to equilibrium. In order to assess the relative magnitudes of the computed errors, each index was divided by its corresponding range of concentrations observed prior to equilibrium.

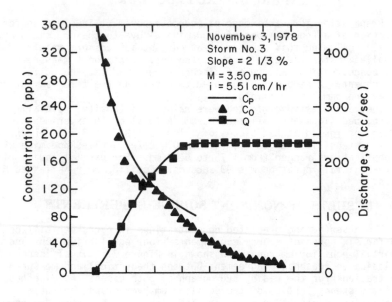

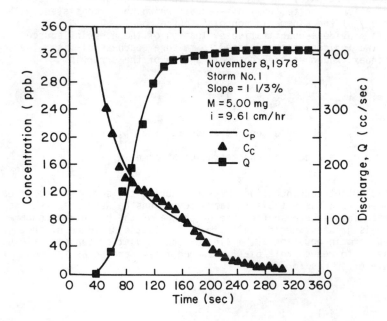

Figure 7.--Predicted concentrations, observed concentrations
and observed discharge versus time for nonpoint
source experiments.

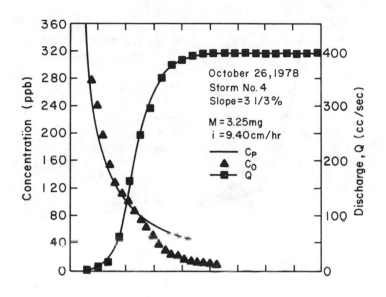

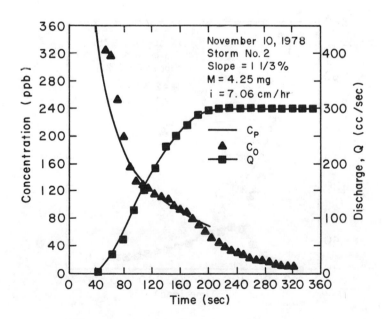

Figure 7.--Predicted concentrations, observed concentrations
and observed discharge versus time for nonpoint
source experiments--Continued

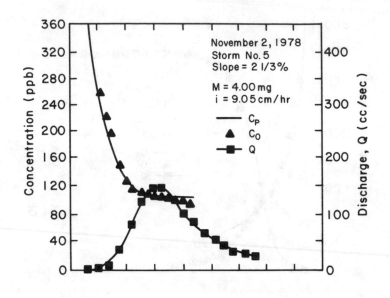

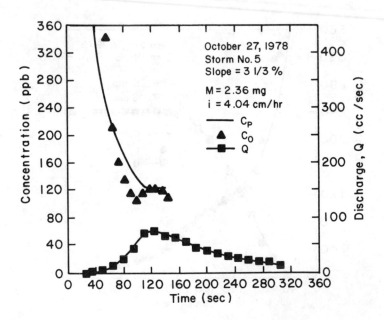

Figure 7.--Predicted concentrations, observed concentrations
and observed discharge versus time for nonpoint
source experiments--Continued

This calculation transformed both indices to relative errors, Δ, expressed as a percent of the total range in observed concentrations.

Average values of ε_1 and ε_2 expressed in parts per billion (ppb) and the percent error were computed for experiments of constant rainfall intensity at each bed slope. These values are shown in Table 1. As expected, ε_2 is consistently less than ε_1. The average percent error computed from results of all 26 experiments are 12.9 percent and 11.5 percent for ε_1 and ε_2, respectively.

Table 1.--Root mean square of the error, mean absolute error, and relative errors between predicted and observed nonpoint source concentrations.

bed slope, S_o	Average rainfall intensity, i (cm/hr)					
	4.7		7.2		9.2	
	ε_1(ppb)	%Δ	ε_1(ppb)	%Δ	ε_1(ppb)	%Δ
1 1/3%	36	16	26	10	12	5
2 1/3%	24	11	40	17	23	8
3 1/3%	37	16	64	21	29	10
	ε_2(ppb)	%Δ	ε_2(ppb)	%Δ	ε_2(ppb)	%Δ
1 1/3%	33	14	20	8	10	4
2 1/3%	22	10	38	16	19	7
3 1/3%	32	14	62	20	24	8

The magnitudes of the percent errors shown in Table 1 are certainly within the range of expected experimental error. Nevertheless, it is possible that deviations between observed and predicted concentrations result from limitations of the assumptions or limitations of the theory. Concerning the assumptions, recall that the pollutant is considered to be mixed instantaneously and completely in the rainfall runoff. If the pollutant is released gradually from the catchment surface, observed concentrations will be less than predicted concentrations. This conjecture, however, is not supported by the results given in Figure 7 where observed concentrations are located above and below predicted values. Concerning the theory, the accuracy of the nonpoint source model depends on the adequacy of the kinematic wave equations for overland flow. Additional insight into the deviation between predicted and observed concentrations, then, is obtained by comparing theoretical and observed runoff hydrographs. According to kinematic wave theory, time of equilibrium signals an abrupt change from the rising hydrograph to steady conditions. The observed hydrographs, however, exhibit a gradual transition from unsteady to steady conditions. This discrepancy between theoretical and observed hydrographs reflects the approximation which results when kinematic wave theory is used to describe the dynamics of overland flow. Extending this limitation to the kinematic nonpoint source model, then, deviations between observed and predicted concentrations may be attributed in part to the approximation inherent in the kinematic wave equations.

423

This shortcoming does not preclude kinematic wave theory as a suitable approach for modeling the transport and concentration of conservative soluble nonpoint source pollutants in rainfall runoff. On the contrary, the results of the nonpoint source experiments are very encouraging and research is continuing at Colorado State University on pollutant transport in overland flow on pervious surfaces. Furthermore, these results provide additional confirmation of the adequacy of the kinematic wave equations for overland flow.

SUMMARY AND CONCLUSIONS

The fundamental equation which describes the one-dimensional convective and diffusive transport of conservative pollutants in gradually varied unsteady overland flow has been derived. Applying principles of kinematic wave theory, the pollutant transport equation reduces to a deterministic model which predicts the concentration of soluble nonpoint source pollutants moving in rainfall runoff during the rising hydrograph. Predicted concentrations are compared against experimental data collected during laboratory runs with simulated rainfall. Rhodamine WT was used to simulate a soluble, conservative pollutant. The variables investigated were rainfall intensity, rainfall duration, bed slope and pollutant mass.

The results presented in this paper demonstrate that kinematic wave theory--ordinarily used to predict the hydrograph of overland flow--also can be used to simulate the transport of soluble pollutants in rainfall runoff. The pollutant transport equation shows that besides salient differences in flow conditions, there are other important distinctions between the regions of steady nonuniform and unsteady uniform flow. In steady nonuniform rainfall runoff, the concentration of soluble nonpoint source pollutants is influenced by dilution, convection, and diffusion. In unsteady uniform rainfall runoff, dilution is the predominant influence on the concentration of soluble nonpoint source pollutants. A kinematic nonpoint source model, based on a simple dilution scheme, provides reasonably accurate predictions of pollutant concentrations at the outfall of an impervious catchment during the rising hydrograph.

REFERENCES

Asmussen, L. E.; White, A. W. Jr.; Hauser, E. W.; and Sheridan, J. M.
    1977. Reduction of 2,4-D Load in Surface Runoff Down a
           Grassed Waterway. Journal of Environmental Quality,
           Vol. 6, No. 2, pp. 159-162.

Brazil, L.E.
    1976. A Water Quality Model of Overland Flow. M.S. Thesis.
           Colorado State University, Fort Collins, Colorado.
           141 p.

Brazil, L. E.; Sanders, T.G.; and Woohiser, D.A.
    1979. Kinematic Parameter Estimation for Transport of
           Pollutants in Overland Flow. Surface and Subsurface
           Hydrology, Water Resources Publications. Fort Collins,
           Colorado. pp. 555-568.

Buchberger, S. G.
    1979. The Transport of Soluble Nonpoint Source Pollutants
           During the Rising Hydrograph. M.S. Thesis. Colorado
           State University, Fort Collins, Colorado. 154 p.

Colston, N. V., Jr.
   1974. Characterization and Treatment of Urban Land Runoff.
          EPA-670/2-74-096, Washington, D.C. 157 p.

Crawford, N. H., and Donigian, A. S.
   1973. Pesticide Transport and Runoff Model for Agricultural
          Lands. EPA-660/2-74-013, Washington, D.C. 221 p.

Donigian, A. S., and Crawford, N. H.
   1976. Modeling Nonpoint Pollution from the Land Surface.
          EPA-660/3-76-083, Washington, D.C. 279 p.

Henderson, F. M., and Wooding, R. A.
   1964. Overland Flow and Groundwater Flow From a Steady
          Rainfall of Finite Duration. Journal of Geophysical
          Research, Vol 69, No. 8, pp. 1531-1540.

Liggett, J. A.
   1975. Basic Equations of Unsteady Flow. In Unsteady Flow
          in Open Channels - Volume 1, K. Mahmood and
          V. Yevjevich (editors). Water Resources Publications,
          Fort Collins, Colorado. pp. 29-62.

Lighthill, M. J., and Whitham, G. B.
   1955. On Kinematic Waves, I: Flood Movement in Long Rivers.
          Proceedings of the Royal Society of London, Vol. 229,
          pp. 281-316.

Oehler, J. R.
   1979. Operator's Manual for the Indoor Rainfall Facility.
          Independent Study. Department of Technical Journalism,
          Colorado State University, Fort Collins, Colorado.
          34 p.

Ponce, S. L., and Hawkins, R. H.
   1978. Salt Pickup by Overland Flow in the Price River Basin,
          Utah. Water Resources Bulletin, Vol. 14, No. 5,
          pp. 1187-1200.

Roesner, L. A. et al.
   1974. A Model for Evaluating Runoff Quality in Metropolitan
          Master Planning. American Society of Civil Engineers,
          Urban Water Research Program, Technical Memorandum
          No. 23, 73 p.

Roffman, H. K.; Roffman, A.; McFeaters, B. D.; and Norris, J. R.
   1975. The Effects of Large Shopping Complexes on Quality of
          Storm Water Runoff. In National Symposium on Urban
          Hydrology and Sediment Control. University of Kentucky,
          Lexington, Kentucky. pp. 245-256.

Sartor, J. D., and Boyd, G. B.
   1972. Water Pollution Aspects of Street Surface Contaminants.
          EPA-R2-72-081, Washington, D.C. 236 p.

Shaheen, D. G.
   1975. Contributions of Urban Roadway Usage to Water Pollution.
          EPA-600/2-75-004, Washington, D.C. 346 p.

Stewart, B. A.; Woolhiser, D. A.; Wischmeier, W. H.; Caro, J. H.; and Frere, M. H.
   1975. Control of Water Pollution from Cropland: Volume I - A Manual for Guideline Development. EPA-600/2-75-026a, Washington, D.C. 111 p.

United States Environmental Protection Agency
   1977. Nonpoint Water Quality Modeling in Wildlife Management: A State-of-the-Art Assessment. EPA-600/3-77-036, Washington, D.C. 145 p.

Woolhiser, D. A.
   1969. Overland Flow on a Converging Surface. Transactions of the American Society of Agricultural Engineers, Vol. 12, No. 4, pp. 460-462.

Woolhiser, D. A.
   1975. Simulation of Unsteady Overland Flow. In Unsteady Flow in Open Channels - Volume II, K. Mahmood and V. Yevjevich (editors). Water Resources Publications, Fort Collins, Colorado. pp. 485-508.

Woolhiser, D. A., and Liggett, J. A.
   1967. Unsteady, One-dimensional Flow Over a Plane - The Rising Hydrograph. Water Resources Research, Vol. 3, No. 3, pp. 753-771.

Wulkowicz, G. M., and Saleem, Z. A.
   1974. Chloride Balance of an Urban Basin in the Chicago Area. Water Resources Research, Vol. 10, No. 5, pp. 974-982.

MODELING NONPOINT POLLUTION DUE TO
ANIMAL WASTE WITH ARM II

K. R. Reddy
Assistant Professor
Agricultural Research and Education Center
University of Florida
Sanford, FL 32771

R. Khaleel
Assistant Professor of Hydrology
New Mexico Institute of Mining and Technology
Socorro, NM 87801

M. R. Overcash
Professor
Department of Chemical Engineering
North Carolina State University
Raleigh, NC 27650

ABSTRACT

An animal waste model was coupled with Agricultural Runoff Management (ARM-II) Model to describe the runoff water quality from watersheds receiving animal wastes. Animal waste model consists of a series of submodels that have been developed to describe the fate of pollutants (nitrogen, phosphorus, carbon, and bacteria) at the soil surface in land areas receiving animal wastes. These submodels were incorporated into the existing computer code of ARM-II to provide for the waste transformations and transport for soils treated with animal wastes. In developing the new computer code and interfacing with the existing code, the same degree of functional approach as in the existing ARM-II model was used. A series of computer runs were made to verify the computer code and study the effects of various swine waste application practices on runoff water quality. Although no experimental verification of the simulated values are available at this time, the predicted values for runoff losses of nitrogen, phosphorus, carbon, and fecal coliforms from a watershed receiving swine wastes were of the same order of magnitude as reported in the literature. The reaction rate coefficients for exchange rate of phosphorus (adsorption-desorption), mineralization of organic nitrogen, and adsorption coefficient of bacteria on sediment were found to be more sensitive compared to other reaction rate co-efficients.

INTRODUCTION

Almost one-third of the pollutants entering the national waterways are derived from nonpoint sources (Agee, 1974). The Federal Water Pollution Control Act Amendments of 1972 (Public Law 92-500) require each state to submit a plan which shall identify agriculturally related nonpoint sources of pollution and define procedures and methods to control, to the extent feasible, these sources of pollution. Concerns about nonpoint source pollution are based on concentrations of pollutants such as nitrogen (N), phosphorus (P), oxygen demanding compounds (C), and microorganisms in runoff water. Since animal wastes contain decomposable organic matter and nutrients, they are considered a major potential source of water quality degradation (ASCE, 1977).

For animal waste land use category, a substantial research effort to evaluate the watershed impact and controlling factors has not been undertaken, although there is considerable information on the practices of animal waste land application. The research for land areas receiving animal wastes should contribute substantially to predictive tools for land application sites of municipal and industrial effluents and sludges -- an inevitably increasing technology.

The animal waste land use category has characteristics that are dissimilar from agricultural cropland segment, although both are rural land uses. First, in addition to high levels of nutrients, animal wastes contain significantly higher levels of organics or oxygen demanding compounds. Secondly, rainfall-runoff transport from areas receiving animal wastes might contain substantially greater levels of coliforms, and pathogenic microorganisms, compared to those from cropland. Thirdly, animal wastes can substantially alter the soil physical properties and the soil-plant system leading to necessary corrections in hydrologic models.

A major impetus for initiating the modeling of animal waste land application is the impending initiation of Best Management Practices (BMP) for controlling the nonpoint source pollution from such areas. A unified understanding of such land applications is essential to determine which if any such BMP's are cost effective from a water quality standpoint.

Research on various land use categories has progressed on a stage-wise basis to allow successful reinforcement of the modeling efforts and in response to the Congressional priorities and financial limitations. For the cropland segment, the first effort was for pesticides in the Pesticide Transport Model (PTR) (Crawford and Donigian, 1973). Following the PTR, several models have been developed for crop land areas receiving fertilizers, primarily the N and P components. Examples are the nonpoint pollution model (Donigian and Crawford, 1976a); modeling pesticides and nutrients from agricultural land (Donigian and Crawford, 1976b); ACTMO, an Agricultural Chemical Transport Model (Frere et al., 1975); and ARM II, Agricultural Runoff Management Model (Donigian et al., 1977). None of these models provide an option to describe the pollutant transport from land areas receiving animal wastes.

Modeling of pollutant transport and transformations, on a continuous basis, for land areas receiving animal wastes requires a number of inputs. These include data on chemical and biochemical reaction kinetics for N, P, C, and bacteria. Generalized conceptual models based on the state-of-the-art approach have been developed for N, P, C, and microorganisms (Overcash et al., 1981; Reddy et al., 1979a,b; Reddy et al., 1980; Reddy et al., 1981). The objectives of this paper are to incorporate the available information and the detailed conceptual submodels developed earlier into the existing computer code of ARM II to provide for the waste transformation and transport for areas receiving animal wastes. In developing the new computer code and interfacing with the existing code, we will seek the same degree of functional approach as in the existing ARM II model. Finally, using the available rainfall-runoff data for Watkinsville,

Georgia, watershed and the revised ARM II: animal waste version, a series of computer runs will be made to verify the computer code and study the effects of various factors, such as animal waste loading rate, timing and frequency of applications, waste types and changes in kinetic reaction rates. The effects of these factors on runoff water quality will be discussed.

ARM II: Animal Waste Model Description

The ARM II Model simulates runoff, sediment, pesticides, and nutrient contributions to stream channels from both surface and subsurface sources. The major components of the model individually simulates the hydrologic response (LANDS) of the watershed, sediment production (SEDT), and nutrient transformations (NUTRNT) Donigian et al., 1977). The NUTRNT submodel will be replaced with the submodel involving the pollutant transformations in land areas receiving animal wastes. The LANDS program simulates all flow components (surface runoff, interflow, groundwater flow), and soil moisture storages by representing the processes of interception, infiltration, overland flow, and percolation. LANDS subroutine is a modification of the Stanford Watershed Model (Crawford and Linsley, 1966), and the Hydrocomp Simulation Program (Hydrocomp, Inc., 1976). Details of the hydrology submodel are reported by Donigian et al. (1977). The SEDT program simulates the erosion processes of soil particle detachment by rainfall and transport by overland flow (Donigian et al., 1977).

Based on the submodels described earlier (Reddy et. al., 1979a,b,c; Reddy et al., 1980; Reddy et al., 1981), the ARM II model was modified to describe the availability of N, P, C, and bacteria at the soil surface for areas receiving animal wastes. The soil-waste complex constituents are represented by relative partitioning between predominantly soluble or water-borne and the sediment or sediment-attached species.

Nitrogen

The contribution of nitrogen species is the sum of contributions from (1) easily mineralizable waste organic N fractions; (2) slowly mineralizable soil plus waste organic N; and (3) inorganic forms of N in soil plus waste. The procedure used to calculate different fractions is shown in Fig. 1. The first step to follow is to characterize the waste material into inorganic and organic-N fractions. For most wastes, inorganic N fraction is in ammonium N form, which is readily available for transport or plant uptake. Ammonium N applied through wastes is also subjected to ammonia volatilization and nitrification. The soil and environmental conditions required for these processes to function at optimum rates were discussed by Reddy et al. (1979a,b). In the modified ARM II model, for surface applied manure, all of the ammonium N was assumed to be lost through volatilization. This assumption is corroborated by the data presented earlier (Reddy et al., 1979a,b). For conditions where waste is incorporated into the soil, only 10% of the added ammonium N was available for nitrification. Ammonium N present in solution phase undergoes nitrification.

Organic N present in the wastes is divided into (1) easily mineralizable; and (2) slowly mineralizable fractions, based on C/N ratio of the waste. Easily mineralizable N is present only in the wastes with C/N ratio less than 23. For wastes with C/N ratio greater than 23, all added organic N was assumed to be in slowly mineralizable fraction and was added to the soil organic N pool, and allowed to mineralize at the

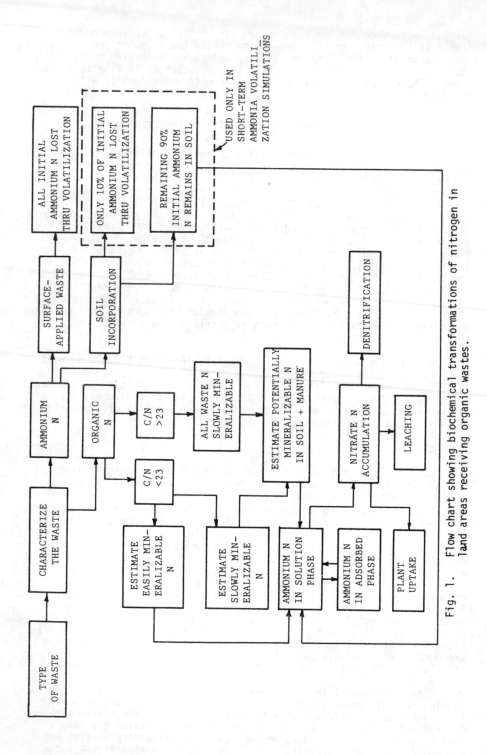

Fig. 1. Flow chart showing biochemical transformations of nitrogen in land areas receiving organic wastes.

same rate as soil organic N. A higher mineralization rate constant was used for easily mineralizable organic N fraction. Ammonium N mineralized from these two fractions was also partitioned into soluble and adsorbed phases. All of the easily mineralizable organic N was assumed to be present in the soluble forms, which can be subjected to leaching and transport in runoff. Procedures for simulation of plant uptake of N and denitrification were same as those described in ARM II model.

Phosphorus

The contribution of phosphorus species is the sum of (1) waste P; and (2) soil P. Phosphorus in both soil and waste exists in organic and inorganic forms. Both these forms can exist in solution and adsorbed phases. The processes considered in simulation of P are shown in Fig. 2. All processes were assumed to follow first order kinetics. A detailed discussion of the processes was presented by Overcash et al. (1981).

Animal waste was first characterized for total P and Ortho P, using literature values. Total P in the waste was also characterized for organic and inorganic P. Using a regression equation obtained from literature data (Overcash et al., 1981), organic P fraction of the waste was calculated and pooled with the soil organic P where it was allowed to mineralize at the same rate as soil organic P. It was assumed that organic P in the soil was about 50% of the total P (Overcash et al., 1981). Mineralization and immobilization rates were assumed to be the same as those for slowly mineralizable organic N. Adsorption-desorption processes were assumed to be time dependent. The rate constants used were similar to those in the ARM II model.

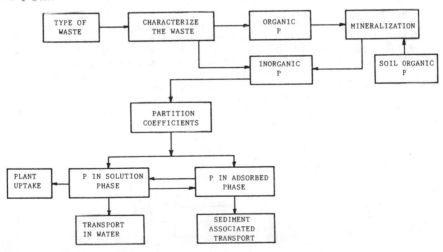

Fig. 2.   Flow chart showing biochemical transformation of phosphorus in land areas receiving organic wastes.

Carbon

The contribution of the total carbon in runoff is the sum of (1) easily decomposible C, and (2) slowly decomposable C fractions. These fractions are decomposed at different rates. A detailed description of the C processes involved in transformations and transport is given by Reddy et al. (1980). The procedure used in the simulation of C transport is presented in Fig. 3. The easily mineralizable fraction of C was assumed to be in water soluble form, whereas the slowly mineralizable

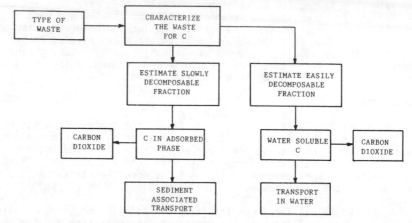

Fig. 3. Flow chart showing biochemical transformations of carbon in land areas receiving organic wastes.

fraction was assumed to be in adsorbed phase. The fraction of C present in the adsorbed phase was allowed to decompose at the same rate as soil organic C. Rate constants used for these two phases are given later.

Bacteria

The contribution of bacteria to runoff water is the sum (1) bacteria transported in water; and (2) sediment associated bacteria. A detailed discussion of bacterial transformation and transport was presented by Reddy et al. (1981). The processes considered in the model are (1) net bacterial dieoff; and (2) bacterial retention to soil particles. Based on literature data, dieoff rate constants were calculated. An average dieoff rate constant for fecal coliforms was used. Only one data set was available for partitioning the fecal coliforms into solution and adsorbed phases. Dieoff of bacteria was assumed to occur at the same rate, both in solution and adsorbed phases. The procedure followed in simulations is shown in Fig. 4.

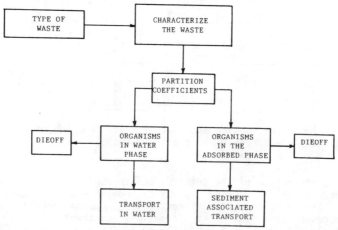

Fig. 4. Flow chart showing pathogen transformations in land areas receiving organic wastes.

The coupled system of differential equations utilized in the simulations, definition of rate constants, and definition of various pollutant forms are given in Tables 1, 2, and 3, respectively.

Table 1. Coupled system of differential equations.

NITROGEN

Easily mineralizable manure N

Organic Nitrogen:

$$\frac{d}{dt} \{ORG\text{-}NM\} = KIM\{NH_4\text{-}SM\} + KKIM\ \{NO_3 + NO_2\text{-}M\} - KAM\ \{ORG\text{-}NM\}$$

Solution Ammonia:

$$\frac{d}{dt} \{NH_4\text{-}SM\} = KAM\ \{ORG\text{-}NM\} - (KSA + K1 + KIM)\ \{NH_4\text{-}SM\} + KAS\ \{NH_4\text{-}AM\}$$

Adsorbed Ammonia:

$$\frac{d}{dt} \{NH_4\text{-}AM\} = KSA\ \{NH_4\text{-}SM\} - KAS\ \{NH_4\text{-}AM\}$$

Nitrate:

$$\frac{d}{dt} \{NO_3 + NO_2\text{-}M\} = K1\ \{NH_4\text{-}SM\} - (KD + KKIM + KPL)\ \{NO_3 + NO_2\text{-}M\}$$

Nitrogen Gas:

$$\frac{d}{dt} \{N_2\text{-}M\} = KD\ \{NO_3 + NO_2\text{-}M\}$$

Plant Nitrogen:

$$\frac{d}{dt} \{PLNT\text{-}NM\} = KPL\ \{NO_3 + NO_2\text{-}M\}$$

Slowly mineralizable soil plus manure N

Organic Nitrogen:

$$\frac{d}{dt} \{ORG\text{-}N\} = KIM\ \{NH_4\text{-}S\} + KKIM\ \{NO_3 + NO_2\} - KAM\ \{ORG\text{-}N\}$$

Solution Ammonia:

$$\frac{d}{dt} \{NH_4\text{-}S\} = KAM\ \{ORG\text{-}N\} - (KSA + K1 + KIM)\ \{NH_4\text{-}S\} + KAS\ \{NH_4\text{-}A\}$$

Adsorbed Ammonia:

$$\frac{d}{dt} \{NH_4\text{-}A\} = KSA\ \{NH_4\text{-}S\} - KAS\ \{NH_4\text{-}A\}$$

Nitrate:

$$\frac{d}{dt} \{NO_3 + NO_2\} = K1\ \{NH_4\text{-}S\} - (KD + KKIM + KPL)\ \{NO_3 + NO_2\}$$

Table 1. (Cont'd.)

Nitrogen Gas:

$$\frac{d}{dt} \{N_2\} = KD \ \{NO_3 + NO_2\}$$

Plant Nitrogen:

$$\frac{d}{dt} \{PLNT-N\} = KPL \ \{NO_3 + NO_2\}$$

SOIL PLUS MANURE P

Organic Phosphorus:

$$\frac{d}{dt} \{ORG-P\} = - KM \ \{ORG-P\} + KIM \ \{PO4-S\}$$

Solution Phosphate:

$$\frac{d}{dt} \{ PO4-S\} = KM \ \{ORG-P\} - (KIM + KSA + KPL) \ \{PO4-S\} + KAS \ \{ PO4-A\}$$

Absorbed and Combined Phosphate:

$$\frac{d}{dt} \{PO4-A\} = KSA \ \{PO4-S\} - KAS \ \{PO4-A\}$$

Plant Phosphorus:

$$\frac{d}{dt} \{PLNT-P\} = KPL \ \{PO4-S\}$$

CARBON

Easily decomposable C (water soluble)

C remaining:

$$\frac{d}{dt} \{CAR-S\} = - KCS \ \{CAR-S\}$$

C loss:

$$\frac{d}{dt} \{CAR-SS\} = KCS \ \{CAR-S\}$$

Slowly decomposable C (absorbed to soil)

C remaining:

$$\frac{d}{dt} \{CAR-A\} = - KCA \ \{CAR-A\}$$

C loss:

$$\frac{d}{dt} \{CAR-AA\} = KCA \ \{CAR-A\}$$

BACTERIA

Bacteria in solution

Survival: $\frac{d}{dt} \{BAC-S\} = - KBS \ \{BAC-S\}$

Table 1. (Cont'd.)

Dieoff: $\dfrac{d}{dt}\{BAC\text{-}SS\} = KBS\ \{BAC\text{-}S\}$

Bacteria retained to soil

Survival: $\dfrac{d}{dt}\{BAC\text{-}A\} = -KBA\ \{BAC\text{-}A\}$

Dieoff: $\dfrac{d}{dt}\{BAC\text{-}AA\} = KBA\ \{BAC\text{-}A\}$

Table 2. Definition of N, P, C and Bacteria Rates

Rate Constant, day^{-1}	Definition
Nitrogen	
K1	Nitrification rate of ammonium in solution to nitrite and nitrate
KD	Denitrification rate of nitrite and nitrate to gaseous nitrogen
KPL	Uptake rate of nitrate (and nitrite) by plants
KAM*	Mineralization or ammonification rate of organic nitrogen to ammonium in solution.
KIM	Immobilization rate of ammonium in solution to organic nitrogen
KKIM	Immobilization rate of nitrate to organic nitrogen
KSA	Exchange rate of ammonium from solution to adsorbed phase
KAS	Exchange rate of ammonium from adsorbed phase to solution
Phosphorus	
KM	Mineralization rate of organic phosphorus to solution phosphate
KIM	Immobilization rate of solution phosphate to organic phosphorus
KPL	Uptake of solution phosphate by plants
KSA	Adsorption or combining of solution to adsorbed or combined phase

*KAM's are different for slowly mineralizable and easily mineralizable fractions.

Table 2. Continued

Rate Constant, day^{-1}	Definition
KAS	Desorption or dissolving or adsorbed or combined phase to solution phosphorus
Carbon	
KCS	Rate constant for easily decomposable C fraction
KCA	Rate constant for slowly decomposable C fraction
Bacteria	
KBS	Rate constant for bacterial dieoff in solution phase
KBA	Rate constant for bacterial dieoff in adsorbed phase

Note: All rates are based on the N, P, C, and Bacteria amounts for various forms.

Table 3. Definition of N, P, C, and Bacteria forms.

Nitrogen	
ORG-NM	Organic N in easily mineralizable manure N
ORG-N	Organic N in slowly mineralizable manure plus soil N
NH_4-SM	Ammonium N in solution derived from easily mineralizable manure N
NH_4-S	Ammonium N in solution derived from slowly mineralizable manure plus soil N
NH_4-AM	Ammonium N in adsorbed phase derived from easily mineralizable manure N
NH_4-A	Ammonium N in adsorbed phase derived from slowly mineralizable manure plus soil N
NO_3+NO_2-M	Nitrate and nitrite N derived from easily mineralizable manure N
NO_3+NO_2	Nitrate and nitrite N derived from slowly mineralizable manure plus soil N
N_2-M	Denitrification of nitrate N derived from easily mineralizable manure N

Table 3. Continued.

N_2	Denitrification of nitrate N derived from slowly mineralizable manure plus soil N
PLNT-NM	Plant uptake of nitrogen derived from easily mineralizable manure N
PLNT-N	Plant uptake of nitrogen derived from slowly mineralizable manure plus soil N

Phosphorus

ORG-P	Organic P derived from soil plus manure
PO_4-S	Soluble ortho-P derived from soil plus manure
PO_4-A	Adsorbed ortho-P derived from soil plus manure
PLNT-P	Plant uptake of P derived from soil plus manure

Carbon

CAR-S	Easily decomposable carbon remaining in solution
CAR-SS	Easily decomposable carbon loss as CO_2 from solution
CAR-A	Slowly decomposable carbon remaining in adsorbed form to soil
CAR-AA	Slowly decomposable carbon loss as CO_2 from adsorbed phase

Bacteria

BAC-S	Bacterial survival in solution phase
BAC-SS	Bacterial dieoff in solution
BAC-A	Bacterial survival in adsorbed phase
BAC-AA	Bacterial dieoff in adsorbed phase

Modifications to ARM II Computer Code

The ARM II computer code was modified to include the waste transformation submodels, i.e., N, P, C, and bacteria submodels. New storages 21 through 26 were created in subroutines NUTRNT and TRANS of ARM computer code to handle rapidly mineralizable waste N components, i.e.,

21 = ORG.NM (see Table 3 for definitions of various constituents);

22 = NH_4-SM,

23 = NH_4-AM,

437

24   = $NO_3 + NO_2 - M$,

25   = $N_2 - M$, and

26   =   PLNT-NM

Nutrient storages 1 through 6 in NUTRNT and TRANS subroutines of ARM are still used for transformations occurring in soil N. The slowly mineralizable waste N fractions are added to the native soil organic N components and the totals are stored in storages 1 though 6, i.e.,

1   =   ORG.N,

2   = NH_4-S,

3   = NH_4-A,

4   = $NO_3 + NO_2$

5   = N_2 gas, and

6   =   PLNT-N.

New storages (i.e., storages C(21,21), C(21,22), C(21,24), etc.) were created in subroutine TRANS for the waste reaction rates and to solve the coupled system of differential equations for the rapidly mineralizable waste N transformations (Table 1). The reaction rates in the coupled system of differential equations for soil N combined with slowly mineralizable waste N fractions are stored in existing storages, i.e., C(1,1), C(1,2), C(1,4) etc. (Table 1).

For bacteria, new storages 7 through 10 were created in the subroutines NUTRNT and TRANS to solve the first order equations for bacterial transformations.

7   =   BAC-S

8   =   BAC-SS

9   =   BAC-A, and

10   =   BAC-AA.

New storages were also created in subroutine TRANS for the bacteria reaction rates, KBS and KBA.

Phosphorus constituents are stored in storages 11 through 14 in subroutines NUTRNT and TRANS of ARM computer code, i.e.,

11   =   ORG.P,

12   = PO_4-S,

13   = PO_4-A, and

14   =   PLNT-P.

No new additional storages were created for P. Instead, P contributions from animal wastes are added to the soil P and stored in storages 11 through 14.

New storages 16 through 19 were created in subroutines NUTRNT and TRANS to solve equations for organic C transformations.

16 = CAR-S,

17 = CAR-SS,

18 = CAR-A, and

19 = CAR-AA.

New storages were also created in subroutine TRANS for the organic C reaction rates, KCS and KCA. The native soil organic C are added to waste organic C and the totals are stored in storages 16 and 18 in subroutine NUTRNT.

Model Inputs

The revised ARM III animal waste version requires data on reaction rates and initial storages for various constituents of N, P, C, and Bacteria. This section describes the sample inputs for a typical swine waste loading rate of 200 kg N/ha. The loading rate is based on N content and is equivalent to 3.1 metric tons/ha (dry basis).

Nitrogen

Forms of N present (TAN/TKN = 0.15):

where TAN = total ammonical nitrogen and TKN = total Kjeldahl nitrogen,

Ammonium N   =   30 kg N/ha,

Organic N   =   170 kg N/ha, and

C/N ratio of swine waste = 8.

According to Reddy et al. (1979a), easily mineralizable fraction of N is given by NX = NW - (0.043) NW (C/N) where NX = potentially mineralizable N, kg/ha; NW = total N applied, kg/ha; and C/N = carbon/nitrogen ratio. Therefore

easily mineralizable N = 112 kg N/ha, and

slowly mineralizable N = 58 kg N/ha.

It is assumed that all easily mineralizable N is in water soluble form. The rate constant for mineralization of easily mineralizable N = 0.021 day^{-1} (at 35°C). The slowly mineralizable waste N was added to the soil organic N pool. The total potentially mineralizable N from this resistant fraction was calculated by the procedures described by Reddy et al. (1979a). The mineralizable rate constant for slowly mineralizable N = 0.0077 day^{-1} (at 35°C). Plant uptake and denitrification simulations were similar to those described in ARM II model. Potential ammonium N loss through volatilization was deleted from the input (Reddy et al., 1979b).

Phosphorus

Phosphorus loading rate = 80 kg P/ha (based on TP/TKN ratio = 0.4) where TP = total phosphorus. Using a ratio of ORG.P/TP = 0.3, total P applied was characterized into:

Organic P = 24 kg P/ha, and

Inorganic P = 56 kg P/ha.

Waste organic P was added to soil organic P and allowed to mineralize at the same rate as that for organic N (i.e., rate constant = 0.0077 day^{-1} at 35°C). Inorganic P applied was partitioned into solution and adsorbed phases using the same ratios as those used for soil P in ARM II. Simulations of adsorption-desorption reactions after waste applications and plant uptake for P were same as those described in ARM II.

Carbon

Carbon loading = 1600 kg C/ha (based on C/N ratio = 8.0). The percentage of easily decomposable C fraction was assumed to be same as that for easily mineralizable organic N (Reddy et al., 1980). Then 66% of the added C is easily decomposable and 34% of the added C is slowly decomposable, i.e.,

easily decomposable C = 1056 kg C/ha;

slowly decomposable C = 544 kg C/ha;

Rates of C decomposition were 0.041 day^{-1} (35°C) for easily decomposable waste C was added to the soil organic carbon, and allowed to decompose at the rate indicated above. The decomposition rate constant was obtained from literature (Reddy et al., 1980).

Bacteria

Bacterial transformations and transport were represented by simulating the fecal coliform behavior in the soil.

Fecal coliform loading = 3.38×10^{13} counts/ha (FC loading is based on the application of 200 kg N/ha, as swine waste). The processes included in the simulation model were (1) bacterial dieoff, and (2) bacterial retention to soil particles. Based on the literature data (Reddy et al., 1981), an average rate constant of fecal coliform dieoff of 1.92 day^{-1} (at 35°C) was used in the simulation model. Only one data set is available on the retention of bacteria to soil particles. Retention coefficients were calculated assuming instantaneous equilibrium, as described by Reddy et al., 1981. A retention coefficient of 1909 ml/g was used to partition the organisms present in solution and adsorbed phases.

Fecal coliforms in solution phase = 1.78×10^{10} counts/ha, and FC retained on soil particles = 3.38×10^{13} counts/ha.

Results and Discussion

Using the available rainfall-runoff data for the Watkinsville, Georgia (P2) watershed, and the revised ARM II: Animal Waste Version, a series of computer runs (Table 4) were made to verify the computer code and to study the effects of various animal waste application practices, such as waste loading rate, timing and frequency of application, and waste types on runoff water quality. Sensitivity analysis

Table 4. Summary of computer runs.

Effects studied	Loading rate KgN/ha	Time of application	Remarks
Control	0	-	-
Loading rate	**200 *600	May May	Manure type swine, KAM = 0.021, KAS = 0.015, 0.0015, 0.005, 0.0 KCS = 0.041, KBS = 1.92, KBA = 1.92
Frequency of applications	*600 200 400 200	May May,Aug,Dec Aug Aug, Dec	Manure type swine, KAM = 0.021, KAS = 0.015, 0.0015, 0.005, 0.0 KCS = 0.041, KBS = 1.92, KBA = 1.92
Time of application	**200 200 200	May Aug Dec	Manure type swine, KAM = 0.021, KAS = 0.015, 0.0015, 0.005, 0.0 KCS = 0.041, KBS = 1.92, KBA = 1.92
Waste types	**200 200 200	May May May	Swine reaction rates same as above, initial storages different Beef for different manure types Poultry
Reaction rates	200 200 200 200	May May May May	Swine type KAM = 0.042, KAS = 0.0 , 0.003, 0.01, 0.0 KCS = 0.082, KBS = 3.8 , KBA = 3.84 KAM = 0.0105, KAS = 0. , 0.015, 0.05, 0.0 KCS = 0.0205, KBS = 0.5 , KBA = 0.96 KAM = 0.0052, KAS = 0.0 , 0.006, 0.02, 0.0 KCS = 0.01025, KBS = 0. , KBA = 0.48 KAM = 0.084, KAS = 0.3, 0.03, 0.10, 0.0 KCS = 0.164, KBS = 2.88, KBA = 2.88

** 200 - same runs
* 600 - same runs.

441.

was also conducted to evaluate the effect of kinetic reaction rate changes on runoff water quality. Most simulations were for soils treated with swine wastes. The kinetic rate constants used in the simulations were obtained from the literature as described in the preceding sections.

Monthly runoff volume and sediment yield for the calibrated P2 watershed at Watkinsville, Georgia, is shown in Fig. 5. Sediment yield and runoff were highest during the month of May followed by June, July, September, and December. The quantity of the nutrients transported in runoff water was expressed in mass units (kg/ha). The data presented are simulated values, and at this time no experimental verification of the model is available.

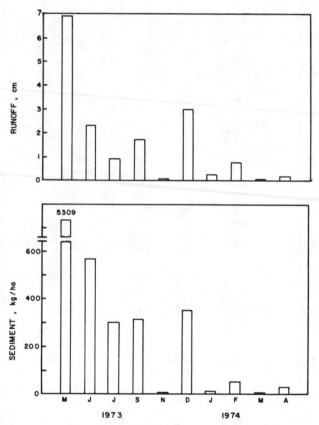

Fig. 5.   Simulated runoff volumes and sediment yields during 1973-1974 for P_2 watershed.

Results of simulation of the effect of swine waste loading rate on nutrient transport are shown in Fig. 6. The simulations are based on three cases, i.e., no waste applied, and 200 and 600 kg N/ha of swine waste applied on May 20, 1973. Waste was assumed to be applied on the surface of the soil. The results show that increased rate of waste application increased the org-N, ammonium N, nitrate N, org-P, soluble PO_4-P, soluble C, and fecal coliforms in the runoff water. These trends are similar to those reported in the literature by several researchers as reviewed by Khaleel et al. (1980).

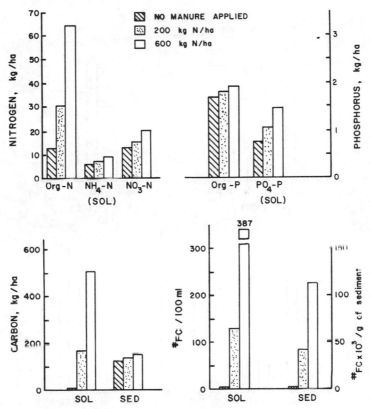

Fig. 6.   Effect of swine waste loading rate on runoff water quality
(May 1973 to April 1974).

Simulated data on the effect of frequency of waste application on
runoff water quality are shown in Fig. 7. Runoff nutrient loadings were
simulated for swine wastes applied at 200 kg N/ha each in May (summer),
August (fall), and December (winter) as compared to one waste loading
rate (600 kg N/ha) applied in summer (May). Nitrogen and carbon concentrations of the runoff water were low in case of treatment receiving
frequent applications of waste, as compared to treatment receiving heavy
applications. Soluble P concentration of the runoff water showed reverse
trend. More fecal coliforms were observed in the treatment receiving
waste applications in summer, fall, and winter compared to one heavy
application in summer. This was primarily due to longer survival periods
of fecal coliforms in winter as compared to summer applications (Fig. 8).

Figure 9 shows the effect of various waste types such as beef,
poultry, and swine on runoff water quality during the period of May
through December 1973. The waste was applied in May 1973 at a rate of
200 kg N/ha. Soluble organic N, ammonium N, and nitrate N transfer into
runoff water were higher for poultry wastes than those for swine and
beef wastes. Soluble carbon losses for poultry wastes were lower than
those for beef and swine wastes, whereas particulate carbon losses for
all three waste types were nearly the same. Fecal coliform transport in
runoff water and in sediment from soils treated with beef and swine
wastes were nearly the same, whereas fecal coliforms transfer from soils
treated with poultry wastes were much lower compared to those for beef

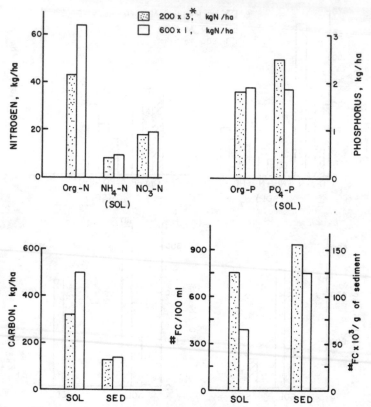

Fig. 7. Effect of frequency of swine waste applications on runoff water quality. Three-time applications of each 200 kg N/ha of swine waste on May 20, Aug. 3, and Dec. 6. One-time application @ 600 kg N/ha on May 20, 1973.

and swine wastes.

The simulated values for nitrogen, phosphorus, carbon and fecal coliforms in runoff water closely agree with some of the literature values on runoff quality as reviewed by Khaleel et al. (1980). Although no experimental verification of the simulated values is available, the revised ARM II model is capable to simulate simultaneously the nitrogen, phosphorus, carbon, and fecal coliforms in runoff water from land areas receiving animal wastes. The model is also applicable to other organic wastes and crop residues.

The results of sensitivity analysis for various reaction rates are shown in Fig. 10. Simulations were run for swine waste applied on May 20, 1973, at a rate of 200 kg N/ha. Following waste application, a highly intense rainfall occurred. The simulation was run for a period of 10 days after waste application. The simulated runoff during this period was 6.98 cm and the total sediment loss was 5309 kg/ha. In Fig. 10, the effects of changes in various reaction rates on total runoff losses during a 10 day period are shown only. The parameter sensitivity lines for peak runoff losses behaved similar to those in Fig. 10; hence they are not included. The reaction rate constant KAM had negligible

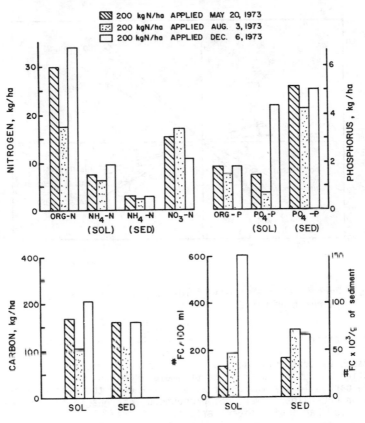

Fig. 8.   Effect of time of swine waste application on runoff water
           quality.

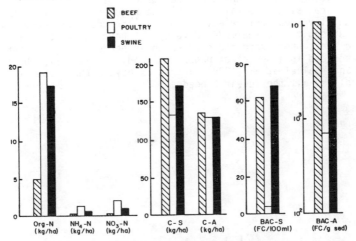

Fig. 9.   Total runoff losses for various waste types and constituents
           during the period (May - Dec., 1973). e.m. = easily
           mineralizable fraction. Loading rate = 200 kg N/ha; time of
           application; May 1973.

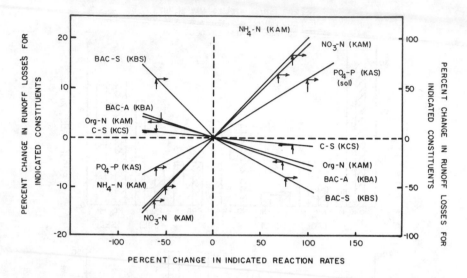

Fig. 10. Percent change in runoff losses, as influenced by the changes in reaction rates.

effect on organic N in runoff water; however, it had a significant effect on ammonium N and nitrate N in runoff water. The exchange rate of phosphorus (desorption), KAS had significant effect on soluble P concentration in runoff water. The parameter KCS has negligible effect on soluble carbon losses in runoff water. Although the effect of changes in KBA on BAC-A losses were not pronounced, changes in KBS on BAC-S in runoff water had a significant effect.

As shown by the simulation results, the ARM-II Animal Waste Model can satisfactorily represent the behavior of pollutants in runoff water. In the future, simulated values should be rigorously tested with experimental values, before any changes in the model are made. The simulated results and the model development have uncovered several specific areas where more field research is needed. These research needs were described in a series of papers by Reddy et al. (1979a,b, 1980, 1981); and Khaleel et al (1980), and it is beyond the scope of this paper to list all the research needs.

Acknowledgements

This research was supported in part by the United States Environmental Protection Agency on grant No. R-805011-0-1-0.

References

Agee, J. L. 1974. The national water quality strategy and the role of agriculture. In: Proc. Workshop on Agricultural Nonpoint Source Water Pollution Control. U. S. E.P.A., Washington, DC. pp. 4-13.

American Society of Civil Engineers. 1977. Quality aspects of agricultural runoff and drainage. ASCE J. of Irrig. and Drainage Div. 103(IR4):475-495.

Crawford, N. H., and R. K. Linsley. 1966. Digital simulation in hydrology: Stanford Watershed Model IV. Dept. of Civil Eng. Stanford Univ., Stanford, CA. Technical Rept. No. 39. 210 p.

Crawford, N. H., and A. S. Donigian, Jr. 1973. Pesticide transport and runoff model for agricultural lands. EPA-660/2-74-013. Environmental Protection Agency, Athens, GA.

Donigian, A. S., and N. H. Crawford, 1976a. Modeling nonpoint pollution from the land surface. EPA-600/3-76-083. Environmental Protection Agency, Athens, GA.

Donigian, A. S., Jr., and N. H. Crawford. 1976b. Modeling pesticides and nutrients on agricultural lands. EPA-600-/2-76-043. Environmental Protection Agency, Athens, GA.

Donigian, A. S., Jr., D. C. Beyerlein, H. H. Davis, Jr., and N. H. Crawford. 1977. Agricultural Runoff Management (ARM) Model Version II: Refinement and testing. EPA-600/3-77-098. Environmental Protection Agency, Athens, GA.

Frere, M. H., C. A. Onstad, and H. N. Holtan. ACTMO - An agricultural chemical transport model. ARS-H-3. Agricultural Research Service, USDA, Hyattsville, MD.

Hydrocomp, Inc. 1976. Hydrocomp simulation programming: Operations Manual. Palo Alto, CA. 2nd Ed.

Khaleel, R., K. R. Reddy, and M. R. Overcash. 1980. Transport of potential pollutants in runoff water from land areas receiving animal wastes: A review - Water Res. 14:421-436.

Overcash, M. R., R. Khaleel, and K. R. Reddy. 1981. Nonpoint source
  model: Watershed inputs from land areas receiving animal
  wastes. EPA - Research Grant No. R 805011-01. Final Rept.

Reddy, K. R., R. Khaleel, M. R. Overcash, and P. W. Westerman. 1979a.
  A nonpoint source model for land areas receiving animal wastes.
  I. Mineralization of organic nitrogen. Trans. ASAE 22:863-873.

Reddy, K. R., R. Khaleel, M. R. Overcash, and P. W. Westerman. 1979b.
  A nonpoint source model for land areas receiving animal wastes.
  II. Ammonia volatilization. Trans. ASAE 22:1398-1405.

Reddy, K. R., R. Khaleel, and M. R. Overcash. 1980. Carbon transformations in the land areas receiving animal wastes in relation to
  nonpoint source pollution: A conceptual model. J. Environ.
  Qual. 9:343-442.

Reddy, K. R., R. Khaleel, and M. R. Overcash. 1981. Behavior and
  transport of microbial pathogens and indicator organisms in soils
  treated with organic wastes. J. Environ. Qual. 10:(in press).

TEMPORAL AND SPATIAL VARIATIONS OF SOLUTE PICKUP DURING RUNOFF GENERATION IN SALINE HILLSLOPES

J. B. LARONNE
DEPARTMENT OF GEOGRAPHY
BEN GURION UNIVERSITY
BEER SHEVA, ISRAEL

H. W. SHEN
DEPARTMENT OF CIVIL ENGINEERING
COLORADO STATE UNIVERSITY
FORT COLLINS, COLORADO 80523

ABSTRACT

Overland and rill flow of constant discharge were generated on highly erodible saline hillslopes of varying inclinations. The runoff contained high concentrations of sediment and solutes with maxima of 64,000 mg/l and 6,900 μmho/cm, respectively. Solute concentration was high in the leading edge of flow due to surface flushing of weathered friable Mancos Shale and soluble minerals contained therein. Solute concentration either decreased after arrival of the leading edge of flow with a later reversal of this temporal trend, or else it increased continuously, reached a peak value and thereafter decreased asymptotically towards a low level.

Solute concentration was determined at 2 to 7 consecutively downslope locations on the studied hillslopes. It increased downslope but temporal trends across the hillslope were identical. Differences of solute pickup across the hillslopes at a given instant of time arose due to spatial differences in the velocities of the leading edges of flow.

Surface flushing and pickup of solutes from underlying less weathered shale is argued to increase with an increase in runoff intensity. One or the other of these two possible high solute pickup phases may dominate depending on antecedent conditions and on total and peak runoff volumes.

INTRODUCTION

Spatial and temporal variations of solute pickup may be determined for individual plots or whole drainage basins by collecting and analyzing runoff samples. Regional and local differences in solute yield may result from climatic and lithologic variations and from the resultant morphologic features, vegetation, and soils. The chemical quality of runoff is known to be a geochemical indicator of rocks and their mineralogic composition (Davis, 1961), and of the change in dominant source during a given runoff event (e.g., Cleaves, Godfrey and Bricker, 1970).

An attempt to describe the temporal and spatial variation of solute pickup is herein presented. To the best of our knowledge this attempt is unique in focusing the attention to the solute producing mechanism proper. Knowledge of the underlying processes of solute pickup is essential to the construction of deterministic water quality models and to the further development of those already in wide use (e.g., Anon, 1970, and Donigian et al., 1977) because they should by definition increasingly rely on correct evaluation of solute producing mechanisms.

Water quality is of particular interest as a pollution indicator where nutrient losses from soils occur, or where agricultural irrigation causes highly saline runoff. It is also an important pollution indicator where herbicides or pesticides are being used, and in situations where discharges of nonbiodegradable materials or heavy metals are possible. Salinity and increased salinity is a problem within the lowlands of the Colorado River Basin, where the major diffuse source of salinity is the Mancos Shale terrain in western Colorado and eastern Utah (Iorns, Hembree and Oakland, 1965). Results presented herein are part of a project undertaken to assess the interaction between sediment and solute transport. This project is one of many recently funded U. S. Government studies (e.g., U. S. Soil Conservation Service, 1975, Laronne and Schumm, 1977, and U. S. Bureau of Reclamation, 1979) endeavoring to solve the water resources and related salinity problems in the arid Southwest.

STUDY AREA AND METHODOLOGY

The study area is located in westernmost central Colorado, north of the
Grand Junction airport and within the Indian Wash and Leach Creek basins (Fig.1).
Nine sites, five of which are referred to in this report, were chosen

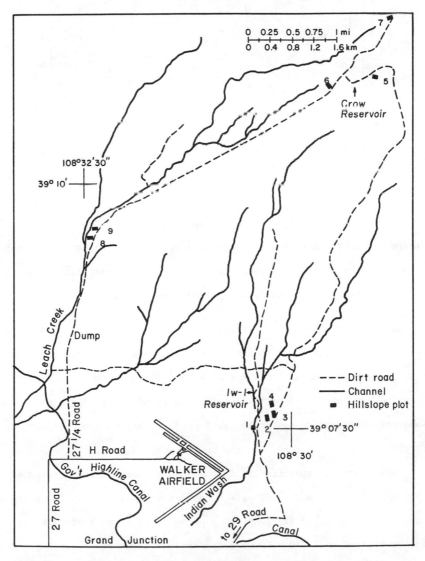

Figure 1.   Map of the Book Cliffs 'desert' north of Grand Junction,
Colorado, showing the location of the study plots.

because they are accessible by a water tank truck, because they were in Mancos Shale terrain, and because they had contours that developed two relatively well defined, though not necessarily rilled, flow conduits near the bottom of the slope. The selected hillslopes ranged in steepness from 16 to 71 percent and were 12.2 - 53.4 m long (Fig. 2).

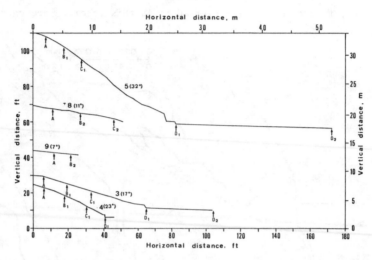

Figure 2.   Vertically unexaggerated long profiles of the hillslopes showing the location of sampling stations.   Note that each of the runoff-generating runs is designated with a symbol (3 - 9).

Knobel et al. (1955) investigated and classified the soils derived from and in association with Mancos Shale. These thin gray silty shale loam soils have a pH of 8.0 and a high salinity. Montmorillonite, Illite, Chlorite, and mica have been identified in these soils. Both fresh and somewhat weathered Mancos Shale swell considerably when wetted with a 25-50 percent volume increase in free swell tests (Schumm, 1964).

Water flow (of specific electrical conductance, herein denoted EC, of 450 μmho/cm @ 25°C and < 100 mg/l suspended solids) was applied using a perforated 3.66 m (13 ft) long PVC pipe and a constant head tank delivering a flow rate of 33.1 l/min (0.0195 ft^3/sec) quite uniformly throughout the length of the pipe. This pipe was situated rather closely to the upper hillslope divide.

Runoff samples were collected 1.52, 4.56 and 9.14 m downslope from the pipe at locations denoted as stations A, B and C, respectively. Samples were also collected at two locations across the hillslope whenever runoff concentrated in more than one clear channel at B_1 and B_2 and at C_1 and C_2. Additionally, samples were collected where the hillslope contacted the channel, stations D_1, and further downstream in the channel proper, stations D_2 (Fig. 2).

The leading edge of flow was sampled with minimum disturbance to the flow by constantly changing the shape of specially prepared aluminum funnels. Broad, flat-shaped funnels were used to collect overland flow and also to decrease sampling time; narrow funnels were used in rills (Shen et al., 1981). The leading edge of flow was collected at each of the stations and sampled again after 2,4,6,8,10,12,15,20,30,45 min and at 15 min intervals thereafter. All odd-numbered samples taken at a particular station underwent immediate EC determination in the field (Lectro Mho Meter, Lab-line Instruments). The precision of the EC meter was 1 percent as compared to a Beckman meter. Sediment concentration was determined by the evaporation method (Vanoni, 1975). Leading edge velocities ranged between 0.9 and 4.6 cm/sec (0.03 and 0.15 ft/sec) and flow depth averaged 1.8 cm with a maximum of 3.8 cm in rills. The flattest slopes produced only overland flow and the steepest only rill flow with a combination of flow types in intermediate slopes. For further discussion of flow type and solute yield refer to Shen et al., 1979.

RESULTS

Figures 3 - 7 show the variation of runoff EC with time from onset of water application at 2 - 7 different stations within each hillslope. Note that sample collection began at each station with the collection of the leading edge of flow and, therefore, the first data point for each station is moved to the right (along the time axis) gradually from station A to station D_2. Flow EC is used to express solute concentration although it is realized that EC depends on indiviual ionic conductances. Stoichiometric concentrations of chemical species in runoff samples (of the Ca, Na (Mg) - SO_4, HCO_3 type) and individual mineral x-ray identification of the soil mantle and of the transported sediment are treated elsewhere (Shen et al., 1981).

The temporal variations in solute load invariably showed distinct patterns. However, the temporal trends in EC cannot be expected to be identical for all plots and studies because each has inherent characteristics of type of encountered flow, soluble mineral content and mineralogy of soluble minerals (Shen et al., 1981). In fact, the trends depicted in Figures 3-7 are as distinct or more so than those presented in other salinity studies (e.g., Ponce and Hawkins, 1978; White, 1977). Solute concentrations were high in the leading edge of flow in comparison with the applied water and at least one maximum occured at the uppermost collection site (stations A). This maximum flattened out with distance downslope. The flattening was especially marked when comparing the EC of the uppermost and lowermost stations. It is best illustrated in Fig. 3, 5 and 6. Note that the flattening of these maxima was sufficiently considerable to cause their complete disappearance at the bottom of the hillslope and

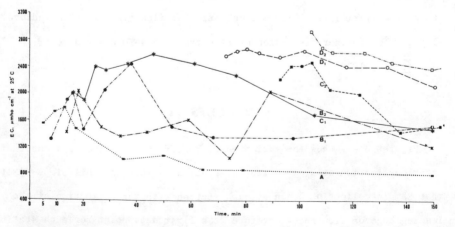

Figure 3. Spatial and temporal variation of runoff conductivity during run 3.

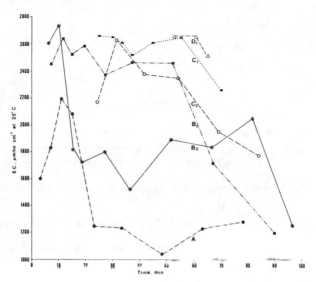

Figure 4. Spatial and temporal variation of runoff conductivity during run 4.

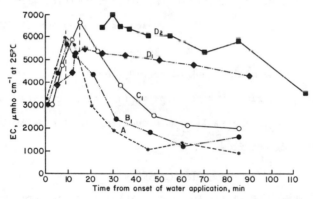

Figure 5. Spatial and temporal variation of runoff conductivity during run 5.

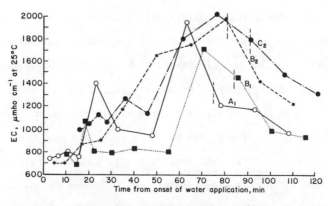

Figure 6. Spatial and temporal variation of runoff conductivity during run 8.

455

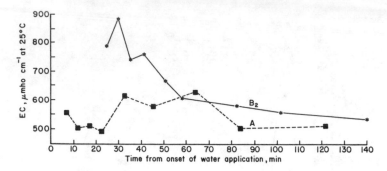

Figure 7. Spatial and temporal variation of runoff conductivity during run 9.

further downstream in the channel. This means that solute concentration was highest in the leading edge of flow at these lower collection sites and consistently decreased with time. Such a decrease is exemplified by the data for stations D_1 and D_2 (Fig.3) and station D_1 (Fig. 4).

After reaching a maximum, EC decreased as a decay function. This occured in all the runs (EC maxima in the range 900 - 6000 μmho/cm) irrespective of the time when the maxima took place (10 - 60 min) after the arrival of the leading edge of flow. Cross slope variations in EC (stations B_1 versus B_2 and C_1 versus C_2) were small when accounting for the delay in arrival of the leading edge of flow.

DISCUSSION

Spatial variations in solute concentration of runoff are typically related in the literature to the soluble mineral content and mineralogy of the dissolving surface (e.g., Miller, 1961) and to biological effects (e.g., Bormann, Likens and Eaton, 1969). Temporal variations may be caused by surface flushing (Pionke and Nicks, 1970), groundwater and inter-flow mixing and straightforward dilution (Lane, 1975). Detailed determination of soluble mineral content of surficial Mancos Shale and associated alluvium in the vicinity of the study area has been described and summarized by Laronne and Schumm (1977). The surface crust of Mancos Shale was shown to be leached and to contain 1/4 - 1/2 as much soluble minerals as underlying layers. It has been demonstrated (Shen et al., 1981) that crust leaching is also typical on the hillslopes reported herein. Kinetics experiments undertaken to determine soluble mineral content showed that this parameter varies within and between hillslope surface materials and does not necessarily increase with hillslope steepness. In fact, both run 5 and run 8 (Fig. 5 and 6) took place on leached crusts but although the crust on the steeper hillslope was 50 percent more saline than the crust developed on the milder hillslope, the latter had underlying material twice as saline as in the steep hillslope (Shen et al., 1981).

Mancos Shale develops a surface of very limited permeability upon impingement of raindrops due to their impact and to the high content of Montmorillonite in the shale; moisture may, however, penetrate deeper into the shale via dessication and freeze-thaw cracks primarily during spring (Schumm, 1964). No deep cracks and very few shallower cracks were present on the hillslopes at the time this study was conducted. Wetting front depths did not exceed 5 cm (2.0 in) in areas where overland flow was generated nor did they exceed 25 cm (9.8 in) in rills. Therefore, no interflow was generated in the runs except some shallow subsurface flow in run 6 (not included in this analysis).

Figures 8 and 9 show the spatial and temporal variations in sediment concentration for runs 5 and 8 (Fig. 5 and 6). The abcissas for comparable figures are different: the first five (Fig. 3 - 7) refer to the time from the moment water was discharged from the pipe (common time), while those for Fig. 8 and 9 refer to the time since the leading edge of flow had reached each station.

Comparison of Fig. 5 and 8 shows that the highest sediment concentrations (33,000 - 64,000 mg/l) occured in the leading edge of flow and that very high solute yields (3,000 - 3,800 µmho/cm and 6,400 µmho/cm at D_2) characterized this initial runoff. High sediment yields in the leading edge of flow of run 8 (Fig. 9), though substantially lower than in run 5, coincided with exceedingly low solute yields. The steeper hillslope generated velocities of flow twice as high as on the flatter hillslope. This difference and

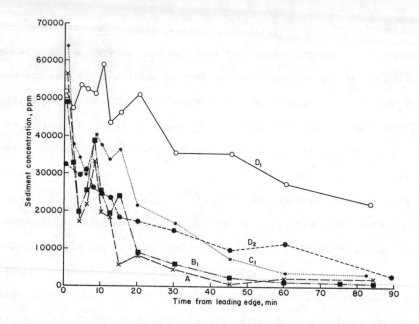

Figure 8. Spatial and temporal variation of sediment concentration during run 5.

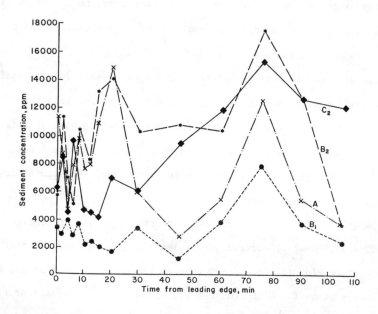

Figure 9. Spatial and temporal variation of sediment concentration during run 8.

the higher soluble mineral content in the crust of the steeper hillslope combined to generate a high sediment and solute surface flushing effect at the beginning of run 5. Lower sediment pickup and very low solute pickup characterized the leading edge of flow during run 8. The greater erodibility of the shallow weathered surface layer of Mancos Shale was, therefore, conducive to the occurrence of high sediment yields where rapid flowing runoff was generated. When the runoff was sufficiently shallow and slow it entrained little sediment, either from the surface layer or from a slightly deeper-lying more erodible layer (compare the maxima at 20 min in Fig. 9 to the compatible maxima in Fig. 6).

Sediment concentration maxima during run 5 occured approximately 10 min after the arrival of the leading edge of flow (Fig. 8) and simultaneously with EC maxima (Fig. 5). This correspondence also occured 75 min after the arrival of the leading edge of flow during run 8 (Fig. 6 and 9). This temporal correspondence between sediment and solute concentrations is causal but cannot be a surface flushing effect, the weathered surface already having been flushed or completely eroded previously. Rather, rills began to develop 48 min after arrival of the leading edge of flow during run 8 and rill entrenchment began decreasing 10 min after the arrival of the leading edges of flow in run 5, rills having been present prior to runoff generation on this latter hillslope. Sediment was continuously and increasingly supplied due to downcutting within the rills during these initial 10 min and more soluble minerals within the transported sediment and within the freshly exposed rill bed were thus dissolving (Laronne and Shen, 1982). A similar increase in sediment supply brought forth a con-comittant but lower increase in solute yield on the flatter hillslope. It should be noted that the bulk of the solute load originated in the rill-entrenching and rill-formation sediment pickup phase during both runs.

The downslope flattening and gradual disappearance of the EC maxima (Fig. 3 - 7) occured due to the increase in sediment concentration downslope (well exemplified in Fig. 8 but not applicable to run 8, Fig. 9) and the resultant and concomittant downslope increase in solute concentration. This downslope increase implies encroachment towards equilibrium arising from smaller ionic concentration gradients between the bulk solution and the dissolving minerals, primarily Gypsum and Na, Mg hydrated sulfates (Laronne and Shen, 1982). Moreover, progressively more of the transported soluble minerals dissolved with distance downslope, leaving less soluble matter for further dissolution while runoff was undersaturated.

The downslope increase in EC is well represented by the data for run 4 (Fig. 4) though with a few data point exceptions, and it is documented in all the runs towards their termination after massive sediment pickup ceased. The apparent anomaly of downslope decrease in EC in several instances, particularly after the arrival of the leading edge of flow, may be due to short-lived pulses of high sediment and/or high solute concentrations at a given station and not downslopewards. Alernatively though unlikely, it may indicate that incongruent dissolution took place with concurrent precipitation. This phenomenon is presently being investigated.

CONCLUSIONS

Temporal variations in solute pickup on saline erodible hillslopes arise from an initial phase of weathered surface layer erosion and concurrent dissolution as a simple flushing effect. A second, later phase of major solute pickup, is produced by rill entrenchment and formation. Low sediment concentrations in surface runoff occur when overland flow (rather than rill flow) predominates (note this type of situation in run 9, Fig. 7) in which case the available power per unit width of overland flow is smaller than in rill flow.

Synthesizing the results presented herein in a magnitude-frequency-duration analysis of flow events provides the following model of solute pickup: The yield of solutes on hillslopes becomes increasingly a down-cutting and rilling-dependent phenomenon rather than a mere surface flushing effect as runoff intensity increases irrespective of the duration and magnitude of preceding overland flow events. Frequent runoff events of low intensity and characterized by overland flow may deplete the surface layer of most of the soluble minerals and may erode most of the friable crust but they will not substantially decrease the salinity hazard from such diffuse sources. Continued weathering of the newly exposed bedrock creates an additional supply of friable material. The less intensely leached bedrock is prone to attack by rill flow and when so attacked will produce very large solute yields.

There is need to compare results from fully simulated runs (sprinkling experiments) to the results presented herein (e.g., Laronne, 1982). Moreover, a quantitative expression relating solute yield produced by the two major solute producing phases to runoff intensity, duration and frequency is yet to be determined.

ACKNOWLEDGEMENTS

We wish to thank the U. S. Department of Interior, Office of Water Resources Technology, for sponsoring this study under grant No. 14-34-001-6063 B-137. J. Hidore reviewed an early draft of this manuscript. The critical comments of the reviewers are greatly appreciated.

REFERENCES

Anon, 1970. DOSAG 1, Simulation of water quality in streams and canals. Texas Water Quality Development Board, Austin.

Bormann, F. H., Likens, G. E. and Eaton, J. S., 1969. Biotic regulation of particulate and solution losses from a forest ecosystem. Bioscience, Vol. 19, pp. 600-610.

Cleaves, E. T., Godfrey, A. E. and Bricker, O. P., 1970. Geochemical balance of a small watershed and its geomorphic implications. Geol. Soc. of Am. Bull., Vol. 81, pp. 3015-3032.

Davis, G. H., 1961. Geologic control of mineral composition of stream waters of the eastern slope of the Southern Coast Ranges, California. U. S. Geol. Survey Water Supply Paper 1535-B, B-1 to B-30.

Donigian, A. S., Jr., Beyerlein, D. C., Davis, H. A. Jr., and Crawford, N. H., 1977. Agricultural runoff management (ARM) model. Version II: Refinement of research and development. U. S. Environmental Protection Agency, EPA 600/3 - 77 - 098, 293p.

Iorns, W. V., Hembree, C. H. and Oakland, G. L., 1965. Water Resources of the Upper Colorado River Basin -- Technical report. U. S. Geological Survey Professional Paper 441, 370 p., and Paper 442, 1036 p.

Knobel, E. V., Dandsdill, R. K., and Richardson, M. L., 1955. Soil Survey of the Grand Junction area, Colorado. U. S. Dept. Agr., Soil Survey Series 1940, No. 19, 118p.

Lane, W. L., 1975. Extraction of information on inorganic water quality. Colorado State University Hydrology Paper 73, 73p.

Laronne, J. B., 1982. Sediment and solute yield from Mancos Shale Hillslopes, Colorado and Utah. in Yair, A., and Bryan, R. B. (eds.), Badland Geomorphology and Pipe Erosion, Geoabstracts, Norwich.

Laronne, J. B. and Schumm, S. A., 1977. Evaluation of the storage of diffuse sources of salinity in the Upper Colorado River Basin. Colorado State University Environmental Resources Center Completion Report No. 79, 111p.

Laronne, J. B. and Shen H. W., 1982. The effect of erosion on solute pickup from Mancos Shale Hillslopes, Colorado. Submitted for publication.

Miller, J. P., 1961. Solutes in small streams draining single rock types, Sangre de Cristo Range, New Mexico. U. S. Geological Survey Water Supply Paper 1535-F, F-1 to F-23.

Pionke, H. B. and Nicks, A. D., 1970. The effect of selected hydrologic variables in stream salinity. International Association of Scientific Hydrology Bulletin, Vol. 15 (4), pp. 13-21.

Ponce, S. L. and Hawkins, R. H., 1978. Salt pickup by overland flow in the Price River Basin. Water Resources Bulletin, 14 (5), pp. 1187-1200.

Schumm, S. A., 1964. Seasonal variations of erosion rates and processes on hillslopes in Western Colorado. Zeitschrift für Geomorphologie, Vol. 5, pp. 215-218.

Shen, H. W., Enck, E. D., Sunday, G. K. and Laronne, J. B., 1979. Salt loading from hillslopes. International Association of Hydraulic Research 18th Congress: Hydraulic engineering in water resources development and planning, Cagliari. Proceedings, Vol. 5, pp. 99-105.

Shen, H. W., Laronne, J. B., Enck, E. D., Sunday, G. K., Tanji, K. K., Whittig, L. D. and Biggan, J. W., 1981. The role of sediment in non-point source salt loading within the Upper Colorado River Basin. Colorado State University Environmental Resources Center Completion Report, 213 p.

U. S. Bureau of Reclamation, 1979. Quality of water, Colorado River Basin. U. S. Bureau of Reclamation Rpt. 9, 206p.

U. S. Soil Conservation Service, 1975. Erosion, sediment and related salt problems and treatment opportunities. U. S. Soil Conservation Service, Special Projects Div., Golden, Colo., unnumbered report, 152 p.

Vanoni, V. A., (ed.), 1975. Sedimentation Engineering. Prepared by the ASCE Hydraulics Division, Task Sedimentation Committee. American Society of Civil Engineers, New York, 745 p.

White, R. B., 1977. Salt production from micro-channels in the Price River Basin, Utah. Unpubl. M.Sc., Dept. of Civil and Environmental Engineering, Utah State Univ., Logan, 121p.

MODELING THE RELEASE OF PHOSPHORUS AND RELATED ADSORBED CHEMICALS FROM SOIL TO OVERLAND FLOW

L. R. Ahuja, A. N. Sharpley, R. G. Menzel, and S. J. Smith
Soil Scientists, USDA-ARS, Southern Plains Watershed and
Water Quality Laboratory, Durant, OK 74701

ABSTRACT

This paper deals with two major problems that have been encountered in modeling the release of soil chemicals to runoff: (1) The extent and dynamics of the zone of rainfall-runoff-soil interaction, and the (2) the partitioning mechanism between solution and solid phases of an adsorbed chemical (P, in this case). Current state of the art and new information available on these processes, especially some recent work done in our laboratory, are reviewed and analyzed. A nonlinear diffusion model of P release from soil to runoff, based on the new information, is tested on some experimental data obtained under controlled conditions, and compared with the concepts used in two leading current models -- the ARM and the CREAMS models. The latter two models used the assumption of first-order kinetics and an empirical relationship, respectively, for P desorption combined with a fitted value for the zone of interaction to compute the P released to runoff. The diffusion model gave closer predictions of the soluble P concentration of runoff than the ARM and CREAMS models. The diffusion model can account for the effect of rainfall intensity and overland flow depth (which determine the water to soil ratio) on P release. For application under field situations, where information on the physical parameters is not currently available, some conceptual simplifications are made. The simpler model is then applied to experimental data from two watersheds, to evaluate the relative importance of the major factors involved. The important factors in determining the mean solution P concentration of a runoff event will be storm size, the portion of storm that infiltrates into the soil before the runoff begins, and initial amount of desorbable P in soil. Finally, applicability of the diffusion model to the release of pesticides from soil to runoff is demonstrated.

INTRODUCTION

The transport of phosphorus (P) in runoff from agricultural land is commonly regarded as one of the major factors in accelerating the biological productivity of natural waters (Vollenweider, 1968; Loehr, 1974; Schindler, 1977). This transport has been a subject of several recent and ongoing studies (Taylor et al., 1971; Kunishi et al., 1972; Schuman et al., 1973; Gburek and Heald, 1974; Duffy et al., 1978; Menzel et al., 1978; Sharpley and Syers, 1979). Other studies have specifically evaluated the effect of different agricultural practices on soil, water and nutrient runoff (Romkens et al., 1973; Burwell

et al., 1975; Johnson et al., 1979; Wendt and Corey, 1980). Although
the major proportion of P transported in runoff is usually attached to
eroded sediment, the solution component can be appreciable under certain conditions of soils, times and methods of fertilizer application,
and tillage systems (Schuman et al., 1973; Romkens and Nelson, 1974;
Baker and Laflen, 1979).

Increasing efforts are being directed towards mathematically modeling the processes of chemical transport from land surface, in conjunction with existing models for water and sediment transport (Crawford
and Donigian, 1973; Bruce et al., 1975; Frere et al., 1975; Donigian and
Crawford, 1976; Donigian et al., 1977; Williams and Hann, 1978;
Knisel, 1980). Two recently developed models, which include P and
other chemicals, are the "ARM" model, developed for the U.S. Environmental Protection Agency by Hydrocomp, Inc. (Donigian et al., 1977),
and the "CREAMS" model, developed by a group of scientists in U.S.
Department of Agriculture (Knisel, 1980). These models incorporate
physically-based descriptions of the various processes where possible.
However, due to the absence of needed information, some of the mechanisms are oversimplified. Limited field tests of both the ARM and
CREAMS models have shown that the prediction of both solid- and
solution-phase chemicals is not satisfactory (Donigian et al., 1977;
Davis and Donigian, 1979; Frere et al., 1980; Leonard and Wauchope,
1980). These studies indicated a need to further investigate the
extent and dynamics of the mass of surface soil that interacts with
rainfall and runoff, and the nature of the partitioning mechanisms
between solid- and liquid-phase portions of an adsorbed chemical.

The purpose of this study is to carefully examine the above
processes with a view to improving the models. We will review and
analyze the current state of the art and the new information available on these processes, especially some recent work done in our
laboratory. A physical model based on this information will then be
tested and compared with the concepts of ARM and CREAMS models on some
experimental data obtained under controlled conditions. We will also
attempt to make some conceptual simplifications in the physical model
for application to field situations, where information on physical
parameters is not currently available. Finally, we will examine the
applicability of the physical P model for describing the release of
some pesticides to runoff water.

THE ZONE OF RAINFALL-RUNOFF-SOIL INTERACTION

The the ARM and CREAMS models assume that rainfall and runoff
interact with a thin zone of surface soil in removing the soluble
material present in the zone. In the ARM model, the rainfall is
assumed to mix completely and uniformly with the soil in the zone.
The amount of a soluble chemical in the zone of interaction is
sequentially apportioned among infiltration, runoff, and water retained
in the soil, in proportion to the amounts of water in each mode. The
thickness of this zone is obtained by model calibration of the experimental data. Values ranging between 1.6 and 6.0 mm have been used,
with the smaller value giving better results in some field testing
(Donigian et al., 1977).

In the CREAMS model, the concept of the interacting zone is
essentially the same as in the ARM model. The zone depth is fixed
at 1.0 cm, but it is assumed that only a fraction of the chemical
present in this zone interacts with the rainfall water. This fraction, termed as an extraction coefficient, is obtained by model cali-

bration. Different coefficients are provided for the infiltration
and runoff loss components (Frere et al., 1980). Limited field tests
of the CREAMS model, as well as of the ARM model, have indicated that
the simulation of P transport in runoff water was poor. For examining
the chemical release mechanisms, the extent, nature, and dynamics of
the interacting zone needed to be understood first.

In order to determine the properties of the interacting zone more
directly, ^{32}P was used as a tracer in packed soil boxes, 100 x 30 x
15 cm dimensions, under a simulated rainfall (Ahuja et al., 1981a).
A known amount of the tracer was placed at the soil surface and at
successive 0.5 cm intervals in separate soil boxes, with permeable
bottoms. A certain amount of ordinary P was mixed uniformly with all
the soil in the box. The concentration of ^{32}P in runoff from the
boxes was measured several times during two 30-min simulated rainfalls. Results for one of three soils investigated, Ruston fine sandy
loam, are shown in Fig. 1. The data points are values adjusted to a
constant amount of ^{32}P initially present at each depth, and are from

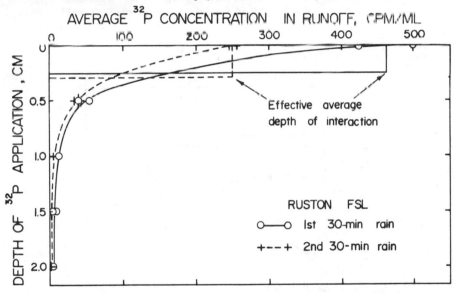

Fig. 1. Average ^{32}P concentrations in runoff as a function of the
depth of ^{32}P placement for two 30-min rain periods on Ruston
soil. (Made from data of Ahuja et al., 1981a)

two replicate boxes, at 4% slope and under 6.5 cm/hr rainfall intensity. The degree of interaction between soil and rainfall-runoff
water was maximum at the soil surface, and decreased very rapidly with
depth. Results for two other soils, Bernow fine sandy loam and
Houston Black clay, were similar. Thus, a uniform zone of interaction
cannot be assumed as in the ARM and CREAMS models.

For practical purposes, we assumed an effective average depth of
interaction (EDI) as the thickness of surface soil in which the degree
of interaction is uniform and equal to that at the soil surface, and
such that the total interaction is equal to that of the real case.
The EDI was obtained by dividing the area under the curve of ^{32}P concentration in runoff versus the depth of ^{32}P placement, by the value
of ^{32}P concentration in runoff for the 0.0 cm (soil surface) place-

ment (Fig. 1). For each rainfall event, the area under the curve is
equal to that of the rectangle shown in Fig. 1. The EDI, thus calculated, for the three soils was between 2.0 and 3.0 mm. The EDI for
the second 30-min rainfall event was slightly greater than that for
the first event (Ahuja et al., 1981a). In order to test the validity
of the EDI hypothesis, we used these values to predict the concentration of ordinary P in runoff during a rainstorm. For this purpose the
kinetics of ^{32}P release to runoff, where ^{32}P was applied at the soil
surface, was used in conjunction with the EDI for each case. The predicted P concentrations of runoff agreed quite well with the experimental values in all cases (Ahuja et al., 1981a). The assumption of
an effective depth of uniform interaction was, thus, shown to be
workable for practical purposes.

Transient changes in EDI during a storm were obtained by comparing
the fractions of the surface-applied ^{32}P appearing in runoff at
different times during a rainfall event, with the amounts of uniformly-
mixed P appearing at the same time. The values, thus determined, for
the three soil types are plotted in Fig. 2. The 30-min rainfall events
shown were separated by an interval of 2 to 3 hours. Although the
EDI generally increased with duration of rainfall, due probably to
increased rilling, similar EDI values were obtained for the three
different soil types (Fig. 2).

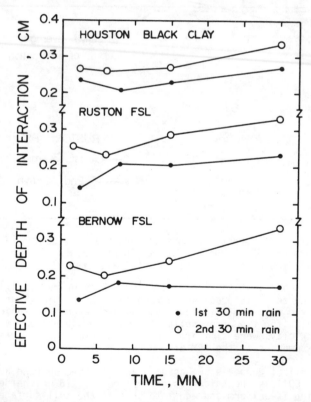

Fig. 2. Changes in the effective depth of interaction during the two
30-min rainstorms. (The graphs are made from numerical data
of Ahuja et al., 1981a)

FACTORS INFLUENCING THE EDI

Ingram and Woolhiser (1980) showed that the release of calcium sulfate to runoff was influenced by soil slope and rainfall kinetic energy. The effect of these factors on the release of soil P to run-off was investigated by Ahuja et al., (1981) using simulated rainfall. The soluble P concentration increased approximately linearly with an increase in soil slope. Between 4 and 16% slopes, the rate of increase was nearly 0.02 mg/l per degree of slope. With an increase in the relative kinetic energy of the raindrops, an increase in soluble P concentration of runoff was observed (Fig. 3). The relationship between kinetic energy and P concentration was approximately linear, with

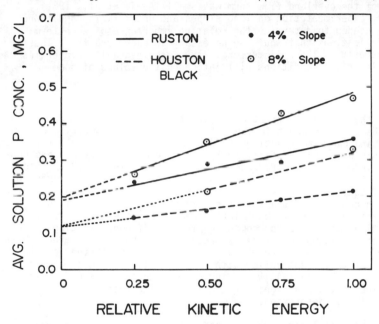

Fig. 3. Average solution P concentration in runoff as a function of relative kinetic energy of the simulated rainfall for Ruston and Houston Black soils at 4 and 8% slopes. The average values are means of two 30-min rainfalls. (Ahuja et al., 1981)

the rate of increase in P concentration slightly greater at the higher soil slope. The relative kinetic energy of the simulated rainfall was varied by covering the soil surface with screens of different mesh sizes. Approximate quantitative ratings of the screens were determined from a separate measurement of water, sediment, and P in splash from the soil boxes (Ahuja et al., 1981). The increase in P release from the soil with an increase in both soil slope and raindrop kinetic energy, was attributed to an increase in the effective depth of interaction.

As a first approximation, we assume that the EDI depends upon the soil slope, rainfall, and runoff characteristics in a manner similar to the influence of these factors on potential erodibility of a soil. Our experimental results are then explained by the following conceptual model, which is analogous to the physically-based soil loss model of Foster et al., (1977):

$$EDI = (aS + b) E + cL^m S^e Q^n \qquad (1)$$

where S is the slope (%), E is the kinetic energy of the rainfall per unit area per unit time, Q is the runoff rate, L is the slope length, and a, b, c, e, m, and n are constants for a given soil. The first term on the right hand side of Eq. (1) represents the effect of raindrop impact and the second term is the effect of overland flow. For our experiments with short slope lengths of 33, 66, and 100 cm, the effect of the overland flow term was small (Ahuja et al., 1981). However, experimentation on long slopes and under field conditions will be needed to determine the relative importance of each term, the modifications that would be necessary in each, and the values of the parameters. The soil texture, structure, hydraulic properties, and soil surface conditions will influence the values of some parameters.

KINETICS OF P DESORPTION

On the ARM model, first-order kinetics is assumed to operate between the adsorbed and solution P forms in soil. Computation of the forward and backward reactions is done numerically for small time steps (Donigian and Crawford, 1976). The reaction constants are considered independent of the water to soil ratio. In the CREAMS model, the kinetics of P release is not invoked, per se. It is assumed that the soil contains a certain amount of readily soluble P, and that the buffering capacity of the soil for P maintains a certain fixed concentration in runoff, percolating water, and the soil water. Mathematical equations for computing the runoff concentrations based on the above assumptions will be presented in the next section. Here, we briefly describe a theoretical and an experimental study of the kinetics of soil P desorption with water for five different soils (Sharpley et al., 1981a), which indicated a departure from the first-order kinetics, and a significant effect of the water to soil ratio on the kinetics of desorption.

A review of literature (Kuo and Lotse, 1974; Evans and Jurinak, 1976) indicated that the desorption of soil P in the short term is a low-activation energy, diffusion-controlled process, which is non-linear (i.e., a case of second or higher-order kinetics). To describe the rate of desorption we assumed the following equation:

$$dP_d/dt = A \cdot a\left(\frac{P_o - P_d}{P_o}\right)^m \cdot \left[\frac{P_o - P_d}{V} - \frac{P_d}{W}\right] \qquad (2)$$

where P_d is the amount (g) of P desorbed per gram of soil at any time, t, dP_d/dt is the rate of desorption, A is the specific surface area of soil (cm/g), P_o is the initial amount of desorbable P, extractable by water, present per gram of soil, the expression $a[(P_o - P_d)/P_o]^m$, with a and m parameters, is the variable diffusivity function (cm/sec), V is the volume of the diffusion layer (cm^3/g) on the surface and/or inside the particle that is envisioned to contain the P on the soil, and W is the water:soil ratio (cm^3/g). The definition of V is rather abstract, in the present analysis it is considered as a constant parameter with units of volume (cm^3). The term within the brackets in Eq. (2) represents the P concentration gradient between soil and water.

The diffusivity has been taken as a power function of the fraction of the initial P remaining on the soil. As this fraction decreases from the initial value of 1, the diffusivity decreases. The diffusivity is, thus, assumed to be dependent on the distance from equilibrium or the final state. The specified function will allow the relationship to be linear or nonlinear, as determined by the value of m. An important assumption is that the initial desorbable P, P_0, is distributed among the various sites on the surface and inside the soil particle in a certain proportion, at any give level of P_0. This leads to the diffusivity being a function of the fraction $(P_0-P_d)/P_0$, rather than of the absolute amount desorbable (P_0-P_d). Our experimental results reported earlier (Sharpley et al., 1981a) showed the validity of the above assumption for diffusivity. The desorption rate was not a unique function of (P_0-P_d), the amount remaining to be desorbed at anytime, for different levels of P_0 in a given soil, but did depend on the term $(P_0-P_d)/P_0$.

For very large water:soil ratio or an infinite sink, Eq. (2) was integrated to obtain:

$$P_d/P_0 = 1 - (1 + mBt)^{-1/m} \qquad (3)$$

where B is a constant $= Aa/V$. It can be shown by graphing that for short to intermediate time periods, Eq. (3) can be approximated by a simplified expression:

$$P_d/P_0 = Ct^n \qquad (4)$$

where C and n are constants.

For small-to-intermediate water:soil ratios, or a finite sink, a closed form solution of Eq. (2) can be obtained for integer values of m. Within a limited range of W values the solution can be simplified as:

$$P_d/P_0 = Kt^\alpha W^\beta \qquad (5)$$

where K, α, and β are constants for a given soil.

We tested the model of Eq. (5) on laboratory batch desorption data for five different soils, at different rates of applied fertilizer P, for times up to 180 min (relevant to runoff situations), and for water:soil ratios (W) ranging from 10:1 to 1000:1 (Sharpley et al., 1981a). Detailed results for two soils, not presented before, are given in Fig. 4. The value of constants for selected cases are shown in Tables 1 and 2. Equation (5) adequately described the desorption of P from surface soil, for times and water:soil ratios relevant to the runoff-surface soil interaction. The power-form model of Eq. (5) was consistent with the experimental results on P desorption reported in the literature (Amer et al., 1955; Romkens and Nelson, 1974; Kuo and Lotse, 1974; Evans and Jurinak, 1976; Barrow, 1979).

The simplified desorption equations (4) and (5) may also be useful for describing desorption of some pesticides. A review of literature

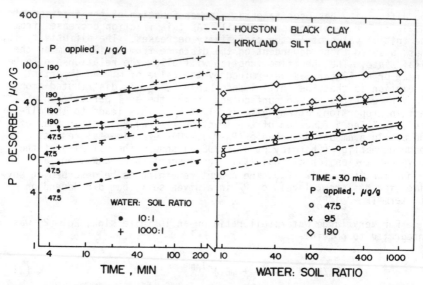

Fig. 4. Logarithm of P desorbed as a function of time at a given
water:soil ratio, and as a function of logarithm of water:
soil ratio at a given time, for two different soils.

Table 1. Slope values for the regression of logarithm P desorbed against
logarithm time at several water:soil ratios (α) and against
logarithm water:soil ratio at several contact times (β).
(Extracted from the data of Sharpley et al., 1981a).

Soil	P added (μg/g)	$\alpha^\dagger$ Water:soil ratio		$\beta^\dagger$ Contact time(min)	
		10:1	1000:1	5	180
Bernow fine	0	0.111^a	0.134^a	0.194^e	0.150^f
sandy loam	47.5	0.124^a	0.127^a	0.144^f	0.145^f
	95	0.122^a	0.126^a	0.146^f	0.162^f
	190	0.121^a	0.117^a	0.148^f	0.145^f
Houston	0	0.101^a	0.159^b	0.405^g	0.432^g
Black clay	47.5	0.259^c	0.247^c	0.216^e	0.211^e
	95	0.297^c	0.253^c	0.216^e	0.205^e
	190	0.174^b	0.193^b	0.200^e	0.184^e
Kirkland	0	0.096^a	0.084^a	0.261^e	0.207^e
silt loam	47.5	0.086^a	0.088^a	0.181^f	0.182^e
	95	0.085^a	0.126^a	0.139^f	0.166^f
	190	0.111^a	0.146^a	0.122^f	0.149^f
Pullman	0	0.036^d	0.108^a	0.373^g	0.393^g
silty clay loam	95	0.049^d	0.076^a	0.189^e	0.199^e
Woodward	0	0.156^b	0.090^a	0.439^g	0.415^g
loam	95	0.083^a	0.101^a	0.187^e	0.195^e

†All relationships are significant at the 0.1% level. The values of
α and β followed by the same letters are not significantly different
at the 5% level.

470

Table 2. Values of constant K, calculated from the slope of the relationship between P desorbed and initial desorbable P. (Extracted from the data of Sharpley et al., 1981a).

| | Contact time (min) | K^{+} Water:soil: ratio | | |
		10:1	100:1	1000:1
Bernow	5	0.148[a]	0.151[a]	0.161[a]
	30	0.135[a]	0.147[a]	0.141[a]
	180	0.138[a]	0.142[a]	0.135[a]
Houston Black	5	0.062[b]	0.064[b]	0.058[b]
	30	0.060[b]	0.061[b]	0.051[b]
	180	0.058[b]	0.052[b]	0.058[b]
Kirkland	5	0.273[c]	0.304[c]	0.288[c]
	30	0.280[c]	0.291[c]	0.276[c]
	180	0.265[c]	0.284[c]	0.292[c]

+Relationships are significant at the 0.1% levels. The values of α and β followed by the same letters are not significantly different at the 5% level.

indicated that diffusion is generally the rate limiting factor for adsorption and desorption of several weakly-to-moderately adsorbed pesticides in soils (Haque and Sexton, 1968; Haque et al., 1968; Lindstrom et al., 1970; Leenheer and Ahlrichs, 1971). Haque and Sexton (1968) found that their experimental data for net adsorption of 2, 4-D on several different soil minerals, under conditions of simultaneous opposing reactions of adsorption and desorption, were quite well described by the following equation of Faya and Eyring (1956):

$$d\phi/dt = 2K'(1-\phi) \text{ Sinh } b(1-\phi) \tag{6}$$

where ϕ is the fraction of the total adsorbable amount that has been adsorbed at any t, and K' and b are constants. Faya and Eyring (1956) showed earlier that the above equation described both adsorption and desorption of a detergent on cotton. Equation (6) is very similar in form to Eq. (2). Both these equations are based on the concept that the rate of desorption depends upon the distance from equilibrium or the final state. The hyperbolic sine function is similar in shape to a power function. We found that our Eq. (5) was consistent with the experimental data on net cumulative adsorption with time for 2, 4-D, isocil, or bromocil on different mineral surfaces (Haque and Sexton, 1968; Lindstrom et al., 1970), and for carbaryl and parathion on organic matter (Leenheer and Ahlrichs, 1971).

COMPARISONS OF ARM, CREAMS, AND DIFFUSION MODELS OF P RELEASE TO RUNOFF

In the ARM model, transformations between adsorbed and solution P are assumed to obey a first-order kinetics (Donigian and Crawford, 1976):

$$dP_d/dt = K_d [P_s] - K_s [P_d] \tag{7}$$

where P_d is the desorbed or solution P (g/g), P_s is the adsorbed P (g/g), $[P_s]$ and $[P_d]$ are the respective concentrations, K_d is the first-order rate constant for the forward reaction (desorption), and K_s is the rate constant for the backward reaction. During a rainfall event, the desorbed P will be largely removed from the site of desorption with the percolating and runoff waters, and its concentration greatly diluted. We may, thus, assume for simplicity, that the backward reaction is small and negligible. If the resulting equation is integrated and combined with the condition of mass balance during a rainstorm, of constant rainfall rate, we obtain:

$$RC_w = -EDI \cdot \Theta_s \cdot dC_w/dt + EDI \cdot BD \cdot K_d P_o \ \exp(-K_d t) \qquad (8)$$

where R is the rainfall rate, C_w is the concentration of P in runoff on soil water, Θ_s is the volumetric soil moisture content at saturation, BD is bulk density of the surface soil layer, and P_o is the initial desorable P. It is assumed that the water percolating through the EDI has the same concentration as the runoff water. The first term on the right-hand side of Eq. (8) will be generally much smaller than the second term. We may, therefore, write Eq. (8) as:

$$C_w = EDI \cdot BD \cdot K_d P_o \exp(-K_d t)/R \qquad (9)$$

If we used the power-form Eq. (5), instead of the first-order kinetic equation, we can derive:

$$C_w = EDI \cdot BD \cdot KP_o \ t^{\alpha - 1} W^\beta/R \qquad (10)$$

This equation incorporates the higher-order kinetics of P desorption and the dependence of desorption on water to soil ratio. Desorption under a rainfall is assumed to occur at a certain effective water:soil ratio.

The equation of P concentration in runoff as used in the CREAMS model (Frere et al., 1980), modified for a constant rainfall rate is:

$$C_w = (C_o - C_b)\exp(-R[K_1 t_1 - K_2 t_1] - RK_2 t) + C_b \qquad (11)$$

where C_o is the initial concentration of soluble P in soil water, C_b is the base concentration maintained by buffering action of the adsorbed soil P, K_1 is the extraction coefficient for water percolating below the zone of interaction, K_2 is the extraction coefficient for runoff, and t_1 is the time when the infiltration ends and the runoff begins. This empirical model resembles Eq. (9) based on the first-order kinetics, but involves different assumptions. It is assumed that the contribution of adsorbed soil P is only by way of maintaining a certain base concentration, C_b. All the infiltration of water is assumed to occur before the runoff begins, and all the rainfall is converted to runoff thereafter. Depth of interaction is implicitly taken equal to 1.0 cm, but empirical extraction coefficients are introduced to adjust this value. For the purpose of our testing, Eq.

(11) may be written as:

$$(C_w - C_b) = G \; \exp(-RK_2 t) \qquad (12)$$

where G is a lumped constant, equal to $(C_0 - C_b) \; \exp(-R[K_1 t_1 - K_2 t_1])$, for a given case.

The applicability of Eq. (9), (10), and (12) is tested on the experimental data for soil P release to runoff, for five different soil types, reported earlier (Sharpley et al., 1981). These experiments were conducted in packed soil boxes, under simulated rainfall, as described already in connection with the ^{32}P tracer studies. For each soil, several levels of applied P were studied. Runoff from each treatment was monitored for several successive 30-min rainfall events with an interval of 3 to 24 hours. In order to test Eq. (9) for the different P_0 levels, resulting from the different applied P levels and from the decrease of soil P in the successive rainfall events, we first normalized the runoff concentrations C_w with respect to the initial desorbable P, P_0. The value of P_0 was determined on a sample of surface soil taken before each rainfall event. Two 1-hour water extractions at a water:soil ratio of 40:1 were used to estimate P_0 (Sharpley et al., 1981). The log of (C_w/P_0) was then plotted against time, t, for different P_0 levels (Fig. 5) and for successive runoff events (Fig. 6). The figures indicate

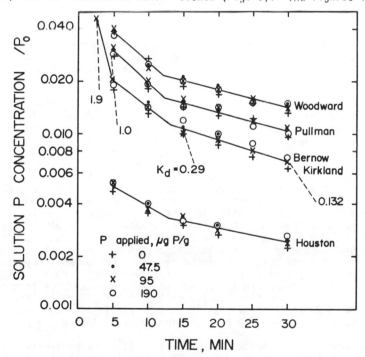

Fig. 5.   Semilog plots of the normalized solution P concentration of runoff as a function of time during the first 30-min rainfall event from five different soils, at four different levels of applied P, as a test of the ARM model. (Experimental data of Sharpley et al., 1981)

473

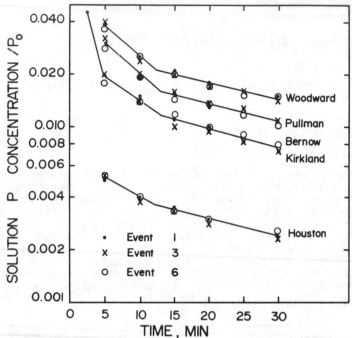

Fig. 6. Semilog plots of the normalized solution P concentration of runoff as a function of time during first, third, and sixth rainfall events from five soils, as a test of the equation for ARM model. (Experimental data of Sharpley et al., 1981)

that the normalized data points for different P_O levels, which occurred either as a result of different rates of applied P or successive desorption of P, coalesce together fairly well. Thus, the transformations occurring before and in between the rainfall events influence the parameter P_O of the model. Consequently, P_O can serve a characterization parameter for field conditions. However, the data points in Fig. 5 and 6 do not fall along a straight line as required by Eq. (9). Most of the data presented above were for times between 5 and 30 minutes after the start of simulated rainfall on pre-wetted soil boxes. For one of the soils, we had a measurement of P concentration for the first minute of runoff. Adding that first point shows more clearly the curvature of the plots (Fig. 5 and 6). The curvature indicates that first-order kinetics do not apply, as K_d is not a constant. For illustration, we determined the values K_d at several different points along the curve for Bernow soil, by fitting Eq. (9) exactly at a given point. The EDI determined independently from the ^{32}P study, discussed in an earlier section, was used for this purpose. The K_d decreased from a value of 1.9 at 2.5 min during the rainfall event to 0.132 at 30 minutes (Fig. 5). Within a narrow range of K_d values, the P release may be approximated as a first-order process.

Equation (10) was tested by replotting all the experimental data, presented in Fig. 5 and 6, as log (C_W/P_O) versus log (time) (Fig. 7 and 8, respectively). The relationships are now linear, in accordance with Eq. (10). Solid lines in the plots are the predicted values of C_W/P_O, obtained by using the independently determined values of the parameters in Eq. (10). The EDI values were from the ^{32}P study dis-

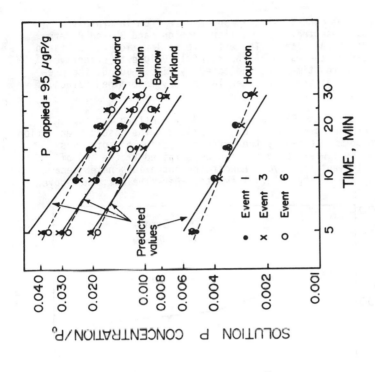

Fig. 7.

Log-log plots of the data shown in Fig. 5, as a test of the kinetic P release model. The solid lines represent the values predicted by the kinetic equation, using independently measured parameters.

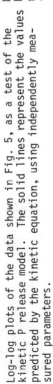

Fig. 8.

Log-log plots of the data shown in Fig. 6, as a test of the kinetic P release model. The solid lines represent the relationships predicted by the kinetic equation, using independently measured parameters.

cussed earlier. The constants α, β, and K were from laboratory kinetic studies, presented in an earlier section, and BD and R were measured for individual cases. The water to soil ratio, W, was taken equal to R/(EDI·BD), the ratio of the rainfall volume to effective soil mass at any given time. As indicated by Fig. 7 and 8, the predictions were satisfactory, even when laboratory determined values of α, β, and K are used.

Equation (12) of the CREAMS was tested by plotting log (C_W-C_b) versus time, for some selected cases, in Fig. 9. The base concentration C_b, assumed to be maintained by the buffering action of soil, was taken equal to the lowest concentration we obtained at the end of the 6th rainfall event. However, runoff concentrations never became constant during any of the rainfall events. Nevertheless, the above

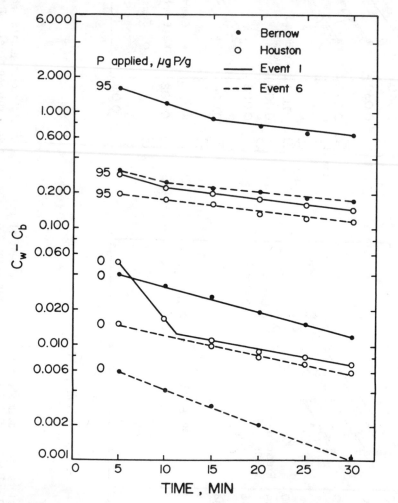

Fig. 9. Semilog plots of the reduced solution P concentration of runoff, $C_w - C_b$, as a function of time during the first and sixth 30-min rainfall events from two different soils, at two applied P levels, as a test of the equation for CREAMS model.

C_b values did appear to make the semilog plots of C_w versus time more linear, especially for soils of low P_o values.

With an increase in rainfall intensity from 6 to 12 cm/hr and consequent increase in W, a decrease in the concentration of soluble P of runoff was measured (Sharpley et al., 1981). Doubling the rainfall rate R in Eq. (10) will change the term W^β/R by a factor of $2^\beta/2$. Runoff concentrations for the 12.0 cm/hr intensity predicted from those of 6.0 cm/hr intensity, by using the above factor, were compared with the actual data (Sharpley et al., 1981). A close agreement was obtained. Thus, it is possible to account for the effect of rainfall intensity on soluble P concentrations in runoff, as described by the kinetic model, in terms of W.

Equation (10) describes the change in P concentration of runoff with time during a rainstorm. For general practical applications, however, the average concentration of runoff for a given rainstorm may be sufficient. For this purpose, Eq. (10) can be integrated, and the time variable may be replaced by ratio of storm size or rainfall volume, V, to average intensity, $I(t = V/I)$. The storm-average P concentration in runoff, $\overline{C}$, is then approximated as (Ahuja et al., 1981):

$$\overline{C} = KP_o(EDI \cdot BD) \left(\frac{I + 0.5LQ}{EDI \cdot BD}\right)^\beta \frac{V^\alpha - V_1^\alpha}{I^\alpha (V-V_1)} \qquad (13)$$

where the fourth term on the right-hand side is an approximation for W^β term, taking into account both the rainfall and runoff waters that interact with soil, and V_1 is the volume of rainfall that infiltrates into the soil before the runoff begins. Equation (13) is very useful in that it specifies the effects of rainfall intensity as well as rainfall volume. A set of field data was not available to rigorously test Eq. (13) and evaluate the relative importance of the major factors involved under field conditions. For a given watershed and soil factors (including P_o), and for rainfall intensity within an order of magnitude, the dominant factor affecting $\overline{C}$ will be the size of the storm, V, and its portion V_1 that infiltrates into the soil before runoff begins. If V_1 is neglected, $\overline{C}$ will be inversely proportional to $V^{1-\alpha}$. For two of our experimental watersheds we plotted $\overline{C}$ as a function of the rainfall volume, V, in Fig. 10. Riesel watershed has 125 hectares of Houston black clay soil (Udic Pellusterts), 60% under grass and 40% under a rotation of cotton, sorghum, and oats. Chickasha watershed has 11 hectares of Renfrow silt loam (Udertic Paleustolls) under grass. The Riesel watershed was fertilized annually (at the rate of 80 kg P/ha for grass, 45 kg P/ha for cotton and sorghum, and 30 kg P/ha for oats), while the Chickasha watershed remained unfertilized. The data shown in Fig. 10 are for individual rainstorms during the years 1976-1979 for Riesel and 1972-1976 for Chickasha. The fluctuations in the data points include the effect of variable P_o status of the soil with time, and that of neglecting V_1, as well as other factors of Eq. (13). While the data have the expected trends, the large fluctuations in the results suggest that the information on P_o and V_1 values for different storms must be available for obtaining a reasonable estimate of P concentrations in runoff.

PESTICIDE RELEASE IN RUNOFF WATER

In the CREAMS model, the release of pesticides from soil is

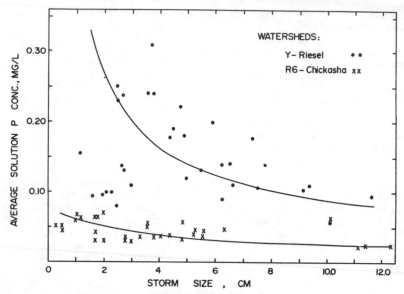

Fig. 10. Storm-average solution P concentrations of runoff as a function of the rainstorm size measured on two watersheds in Texas and Oklahoma.

modeled by assuming a constant partition coefficient between the adsorbed and solution concentrations at any given time (Leonard and Wauchope, 1980; Steenhuis and Walter, 1980). In the ARM model, a nonlinear isotherm is taken as the basis for partitioning (Donigian et al., 1977). In both cases, an instantaneous equilibrium between the adsorbed and solution phases is assumed to exist. The field tests of both these models have indicated problems with the partitioning mechanisms employed. Leonard and Wauchope (1980) found that the partition coefficient increased with time. Studies with the movement of pesticides through soil columns have shown that an instantaneous equilibrium may be assumed at low water flow velocities, but not at high velocities (Van Benuchten et al., 1974; James and Rubin, 1979). The velocities of overland flow are much larger than those of flow through soils. A diffusion controlled kinetic release is more likely to operate. A few laboratory studies on the kinetics of pesticide adsorption and desorption reported in the literature, that were reviewed in an earlier section of this paper, support the above statement. Some detailed experimental studies relating the kinetics of desorption in the laboratory with the release of several pesticides to runoff are needed to verify this. However, to illustrate our point, we present here some data of Baker et al. (1979) on the movement of Propochlor, Atrazine, and Alachlor pesticides in runoff from field plots, under simulated rainfall. Figure 11 presents a semilog plot of the runoff-water concentrations versus time. Instantaneous equilibrium and a constant partition coefficient should have given us a straight line on the semilog plot (Leonard and Wauchope, 1980; Steenhuis and Walter, 1980). Actually, the experimental data show a definite curvature. In Fig. 12, we present the same data as a log-log plot. Since the pesticides were applied at the soil surface and a rainfall of 6.35 cm/hr was applied, the effective water to soil ratio was very large. We, therefore, fitted Eq. (3) to the data. Based on Eq. (3), the instantaneous concentration of a pesticide in runoff is given by:

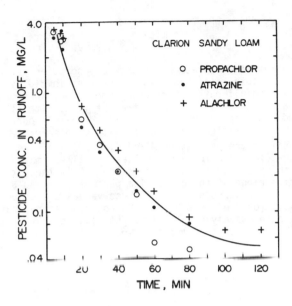

Fig. 11. Semilog plots of the pesticide concentrations in runoff water as a function of time during a rainfall event from Clarion sandy loam soil, for three different pesticides, as a test of the constant-partition coefficient assumption used in the CREAMS model. The solid curve is an eyeball fit. (Experimental data of Baker et al., 1979.)

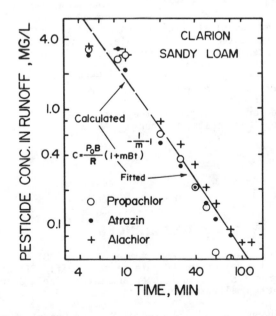

Fig. 12. Log-log plots of the data shown in Fig. 11, as a test of the kinetic pesticide release model indicated in the figure.

$$C = \frac{1}{R} \frac{dP_d}{dt} = \frac{BP_o}{R} (1 + mBt)^{-\frac{1}{m} - 1} \qquad (14)$$

where dP_d/dt represents the rate of pesticide desorption per unit area of soil, P_o is the total desorbable amount present initially per unit area, and R the rainfall rate. For large times, i.e. when $mBt \gg 1$, a plot of log C against log t will be a straight line, with slope = $-(1 + 1/m)$, and intercept = log $[BP_o (mB)^{-1-1/m}/R]$. In our graph of Fig. 12, we fitted a straight line to the data points for t greater than 20 min, and obtained the parameters m and B, as P_o and R were known. We then used Eq. (14) to calculate C values for times smaller than 20 minutes. The solid-dashed curve shown in Fig. 12 is the fit of Eq. (3) to the entire range of the data. The parameters m and B are equal to 2.0 and 1.023, respectively. These values are comparable with values of m and B for P desorption from Bernow fine sandy loam soil (3.0 and 0.82, respectively), calculated from laboratory kinetics measurements (Sharpley et al., 1981a).

CONCLUDING REMARKS

The experimental evidence and analyses presented in the proceeding sections enhance our basic understanding of the two important problems that have been encountered in modeling the release of soil chemicals to runoff. These are the extent and dynamics of the zone of rainfall-runoff-soil interaction, and the partitioning mechanism between solution and solid phases of phosphorus and some pesticides. Simplified, yet realistic, mathematical descriptions of the mechanisms involved are presented and tested on a small scale. These simplified models should now be extensively tested under field conditions. Efforts should also be directed towards simple ways of measuring or estimating the required parameters (the initially desorbable P, P_o, the storm size V, the portion of the storm that infiltrates before the runoff begins, V_1 and the exponents α, β, and K). The applicability of the model developed for phosphorus, to the release of pesticides to runoff water, should be investigated further.

REFERENCES

Ahuja, L. R., Sharpley, A. N., and Lehman, O. R. 1981. Further on a simplified model of soil P release to runoff: Effects of slope length, percent slope, soil cover and storm size. Journal of Environmental Quality (in press).

Ahuja, L. R., Sharpley, A. N., Yamamoto, M., and Menzel, R. G. 1981a. The depth of rainfall-runoff-soil interaction as determined by 32P. Water Resources Research (in press).

Amer, F., Bouldin, D. R., Black, C. A., and Duke, F. R. 1955. Characterization of soil phosphorus by anion-exchange resin and ^{32}P equilibration. Plant and Soil, Vol. 6, pp. 391-408.

Baker, J. L., and Laflen, J. M. 1979. Effect of corn residue and chemical placement on chemical runoff losses. Paper No. 79-2050, 25 pp., American Society of Agricultural Engineers, St. Joseph, Michigan.

Baker, J. L., Laflen, J. M., and Hartwig, R. O. 1979. Effects of corn residue and herbicide placement on herbicide runoff losses. Paper No. 79-2050a, 22 pp., American Society of Agricultural Engineers, St. Joseph, Michigan.

Barrow, W. J. 1979. The description of desorption of phosphate from soil. Journal of Soil Science, Vol. 30, pp. 259-281.

Bruce, R. R., Harper, L. A., Leonard, R. A., Snyder, W. M., and Thomas, A. W. 1975. A model for runoff of pesticides from small upland watersheds. Journal of Environmental Quality, Vol. 4, pp. 541-548.

Burwell, R. E., Timmons, D. R., and Holt, R. F. 1975. Nutrient transport in surface runoff as influenced by soil cover and seasonal periods. Soil Society of America Proceedings, Vol. 39, pp. 523-628.

Crawford, N. H., and Donigian, A. S., Jr. 1973. Pesticide transport and runoff model for agricultural lands. Report No. EPA 660/2-74-013, 211 pp., Environmental Research Laboratory, U.S. Environmental Protection Agency, Athens, Georgia.

Davis, H. H., Jr., and Donigian, A. S., Jr. 1979. Simulating nutrient movement and transformations with the ARM model. Transactions of the American Society of Agricultural Engineers, Vol. 22, pp. 1081-1087.

Donigian, A. S., Jr., and Crawford, N. H. 1976. Modeling pesticides and nutrients on agricultural lands. Report No. EPA 600/2-76-043, 317 pp., Environmental Research Laboratory, U.S. Environmental Protection Agency, Athens, Georgia.

Donigian, A. S., Jr., Beyerlein, D. C., Davis, H. H., and Crawford, N. H. 1977. Agricultural Runoff Management (ARM) Model version II: Refinement and testing. Report No. EPA 600/3-77-098, 293 pp., Environmental Research Laboratory, U.S. Environmental Protection Agency, Athens, Georgia.

Duffy, P. D., Schreiber, J. D., McClurkin, D. C., and McDowell, L. L. 1978. Aqueous- and sediment-phase phosphorus yields from five southern pine watersheds. Journal of Environmental Quality, Vol. 7, pp. 45-50.

Evans, R. L., and Jurinak, J. J. 1976. Kinetics of phosphate release from a desert soil. Soil Science, Vol. 121, pp. 205-211.

Faya, A. and Eyring, H. 1956. Equilibrium and kinetics of detergent adsorption--a generalized equilibration theory. Journal of Physical Chemistry, Vol. 60, pp. 890-898.

Frere, M. H., Onstad, C. A., and Holtan, H. N. 1975. ACTMO, An Agricultural Chemical Transport Model. Report No. ARS-H-3, 56 pp., U.S. Department of Agriculture, Washington, D.C.

Frere, M. H., Ross, J. D., and Lane, L. J. 1980. The nutrient submodel. Chap. 4 in: Knisel, W. G. (editor), CREAMS: A Field Scale Model for Chemicals, Runoff, and Erosion from Agricultural Management Systems, Conservation Research Report, No. 26, pp. 65-86, U.S. Department of Agriculture, Washington, DC.

Foster, G. R., Meyer, L. D., and Onstad, C. A. 1977. An erosion equation derived from basic erosion principles. Transactions of the American Society of Agricultural Engineers, Vol. 20, pp. 678-682.

Gburek, W. J., and Heald, W. R. 1974. Soluble phosphate output of an agricultural watershed in Pennsylvania. Water Resources Research, Vol. 10, pp. 113-118.

Haque, R., and Sexton, R. 1968. Kinetic and equilibrium study of the adsorption of 2, 4-Dichlorophenoxy acetic acid on some surfaces. Journal of Colloid and Interface Science, Vol. 27, pp. 818-827.

Haque, R., Lindstrom, F. T., Freed, V. H., and Sexton, R. 1968. Kinetic study of sorption of 2, 4-D on some clays. Journal of Environmental Science and Technology, Vol. 2, pp. 207-211.

Ingram, J. J., and Woolhiser, D. A. 1980. Chemical transfer into overland flow. Proceedings American Society of Civil Engineers, Watershed Management Symposium, Boise, Idaho, pp. 40-53.

James, R. V., and Rubin, J. 1979. Applicability of the local equilibrium assumption to transport through soils of solutes affected by ion exchange. American Chemical Society Symposium Series, No. ACSV061, pp. 225-235.

Johnson, H. P., Baker, J. L. Schrader, W. D., and Laflen, J. M. 1979. Tillage system effects on sediment and nutrients in runoff from small watersheds. Transactions of the American Society of Agricultural Engineers, Vol. 22, pp. 1110-1115.

Knisel, W. J. (editor). 1980. CREAMS: A Field Scale Model for Chemicals, Runoff, and Erosion from Agricultural Management Systems. Conservation Research Report No. 26, 643 pp., U.S. Department of Agriculture, Washington, DC.

Kunishi, H. M., Taylor, A. W., Helad, W. R., Gburek, W. J. and Weaver, R. N. 1972. Phosphate movement from an agricultural watershed during two rainfall periods. Journal of Agricultural and Food Chemistry, Vol. 20, pp. 900-905.

Kuo, S. and Lotse, E. G. 1974. Kinetics of phosphate adsorption and desorption by hematite and gibbsite. Soil Science, Vol. 116, pp. 400-406.

Leenheer, J. A., and Ahlrichs, J. L. 1971. A kinetic and equilibrium study of the adsorption of carbaryl and parathion upon soil organic matter. Soil Science Society of America Proceedings, Vol. 35, pp. 700-705.

Leonard, R. A., and Wauchope, R. D. 1980. The pesticide submodel.
Chapter 5 in: Knisel, W. G. (editor), CREAMS: A Field Scale
Model for Chemicals, Runoff, and Erosion from Agricultural
Management Systems, Conservation Research Report No. 26, pp. 88-112,
U.S. Department of Agriculture, Washington, D.C.

Lindstrom, F. T., Haque, R., and Coshow, W. R. 1970. Adsorption
from solution. III. A new model for the kinetics of adsorption-
desorption processes. Journal of Physical Chemistry, Vol. 74,
pp. 495-502.

Loehr, R. C. 1974. Characteristics and comparative magnitude of
nonpoint sources. Journal of Water Pollution Control Federation,
Vol. 46, pp. 1849-1870.

Menzel, R. G., Rhoades, E. D., Olness, A. E., and Smith, S. J. 1978.
Variability of annual nutrient and sediment discharges in
runoff from Oklahoma cropland and rangeland. Journal of
Environmental Quality, Vol. 7, pp. 401-406.

Romkens, M. J. M., and Nelson, D. W. 1974. Phosphorus relationships
in runoff from fertilized soils. Journal of Environmental
Quality, Vol. 3, pp. 10-13.

Romkens, M. J. M., Nelson, D. W., and Mannering, J. V. 1973. Nitrogen
and phosphorus composition of surface runoff as affected by
tillage method. Journal of Environmental Quality, Vol. 2, pp.
292-298.

Schindler, D. W. 1977. Evolution of phosphorus limitation in
lakes. Science, Vol. 195, pp. 260-262.

Schuman, G. E., Spomer, R. G., and Piest, R. F. 1973. Phosphorus
losses from four agricultural watersheds on Missouri Valley
loess. Soil Science Society of America Proceedings, Vol. 37,
pp. 424-427.

Sharpley, A. N., and Syers, J. K. 1979. Phosphorus inputs into a
stream draining an agricultural watershed II. Amounts contributed
and relative significance of runoff types. Water, Air, and Soil
Pollution, Vol. 11, pp. 417-428.

Sharpley, A. N., Ahuja, L. R., and Menzel, R. G. 1981. The release
of soil phosphorus to runoff in relation to the kinetics of
desorption. Journal of Environmental Quality, Vol. 10, pp. 386-391.

Sharpley, A. N., Ahuja, L. R., Yamamoto, M., and Menzel, R. G. 1981a.
The kinetics of phosphorus desorption from soil. Soil Science
Society of America Journal, Vol. 45, pp. 493-496.

Steenhuis, T. S., and Walter, M. F. 1980. Closed form solution for
pesticide loss in runoff water. Transactions of the American
Society of Agricultural Engineers, Vol. 23, pp. 615.

Taylor, A. W., Edwards, W. M., and Simpson, E. C. 1971. Nutrients
in streams draining woodland and farmland near Coshocton, Ohio.
Water Resources Research, Vol. 7, pp. 81-90.

Van Genuchten, M. Th., Davidson, J. M., and Wierenga, P. J. 1974.
An evaluation of kinetic and equilibrium equations for the

prediction of pesticide movement through porous media. Soil
Science Society of America Proceedings, Vol. 38, pp. 29-35.

Vollenweider, R. A. 1968. Scientific fundamentals of the eutrophication of lakes and flowing waters with particular reference to
nitrogen and phosphorus. Organization for Economic Cooperation
and Development. Report OAS/CSI/68.27, Paris.

Wendt, R. C. and Corey, R. G. 1980. Phosphorus variations in
surface runoff from agricultural lands as a function of land
use. Journal of Environmental Quality, Vol. 9, pp. 130-136.

Williams, J. R., and Hann, R. W., Jr. 1978. Optimal operation of
large agricultural watersheds with water quality constraints.
Technical Report, No. 96, 1520 pp., Water Resources Research
Institute, Texas A&M University, College Station.

ENVIRONMENTAL IMPACT OF RAIN, RUNOFF AND INFILTRATION ON SOLID WASTE PILES

Anand Prakash
Chief Water Resources Engineer
Dames & Moore
1626 Cole Boulevard
Golden, Colorado 80401

Manju Prakash
Engineer
Denver, Colorado

Jagdish Mohan
Irrigation Research Institute
Roorkee, India

ABSTRACT

This paper analyzes the interaction of rain, runoff, infiltration, and evapotranspiration on solid waste piles. As precipitation falls on the surface of a solid waste pile, a portion of it returns to the atmosphere as evapotranspiration, another portion runs off the surface as overland flow, and the remaining portion infiltrates through the pile. The infiltrated portion transmits leachates into the underlying groundwater environment.

An approximate analytical approach is presented to perform a water balance study for solid waste piles. The output of the water balance study includes the rates of surface runoff and leachates released from the pile. The leachate release is approximated as a continuous source of contamination for the underlying saturated porous medium.

Analytical models are presented to determine the spatial distribution of concentrations caused by continuous release of effluents from a point, line, or plane source in a ground water environment with one- or two-dimensional uniform steady flow. The solutions include the effects of radioactive decay and soil adsorption. The contributions of the upper and lower confining boundaries and of a stream fully penetrating the aquifer are accounted for by the method of images.

Illustrative examples are presented to predict the concentration distributions in one-dimensional and two-dimensional flow fields. In all, four source configurations are illustrated: a point source, a vertical line source, a horizontal line source, and a plane rectangular source. A case study illustrating the use of the proposed model for a field situation is also presented. In this case study, isopleths of concentrations are developed for an aquifer subjected to ground water pollution due to leakage of contaminants from an ash pond. Ground water movement in the aquifer is two-dimensional with dominant flow towards a perennial stream which forms a boundary of the flow field.

INTRODUCTION

The interaction of the hydrologic processes of rain, runoff, infiltration, and evapotranspiration with solid wastes stockpiled at different disposal sites is a matter of increasing concern to hydrologists and environmentalists. As precipitation falls on the surface of a solid waste pile, a portion of it returns to the atmosphere as evapotranspiration, another portion runs off the surface as overland flow, and the remaining portion infiltrates through the pile. The evaporated gases may cause air pollution. The overland flow carries with it dissolved and eroded constituents of the solid waste and enters the nearest surface waterbody as a point or non-point continuous or intermittent source of pollution. The infiltrated portion transmits leachates into the groundwater environment through the saturated or unsaturated medium between the surface of the pile and the water table. Such release of effluents into an aquifer can be approximated as a point, line, rectangular, or parallelepiped source.

This paper presents a deterministic approach to analyze the water balance for rain, runoff, infiltration, and evapotranspiration over solid waste piles and simple analytical models to predict the spatial distribution of steady-state concentrations caused by continuous release of contaminants from a point, line, rectangular or parallepiped source in a groundwater environment with one- or two-dimensional uniform flow.

WATER BALANCE ANALYSIS FOR SOLID WASTE PILES

Computerized deterministic approaches that can be used to compute the water balance for solid waste piles include continuous rainfall-runoff simulation models like the HYDROCOMP (Crawford and Linsley, 1966) and its revised versions, HSP and HSPF; Sacramento Streamflow Simulation Model (Burnash, Ferral and McGuire, 1973); ILLUDAS (Terstriep and Stall, 1974; Wenzel and Voorhees, 1980); and Hydrologic Model of Storm Drainage Systems for Urban Areas (Shih, Israelsen, Parnell, and Riley, 1976). Algorithms used to route the surface runoff components for urban, agricultural, or forested watersheds have included the kinematic wave, diffusion analogy, and complete dynamic wave models (Weinmann and Laurenson, 1979). These models, or appropriate versions thereof, have been used extensively to generate continuous runoff sequences for mine complexes, urban watersheds, and agricultural or forested catchments.

A simplified hydrologic simulation model for solid waste piles has been developed by the U.S. Environmental Protection Agency (Perrier and Gibson, 1980). This model uses the Soil Conservation Approach, 1972) to develop the rainfall-runoff relationship, the modified Penman method (Penman, 1948) to estimate the evapotranspiration, and a simplified soil storage routing technique to estimate the soil drainage. The proposed method follows the same basic equations. Thus,

$$S_{max} = 1.2 \left(\frac{1000}{CN} - 10 \right) \tag{1}$$

$$S = S_{max} \left(1 - \frac{SM}{UL} \right) \tag{2}$$

$$Q = \frac{(P - 0.25)^2}{P + 0.8S} \tag{3}$$

in which S_{max} = potential soil moisture retention; S = actual soil moisture retention; CN = SCS curve number for antecedent soil moisture condition I; SM = soil water content; UL = upper limit of soil moisture storage; P = daily rainfall; and, Q = daily runoff, all the parameters being expressed in inches.

Soil moisture entering the solid waste pile is given by,

$$F = P - I_a - Q \tag{4}$$

in which, I_a = initial abstraction = 0.2S

As water enters the solid waste pile, it undergoes evapotranspiration, storage, and drainage below the pile. The water balance among these processes is represented by,

$$SM_i = SM_{i-1} + F_i - E_i - D_i \tag{5}$$

in which SM_i = soil water storage on day i; F_i = water enturing the soil on day i; E_i = evapotranspiration on day i; and D_i = drainage below the pile on day i.

The evapotranspiration, E_i, is the sum of soil evaporation, E_s, and plant transpiration, E_p, and can be approximated as follows:

$$E_i = E_s + E_p \tag{6}$$

$$E_s = E_0 \exp (0.4 L_f)$$

$$= \frac{1.28 \Delta H_0}{\Delta + \Gamma} \exp (0.4 L_f) \tag{7}$$

$$E_p = \frac{E_0 L_f}{3} \tag{8}$$

in which E_0 = potential evaporation; L_f = area of plant leaves relative to soil surface; Δ = slope of the saturation vapor pressure curve at the mean air temperature; H_0 = net solar radiation; and, Γ = psychrometric constant. A simplified equation to estimate the drainage rate per day, D_i, is given by,

$$D_i = \sigma (F_i + \frac{SM_i}{\Delta t}) \tag{9}$$

in which $\sigma = \frac{2 \Delta t}{2t + \Delta t}$; Δt = routing inverval; $t = \frac{SM_i - FC}{\kappa}$; FC = field capacity; and κ = hydraulic conductivity of the material of the solid waste pile. If the waste pile is formed of stratified layers of different hydrologic characteristics, the above-mentioned equations may be used to analyze the water balance for each successive layer.

Using the aforementioned methodology, the rate of release of leachates from a solid waste pile into the underlying groundwater medium is estimated and averaged over a long period of time. The dispersion and movement of such leachates for steady-state drainage is described in the following paragraphs. Non-steady-state transport and dispersion of surface runoff are beyond the scope of this study.

LEACHATE TRANSPORT THROUGH SATURATED POROUS MEDIA

The convective-dispersion equation for the steady-state transport of contaminants in an adsorbing proous medium with uniform flow can be written as (Bear, 1972),

$$u \frac{\partial C}{\partial x} = D_x \frac{\partial^2 C}{\partial x^2} + D_y \frac{\partial^2 C}{\partial y^2} + D_z \frac{\partial^2 C}{\partial z^2} - R_d \lambda C - v \frac{\partial C}{\partial y} - w \frac{\partial C}{\partial z} \qquad (10)$$

in which, u, v, w = seepage velocities in the x, y, and z directions, respectively; C = concentration of the contaminant per unit volume of fluid; D_x, D_y, D_z = coefficients of dispersion in the x, y, and z directions, respectively; R_d = a retardation factor (Prakash, 1976); λ = radioactive decay factor, and x, y, z = Cartesian coordinates in the longitudinal, lateral and vertical directions, respectively.

For this study, it is assumed that v = w = o, $D_x \cong o$ (Hunt, 1973; 1978), and the principal axes of the dispersion tensor coincide with the principal axes of groundwater flow so that

$$D_y = D_T + D_d T \qquad (11a)$$

and

$$D_z = D_T + D_d T \qquad (11b)$$

in which D_T = coefficient of lateral dispersion; D_d = coefficient of molecular diffusion; and, T = scalar tortuosity.

With these simplications, Equation 1 reduces to:

$$\frac{\partial C}{\partial x} = D_z^* \frac{\partial^2 C}{\partial z^2} + D_y^* \frac{\partial^2 C}{\partial y^2} - \lambda^* C \qquad (12)$$

in which $D_z^* = \frac{D_z}{u}$; $D_y^* = \frac{D_y}{u}$; and $\lambda^* = \frac{\lambda R_d}{u}$. For a point source of strength M located at (o, y_o, z_o) in a semi-infinite porous medium, the boundary conditions are:

$$C (o, y_o, z_o) = M \delta (y - y_o) \delta (z - z_o) ;$$

$$C (x, \pm \infty, z) = o, x > o ; \text{ and } C (x, y, \pm \infty) = o , x > o.$$

Using Fourier transforms and the product law, the solution to Eq. 4 is:

$$C(x, y, z) = \frac{M}{4 \pi x \sqrt{D_y^* D_z^*}} \exp \left[-\frac{(y - y_0)^2}{4 D_y^* x} - \frac{(z - z_0)^2}{4 D_z^* x} - \lambda^* x \right] \quad (13)$$

So far as the transport of liquid contaminants is concerned, the bedrock and upper confining soil layer in a confined aquifer, and the bedrock and the free surface in an unconfined aquifer are non-flux boundary conditions, and can be simulated by the method of images. For the sake of simplicity, it is assumed that the upper confining soil layer or the free surface, as the case may be, is parallel to the bedrock.

Treating the origin of coordinates to be at the center of the vertical cross section of the aquifer at $x = 0$, solutions for a point, line, and rectangular source are:

(i) Point Source at $x = 0$, $y = y_0$, $z = z_0$

$$C(x,y,z) = \frac{M}{4 \pi x \sqrt{D_y^* D_z^*}} \exp \left[-\frac{(y - y_0)^2}{4 D_y^* x} - \lambda^* x \right]$$

$$\left[\sum_{m = -\infty}^{\infty} \exp \left\{ -\frac{[z - mD - (-1)^m z_0]^2}{4 D_z^* x} \right\} \right] \quad (14)$$

in which $M = \frac{C_0 Q_0}{u}$; D = saturated thickness of the aquifer; C_0 = concentration of contaminant in the effluent discharge; Q_0 = rate of release of the effluent; and M = index for the number of images considered.

(ii) Horizontal line source extending from $y = y_1$ to $y = y_2$ at $x = 0$, $z = z_0$,

$$C(x,y,z) = \frac{C_0 Q_0}{4 u (y_2 - y_1) \sqrt{\pi x D_z^*}} \exp (-\lambda^* x) \left[\text{erf} \frac{y - y_1}{\sqrt{4 D_y^* x}} - \text{erf} \frac{y - y_2}{\sqrt{4 D_y^* x}} \right]$$

$$\left[\sum_{m = -\infty}^{\infty} \exp \left\{ -\frac{[z - mD - (-1)^m z_0]^2}{4 D_z^* x} \right\} \right] \quad (15)$$

(iii) Vertical line source extending from $z = z_1$ to $z = z_2$ at $x = 0$, $y = y_0$,

$$C(x, y, z) = \frac{C_o Q_o}{4 u (z_2 - z_1) \sqrt{\pi x D_y^*}} \exp \left\{ -\frac{(y - y_0)^2}{4 D_y^* x} - \lambda^* x \right\}$$

$$\left[\sum_{m = -\infty}^{\infty} \left\{ \mathrm{erf} \frac{z - m D - (-1)^m z_1}{\sqrt{4 D_z^* x}} - \mathrm{erf} \frac{z - m D - (-1)^m z_2}{\sqrt{4 D_z^* x}} \right\} \right] \qquad (16)$$

(iv) Plane vertical source from $y = y_1$ to $y = y_2$ and $z = z_1$ to $z = z_2$ at $x = 0$,

$$C(x, y, z) = \frac{C_o Q_o}{4 u (y_2 - y_1)(z_2 - z_1)} \left[\mathrm{erf} \frac{y - y_1}{\sqrt{4 D_y^* x}} - \mathrm{erf} \frac{y - y_2}{\sqrt{4 D_y^* x}} \right]$$

$$\left[\sum_{m = -\infty}^{\infty} \left\{ \mathrm{erf} \frac{z - m D - (-1)^m z_1}{\sqrt{4 D_z^* x}} - \mathrm{erf} \frac{z - m D - (-1)^m z_2}{\sqrt{4 D_z^* x}} \right\} \right] \left[\exp(-\lambda^* x) \right] \qquad (17)$$

SIMULATION OF A STREAM-AQUIFER SYSTEM

Eqs. 14, 15, 16 and 17 are valid for an aquifer extending over the domain $0 < x < \infty$, $-\infty < y < \infty$, and $-D/2 < z < D/2$. In many field situations, the aquifer extends from a groundwater divide at $x = 0$ to a perennial stream at the other end. A typical case is illustrated in Fig. 1. For the transport of liquid contaminants through such a system, the stream forms a Dirichlet boundary, where, because of large dilution provided by the fresh water of the stream, $C(L, y, z) \simeq 0$ in which L = distance of stream from source. As shown in Fig. 1, this can be simulated by introducing imaginary sinks, P, P_1, P_2, etc. as mirror images of the real source, S, and the virtual sources, S_1, S_2, etc., and superposing the concentrations at any point due to these two series of sources and sinks. Thus,

$$C(x, y, z) = C_s(x, y, z) + C_p(x, y, z) \qquad (18)$$

$$C_p(x, y, z) = -C_s(2L - x, y, z) \qquad (19)$$

in which $C_s(x, y, z)$ = concentration at (x, y, z) due to the source, S,

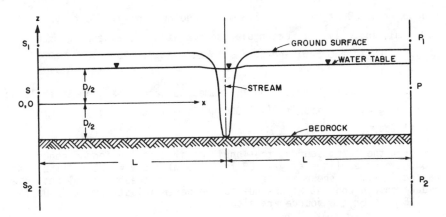

Fig. I. Cross–Section of a Typical Stream – Aquifer System

at $x = 0$, and C_p (x, y, z) = concentration at (x, y, z) due to a sink, P, at $x = 2L$. The concentrations C_s (x, y, z) and C_p (x, y, z) are given by Eq. 14,15,16 or 17. If the source, point, line or plane, extends from $x = 0$ to $x = L'$, the resulting concentrations, C_0, can be estimated by the principle of superposition. Thus,

$$C_0 \ (x, y, z) = \int_0^{L'} [C \ (x - x'), y, z] \ dx' \qquad (20)$$

in which C (x, y, z) = concentrations due to a point, line or plane source at $x = 0$. The expression on the right hand side of Eq. 20 is the convolution integral and can be evaluated numerically (Prakash, 1977).

APPLICATIONS

One-dimensional Flowfield

To demonstrate the applicability of the aforementioned methods to field problems, the following numerical examples are considered.

An underground source releases an effluent in a semi-infinite saturated porous medium. The ambient-groundwater velocity is 2 ft/day (0.61 m/d) from the source towards a perennial stream which is 7,000 feet (2,135 m) away. The saturated thickness of the aquifer is 100 feet (30.5 m). The coefficients of dispersion D_y and D_z are both 20 feet^2/day (1.86 m^2/d). The analyst is required to predict the spatial distribution of concentration in the porous medium with and without the stream.

For illustrative purposes, four source configurations are considered: (i) a point source located at $x = 0$, $y_0 = 100$ ft (30.7 m), $z_0 = 25$ ft (7.6 m); (ii) a vertical line source at $x = 0$, $y_0 = 100$ ft (30.7 m) extending from $z_1 = 0$ to $z_2 = 25$ ft (7.6 m); (iii) a horizontal line

491

source at $x = 0$, $z_o = 25$ ft (7.6 m) extending from $y_1 = 0$ to $y_2 = 100$ ft (30.5 m); and, (iv) a rectangular source at $x = 0$ extending from $y_1 = 0$ to $y_2 = 100$ ft (30.5 m) and $z_1 = 0$ to $z_2 = 25$ ft (7.6 m).

In each case, the source is assumed to be discharging an effluent containing 4,000 units/ft^3 (143,000 units/m^3) at a steady rate of 100 ft^3/day (2.8 m^3/d).

For the point source, the longitudinal distribution of concentration along two streamlines at $y = 100$ ft (30.5 m) and $y = 500$ ft (152.5 m), respectively, is shown in Fig. 2. For each of these streamlines, the reductions in concentrations due to the perennial stream at 7,000 ft (2,135 m) from the source are also shown.

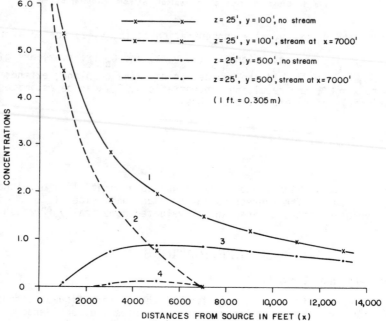

Fig. 2   Long. Distribution of Concentration
Due to a Point or Vertical Line Source

Analyses of the results of the aforementioned cases indicated that with $D_z = 20$ ft^2/day (1.86 m^2/d), there was almost uniform distribution of concentration along the entire aquifer depth for all the four source configurations considered. Thus, as one would expect, the isopleths of concentrations for the vertical line source were identical to those for the point source. Similarly, the isopleths for the rectangular source were identical to those for the horizontal line source.

To examine the sensitivity of the concentration distribution to the source configuration, the longitudinal distribution of concentration

along a streamline at y = -100 ft (30.5 m) for the vertical line source
(same as for the point source) and for the rectangular source (same as
for the horizontal line source), respectively, are plotted in Fig. 3.
The corresponding concentrations with the perennial stream at 7,000 ft
(2,135 m) from the respective sources are also plotted in Fig. 3. Note
the reductions in concentrations caused by the presence of the perennial
stream. Evidently, the nearer the stream is to the source, the larger
will be the reductions in concentrations. The slightly lower concentrations near the source, x < 2,000 ft (610 m), seen in Curve 3 of Fig. 2
and Curve 2 of Fig. 3 merely indicate that, with D_y = 20 ft^2/day
(1.86 m^2/d), lateral dispersion becomes pronounced only beyond
x = 2,000 ft (610 m). Along a streamline at y = 100 ft (30.5 m), the
logitudinal distributions of concentration for the four source configurations considered were almost indistinguishable from one another. This
distribution is shown in Fig. 2.

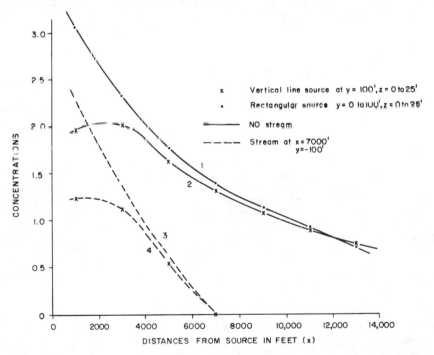

Fig. 3 - Long. Distribution of Concentration
Due to Vertical Line and Rectangular Sources

Two-dimensional Flowfield

To examine the effect of lateral flow, test cases were run with
u = 6 ft/day (1.83 m/d), v = 1 ft/day (0.305 m/d); and u = 2 ft/day
(0.61 m/d), v = 0.2 ft/day (0.061 m/d) for D_y = D_z = 20 ft^2/day
(1.86 m^2/d) and D_y = D_z = 84 ft^2/day (7.8 m^2/day), respectively. The
resulting lateral distributions of concentration are shown in Fig. 4
Curves 1 and 2 of Fig. 4 show the effect of lateral flow on the lateral
distribution of concentrations at two transects 300 feet (91.5 m)

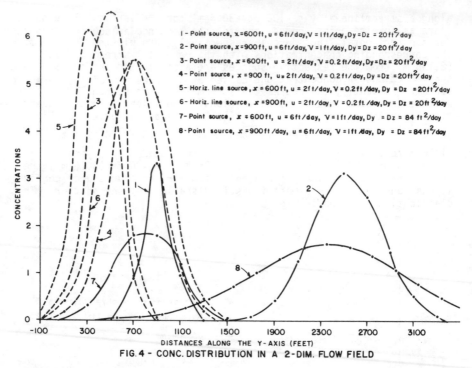

FIG. 4 - CONC. DISTRIBUTION IN A 2-DIM. FLOW FIELD

apart in the direction of dominant groundwater flow due to the point (or vertical line) source. Note the skew in concentration distribution introduced by lateral flow. Curves 3 and 4 of Fig. 4 also show the effect of lateral flow but with the groundwater velocities, u and v, reduced from 6 ft/day (1.83 m/d) and 1 ft/day (0.305 m/d) to 2 ft/day (0.61 m/d and 0.2 ft/day (0.061 m/d), respectively. Note that the dispersion and attenuation of peak concentrations are slower in the slower flowfield. Curves 5 and 6 of Fig. 4 show the effect of lateral flow on the lateral distribution of concentration due to the horizontal line (or rectangular) source. The slightly larger spread of concentration in the negative direction of y discernible in Curves 5 and 6 as compared to Curves 3 and 4, respectively, reflects the effect of the change in source configuration. Comparisons of Curves 1 and 2 with 7 and 8, respectively, indicate the effect of increasing the coefficients of dispersion, D_y and D_z, from 20 ft^2/day (1.36 m^2/d) to 84 ft^2/day (7.8 m^2/d). As expected, with increased coefficients of dispersion, the attenuation of concentration peaks is increased and the lateral spread of concentrations is elongated.

A CASE STUDY

The proposed model is now applied to a field situation where the objective is to plot the isopleths of concentration due to leakage of contaminants from an ashpond into a two-dimensional groundwater flow-field with a perennial stream about 1,500 feet (457 m) from the edge of the ashpond. The groundwater velocities and coefficients of dispersion are u = 6 ft/day (1.83 m/d), v = 1 ft/day (0.305 m/d) and $D_y = D_z = 84$ ft^2/day (7.8 m^2/d), respectively. The shape of the

494

ashpond is an irregular pentagon. The analyst has to predict if groundwater in a nearby community will be contaminated.

The ashpond is modeled as a plane rectangular source extending from y_1 = -900 feet (-274.5 m) to Y_2 = 100 feet (30.5 m) and x_1 = 800 feet (-244 m) to s_2 = 0. Note that a parallelepiped typed ashpond can be treated equally conveniently. Isopleths obtained from the present model are shown in Fig. 5. Note that no significant contamination is indicated beyond about 4,000 feet (1,220 m) from the right edge of the source. The isopleths of Fig. 5 compare very well with those obtained

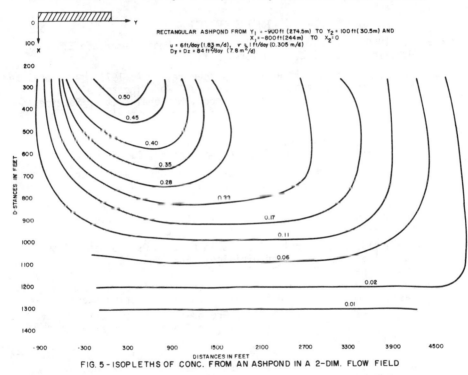

FIG. 5 - ISOPLETHS OF CONC. FROM AN ASHPOND IN A 2-DIM. FLOW FIELD

from a finite-element study of the same field problem. Details of these comparisons are described in a separate publication (Prakash, 1982). The model has also been used to predict the concentrations of leachates from fly ash deposits at the Brandywine and Faulkner facilities of the Potomac Electric Power Company. The average time for a computer run for 390 computational points with 11 images for a horizontal line source was 0.851 CP seconds for compilation and 3.489 CP seconds for execution on the United Computing Service Facilities used by Dames & Moore. Computing times for other cases were of the same order of magnitude.

CONCLUSION

A simple approximate method is presented to analyze the water balance in response to rain, runoff, evapotranspiration, and infiltration through a solid waste pile. The output of the water balance study provides the rates of runoff and leachates from the pile. The release of leachates is approximated by a steady-state source of contaminants for the underlying groundwater environment.

Simple analytical models are presented to simulate groundwater contamination due to point, line, plane or parallelepiped sources in a confined or unconfined aquifer. Adaptation of the method to compute the concentrations in a stream-aquifer system is described. Methods are presented to analyze groundwater contamination in an aquifer or stream-aquifer system with two-dimensional uniform groundwater flow also. The applicability of the proposed models to field situations is demonstrated by illustrative examples. The analytical model is simple, requires much smaller computer time and is free from convergence and stability problems which are common with most numerical models.

REFERENCES

Bear, J. 1959. Dynamics of Fluids in Porous Media, American Elsevier Publishing Co., Inc., New York, N.Y.

Burnash, R. J. C.; Ferral, R. L.; and McGuire, R. A. 1973. A Generalized Streamflow Simulation System Conceptual Modeling for Digital Computers, U. S. Department of Commerce, National Weather Service and State of California, Department of Water Resources.

Crawford, N. H., and Linsley, R. K. 1966. Digital Simulation in Hydrology; Stanford Watershed Model IV, Technical Report 39, Civil Engineering Department, Stanford University, California.

Penman, H. L. 1948. Natural Evaporation from Open Water, Bare Soil, and Grass, Proc. Royal Soc. (London), Ser. A, Vol. 193.

Hunt, B. W. 1973. Dispersion From Pit in Uniform Seepage, Journal of the Hydraulics Division, ASCE, Vol. 99, No. HY1, Proc. Paper 9474, pp. 13-21.

Hunt, B. W. 1978. Dispersive Sources in Uniform Ground-Water Flow, Journal of the Hydraulics Division, ASCE, Vol. 104, No. HY1, Proc. Paper 13467, pp. 75-85.

Perrier, E. R., and Gibson, A. C. 1980. Hydrologic Simulation on Solid Waste Disposal Sites, U.S. Environmental Protection Agency.

Prakash, A. 1982 (to be published), Groundwater Contamination Due to Vanishing and Finite-Size Continuous Sources, Journal of the Hydraulics Division, ASCE.

Prakash, A. 1976. Radial Dispersion Through Adsorbing Porous Media, Journal of the Hydraulics Division, ASCE, Vol. 102, No. HY3, pp. 379-396.

Prakash, A. 1977. Convective-Dispersion in Perennial Streams, Journal of the Environmental Engineering Division, ASCE, Vol. 103, No. EE2, Proc. Paper 12891, pp. 321-340.

Shih, G. B.; Israelsen, E. K.; and Riley, J. P. 1976. Application of a Hydrologic Model to the Planning and Design of Storm Drainage Systems for Urban Areas, Utah Water Research Laboratory, Utah State University, Logan, Utah.

Terstriep, M. L., and Stall, J. B. 1974. The Illinois Urban Drainage Area Simulation, ILLUDAS, Illinois State Water Survey Bulletin 58.

U. S. Department of Agriculture, Soil Conservation Service 1972. National Engineering Handbook, Section 4, Hydrology.

Weinmann, P. E., and Laurenson, E. M. 1979. Approximate Flood Routing Methods; A Review, Journal of the Hydraulics Division, ASCE, Vol. 105, No. HY12.

Wenzel, H. G. Jr., and Voorhees, M. L. 1980. Adaptation of ILLUDAS for Continuous Simulation, Journal of the Hydraulics Division, ASCE, Vol. 106, No. HY11.

Section 7
SEDIMENT YIELD

MATHEMATICAL MODELING OF WATERSHED SEDIMENT YIELD

J. R. Williams
Hydraulic Engineer
U.S. Department of Agriculture
Grassland, Soil & Water Research Laboratory
P. O. Box 748
Temple, TX 76501, U.S.A.

INTRODUCTION

Estimates of watershed sediment yield are required to solve a variety of soil and water resources problems. Usually a mathematical model of the sediment yield process is used in making such estimates. Sediment yield models vary considerably to accommodate a wide range in soil and water resources problems. These problems can be divided into three broad categories--(1) erosion control planning, (2) water resources planning and design, and (3) water quality modeling. Some of these problems can be solved with simple models but others require quite complex models depending on temporal and spatial scales, cost and expected life of the project, risk involved if failure occurs, etc.

Generally the simplest models are used in erosion control planning for agricultural fields, construction sites, reclaimed mines, etc. The only estimate necessary in such applications is the average annual soil loss for various erosion control systems. Although the estimates must be realistic and provide relative rankings of practices, a high degree of accuracy is not required (annual time scale, small areas, low cost/short life projects, and little risk from failure). Since thousands of these estimates are made annually (usually on site), a simple computationally efficient model is required.

Model requirements for water resources planning and design depend mainly on cost and life of the project and risk of failure. Sediment yield estimates are needed for designing structures ranging from temporary sediment basins at construction sites to large dams that control the flow from thousands of km². Estimates need to be fairly accurate for sediment pool design to prevent early filling or excessive cost. Besides designing structures, sediment yield models are also used in evaluating the effects of structural systems on floodplain and channel degradation and deposition.

Sediment yield models used in simulating water quality also vary in complexity depending upon the nature of the water quality parameter being modeled. Some chemicals that are highly toxic may require short time steps (<1 day) to adequately define the change in concentration within individual runoff events. The long-term accumulation is of particular importance for other chemicals like nutrients and, therefore, longer time steps (≥1 day) are appropriate.

499

Recently developed sediment routing models are superior to traditional approaches particularly for modeling water quality. In simulating water quality, sediment is treated as a pollutant and as a carrier of other pollutants. Thus, other sediment properties like particle size distributions must be calculated as well as total sediment yield. Some of the routing models predict sediment graphs (time distribution of sediment for an individual event) in addition to total sediment yield for the event. In routing, basins are subdivided into a manageable number of relatively homogeneous sub-basins. Models for continuously simulating sub-basin sediment yields have been developed recently by linking erosion-sedimentation models to hydrology models. Although these modern routing-simulation models are more time consuming (assembling inputs and computing) than traditional methods, the potential accuracy is greater.

The objective here is to determine the state-of-the-art in modeling sediment yield. A brief review of the literature describing accomplishments in model development is presented. Also suggestions for future research in long-term simulation, sediment routing, gully erosion, and average annual methods are described.

LITERATURE REVIEW

Recent literature has provided excellent descriptions of the state-of-the-art of various phases of sediment yield modeling. For example, a comprehensive review of erosion process modeling was given by Foster (1981). Earlier, Renard (1977) reviewed the available techniques for estimating erosion rates. In further work, Renard (1980) discussed progress in estimating erosion and sediment yield from rangeland. Woolhiser and Renard (1980) provided insight to stochastic aspects of watershed sediment yield. Alonso (1980) evaluated a number of sediment transport formulas, both bedload and total load.

A review of the most pertinent literature describing model development for erosion control, water resources planning, and water quality modeling follows.

Erosion Control Planning

One of the earliest and most successful equations for use in erosion control planning was developed by Musgrave (1947). The Musgrave equation predicts annual soil loss from sheet and rill erosion as a function of soil erodibility, cover, land slope, slope length, and rainfall intensity. In 1960 the Universal Soil Loss Equation (USLE) (Wischmeier and Smith, 1960) was developed using 10,000 plot years of data from various research facilities throughout the U.S. Although the USLE is similar to the Musgrave equation, it has a different form, includes additional factors, and is more generally applicable (based on 10,000 plot years of data). Like the Musgrave equation, the USLE predicts annual soil loss from sheet and rill erosion as a function of soil erodibility and slope length and steepness. However, the USLE includes rainfall energy as well as intensity, a crop management factor instead of a cover factor, and an erosion-control practice factor.

Currently, the USLE is the most widely used and accepted model for planning erosion control practices in the U.S. Two handbooks (Wischmeier and Smith, 1965; Wischmeier and Smith, 1978) provide user convenience for a variety of on-site applications.

Recently, more complex models have been developed for predicting sediment yield from sheet and rill erosion caused by individual storms.

Some of the more promising models include those developed by Onstad and Foster (1975), Foster and Meyer (1975), Beasley et al. (1977), Shirley and Lane (1978), and Foster et al. (1980). These models provide estimates of locations and amounts of deposition as well as erosion. They include soil particle detachment and transport by runoff in addition to rainfall. Because of the transport components, these models are capable of evaluating erosion control practices implicitly without using the empirically derived USLE erosion control practice factor. Since these models operate on individual storms, long-term simulations (30 years or more) are required to estimate the average annual sediment yield. Also they must be linked to an appropriate hydrology model because they require runoff inputs. Thus, these more complex erosion simulation models are not nearly so computationally efficient as the USLE. However, they do provide considerably more information (locations and amounts of erosion and deposition, sediment yield frequency distributions as well as the annual yields, etc.). Thus, for some critical erosion problems, the increased input detail and computing cost may be justified. Generally, however, these simulation models are probably better suited for other applications like detailed studies to determine the effect of soil erosion on soil productivity, water quality modeling, etc.

Water Resources Planning

Two of the most widely used approaches in predicting sediment yield for water resources planning are the gross-erosion/delivery-ratio technique and the sediment rating curve procedure. Two other more modern approaches are sediment yield simulation models and sediment yield routing models. A brief discussion of the literature describing each of the four methods follows.

Gross-erosion/delivery-ratio method

This method is generally used in planning small to medium water resources projects. One of the most important applications is the Soil Conservation Service's flood control program that involves planning, designing, and evaluating flood water retarding structures. Gross erosion is composed of total sheet, rill, channel and gully erosion on a watershed. The delivery ratio is the sediment yield at the watershed outlet divided by the gross erosion. The USLE (Wischmeier and Smith, 1978) is currently the most accepted method for predicting sheet and rill erosion. The Soil Conservation Service (1966, 1971) has developed guidelines for estimating gully and channel erosion. These methods of estimating gross erosion provide a means for evaluating the effects of various land management strategies on sediment yield.

Applying a delivery ratio to estimated gross erosion can be a fairly accurate technique if delivery ratios can be predicted accurately. Traditionally, delivery ratios have been estimated by comparing limited amounts of measured sediment yield data with predicted gross erosion. These delivery ratios have been related to basin characteristics to develop delivery ratio prediction equations for use on ungaged basins (Gottschalk and Brune, 1950; Maner, 1958; Maner, 1962; Roehl, 1962; Williams and Berndt, 1972). However, prediction equations have been developed for only a few regions of the U.S. because of limited sediment yield data. This deficiency can be partially overcome by using simulated sediment yields (Williams, 1977) for determining delivery ratios. Long-term average annual simulated sediment yields are divided by the gross erosion to calculate delivery ratios. These simulated delivery ratios are related to basin characteristics to develop equations for predicting delivery ratios for nearby ungaged basins.

501

Sediment rating curve method

The sediment rating curve method (Campbell and Bauder, 1940) is
normally used for planning and designing large water resources projects
like reservoirs on major streams. Sediment rating curves (the relationship
between flow rate and sediment discharge rate) can be constructed by
sampling stream flow or by applying a sediment transport equation.
Sediment yield frequency distributions can be established using flow
frequency distributions and sediment rating curves. Average annual
sediment yield can be computed by integrating the sediment frequency
distribution (Task Committee, 1970; Williams, 1974).

Sampling to establish sediment rating curves is costly and time
consuming (several years of data collection may be required to develop
a representative rating curve). Also changing land uses and management
practices may cause instability in the rating curve and thus prevent the
establishment of a reliable relationship.

Sediment transport equations can be used to calculate sediment
rating curves, but the equations usually require calibration with measured
data to obtain reliable results. Some of the early work in developing
total sediment load transport relationships includes equations by
Einstein (1950), Colby and Hembree (1955), and Laursen (1958). Recently
developed equations by Ackers and White (1973), Engelund and Hansen
(1967), and Yang (1973) appear promising, although they have not been
tested extensively.

An advantage of sediment rating curves over annual prediction
methods is that sediment yield frequency distributions can be obtained
(not just average annual sediment yield). The main disadvantage is that
land management strategies other than existing conditions cannot be
evaluated.

Sediment yield models

Sediment yield models can be used to estimate sediment yield from a
basin without delivery ratios or sediment rating curves. These models
can be divided into two groups. Models in the first group (annual
sediment yield models) estimate long-term average annual sediment yield;
use simple, easy to obtain inputs; and are very efficient computationally
(do not require computer solutions). Models in the second group (continuous simulation models) simulate individual storm or daily sediment
yields, use fairly detailed inputs, and require computer solutions.

(a) Annual sediment yield models. A number of annual sediment yield
models have been developed (generally empirically) for several areas of
the U.S. These models have been used mostly in the Western states to
estimate sediment yield from range and forest lands. Some of the most
widely used and accepted models include Anderson's model (Anderson, 1954)
for use mainly on forested basins; Flaxman's model (Flaxman, 1972) for
use mainly on rangeland basins; and the Pacific Southwest Inter-Agency
Committee (PSIAC) method (Pacific Southwest Inter-Agency Committee, 1968)
for use in broad planning for a variety of conditions encountered in
the southwestern U.S. Recently a more generally applicable sediment
yield model was developed by relating sediment yield to average annual
runoff and basin area (Dendy and Bolton, 1976). This method may not be
appropriate for application to specific basins because it does not
account for variations in many local factors like slope, cover, and
management. However, it should be a valuable tool in preliminary and
large area planning.

(b) Continuous simulation models. Recently developed continuous sediment yield simulation models offer several advantages for solving a range of water resources problems. Some of the most important attributes of the simulation models include:

(1) They provide a means for evaluating the effects of agricultural management on sediment yield;

(2) Since they are driven by inputs from hydrologic models, accuracy is generally better than models that consider only rainfall energy;

(3) Accuracy is also improved by simulating individual storm sediment yields instead of estimating average annual amounts;

(4) They are capable of simulating crop growth and residue decay for use in estimating cover effects on infiltration and erosion;

(5) They produce sediment yield frequency distributions, not just annual amounts; and

(6) Sediment delivery is determined directly in the model--delivery ratio estimates are not necessary.

Disadvantages of the simulation models include increased input data requirements and computing costs. For some applications, however, the simulation model results can easily justify the additional data requirements and computing time.

Two recently developed deterministic simulation models that have proven to be reliable in some areas of the U.S. are Negev's model (1967) and the Modified Universal Soil Loss Equation (MUSLE) (Williams, 1975; Williams and Berndt, 1977). Negev's model is linked to the Stanford Watershed Model (Crawford and Linsley, 1966) to supply necessary hydrologic inputs. The Negev model simulates surface and channel erosion using hourly time steps. The Williams model was developed by linking MUSLE with a water yield model (Williams and LaSeur, 1976) to supply estimates of runoff volume and peak rate. The combined runoff-sediment yield model simulates surface sediment yield using a daily time step.

There are also a number of recently developed stochastic sediment yield simulation models. Several of these models were developed by linking MUSLE with various stochastic runoff models (Smith et al., 1977; Simons et al., 1977; Fogel et al., 1977; Mills, 1981). Renard and Lane (1975) linked a stochastic runoff model to a model for simulating sediment yield from channels.

Although the Renard and Lane model can be used to simulate sediment yield, there is no mechanism for evaluating the effects of agricultural management on sediment yield. Actually, the method is quite similar to a sediment rating curve approach because it uses a sediment transport equation (Laursen, 1958) to simulate sediment discharge for individual runoff events.

Sediment routing models

Sediment routing for agricultural basins has received considerable attention recently mainly because of increased concern for non-point source water pollution. Several of the sediment yield prediction techniques described in the Water Resources Planning section of this paper perform

quite well even on large basins if the sediment sources are uniformly
distributed over the basin and if the major basin tributaries are hydrau-
lically similar. However, these conditions do not generally exist on
large agricultural basins. Thus, sediment routing is required to produce
satisfactory results for many applications.

Routing increases prediction accuracy on large basins and allows
determination of sub-basin contributions to the total sediment yield.
Also, the locations and amounts of channel and floodplain deposition and
degradation can be determined. Besides predicting sediment yields,
sediment routing also calculates the change in particle-size distribution
as sediment moves from its source to the basin outlet.

An early version of a sediment routing model (Williams, 1975) routed
the sediment yield for an individual storm from each sub-basin to the
basin outlet considering the travel time and the median particle size.
The model included deposition and degradation components, but the
deposition process was emphasized. Onstad and Bowie (1977) simplified
the routing model for use in routing average annual sediment yields.
The Onstad-Bowie model neglected particle size and did not include a
degradation component. However, it is quite convenient and represents a
significant improvement over gross-erosion/delivery-ratio methods.

The Williams sediment routing model was refined (Williams and Hann,
1978) by replacing the median particle size with the entire particle-size
distribution. Also, a technique was developed for determining the
deposition parameter for each routing reach, instead of using one value
of the parameter for the entire basin. Further refinement (Williams,
1978) included the development of a new degradation component based on
stream power.

Li et al. (1977) developed a sediment routing model for application
to small basins (channel processes were not considered). The Li model
considers suspended and bed loads separately, contains both deposition
and degradation components, and routes sediment graphs--not just total
yield for an event. Alonso et al. (1978) modified the infiltration and
water and sediment routing schemes of the Li model and included a
channel routing component. Both the Li and Alonso models require calibration with measured sediment data. However, they both provide estimates
of sediment graphs which are necessary for solving some problems.

Sediment graphs can also be estimated for models that output only
total sediment yield using an independent sediment graph model. Williams
(1978) developed a model for simulating sediment graphs for ungaged
basins given the total yield. Storm sediment graphs are predicted by
convolving rainfall excess with an instantaneous unit sediment graph
(IUSG). The IUSG is the distribution of sediment from an instantaneous
burst of rainfall producing one unit of runoff.

Water Quality Modeling

Recently several erosion-sedimentation models have been developed
or adapted for use in simulating non-point source water pollution.
Because sediment is a pollutant and transports other pollutants, it is
important to estimate sediment yield accurately for individual storms
(annual estimates are generally not adequate because chemical concentrations
in soils vary during a year). Besides sediment yield, a model should
also simulate the particle size distribution for use in determining the
pollutant carrying capacity of the sediment. One of the most important
requirements of a water quality model is the ability to accurately

estimate the effects of management on model outputs.

Water quality models that rely heavily on erosion-sedimentation components can be divided into two groups--continuous simulation and event models. Two of the most popular continuous simulation models are CREAMS (Knisel, 1980) and ARM (Donigian et al., 1977). The CREAMS erosion-sedimentation model (Foster et al, 1980) represents the most advanced USDA erosion modeling. The ARM model relys on the work of Negev (1967). The CREAMS model uses a daily time step and the ARM model uses a 5 minute time step. Both models are designed for application to small field size basins. Another field scale model was developed by Williams (1979) for simulating daily sediment, phosphorus, and nitrogen yields. The Williams model uses MUSLE (Williams, 1975) to estimate sediment yield. A continuous simulation model for application to larger basins was developed by Holtan (1979). Holtan's model is based on the USDAHL model (Holtan et al., 1975) and MUSLE.

Two well-known event water quality models are ANSWERS (Beasley et al., 1977) and SPNM (Williams, 1980). The ANSWERS model is designed for use in planning conservation practices on farms to reduce erosion and improve water quality. It allows a detailed description of the basin using a grid system. Since grid elements are usually relatively small (5-10 ha) and time steps are short ($\sim$15 minutes), computing cost may be excessive for many applications, especially on large areas. The SPNM model is written in the form of a problem-oriented computer language for user convenience in solving water quality problems on large basins (<2500 km^2). Sediment yield for an individual storm is predicted with MUSLE for each sub-basin and routed to the basin outlet.

Leytham and Johanson (1979) developed the Watershed Erosion and Sediment Transport Model (WEST) by linking the ARM model to a channel routing model called CHANL. The WEST model is unique because it combines continuous simulation and channel routing. Continuous simulation and channel routing have not been combined previously because of excessive computing cost (continuous simulation expands the temporal scale and routing expands the spatial scale). Actually, the ARM and CHANL models run separately--CHANL uses ARM output files as input. As the authors point out, this causes real data handling problems that may limit the models usefulness. Considerable testing is needed to determine the usefulness and realiability of the WEST model.

FUTURE RESEARCH

The major emphasis in future research should be placed on developing and refining long-term simulation and sediment routing models because: (a) These approaches offer the greatest promise for improved understanding of erosion-sedimentation processes and for increased prediction accuracy; (b) They have great potential as a management tool for maximizing long-term agricultural production and maintaining a safe environment; and (c) Many problems particularly those involving water quality require considerably more information than just sediment yield (particle size distribution, degradation-deposition amounts and locations, the effect of erosion on soil productivity, etc). Another extremely important area of future research is gully erosion (actually a component of the long-term simulation model). Gully erosion modeling research should reap tremendous benefits because: (a) The state-of-the-art is not nearly as advanced as that of sheet and rill erosion; and (b) Gully erosion is extremely damaging to cropland and is a major sediment source on many basins. A fourth area that also deserves future research attention is the refinement of efficient methods for estimating the long-term average

annual sediment yield. For designing many present and future low cost/
risk projects, a good estimate of the average annual sediment yield is
the only information necessary. Also, some applications require a large
number of these long-term sediment yield estimates annually (usually in
the field without access to a computer).

Suggestions for future research in each of the four indicated areas
(long-term simulations, sediment routing, gully erosion, and average
annual methods) follows.

Long Term Simulation Models

Usually the purpose of a long-term simulation model is to establish
a sediment yield frequency distribution for use in evaluating the effects
of management strategies. Long-term simulations (20-100 years) are
almost essential in establishing frequency distributions for complex
systems with many interacting processes. The long term usually insures
a wide range in weather variables which causes wide ranges in other
important variables. For increased computing efficiency, the spatial
scale is generally limited to a relatively homogeneous area and time
steps are relatively long ($\sim$1d).

Long-term simulation models should be expanded to include most of
the important processes that affect erosion-sedimentation (evapotranspiration, crop growth, nutrient cycling, vegetative cover, residue cover,
tillage, surface runoff, and animal uptake for range and pasture). The
largest potential improvement in present sediment simulation models is
in the addition of crop growth and residue simulation components. Since
erosion is so strongly affected by crop and residue cover, correctly
simulating growth and decay will greatly improve sediment yield predictions.
Another major advancement could come through the addition of tillage
components. Most sediment simulation models currently ignore tillage
effects on erosion. Of course, crop growth is also affected by tillage
as well as water and nutrient availability.

Better climatic inputs could also produce significant improvement
in sediment yield simulation. Three climatic inputs frequently required
by hydrologic models are rainfall, temperature, and solar radiation.
Solar radiation is generally not available for long-term simulations and
rainfall and temperature may also be scarce in some areas (or have
missing records). Thus a good model for generating climatic inputs
(rainfall, temperature, and solar radiation) would improve sediment
prediction accuracy through more realistic crop growth, better rainfall
and runoff energy estimates, etc. Besides being generally superior to
recorded climatic data because of missing records, the generation
approach also has the advantage of providing any number of long-term
(hundreds of years) climatic sequences. This is particularly important
in developing frequency distributions.

Sediment Routing

The main reasons for needing sediment routing models are: (a) They
increase sediment yield prediction accuracy for large complex basins;
(b) They determine locations and amounts of degradation and deposition;
and (c) They provide a means for determining sub-basin contributions to
the basin sediment yield, and thus a way to evaluate management strategies
for sub-basins. Sediment routing is used to simulate detachment of
particles by rainfall, deposition, reentrainment of previously deposited
particles, and degradation by overland and channel flow. Sediment
routing models allow detailed descriptions of topography, channel geometry,

etc. and are capable of operating on large complex basins. Generally the spatial scale is much larger (a basin may be divided into many sub-basins) than that of the long-term simulation models and the time steps are shorter (0.1h to 1d). Therefore, sediment routing models are usually designed to operate on individual storms given initial conditions. Frequency distributions can be estimated by relating sediment yield frequency to runoff frequency--routing several events with a range of runoff frequencies.

Sediment routing can be divided into two phases, the upland phase (sometimes called overland flow) and the channel phase. Although valuable research has been reported recently on both phases, much work is needed to refine existing models, establish principles, and develop new and better models. In modeling the upland phase, an improved more physically based rainfall energy component is needed. Also, runoff must be modeled more realistically. Overland flow does not generally occur in uniform sheets, but in small concentrated flow streams. Upland runoff concentrates into flow patterns similar to channel patterns on large basins. In cultivated fields these concentrated flow areas are tilled over and thereby hidden between storms. If they are not plowed over, they may become gullies when subjected to frequent runoff events. Models need to have the ability to simulate the concentrated flow patterns and to route sediment through the network to the channel system. To accomplish this upland routing, a model should include the processes of detachment, deposition, reentrainment, and degradation and their interactions as affected by rainfall and runoff, soils characteristics, conservation practices, cover, and topography.

Sediment routing in channels and floodplains is based on similar principles to those of upland routing. The major differences are in the flow depths and the rainfall energy effect (generally rainfall energy can be ignored in channel and floodplain routing). Also the geometry of the channel and floodplain network is much better defined than that of the concentrated flow networks in the uplands. Much research is needed to develop physically based model components for simulating detachment, deposition, reentrainment, and degradation both on the uplands and in the channels and floodplains (physically based components should serve both purposes).

Particle size distributions of detachment material and the dynamics of the distributions as affected by flow are extremely important in both phases of sediment routing. Not only is this an important subject, but it is a complex one that needs much research attention. Particle size distributions are affected by soil aggregates. The degree of aggregation, of course, varies with soil types and also within soil types as a function of soil management. Soil aggregates have different transport character-istics than soil particles of the same size, but little is known about aggregate transport. Also, the stability of the aggregates in flowing water varies, although little is known about the variation or the stability.

Soil erodibility is another subject of great importance in sediment routing research. Although much research has been devoted to developing a soil erodibility factor for upland erosion (Wischmeier and Smith, 1978), little is known about the soil erodibility of channel material subjected to deep flows. Also, little information is available concerning soil erodibility of floodplains under a range of flows with various vegetative covers.

Gully Erosion

Although gully erosion is extremely damaging to cropland and is a major sediment source, it has received much less research attention than sheet and rill erosion. Gully erosion modeling has not been approached in a physically based process oriented manner. Research is badly needed to identify and describe mathematically the major processes involved in gully erosion. A long-term simulation model with a good upland sediment routing component should be quite helpful in studying the problem. Gullies may be initiated as part of an upland flow network that is not repaired between storms. In non-cultivated areas, the flow network may follow animal paths or roads to initiate gullies. Gullies also form in floodplains when extreme events occur and some disturbance (poor vegetative cover, animal paths, man's activities, etc.) provides the opportunity. Again, a long-term simulation model with a good channel and floodplain routing component should aid these floodplain gully studies. Thus, by using a combination of long-term simulation and sediment routing it may be possible to construct a reliable gully erosion research model. Although as stated earlier, computing costs are large for the continuous simulation/ routing combination, the high cost may be justified for a research model (generally such cost would be excessive for management models). By simulating crop growth, management practices, weather information, etc., the initiation of gully erosion should be predictable. However, much work is needed to describe the processes that cause gullies to expand. Examination of the flow-energy relationships at the gully head cut offers tremendous potential modeling advances. Also, the variation in soil erodibility between top soil and subsoil should be studied thoroughly.

Average Annual Methods

For designing many present and future low cost/risk projects, a good estimate of the average annual sediment yield is the only information necessary. Also, some of these estimates must be made many times annually in the field without access to a computer. Thus, it is important to continue research to refine existing methods and to develop new efficient sediment yield equations. These equations should predict the annual sediment yield realistically based on easy to obtain inputs. Rainfall inputs have a big advantage over runoff inputs for user convenience. However, runoff is strongly related to sediment yield and should not be eliminated from consideration as an energy factor. Another important requirement of an annual equation is that it be sensitive to agricultural management decisions.

Generally, the approach used in developing annual equations must be empirical at least to a degree. Annual sediment yield prediction schemes usually fall into two groups: (a) Gross erosion estimates adjusted with a sediment delivery ratio; and (b) Regression equations that predict sediment yield directly. The gross-erosion/delivery-ratio approach has the advantage of the availability of the widely used and accepted USLE (Wischmeier and Smith, 1978). However, delivery ratios need considerable work to improve accuracy and general applicability. A sediment yield equation like the Dendy-Bolton equation (Dendy and Bolton, 1976) has the advantage of considering annual runoff volume (generally a better sediment yield predictor than rainfall). However, the Dendy-Bolton equation needs modification to include factors that consider cover, topography, soils, and management (perhaps the appropriate USLE factors).

A shortage of data is a serious problem for developing delivery ratio relationships or regression equations for predicting sediment yield. For areas where little data is available, generated data (using reliable long-term sediment yield simulation models and sediment routing models) may provide a satisfactory substitute. Although the complex simulation and routing models are too costly for many simple applications, they can be operated on a few representative basins throughout the U.S. to simulate data. If the simulation and routing models have been properly validated, the output should give a reasonable estimate of measured sediment yield data (at least it can be used to fill in badly needed data gaps).

SUMMARY AND CONCLUSIONS

Sediment yield models vary considerably to accommodate a wide range in soil and water resources problems. These problems can be divided into three categories--(1) erosion control planning, (2) water resources planning and design, and (3) water quality modeling. Generally the simplest models are used in erosion control planning because the only estimate necessary is the average annual soil loss. Currently the USLE is the most widely used and accepted model for planning erosion control practices in the U.S. However, recently more complex models have been developed for predicting sediment yield from sheet and rill erosion caused by individual storms. Model requirements for water resources planning and design depend mainly on cost and life of the project and risk of failure. Two of the most widely used approaches in predicting sediment yield for water resources planning are the gross-erosion/delivery-ratio technique and the sediment rating curve procedure. Two other more modern approaches are sediment yield simulation models and sediment yield routing models. Recently, several erosion-sedimentation models have been developed or adapted for use in simulating non-point source water pollution. Although the water quality models may be continuous or event oriented, generally they use relatively short time steps (0.1h to 1d), simulate changes in particle size distribution, and estimate the effects of management on model outputs.

The major emphasis in future research should be placed on developing and refining long-term simulation and sediment routing models. These approaches offer the greatest promise for improved understanding of erosion-sedimentation processes and for increased prediction accuracy. Gully erosion research, another extremely important area of future research, should reap tremendous benefits because it is a major sediment source and the governing principles are relatively undeveloped. A fourth area that also deserves future research attention is the refinement of efficient methods for estimating the long-term average annual sediment yield from small watersheds. For designing many present and future low cost/risk projects, a good estimate of the annual sediment yield is the only information necessary.

REFERENCES

Ackers, P. and White, W.R. 1973. Sediment Transport: New Approach and Analysis. Journal of the Hydraulics Div. ASCE, Vol. 99, No. HY11, pp. 2041-2060.

Alonso, C.V. 1980. Selecting a Formula to Estimate Sediment Transport Capacity in Nonvegetated Channels. in: CREAMS: A Field Scale Model for Chemicals, Runoff, and Erosion from Agricultural Management Systems, Vol. 3. Supporting Documentation, Chapter 5, USDA Conservation Research Report No. 26, pp. 426-439.

Alonso, C.V., DeCoursey, D.G., Prasad, S.N., and Bowie, A.J. 1978. Field Test of a Distributed Sediment Yield Model. Proceedings of the Specialty Conference on Verification of Mathematical and Physical Models in Hydraulic Engineering, ASCE, College Park, Maryland, pp. 671-678.

Anderson, H.W. 1954. Suspended Sediment Discharge as Related to Stream-flow, Topography, Soil and Land Use. Trans. American Geophysical Union, Vol. 35, No. 2, pp. 268-281.

Beasley, D.B., Huggins, L.F., and Monke, E.J. 1980. ANSWERS: A Model for Watershed Planning. Trans. of the Americal Society of Agricultural Engineers, Vol. 23, No. 4, pp. 938-944.

Campbell, F.B. and Bauder, H.A. 1940. A Rating Curve Method for Deter-mining Silt Discharge of Streams. Trans. American Geophysical Union, Part 2, pp. 603-607.

Colby, B.R. and Hembree, C.H. 1955. Computations of Total Sediment Discharge, Niobrara River near Cody, Nebraska. U.S. Geological Survey Water Supply Paper 1357.

Crawford, N.H. and Linsley, R.K. 1966. Digital Simulation in Hydrology; Stanford Watershed Model IV. Department of Civil Engineering, Stanford University, Tech. Report No. 39, 210pp.

Dendy, F.F and Bolton, G.C. 1976. Sediment Yield-Runoff-Drainage Area Relationships in the United States. Journal of Soil and Water Conservation, Vol. 31, No. 6, pp. 264-266.

Donigian, A.S., Beyerlein, D.C., Davis, H.H., and Crawford, N.H. 1977. Agricultural Runoff Management (ARM) Model Version II: Refinement and Testing. Environmental Protection Agency, EPA-600/3-77-098, 294pp.

Einstein, H.A. 1950. The Bedload Function for Sediment Transportation in Open Channel Flows. USDA Tech. Bull. 1026, 70pp.

Engelund, F. and Hansen, E. 1967. A Monograph on Sediment Transport in Alluvial Streams. Teknisk Vorlag, Copenhagen.

Flaxman, E.M. 1972. Predicting Sediment Yield in Western United States. Journal of the Hydraulics Div., ASCE, Vol. 98, No. 12, pp. 2073-2085.

Fogel, M.M., Hekman, L.H., and Duckstein, L. 1977. A Stochastic Sediment Yield Model Using the Modified Universal Soil Loss Equation. in: Soil Erosion: Prediction and Control, SCSA Special Publ. No. 21, pp. 226-233.

Foster, G.R. 1981. Modeling the Erosion Process. in: ASAE Monograph on Hydrologic Modeling of Small Watersheds. In Press.

Foster, G.R., Lane, L.J., Nowlin, J.D., Laflen, J.M., and Young, R.A. 1980. A Model to Estimate Sediment Yield from Field-Sized Areas: Development of Model. in: CREAMS: A Field Scale Model for Chemicals, Runoff, and Erosion from Agricultural Management Systems, Vol. 1. Model Documentation, Chapter 3, USDA Conservation Research Report No. 26, pp. 36-64.

Foster, G.R. and Meyer. L.D. 1975. Mathematical Simulation of Upland Erosion by Fundamental Erosion Mechanics. in: Present and Pro-spective Technology for Predicting Sediment Yields and Sources, USDA, ARS-S-40, pp. 190-207.

Gottschalk, L.C. and Brune, G.M. 1950. Sediment Design Criteria for the Missouri Basin Loess Hills. USDA, SCS TP-97.

Holtan, H.N. 1979. Procedures Manual for Sediment, Phosphorus and
Nitrogen Transport Computations with USDAHL. Maryland Agricultural
Experiment Station, University of Maryland, College Park, Maryland,
MP-943, 34pp.

Holtan, H.N., Stiltner, G.J., Henson, W.H., and Lopez, N.C. 1975.
USDAHL-74 Revised Model of Watershed Hydrology. USDA Tech. Bull.
No. 1518, 99pp.

Knisel, W.G. 1980. CREAMS, A Field Scale Model for Chemicals, Runoff,
and Erosion from Agricultural Management Systems. USDA Conservation
Research Report No. 26, 643pp.

Laursen, E. 1958. The Total Sediment Load of Streams. Journal of the
Hydraulics Division, ASCE, Vol. 54, No. HY1, Paper 1530.

Leytham, K.M. and Johanson, R.C. 1979. Watershed Erosion and Sediment
Transport Model. Environmental Protection Agency, EPA-600/3-79-028,
357pp.

Li, Ruh-Ming, Simons, D.B., and Carder, D.R. 1977. Mathematical Modeling
of Soil Erosion by Overland Flow. in: Soil Erosion: Prediction
and Control, SCSA Special Publ. No. 21, pp. 210-216.

Maner, S.B. 1958. Factors Affecting Sediment Delivery Rates in the Red
Hills Physiographic Area. Trans. American Geophysical Union,
Vol. 39, No. 4, pp. 669-675.

Maner, S.B. 1962. Factors Influencing Sediment Delivery Ratios in the
Blackland Prairie Land Resource Area. USDA, Soil Conservation Service,
Fort Worth, Texas, 10pp.

Mills, W.C. 1981. Deriving Sediment Yield Probabilities for Evaluating
Conservation Practices. Trans. of the American Society of Agricultural Engineers.

Mulkey, L.A. and Falco, J.W. 1979. Sedimentation and Erosion Control
Implications for Water Quality Management. Proceedings of the
National Symposium on Erosion and Sedimentation by Water, ASAE
Publ. 4-77, 22pp.

Musgrave, G.W. 1947. The Quantitative Evaluation of Factors in Water
Erosion, A First Approximation. Journal of Soil and Water Conservation,
Vol. 2, pp. 133-138.

Negev, M. 1967. A Sediment Model on a Digital Computer. Department of
Civil Engineering, Stanford University, Tech. Report No. 76,
109pp.

Onstad, C.A. and Bowie, A.J. 1977. Basin Sediment Yield Modeling Using
Hydrological Variables. in: Erosion and Solid Matter Transport in
Inland Waters, IAHS-AISH, Publ. No. 122, pp. 191-202.

Onstad, C.A. and Foster, G.R. 1975. Erosion Modeling on a Watershed.
Trans. of the American Society of Agricultural Engineers, Vol. 18,
No. 2, pp. 288-292.

Pacific Southwest Inter-Agency Committee. 1968. Factors Affecting

Sediment Yield and Measures for the Reduction of Erosion and Sediment Yield. 13pp.

Renard, K.G. 1977. Erosion Research and Mathematical Modeling. in: Erosion: Research Techniques, Erodibility, and Sediment Delivery. T.J. Toy editor. GEO Books, Norwich, England, pp. 31-44.

Renard, K.G. 1980. Estimating Erosion and Sediment Yield from Rangeland. in: ASCE Symposium on Watershed Management, Boise, Idaho, pp. 164-175.

Renard, K.G. and Lane, L.J. 1975. Sediment Yield as Related to a Stochastic Model of Ephemeral Runoff. in: Present and Prospective Technology for Predicting Sediment Yields and Sources, USDA, ARS-S-40, pp. 253-263.

Roehl, J.W. 1962. Sediment Source Areas, Delivery Ratios and Influencing Morphological Factors. in: Land Erosion, IAHS Publ. No. 59, pp. 202-213.

Shirley, E.D. and Lane, L.J. 1978. A Sediment Yield Equation from an Erosion Simulation Model. Proceedings of the 1978 Meetings on Hydrology and Water Resources in Arizona and the Southwest, Flagstaff, Arizona, Vol. 8, pp. 90-96.

Simons, D.B., Reese, A.J., Li, Ruh-Ming, and Ward, T.J. 1977. A Simple Method for Estimating Sediment Yield. in: Soil Erosion: Prediction and Control, SCSA Special Publ. No. 21, pp. 234-241.

Smith, J.H., Davis, D.R., and Fogel, M.M. 1977. Determination of Sediment Yield by Transferring Rainfall Data. Water Resources Bulletin, AWRA, Vol. 13, No. 3, pp. 529-541.

Soil Conservation Service. 1966. Procedures for Determining Rates of Land Damage, Land Depreciation, and Volume of Sediment Produced by Gully Erosion. USDA Tech. Release No. 32, Geology.

Soil Conservation Service. 1971. Sediment Sources, Yields, and Delivery Ratios. Soil Conservation Service National Engineering Handbook, Section 3, Sedimentation, Chapter 6.

Task Committee on Preparation of Manual of Sedimentation, Vito A. Vanoni, Chairman. 1970. Sedimentation Engineering, Chapter IV: Sediment Sources and Sediment Yields. Journal of the Hydraulics Division, ASCE, Vol. 96, No. HY6, Proc. Paper 7337, pp. 1283-1329.

Williams, J.R. 1974. Predicting Sediment Yield Frequency for Rural Basins to Determine Man's Effect on Long-Term Sedimentation. in: Effects of Man on the Interface of the Hydrological Cycle with the Physical Environment, Symposium Proceedings IAHS Publ. No. 113, pp. 105-108.

Williams, J.R. 1975. Sediment-Yield Prediction with Universal Equation Using Runoff Energy Factor. in: Present and Prospective Technology for Predicting Sediment Yield and Sources, USDA, ARS-S-40, pp. 244-252.

Williams, J.R. 1975. Sediment Routing for Agricultural Watersheds. Water Resources Bulletin, AWRA, Vol. 11, No. 5, pp. 965-974.

Williams, J.R. 1977. Sediment Delivery Ratios Determined with Sediment and Runoff Models. in: Erosion and Solid Matter Transport in Inland Waters, IAHS-AISH, Publ. No. 122, pp. 168-179.

Williams, J.R. 1978. A Sediment Yield Routing Model. Proceedings of the Specialty Conference on Verification of Mathematical and Physical Models in Hydraulic Engineering, ASCE, College Park, Maryland, pp. 662-670.

Williams, J.R. 1978. A Sediment Graph Model Based on an Instantaneous Unit Sediment Graph. Water Resources Research, Vol. 14, No. 4, pp. 659-664.

Williams, J.R. 1979. Model for Predicting Sediment, Phosphorus, and Nitrogen Yields from Rural Basins. in: Hydraulic Engineering in Water Resources Development and Management, Proceedings of XVIII Congresso, IAHR, Italia, Vol. 5D, pp. 107-116.

Williams, J.R. 1980. SPNM, A Model for Predicting Sediment, Phosphorus, and Nitrogen Yields from Agricultural Basins. Water Resources Bulletin, AWRA, Vol. 16, No. 5, pp. 843-848.

Williams, J.R. and Berndt, H.D. 1972. Sediment Yield Computed with Universal Equation. Journal of the Hydraulics Div., ASCE, Vol. 98, No. HY12, pp. 2087-2098.

Williams, J.R. and Berndt, H.D. 1977. Sediment Yield Prediction Based on Watershed Hydrology. Trans. of the American Society of Agricultural Engineers, Vol. 20, No. 6, pp. 1100-1104.

Williams, J.R. and Hann, R.W. 1978. Optimal Operation of Large Agricultural Watersheds with Water Quality Constraints. Texas Water Resources Institute, Texas A&M University, Tech. Report No. 96, 152pp.

Wischmeier, W.H. and Smith, D.D. 1960. A Universal Soil-Loss Equation to Guide Conservation Farm Planning. 7th International Congress of Soil Science Trans., Vol. 1, pp. 418-425.

Wischmeier, W.H. and Smith, D.D. 1965. Predicting Rainfall-Erosion Losses from Cropland East of the Rocky Mountains. Agriculture Handbook 282, USDA, ARS, 47pp.

Wischmeier, W.H. and Smith, D.D. 1978. Predicting Rainfall Erosion Losses. Agriculture Handbook 537, USDA, SEA, 58pp.

Woolhiser, D.A. and Renard, K.G. 1980. Stochastic Aspects of Watershed Sediment Yield. in: Application of Stochastic Processes in Sediment Transport. H.W. Shen and H. Kikkawa editors. Water Resources Publications, Littleton, Colorado, Chapter 3, 28pp.

Yang, C.T. 1973. Incipient Motion and Sediment Transport. Journal of the Hydraulics Div., ASCE, Vol. 99, No. HY10, pp. 1679-1704.

EXPLICIT SOLUTIONS TO KINEMATIC EQUATIONS FOR EROSION ON AN INFILTRATING PLANE

Vijay P. Singh*
Associate Professor of Civil Engineering
Department of Civil Engineering
Louisiana State University
Baton Rouge, LA 70803, U.S.A.

Shyam N. Prasad
Professor of Civil Engineering
Department of Civil engineering
The University of Mississippi
University, MS 38677, U.S.A.

ABSTRACT

The one-dimensional form of kinematic equations have been utilized in the past for modeling erosion from a sloping plane subject to rainfall or the effective rainfall of finite duration. These equations consist of continuity equation and kinematic approximation to momentum equation for water and continuity equation for sediment. The sediment continuity requires specification of a source term which is frequently considered to be composed of interrill and rill erosion. Interrill erosion is assumed entirely due to rainfall impact. Rill erosion is defined as erosion due to tractive forces and transport capacity of the flow as it occurs in small rills and channels. Solutions to kinematic equations subject to simple initial and boundary conditions, obtained in the past, have been mostly numerical.

This study employs the same kinematic formulation, as used previously, for erosion from a sloping plane subject to infiltration. However, infiltration, runoff and erosion are considered concurrently during and after the occurrence of rainfall. This simultaneous treatment, subject to initial and boundary conditions, gives rise to a free boundary problem not encountered in the case when infiltration is either disregarded or considered through the effective rainfall. Thus the effective rainfall is no longer used. It is assumed that rainfall and infiltration are space-time invariant and occur for finite periods of time. Given these assumptions, explicit solutions are derived for sediment concentration and sediment discharge. The solutions contain a function which is related to a generalization of Dawson's integral. This integral can be easily evaluated and tabulated for ready use. It is shown that the solutions obtained by previous investigators constitute special cases of the solutions derived here.

*Formerly at Mississippi State University, Mississippi State, Mississippi 39762, U.S.A.

INTRODUCTION

In recent years several studies have addressed themselves to modeling of erosion by rainfall of soil from upland areas. An excellent exposition of the present and prospective technology for prediction and control of soil erosion is contained in Bennett (1974), Agricultural Research Service (1975), Water Resources Council (1976), and Soil Conservation Society of America (1977). From a survey of literature it is evident that there are several approaches to modeling erosion from upland watersheds (Foster and Meyer, 1972, 1975; Bennett, 1974; Simons, Li and Stevens, 1975; Curtis, 1976; Komura, 1976; Smith, 1976; Williams, 1978; Rendon-Herrero, 1978; Wischmeier and Smith, 1978; Li, 1979; Foster and Lane, 1980; Knisel, 1980; Ross, Shanholtz and Contractor, 1980). Of these approaches one approach has been to utilize kinematic wave theory (Lighthill and Whitham, 1955) to develop mathematical models (Li, Shen and Simons, 1973; Hjelmfelt, Piest and Saxton, 1975; Curtis, 1976; Smith, 1976; Li, Simons and Shiao, 1977; Shirley and Lane, 1978; Li, 1979; Ross, Shanholtz and Contractor, 1980). In this approach the one-dimensional kinematic equations have been utilized for modeling erosion from upland areas subject to rainfall or the effective rainfall of finite duration. Infiltration has been considered through the effective rainfall. Solutions to these equations have been mostly numerical.

This study employs the same kinematic equations for modeling erosion. However, infiltration, runoff and erosion are considered concurrently during and after the occurrence of rainfall. This simultaneous treatment, subject to initial and boundary conditions, gives rise to a free boundary problem not encountered in the case when infiltration is either disregarded or considered through the effective rainfall. Explicit solutions are derived for sediment concentration and sediment discharge. The solutions contain a function which is related to a generalization of Dawson's integral. This integral can be easily evaluated and tabulated for ready use.

KINEMATIC MODELING

Kinematic modeling requires specification of (1) geometry, (2) kinematic equations, (3) rainfall and infiltration, and (4) initial and boundary conditions. Depending upon the geometric complexity upland watersheds may be represented by a combination of (1) plane, (2) converging section, (3) diverging section, and (4) channel. Thus a sloping plane can be employed to represent an upland area or a portion thereof. This study will be confined to a plane geometry only.

Kinematic Equations

Kinematic equations for modeling erosion from upland areas consist of (1) continuity equation for water, (2) kinematic depth-discharge relation, and (3) continuity equation for sediment. These equations, in one dimensional form, can be written for a plane on a unit width basis (Bennett, 1974; Li, 1979) as

$$\frac{\partial h}{\partial t} + u \frac{\partial h}{\partial x} + h \frac{\partial u}{\partial x} = q - f \tag{1}$$

$$u = \alpha n^{n-1} \; ; \quad Q = \alpha h^n \tag{2}$$

$$\frac{\partial(Ch)}{\partial t} + \frac{\partial}{\partial x}(h\,u_s C) + (1-\lambda)\frac{\partial y}{\partial t} = \frac{\partial h}{\partial x}\,\varepsilon_s\frac{\partial C}{\partial x} + E \tag{3}$$

$$Q_s = CQ \tag{4}$$

where h is depth of flow (L), u velocity of flow (L/T), Q discharge of water per unit width (L^2/T), y local bed elevation (F/L^2), q lateral inflow or rainfall (L/T), f infiltration (L/T), C sediment concentration (F/L^3), Q_s sediment discharge per unit width (F/TL), u_s average velocity of sediment (L/T), x distance in the direction of flow (L), t time (T), n an exponent, α depth-discharge coefficient (L^{2-n}/T), ε_s sediment particle mass transfer coefficient (L^2/T), λ porosity of deposited sediment, and E lateral sediment inflow from adjacent flow units (F/TL^2).

Equations (1)-(3) describe the movement of suspended sediment particles in free surface plane flow. The right side of equation (3) accounts for dispersion of sediment during suspension in the flow, and is normally negligible (Bennett, 1974); as a result, it will be neglected in this study. The second term on the left side of equation (3) contains u_s, the transport velocity of suspended sediment, which may not be equal to u, the flow velocity. However, it will be assumed here that u_s = u, (Smith, 1976; Li, 1969). The third term on the left side of this equation accounts for deposition or erosion from the bed.

Equation (3) can be written in a simplified form (Hjelmfelt, Piest, Saxton, 1975; Borah, 1979) as

$$\frac{\partial(Ch)}{\partial t} + \frac{\partial Q_s}{\partial t} = E_I + E_R \tag{5}$$

where E_R represents the rate of sediment flux, across the control volume surface, by erosion or deposition per unit length (F/TL^2). The amount of sediment available for transport is the sum of the sediment inflow from contributing flow units, initial loose soil depth left from previous storms, the amount of soil detached by rain-drop impact and the amount of soil detached by flow of water. In the context of erosion from upland watersheds, E_I may be considered to represent interrill erosion or sheet erosion which is principally due to the impact of rain. This has been discussed in detail by Foster, Meyer and Onstad (1973). They conclude that E_I varies with rainfall intensity, soil characteristics, vegetative cover and slope. As used by Hjelmfelt, Piest and Saxton, (1975) and Shirley and Lane (1978), E_I, as a first approximation, can be expressed as

$$E_I = Bq^m \tag{6}$$

where B is a coefficient (F/L^3) and m an exponent.

E_R can be considered to represent rill erosion but accounts for deposition also. Based on the work of Foster and Meyer (1972) a relationship for E_R was derived by Hjelmfelt, Piest and Saxton (1975) as

$$E_R = \gamma(Kh^b - CQ) \tag{7}$$

where $\gamma(1/L)$ and $K(F/TL^{1+b})$ are coefficients. b is an exponent

517

which may be identified with the exponent n of equation (2). This appears plausible since it in effect assumes that the sediment exists in the same flow regime as the water in which it is imbedded. Equation (7) was also used by Shirley and Lane (1978). In this study equation (5) will be used in conjunction with equations (6)-(7) for conservation of sediment mass.

Rainfall, q(x,t)

We make the assumption that q is space-time invariant and is represented by

$$q(x,t) = q > 0, \; t < T, \; t > 0$$

$$q(x,t) = 0, \; t \geq T \tag{8}$$

where T is the duration of q.

Infiltration

Let $f(x,t)$ be the rate of infiltration. Then f is dependent on the depth of flow h in the following sense,

$$f(x,t) > 0 \quad \text{if } h(x,t) > 0$$

$$f(x,t) = 0 \quad \text{if } h(x,t) = 0$$

We will assume further that

$$q(x,t) > f(x,t), \; 0 \leq t \leq T, \; 0 \leq x \leq L$$

where L is the length of planar flow. In this study we will consider the case when $f(x,t) = f$, a constant.

Initial and Boundary Conditions

To solve for flow of water and sediment on a sloping plane subject to rainfall and infiltration the following initial conditions can be assumed (Hjelmfelt, Piest, and Saxton, 1975; Shirley and Lane, 1978),

$$h(0,t) = 0, \quad t \geq 0; \quad h(x,0) = 0, \quad 0 \leq x \leq L \tag{9}$$

$$C(0,t) \text{ is bounded for } t > 0 \tag{10}$$

$$C(x,0) \text{ is bounded for } x > 0 \tag{11}$$

The plausibility of the conditions in equations (10) - (11) is seen when one considers the conditions in equation (9) and that at zero depth of water any finite amount of sediment would correspond to a sediment concentration of infinity. Thus at zero depth (no water) there must be zero quantity of sediment, and thus the sediment concentration is indeterminate, hence having no apriori definite value. We may, however, assume that the concentration is bounded since as a physically meaningful quantity it must be so. Therefore, the boundedness condition is imposed on the mathematical derivations. Thus the problem of erosion due to rainfall on a plane is formulated in equations (1) - (2), (5) - (10), and is hopefully well posed.

Solution Domain

It is plausible on physical grounds that there will be a curve $t = t°(x)$ in $t > T$, $0 < x < L$ starting at $x = 0$, $t = T$, and such that $h(x, t°(x)) = 0$. This curve gives the time history of the water edge as it recedes from $x = 0$ to $x = L$. Equations (1) - (2) are satisfied in the region $S \equiv \{0 < t < t°(x), 0 < x < L\}$. Thus $t°(x)$ is a free boundary, and equations (1) - (2), (9), and $h(x, t°(x)) = 0$ constitute a free boundary problem. In the domain above the curve $t = t°(x)$, $h(x,t) = 0$. The determination of free boundary is relatively simple when q and f are constant.

MATHEMATIC SOLUTIONS

The solution of the problem formulated above can be considered in two parts. The first part consists of solving equations (1) - (2) subject to equations (8) - (9). Singh (1976), and Singh and Mahmood (1979) have given solutions for this part. For the sake of completeness we will summarize these solutions here and make some pertinent remarks. The second part consists of solving equations (5) - (7) subject to equations (10) - (11), using the solution of the first part. The method of characteristics can be employed to obtain the solutions. Depending upon the relative disposition of the bounding characteristics of sediment and water (the characteristics issuing from the origin (0,0)) as well as the line $t = T$ five cases can be distinguished, A, B, C, D, and E as shown in figures 1 - 5. We derive solutions for each of these five cases under the assumption that q and f are constant.

For simplicity we introduce the following symbols:

$$q^* = q - f$$

$$\beta = (\tfrac{q}{2})^{1/n}$$

$$\beta^* = (q^*/2)^{1/n}$$

$$\xi = \frac{1}{n}(q/2)^{-(n-1)/n}$$

$$\xi^* = \frac{1}{n}(q^*/2)^{-(n-1)/n}$$

$$\rho = f/q$$

$$r = (n-1)/n$$

$$B_* = Bq^m/q^*$$

Solution for Case A

The solution domain for this case, as shown in figure 1, can be partitioned into domains D_1, D_2 composed of D_{2A} and D_{2B}, and D_3. The domain D_1 is bounded by $t = 0$, $x = L$, and $t = t(x,0)$; $t = t(x,0)$ is the bounding water characteristic. The domain D_3 is bounded by $x = 0$, $t = T$, $x = L$ and $t = t°(x)$. The domain D_2 is bounded by $x = 0$, $t = T$, $x = L$, and $t(x,0)$. It is divided into D_{2A} and D_{2B} by the bounding sediment characteristic $t = t_s(x,0)$.

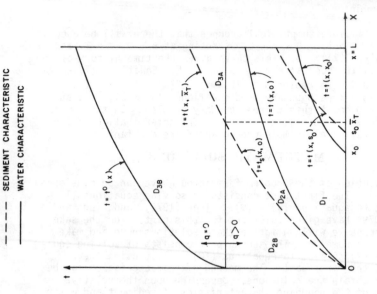

Figure 2. Solution domain for Case B.

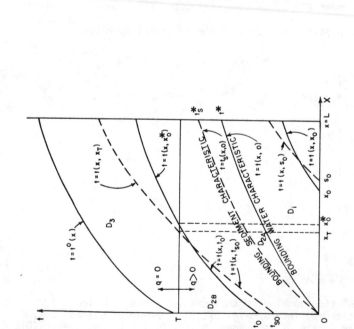

Figure 1. Solution domain for Case A.

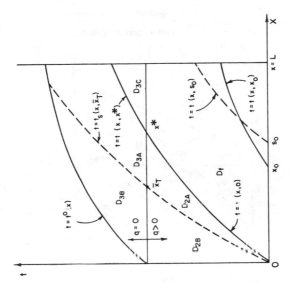

--- SEDIMENT CHARACTERISTIC

—— WATER CHARACTERISTIC

Figure 4. Solution domain for Case D.

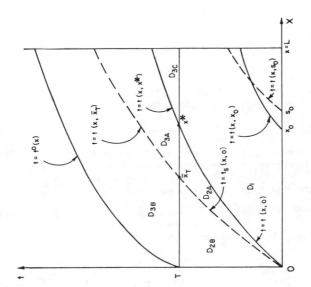

--- SEDIMENT CHARACTERISTICS

—— WATER CHARACTERISTICS

Figure 3. Solution domain for Case C.

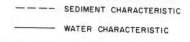
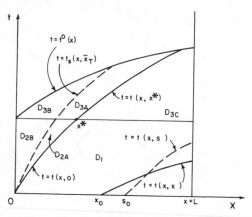

Figure 5.   Solution domain for Case E.

Domain D_1

The solution of equations (1) – (2) subject to equation (9) is given in parametric form by

$$h(x,x_0) = [\frac{q^*(x-x_0)}{\alpha}]^{1/n} \tag{12}$$

$$t(x,x_0) = q^{*^{-r}}(\frac{x-x_0}{\alpha})^{1/n} \tag{13}$$

where x_0, $0 \le x_0 \le L$, is a parameter representing the intersection of a characteristic curve with the x-axis. This parameter may be eliminated to give h as an explicit function of x and t,

$$h(x,t) = q^* t \tag{14}$$

Thus in this domain h is independent of x. In other words the flow is uniform but unsteady, and increases with time ($\partial h/\partial t > 0$).

Substituting equation (14) into equation (4),

$$Q_s = C\alpha q^{*^n} t^n \tag{15}$$

Inserting equation (15) in equation (5)

$$\frac{\partial C}{\partial t} + \alpha q^{*^{n-1}} t^{n-1} \frac{\partial C}{\partial x} t(\frac{1}{t} + \gamma\alpha q^{*^{n-1}} t^{n-1}) C = \frac{B_*}{t} + K\gamma q^{*^{n-1}} t^{n-1}$$

By applying the method of characteristics the characteristic base curves are thus seen to satisfy

$$\frac{dx}{dt} = \alpha q^{*n-1} t^{n-1} \qquad (16)$$

The concentration C along these curves satisfies

$$\frac{dC}{dt} + (\frac{1}{t} + \gamma\alpha q^{*n-1} t^{n-1})C = \frac{B_*}{t} + \gamma Kq^{*n-1} t^{n-1} \qquad (17)$$

The solution to equation (16) is

$$(x-S_0) = \frac{\alpha}{n} q^{*n-1} t^n \qquad (18)$$

where S_0, $0 \le S_0 \le L$, is the point of intersection of the particular curve with $t = 0$ axis. Equation (18) may be solved to get

$$t = (n)^{1/n} q^{*-1} [\frac{x - y_0}{\alpha}]^{1/n} \qquad (19)$$

which is seen to be a constant multiple of equation (13). Since n is invariably greater than unity, it is clear that these characteristic lines intersect those given by equation (13).

Because equation (17) is linear, its solution can be obtained using standard techniques. Its complementary solution C_0 is

$$C_0 = \frac{A}{t} \exp (- \frac{\gamma\alpha q^{*n-1} t^n}{n}) \qquad (20)$$

Where A is the constant of integration. The particular integral may be found, after some manipulation, to be

$$C_p = \frac{K}{\alpha} + (B_* - \frac{K}{\alpha}) \; F_n [(\frac{\gamma\alpha}{n})^{1/n} q^{*r} t] \qquad (21)$$

where the function $F_n(x)$ is defined by

$$F_n(x) = \frac{1}{x \exp (x^n)} \int_0^x \exp (s^n)ds \qquad (22)$$

This function is considered as a generalization of Dawson's integral (Abramowitz and Stegun, 1964). Some pertinent properties of the function may be shown to be:

(a) $F_n(x)$ is a monotonically decreasing function of x for values of n > 0.

(b) $F_n(x)$ is positive for all x > 0.

(c) $F_n(x) = 1$.

(d) $F_n(\infty) = 0$, n > 0.

Thus along the characteristic curves given by equation (16) the sediment concentration $C_0 + C_p$ subject to the condition that at $t = 0$, C is bounded,

$$C = \frac{K}{\alpha} + (B_* - \frac{K}{\alpha}) \; F_n \; [(\frac{\gamma\alpha}{n})^{1/n} \; q^{*r} \; t] \qquad (23)$$

Taking into account the properties of $F_n(x)$ it is seen that

$$C(x,0) = B_* \; ; \quad C(x,\infty) = K/\alpha \qquad (24)$$

Thus the initial sediment concentration, which had previously been indeterminate, is seen to be equal to $B_* = Bq^m/q^*$, while the concentration approaches K/α after a long time. Further, equation (23) is independent of S_0 and thus independent of x indicating that in domain D_1 the temporal growth (or decay) of sediment concentration is the same for all points on the watershed surface. This implies that C is uniform but unsteady with $dC/dt < 0$ if $(B_* - K/\alpha) < 0$.

Domain D_2

The solution of equations (1) - (2) subject to equation (9) is given in parametric form by

$$h(x,t_0) = (q^* x/\alpha)^{1/n} \qquad (25)$$

$$t(x,t_0) = t_0 + q^{*-r} \; (x/\alpha)^{1/n} \qquad (26)$$

where the parameter t_0, $0 \leq t_0 \leq T$, represents the intersection of a characteristic curve with the t-axis. Here h is independent of t_0 and thus independent of t, that is,

$$h(x,t) = (q^* x/\alpha)^{1/n} \qquad (27)$$

This implies that in this domain the flow of water is steady but nonuniform with $\partial h/\partial t = 0$.

In domain D_2 equation (5) must be satisfied with the corresponding h given by equation (27) and subject to the condition that on the bounding water characteristic $t = t(x,0)$ between domains D_1 and D_2 the concentration C must assume the same values as those of equation (23) evaluated on that characteristic and that on the $x = 0$ axis C is bounded.

In this domain equation (5) becomes

$$\frac{\partial C}{\partial x} + (\frac{1}{\alpha})^{1/n} \; (q^* x)^{-r} \; \frac{\partial C}{\partial t} = \frac{1}{x} (B_* - C) + \frac{\gamma}{\alpha} (K - C\alpha) \qquad (28)$$

The characteristic base curves thus satisfy

$$\frac{dt}{dx} = (1/\alpha)^{1/n} \; (q^* x)^{-r} \qquad (29)$$

and C along these characteristic satisfies

$$\frac{dC}{dx} + (\gamma + \frac{1}{x})\, C = \frac{K\gamma}{\alpha} + \frac{B_*}{x} \qquad (30)$$

Equation (29) can be integrated to get the characteristic base curves as

$$t = t_{s0} + nq^{*-r}\,(\frac{x}{\alpha})^{1/n} \qquad (31)$$

where t_{s0}, $x \leq t_{s0} \leq T$, is the point of intersection of this characteristic with the $x = 0$ axis. It is seen that the characteristic curves are given by an expression that is n times the flow characteristic given by equation (26). The bounding sediment characteristic is obtained by putting t_{s0} equal to 0,

$$t_s(x,0) = nq^{*-r}\,(\frac{x}{\alpha})^{1/n} \qquad (32)$$

This characteristic divides the domain D_2 into two subdomains D_{2A} and D_{2B} as shown in figure 1. Equation (30) can be solved to give

$$C = \frac{K}{\alpha} + (B_* - \frac{K}{\alpha})\, F_1(\gamma x) + \frac{A}{x}\,\exp\,(-\gamma x) \qquad (33)$$

where A is a constant of integration, and F_1 (.) is a function defined by equation (22) with n = 1,

$$F_1(\gamma x) = \frac{\exp\,(\gamma x) - 1}{\gamma x\,\exp\,(\gamma x)} \qquad (34)$$

The constant A is to be evaluated for each of the subdomains D_{2A} and D_{2B} by the condition that at $x = 0$, C is bounded, and that along the bounding flow characteristic C must match the solution for domain D_1. Thus in subdomain D_{2B},

A = 0

In subdomain D_{2A} the characteristic curves given by equation (32) are characterized by the fact that t_{s0} is negative in this region. Indeed in this region,

$$(1-n)\, q^{*-r}\,(\frac{L}{\alpha})^{1/n} \leq t_{s0} \leq 0$$

Since $n > 1$, $t_{s0} < 0$ in subdomain D_{2A}. Any value of t_{s0} within these limits will define a sediment characteristic that intersects the bounding flow characteristic at a point where the coordinates are

$$t = \frac{t_{s0}}{1-n} \qquad (35)$$

$$x = \alpha q^{*1-n}\,(\frac{t_{s0}}{1-n})^n \qquad (35b)$$

At such a point the value of C given by equation (33) must equal that given by equation (23). Toward this end we introduce the dimensionless parameter τ_0 defined by

$$\tau_0 = \frac{\alpha q_*^{n-1}}{L} \left(\frac{t_s 0}{1-n}\right)^n \tag{36}$$

where in subdomain D_{2A}, $0 \le \tau_0 \le 1$. Writing equation (23) in terms of τ_0,

$$C = \frac{K}{\alpha} + (B_* - \frac{K}{\alpha})\, F_n\, [(\frac{\gamma L \tau_0}{n})^{1/n}] \tag{37}$$

Writing equation (33) in terms of τ_0,

$$C = \frac{K}{\alpha} + (B_* - \frac{K}{\alpha})\, F_1\, (\gamma L \tau_0) + \frac{A}{\gamma L \tau_0}\, \exp(-\gamma L \tau_0) \tag{38}$$

Determining A by equation (37) with equation (38),

$$A = \frac{1}{\gamma} (B_* - \frac{K}{\alpha}) \; [1 - \exp(\gamma L \tau_0) \{1 - \gamma L \tau_0\, F_n \left(\frac{\gamma L \tau_0}{n}\right)^{1/n} \}]$$

Thus in domain D_{2A} the sediment concentration is given by

$$C = \frac{K}{\alpha} + (B_* - \frac{K}{\alpha})\, F_1\, (\gamma x) + \frac{\exp(-\gamma x)}{\gamma x} (B_* - \frac{K}{\alpha}) \; [1 - \exp(\gamma L \tau_0)$$

$$\{1 - \gamma L \tau_0\, F_n\, (\gamma L \tau_0/n)^{1/n} \}] \tag{39}$$

The parameter τ_0 may be eliminated by using equations (36) and (31) to give C as an explicit function of x and t, which can be conveniently expressed as

$$C = \frac{K}{\alpha} + (B_* - \frac{K}{\alpha}) \; [F_1\, (\gamma x) + \frac{\exp(-\gamma x)}{\gamma x} \{1 - \exp(\mu(x,t))$$

$$(1-\mu(x,t)\, F_n(\frac{1}{n}\mu(x,t)))\}] \tag{40}$$

where

$$\mu(x,t) = \gamma \alpha q_*^{n-1} \; [\frac{nq_*^{-r}\, (x/\alpha)^{1/n} - t}{n-1}]^n \tag{41}$$

In subdomain D_{2B},

$$C = \frac{K}{\alpha} + (B_* - \frac{K}{\alpha}) \; [\frac{\exp(\gamma x) - 1}{\gamma x \exp(\gamma x)}] \tag{42}$$

It is thus seen that C depends on both x and t (unsteady and nonuniform) in subdomain D_{2A} whereas it is dependent on x only in D_{2B}. Therefore, $\partial C/\partial t = 0$ in D_{2B}, and $\partial C/\partial t < 0$ in D_{2A}.

Domain D_3

In this domain q = 0. The solution for both water and sediment can, however, be obtained in a similar manner. The solution for flow of water is given by

$$h(x,x_0^*) = \left[\frac{qx_0^* - fx}{\alpha} \right]^{1/n} \tag{43}$$

$$t(x,x_0^*) = T + \frac{1}{f} \left[\frac{-(qx_0^* - fx)^{1/n} + ((q-f) x_0^*)^{1/n}}{\alpha^{1/n}} \right] \tag{44}$$

where $x_0^*, 0 \leq x_0^* \leq L$, is the intersection of a characteristic $t - t(x,x_0^*)$, with the line $t = T$, $t(x,x_0^*)$ is an elongation of $t(x,x_0)$ beyond t = T. It is to be noted that the characteristics in this domain are not straight lines and that h is not constant along a characteristic.

The parameter x_0^* may be eliminated to give an explicit expression (in inverted form) for h as a function of x and t,

$$(t - T) = -\frac{h}{f} + \frac{1}{f} \left[\frac{(q-f) (\alpha h^n + fx)}{\alpha q} \right]^{1/n}$$

or

$$(t - T) = -\frac{h}{f} + \frac{1}{f} \left(\frac{1}{\alpha}\right)^{1/n} \left[(1 - \rho) (\alpha h^n + \rho x) \right]^{1/n} \tag{45}$$

It is evident that h depends on both x and t (unsteady, nonuniform flow). The free boundary $t = t^\circ(x)$ is determined by

$$qx_0^* - fx = 0 \tag{46}$$

and equation (44). Eliminating x_0^* between equations (44) and (46),

$$t^\circ(x) = T + \frac{f^{(1-n)/n}}{\alpha^{1/n}} \left[x - \frac{fx}{q} \right]^{1/n}$$

or

$$t^\circ(x) = T + f^{(1-n)/n} \left[x(1 - \rho)/\alpha \right]^{1/n} \tag{47}$$

It may be appropriate to summarize the results of water flow at this point. Figure 6 is a sketch of the characteristic curves in the various domains. Figure 7 shows solution domain for water

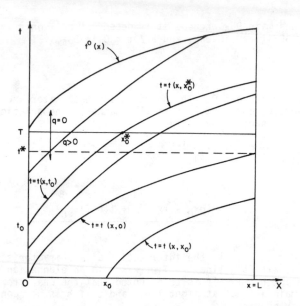

Figure 6. Characteristic diagram for water flow in
Case A with infiltration.

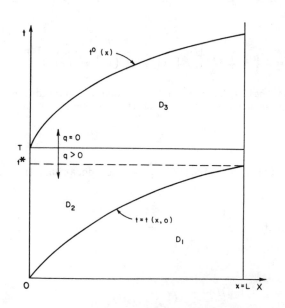

Figure 7. Solution domain for water flow in
Case A with infiltration.

528

flow problem. For a fixed value of x, a hydrograph may be developed from these expressions showing the depth of water h as a function of time. Figure 8 is a sketch of such a system of hydrographs for various values of x. In the figure $\bar{t}(x_j)$ is the time taken by water to travel from the upper end of the watershed to location x_j. For any value of x the maximum depth $\bar{h}(x)$ is achieved at time $\bar{t}$ given by

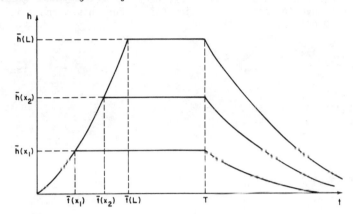

Figure 8.   Depth hydrographs as a function of x in Case A with infiltration; $\bar{h}(x)$ is maximum depth at x; and $\bar{t}(x)$ is time to minimum depth at x.

$$\bar{t}(x) = q^{*-r} \left(\frac{x}{\alpha}\right)^{1/n} \qquad (48)$$

$$\bar{h}(x) = \left(\frac{q^* x}{\alpha}\right)^{1/n} \qquad (49)$$

For sediment concentration, equation (5) needs to be modified. From equation (1),

$$\frac{\partial h}{\partial t} = -\frac{\partial Q}{\partial x} - f$$

Thus equation (5) can be modified to

$$\frac{1}{\alpha h^{n-1}} \frac{\partial C}{\partial t} + \frac{\partial C}{\partial x} = \frac{Cf}{\alpha h^n} + \gamma \left(\frac{K}{\alpha} - C\right) \qquad (50)$$

The characteristic base curves thus satisfy

$$\frac{dt}{dx} = \frac{1}{\alpha h^{n-1}} \qquad (51)$$

The sediment concentration along these characteristics satisfies

$$\frac{dC}{dx} = \frac{Cf}{\alpha h^n} + \gamma \left(\frac{K}{\alpha} - C\right) \tag{52}$$

Equation (50) must be solved using the expression for h in domain D_3 given by equation (45). However, it is more convenient to use the parametric equations (43) - (44) to define h in this region than the explicit but inverted form of equation (45). For this purpose it is convenient to define a new variable as

$$y = \frac{qx_0^* - fx}{\alpha} \tag{53a}$$

Writing equation (43) as

$$h(x, x_0^*) = y^{1/n} \tag{53b}$$

and equation (44) as

$$x = \frac{\alpha q}{f(q - f)} \left[(t - T) f + y^{1/n}\right]^{n-1} - \frac{\alpha}{f} \tag{54}$$

Equation (51) becomes

$$\frac{dt}{dx} = \frac{1}{\alpha y^{(n-1)/n}} \tag{55}$$

To solve equation (55) we need to obtain dt/dy which can be obtained as

$$\frac{dt}{dy} = \frac{dt}{dx} \frac{\partial x}{\partial y} \left(1 - \frac{\partial x}{\partial t} \frac{dt}{dx}\right)^{-1} \tag{56}$$

Using equation (54) we can write

$$\frac{\partial x}{\partial y} = \frac{\alpha q y^{(1-n)/n}}{f(q - f)} \left[(t - T) f + y^{1/n}\right]^{n-1} - \frac{\alpha}{f}$$

$$\frac{\partial x}{\partial t} = \frac{\alpha n \, q}{(q - f)} \left[(t - T) f + y^{1/n}\right]^{n-1}$$

Therefore,

$$\frac{dt}{dy} = - \frac{(q - f) - q y^{(1-n)/n} \left[(t - T) f + y^{1/n}\right]^{n-1}}{f[(q - f) y^{(n-1)/n} - nq\{(t - T) f + y^{1/n}\}^{n-1}]} \tag{57}$$

Equation (57) is nonlinear and its analytical solution does not seem tractable. However, a first approximation can be obtained by assuming f to be so small that the term $(t - T)f$ can be neglected in both numerator and denominator,

$$\frac{dt}{dy} = - \frac{f}{y^{(n-1)/n} \, [(n-1) \, q - f]} \tag{58}$$

On integrating,

$$t = - \frac{nf}{[(n-1) \, q + f]} \, y^{1/n} + A \tag{59}$$

where A is a constant of integration. Transforming back to the original variables,

$$t = - \frac{fn \, [(qx_0^* - fx)/\alpha]^{1/n}}{[(n-1) \, q + f]} + A$$

where x_0^* is given by equation (44). Denoting the point of intersection of a sediment characteristic with the line $t = T$ by x_T, A may be expressed in terms of x_T,

$$A = T + \frac{fn \, [(q - f) \, x_T/\alpha]^{1/n}}{(n-1) \, q + f}$$

The characteristic base curves for sediment are thus given by

$$t = T + \frac{fn}{(n-1) \, q + f} \, (\frac{1}{\alpha})^{1/n} \, [((q - f) \, x_T)^{1/n} - (qx_0^* - fx)^{1/n}] \tag{60}$$

Invoking the same assumption equation (52) can be simplified as

$$\frac{dC}{dt} = \gamma \, (\frac{K}{\alpha} - C) \tag{61}$$

whose solution is

$$C = A \exp (-\gamma x) + \frac{K}{\alpha} \tag{62}$$

The constant of integration A is to be evaluated using the condition that at $t = T$, C given by equation (62) must equal that given by equation (42),

$$A = (B_* - \frac{K}{\alpha}) \, [\frac{\exp (\gamma x_T) - 1}{\gamma x_T}]$$

531

Therefore,

$$C = \frac{K}{\alpha} + (B_* - \frac{K}{\alpha}) \exp(-\gamma x) [\frac{\exp(\gamma x_T) - 1}{\gamma x_T}] \tag{63}$$

It is seen from equation (63) that C, and consequently Q_s, depends on both x and t (nonuniform and unsteady).

Solution for Case B

The solution domain for this case, as shown in figure 2, is divided into domains D_1, D_2 composed of D_{2A} and D_{2B}, and D_3 composed of D_{3A} and D_{3B}. The domains D_1 and D_2 are as defined before. The subdomain D_{3A} is bounded by $t = T$, $x = L$ and $t = t_s(x, \bar{x}_T)$; and D_{3B} by $t = t^0(x)$, $t = T$, $t = t_s(x, \bar{x}_T)$ and $x = L$.

Domain D_1

The solution in this domain is given by equations (12) - (13) for water and equations (19) and (23) for sediment.

Domain D_2

The solution for water in this domain is given by equations (25) - (26). The solution for sediment in subdomain D_{2A} is given by equations (31) and (40) and in subdomain D_{2B} by equations (31) and (42).

Domain D_3

The solution for water in this domain is given by equations (43) - (44) with the free boundary given by (47). For $0 < x \leq \bar{x}_T$ the solution in subdomain D_{3B} is given by equations (60) and (63). For $\bar{x}_T \leq x < L$ the solution in subdomain D_{3A} remains yet to be determined. Here the solution will be given by equation (62) in which the constant A is to be evaluated using the condition that at $t = T$, C given by equation (62) must equal that given by equation (40),

$$A = (B_* - \frac{K}{\alpha}) \exp(\gamma x_T) \{F_1(\gamma x_T) + \frac{\exp(-\gamma x_T)}{\gamma x_T} [1 - \exp(\mu(x_T, T))$$

$$(1 - \mu(x_T, T) F_n(1/n \, \mu(x_T, T)))] \tag{64}$$

where

$$\mu(x_T, T) = \gamma \alpha q^{*(n-1)} [\frac{nq^{*-r} (x/\alpha)^{1/n} - 1}{n-1}]$$

Therefore,

$$C = \frac{K}{\alpha} + A \exp(-\gamma x) \tag{65}$$

where A is given by equation (64). The quantity $\bar{x}_T$ in $t = t_s\,(x,\bar{x}_T)$ is given by

$$\bar{x}_T = \alpha \left(\frac{T}{n}\right)^n q^{*(n-1)} \tag{66}$$

It is thus seen that C and consequently Q_s depends on both x and t.

Solution for Case C

The solution domain for this case is shown in figure 3. It is divided into domains D_1, D_2 composed of D_{2A} and D_{2B}, and D_3 composed of D_{3A}, D_{3B} and D_{3C}. The domains D_1 and D_2 are as defined before. The subdomain D_{3B} is bounded by $t = T$, $t = t_0(x)$, $t = t_s(x, \bar{x}_T)$ and $x = L$; D_{3A} by $t \triangleq t_s(x, \bar{x}_{T*})$, $t = T$, $t = t(x,x^*)$ and $x = {}^sL$, and D_{3C} by $t = T$, $x \triangleq L$ and $t \triangleq t(x,x^*)$.

Domain D_1

The solution in this domain is given by equations (12) - (13) for water and equations (14) and (23) for sediment.

Domain D_2

The solution for water in this domain is given by equations (25) - (26). The solution for sediment in subdomain D_{2A} is given by equations (31) and (40) and in subdomain D_{2B} by equations (31) and (42).

Domain D_3

The solution for water in this domain is given by equations (43) - (44) with the free boundary given by equation (47). For $0 \leq x \leq \bar{x}_T$ the solution for sediment in subdomain D_{3B} is given by equations (60) and (63). For $\bar{x}_T \leq x < x^*$ the solution for sediment in subdomain D_{3A} is given by equations (60) and (65) with $\bar{x}_T$ given by equation (66) and x^* given by

$$x^* = \alpha T^n q^{*(n-1)} \tag{67}$$

For $x^* \leq x_T \leq L$ the solution in subdomain D_{3C} remains yet to be determined. The solution for water is

$$h(x;x_0^*,x_0) = \left[\frac{q\,(x_0^* - x_0) + f\,(x_0 - x)}{\alpha}\right]^{1/n} \tag{68}$$

$$t(x;x_0^*,x) = T + \frac{1}{f}\left(\frac{1}{\alpha}\right)^{1/n}\left[\{(q - f)\,(x_0^* - x_0)\}^{1/n}\right.$$

$$\left. - \{q\,(x_0^* - x_0) + f\,(x_0 - x)\}^{1/n}\right] \tag{69}$$

where x_0 and x_0^* are related by

533

$$T = (q - f)^{(1-n)/n} \left[\frac{(x_0^* - x_0)}{\alpha} \right]^{1/n} \tag{70}$$

Eliminating the parameters in equations (68) and (69),

$$h = Tq - ft \tag{71}$$

It is clear that h depends on t only (unsteady but uniform).

For sediment we solve equation (51) using equation (71) with the condition that at $x = x_T$, $t = T$,

$$t = \frac{1}{f} \left[qT - \{ T^n (q - f)^n - \frac{fn}{\alpha} (x - x_T) \}^n \right] \tag{72}$$

To solve equation (52) we invoke the same assumption that allows dropping of the first term on the right side of equation (52). The solution then is the same as equation (65) in which the constant A is to be evaluated by matching the solutions at $x = x_T$, $t = T$,

$$A = (B_* - \frac{K}{\alpha}) \exp (\gamma x_T) F_n \left[(\frac{\gamma\alpha}{n})^{1/n} q^{*r} T \right] \tag{73}$$

Therefore,

$$C = \frac{K}{\alpha} + (B_* - \frac{K}{\alpha}) \exp (\gamma (x_T - x)) F_n \left[(\frac{\gamma\alpha}{n})^{1/n} q^{*r} T \right] \tag{74}$$

A close scrutiny of equations (72) - (74) indicates that C is independent of x but depends on t only. The same, of course, will be true of sediment discharge.

Solutions for Cases D and E

The solutions in case of D are the same as in case C. Therefore, we will not repeat them here. There is, however, one difference. Part of the boundary of subdomain D_{3A} is a free boundary and is given by $t^0(x)$. The point at which $t_s(x, \bar{x}_T)$ and $t^0(x)$, intersect is given by equating the expressions for these two quantities.

The solution in case of E is obtained in a similar manner. The difference between this case and the case C lies in the boundaries of subdomain D_{3A}, D_{3B} and D_{3C}. Since $t(x, x^*)$, $t_s(x, \bar{x}_T)$ and $t^0(x)$ are known, these boundaries are easily defined.

CONCLUDING REMARKS

Explicit solutions to kinematic equations for erosion occurring on a plane subject to space-time invariant rainfall and runoff are derived.

If rainfall and infiltration vary in time, explicit solutions are intractable. However, if these vary in space, explicit solutions can be obtained with only a modest increase in mathematical complexity. The solutions obtained in this study may be useful in determining the effect of land use management practices on erosion in upland areas.

ACKNOWLEDGEMENTS

This study was supported in part by funds provided by the National Science Foundation under the project, 'Free Boundary Problems in Water Resource Engineering,' NSF-ENG-CME-79-23345.

REFERENCES

Abramowitz, M. and Stegun, I. 1964. Handbook of Mathematical Functions with Formulas, Graphs and Mathematical Tables. National Bureau of Standards, Applied Mathematics Series 55, Washington, D.C., 1046 p.

Agricultural Research Service. 1975. Present and Prospective Technology for Predicting Sediment Yields and Sources. ARS-S-40, U.S. Department of Agriculture, Washington, D.C.

Bennett, J. P. 1974. Concepts of Mathematical Modeling of Sediment
Yield. Water Resources Research, Vol. 10, No. 3, pp. 485-492.

Borah, D. K. 1979. The Dynamic Simulation of Water and Sediment in
Watersheds. Unpublished Ph.D. dissertation, The University of
Mississippi, University, Mississippi, 224 p.

Curtis, D. C. 1976. A Deterministic Urban Storm Water and Sediment
Discharge Model. Proceedings, National Symposium on Urban
Hydrology, Hydraulics and Sediment Control, University of Kentucky,
Lexington, Kentucky.

Foster, G. R. and Lane, L. J. 1980. Simulation of Erosion and Sediment Yield from Field Sized Areas. Proceedings of the International Conference on Watershed Management and Land Development
in Tropics, John Wiley, London.

Foster, G. R. and Meyer, L. D. 1975. Mathematical Simulation of
Upland Erosion by Fundamental Erosion Mechanics. In: Present and
Prospective Technology for Predicting Sediment Yields and Sources,
Agricultural Research Service, U.S. Department of Agriculture,
Washington, D.C.

Foster, G. R., Meyer, L. D., and Onstad, C. A. 1973. Erosion Equations
Derived from Modeling Principles. Paper No. 73-2550 presented at
the National Meeting of the American Society of Agricultural
Engineers, Chicago, Illinois.

Foster, G. R. and Meyer, L. D. 1972. A Closed Form Soil Erosion
Equation for Upland Areas. In: H. W. Shen (editor), Sedimenta-
tion, pp. 12.2 - 12.9, Water Resources Publications, Fort Collins,
Colorado.

Hjelmfelt, A. T., Piest, R. P., and Saxton, K. E. 1975. Mathematical
Modeling of Erosion on Upland Areas. Proceedings, XVI Congress
of the International Association for Hydraulic Research, San
Paulo, Brazil, Vol. 2, pp. 40-47.

Knisel, W. G., editor. 1980. A field-scale Model for Chemicals, Runoff, and Erosion from Agricultural Systems. U.S. Department of
Agriculture, Conservation Research Report No. 26, 840 p.

Komura, S. 1976. Hydraulics of Slope Erosion by Overland Flow.
Journal of the Hydraulics Division, Proc. ASCE, Vol. 102, pp HY10,
pp. 1573-1586.

Li, R. M. 1979. Water and Sediment Routing from Watersheds. In: H. W. Shen (editor), Modeling of Rivers, pp. 9-1-9-83, Wiley-Interscience, New York.

Lighthill, M. M., and Whitham, G. B. 1955. On Kinematic Waves: Flood Movement in Long Rivers. Proceedings, Royal Society (London), Series A, Vol. 229, pp. 281-901.

Rendon-Herrero, O. 1978. Unit Sediment Graph. Water Resources Research, Vol. 14, No. 5, pp. 889-901.

Ross, B. B., Shanholtz, V. O., and Contractor, D. N. 1980. A Spatially Responsive Hydrologic Model to Predict Erosion and Sediment Transport. Water Resources Bulletin, Vol. 16, No. 3, pp. 538-545.

Shirley, E. D., and Lane, L. J. 1978. A Sediment Yield Equation From an Erosion Simulation Model. Proceedings of the 1978 Meetings of the Arizona Section of the American Water Resources Association and the Hydrology Section of the Arizona Academy of Science, Hydrology and Water Resources in Arizona and the Southwest, Vol. 8, pp. 90-96.

Simons, D. B., Li, R. M., and Shiao, L. Y. 1977. Formulation of Road Sediment Model. U.S. Department of Agriculture, Forest Service, Rocky Mountain Forest and Range Experiment Station, Flagstaff, Arizona, CER 76-77 DBS-RML-LySSO.

Simons, D. B., Li, R. M., and Stevens, M. A. 1975. Development of Models for Predicting Water and Sediment Routing and Yield from Storms on Small Watersheds. U.S. Department of Agriculture, Forest Service, Rocky Mountain Forest and Range Experiment Station, Flagstaff, Arizona.

Singh, V. P. and Mahmood, K. 1979a. Kinematic Modeling of Watershed Runoff: 1. Equilibrium Hydrograph. Proceedings of the Third World Congress on Water Resources held in Mexico City, Mexico, Vol. 4, pp. 2052-2073.

Singh, V. P. and Mahmood, K. 1979b. Kinematic Modeling of Watershed Runoff: 2. Partial Equilibrium Hydrograph. Proceedings of the Third World Congress on Water Resources held in Mexico City, Mexico, Vol. 4, pp. 2074-2086.

Singh, V. P. 1976. Studies on Rainfall-runoff Modeling: 2. A Distributed Approach to Kinematic Wave Modeling. WRRI Report

No. 065, 154 p., New Mexico Water Resources Research Institute, New Mexico State University, Las Cruces, New Mexico.

Smith, R. E. 1976. Simulating Erosion Dynamics with a Deterministic Distributed Watershed Model. Proceedings of the Third Federal Interagency Sedimentation Conference, held March 22-25, 1976, Denver Colorado, pp. 1- 163- 1 -173.

Soil Conservation Society of America. 1977. Soil Erosion: Prediction and Control. 7515 Northeast Ankeny Road, Ankeny, Iowa, 50021.

Water Resources Council. 1976. Proceedings of the Third Federal Inter-Agency Sedimentation Conference, Washington, D.C.

Williams, J. R. 1978. A Sediment Graph Model Based on an Instantaneous Unit Sediment Graph. Water Resources Research, Vol. 14, No. 4, pp. 659-664.

Wischmeier, W. H. and Smith, D. D. 1978. Predicting Rainfall Erosion Losses - A Guide to Conservation Planning. Agricultural Handbook 537, Science and Education Administration, U.S. Department of Agriculture, Washington, D.C., 58 p.

AN INSTANTANEOUS UNIT SEDIMENT GRAPH STUDY
FOR SMALL UPLAND WATERSHEDS

Vijay P. Singh
Associate Professor of Civil Engineering
Department of Civil Engineering
Louisiana State University
Baton Rouge, LA 70803, U S A

Andrez Baniukiwicz
Graduate Student
Department of Civil Engineering
Virginia Polytechnic Institute and State University
Blacksburg, VA 24060, U.S.A.

Victor J. Chen
Graduate Student
Department of Civil Engineering
Mississippi State University
Mississippi State, MS 39762, U.S.A.

ABSTRACT

An instantaneous unit sediment graph (IUSG) procedure is investigated for prediction of sediment yield (wash load) for an upland watershed in Northwestern Mississippi. An analysis of rainfall, runoff, and sediment data from this watershed shows that the sediment concentration distribution (SCD) for a unit volume of the effective rainfall is almost constant or only a function of time and independent of time characteristics of the effective rainfall. Therefore, SCD remains amenable to apriori specification. However, SCD does change linearly with the volume of the effective rainfall.

The instantaneous unit sediment graph, defined as a product of SCD and the instantaneous unit hydrograph (IUH), is known to depend on the characteristics of the effective rainfall. This dependence seems to arise through the IUH characteristics. Although the relation between sediment discharge and volume of the effective rainfall is not strictly linear, the IUSG procedure is reasonably accurate for prediction of sediment yield (wash load) from upland areas.

Thirteen rainfall-runoff-sediment events were selected for investigating the IUSG procedure. The IUH was derived by the Nash model for each event. The IUH parameters were correlated with the effective rainfall characteristics. The SCD was assumed to be an exponential function for each event and its parameters were correlated with the effective rainfall characteristics. Finally the IUSG was derived for each event and its parameters were correlated with the effective rainfall characteristics. These correlations demonstrate that the sediment yield process is not strictly linear. However, as a practical tool the assumption of linearity is reasonable. The IUH's and IUSG's are non-dimensionalized. Their dimensionless forms are found to be equivalent to each other. Utilizing the derived IUH's and IUSG's discharge hydrographs and sediment discharge graphs are generated for each event and compared with their observed counterparts.

INTRODUCTION

The last few years have witnessed applications of linear systems theory in development of tools for prediction of sediment from upland areas. Perhaps the first attempt in this direction was made by Johnson (1943) where he derived distribution graphs of suspended sediment concentration for East Fork of Deep River, North Carolina. An application of the distribution graph requires stream hydrograph and one measurement of suspended-matter concentration during a rise of the streamflow. Although the work by Johnson (1943) was a significant contribution, it was only recently that linear systems theory began to receive attention in sedimentation engineering.

Following the work by Johnson (1943), a unit sediment graph (USG) study was undertaken by Rendon-Herrero (1974, 1978) where he derived unit sediment graphs corresponding to different storm durations for Bixler Run, a small wash load-producing watershed near Loysville, Pennsylvania. He defined the USG as one unit of sediment for a given duration distributed over a watershed. The USG ordinates are obtained by dividing a sediment discharge graph by its total sediment load. This procedure is completely dependent upon data and cannot be used to assess environmental impact of land use management practices.

Williams (1978) is perhaps the first to have developed an instantaneous unit sediment graph (IUSG) overcoming some of the handicaps of the previous investigations. In the spirit of the instantaneous unit graph (IUH) he defined the IUSG as the distribution of sediment from an instantaneous burst of rainfall producing one unit of runoff. The IUSG is the product of the IUH and the sediment concentration distribution (SCD). Sediment concentration of the IUSG is assumed to vary with the effective rainfall volume. A sediment routing function, using travel time and sediment particle size, was used to determine the SCD. This study holds a great deal of promise in sedimentation technology.

On the other hand, Sharma and his colleagues (Sharma and Dickinson, 1979a, b, 1980; Sharma, Hines and Dickinson, 1979) developed input-output models for runoff-sediment yield processes for daily and monthly time bases. They derived unit step and frequency response functions and studied the noise component in runoff-sediment yield processes. These investigations represent valuable contributions on application of linear systems theory to sedimentation technology.

The above studies require further testing both with regard to their underlying assumptions and their accuracy in prediction of sediment discharge and yield. Thus the objective of this study is threefold: (1) to test the IUSG procedure (Williams, 1978) for an upland watershed in Mississippi, (2) to develop dimensionless IUSG's and IUH's and compare them, and (3) to explore extension of the dimensionless IUSG procedure to ungaged watersheds.

INSTANTANEOUS UNIT SEDIMENT GRAPH

Following Williams (1978) the IUSG is the distribution of sediment due to an instantaneous burst of the effective rainfall having a unit volume. Therefore, for one unit of runoff,

$$u_i = h_i c_i \ , \ i = 1, N \tag{1}$$

where u is the IUSG ordinate or sediment flow rate, h the IUH ordinate, c the SCD ordinate and N the number of IUSG points. A fundamental assumption underlying the IUSG is that c varies linearly with the volume of the effective rainfall. Thus the storm sediment discharge y is computed by convolving the IUSG with the effective rainfall squared,

$$y_i = \sum_{j=1}^{i} R_j^2 \ u_k \ , \ k = i + 1 - j, \ i = 1, M \tag{2}$$

The SCD (Williams, 1978) can be defined as

$$c = c_0 \exp(-bt) \tag{3}$$

where c_0 is the initial value of c, t time and b a parameter.

The IUH can be specified by the Nash model (1957),

$$h(t) = \frac{1}{\Gamma(n)} \ \frac{1}{k} \ \left(\frac{t}{k}\right)^{(n-1)} \exp(-t/k) \tag{4}$$

where n and k are parameters. k represents the watershed lag and n the number of reservoirs. Thus the IUSG procedure consists in specification of the SCD and the IUH.

APPLICATION TO A NATURAL WATERSHED

A small upland watershed, W-5, a part of Pigeon Roost basin, located near Oxford, Marshall County, Mississippi, was selected for testing of the IUSG procedure. It has an area of approximately $4km^2$, is 1288 m long and 128.8 m wide. The watershed consists of a rather flat flood plain with natural channels and rolling severely dissected interfluvial areas. The channels have few straight reaches, and most have banks that scour easily. The average channel width-depth ratio is approximately 2:1 at the gaging station. A detailed description of this watershed is given by Bowie and Bolton (1972).

Thirteen rainfall-runoff-sediment events on the watershed W-5 were selected for this analysis. These are the events where the direct runoff constitutes a significant portion of rainfall. For these events data are available on rainfall hyetograph, discharge, sediment concentration, and sediment discharge. Using a standard hydrograph separation technique the direct runoff and baseflow were determined. These data are available at irregular time intervals, and were expressed at a 5-minute interval. Thus the IUH and IUSG, derived herein would correspond to a 5-minute effective rainfall.

Determination of Effective Rainfall

The effective rainfall was determined by subtracting infiltration from rainfall which was determined by Philip infiltration equation

(Philip, 1957),

$$f = a_0 + a_1 \, t^{-0.5} \qquad\qquad (5)$$

where f is infiltration rate, t time, and a_0 and a_1 are parameters. a_0 is approximately equal to saturated hydraulic conductivity and was fixed at 0.01 cm/hr for the watershed. a_1 is sorptivity and varies from one rainfall event to another; this was determined along with other parameters for each event by optimization to be discussed later.

Parameter Estimation

The parameters to be estimated include c_0, b, a_1, k and n. These parameters were estimated for each rainfall-runoff-sediment event by the modified Rosenbrock-Palmer algorithm (Rosenbrock, 1960; Palmer, 1969) using the objective function E,

$$E = \Sigma \left[\left(\frac{x_0 - x_e}{x_0} \right) \left(\frac{x_0}{x_{0,m}} \right) \right]^2 \Rightarrow \min \qquad (6)$$

where x denotes either discharge or sediment discharge and subscripts 0, e, and (0,m) relate respectively to observed, estimated and maximum observed quantities. In equation (6) the first part is analogous to the least squares criterion; the second part assigns weight, giving the maximum weight to the peak ordinate.

The optimized parameter values for all 13 events are given in Table 1. The values of c_0 and b were plotted against the volume of the effective rainfall R as shown in figures 1-2. It is clear that on an average c_0 decreases with increasing R and becomes asymptatically constant fast. However, b does not appear to have a significant relation with R. The mean values of c_0 and b are respectively 0.025 metric

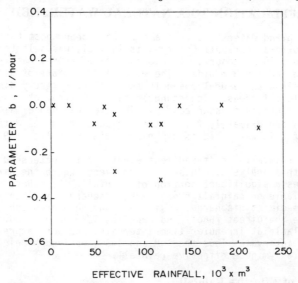

Figure 1. The relation between the parameter b and the effective rainfall volume for watershed W-5, Oxford, Mississippi.

Table 1. Optimized parameter values for 13 rainfall-runoff-sediment events on watershed W-?? Oxford, Mississippi.

Date Year Mo/Day	Rainfall Volume (cm)	(M³)	Runoff Volume (M³)	Sediment Yield (tons) Observed	Computed	K (hrs)	n	a_1 (cm/hr)	b (1/hr)	c_0 ($\frac{\text{metric tons}}{\text{M}^3}$)
1972 11/6-11/7	4.54	183416	183455	482.619	532.000	0.300	2.363	2.89	0	0.0307
1972 12/9-12/10	2.50	101000	101225.3	330.058	363.829	0.228	4.559	1.77	-0.207	0.025
1973 1/21-1/22	5.43	219372	219684.2	975.095	1074.866	0.453	3.577	1.25	-0.093	0.072
1973 3/14-3/15	2.29	92516	92824.9	842.991	929.245	0.147	8.493	2.53	-0.130	0.025
1973 4/19-4/20	0.96	38784	38879.1	272.583	272.583	0.097	10.526	5.65	-0.364	0.036
1973 5/27	2.96	119584	120056.4	520.992	574.300	0.172	17.557	3.33	-0.053	0.021
1973 11/27	0.11	4444	4417.4	7.074	7.798	0.184	8.526	5.30	0	0.0382
1974 7/21-7/23	0.41	16564	16477.4	77.561	85.497	0.25	18.492	11.60	0	0.0305
1974 11/19	1.45	58580	58534.7	205.807	226.865	0.49	12.055	2.52	-0.027	0.022
1975 1/10	1.55	62620	62496.1	374.547	412.870	0.135	7.450	3.38	-0.033	0.032
1975 3/12	2.57	103828	104135.2	421.276	464.381	0.150	6.354	1.96	-0.096	0.021
1976 2/17-2/18	1.52	61408	61830.5	449.891	495.924	0.148	10.497	2.94	-0.308	0.034
1976 3/20-3/21	2.48	100192	100110.8	728.416	802.947	0.206	3.605	3.98	-0.001	0.027

543

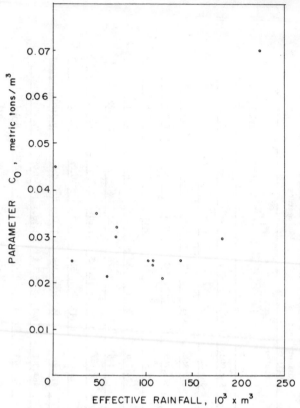

Figure 2. The relation between the parameter C_0 and the effective
rainfall volume for watershed W-5, Oxford, Mississippi.

tons/m^3 and .08/hr, and can be taken as representative values for watershed W-5. These parameters are subject to a small dispersion which may
well be due to data errors.

Likewise, k was plotted against R and the effective rainfall intensity r as shown in figures 3-4. It is clear that k increases with
R and decreases with r, although the latter correlation is not conclu-
sive. Similarly n was related to R and r as shown in figures 5-6.
It is seen that n decreases with R and increases with r. Thus figures
3-6 show the dependence of n and k on r and R.

Determination of the IUH

The IUH was determined by equation (4) for each event. The
moments, cumulants and shape factors (Dooge, 1973) of each IUH were
computed. The shape factors were remarkably close to each other for
all IUH's. The first shape factor varied from -0.88 to 0.93, the
second from 1.65 to 1.83, and the third from -7.26 to -8.05. The
peak IUH ordinate h_p was related to R as shown in figure 7. It is
seen that the peak decreases with increasing R. The IUH's vary considerably from one event to another, and were, therefore, non-
dimensionalized as

544

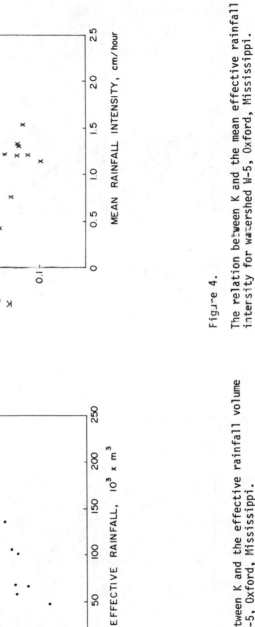

MEAN RAINFALL INTENSITY, cm/hour

K , hours

Figure 4.

The relation between K and the mean effective rainfall intensity for watershed W-5, Oxford, Mississippi.

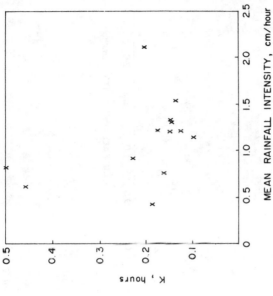

EFFECTIVE RAINFALL, 10^3 x m³

K , hours

Figure 3.

The relation between K and the effective rainfall volume for watershed W-5, Oxford, Mississippi.

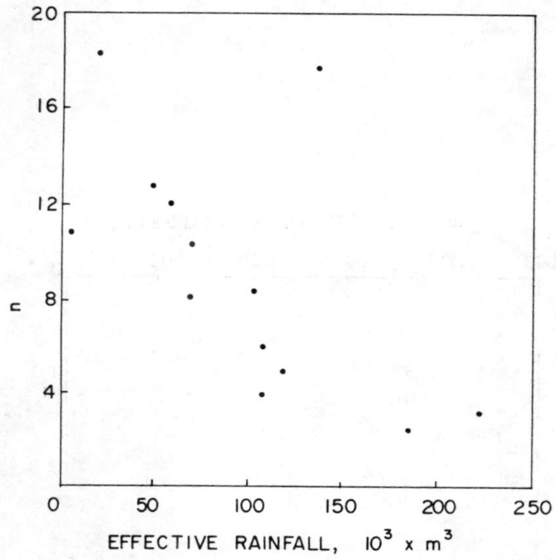

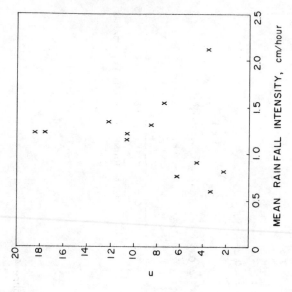

Figure 5.

The relation between the Nash parameter n and the effective
rainfall volume for watershed W-5, Oxford, Mississippi

Figure 6.

The relation between the Nash parameter n and the mean
effective rainfall intensity.

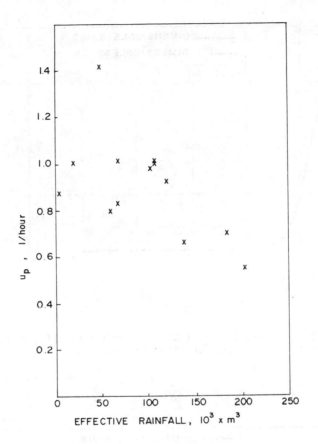

Figure 7. The relation between the IUH peak and the effective rainfall for watershed W-5, Osford, Mississippi.

$$h_*(t) = \frac{h(t)}{h_p} \tag{7}$$

The dimensionless IUH's $h_*(t)$ are similar in shape and experience much variation. For two sample events these are shown in figures 8-9.

Determination of the IUSG

The IUSG was determined by equation (1) for each event. The moments, cumulants and shape factors were computed for each IUSG. These parameters were subject to considerable variation unlike the IUH. The IUSG peak u_p and its time t_u were related to R as shown in figures 10-11. It is seen that t_u decreases with increasing R whereas u_p first decreases and then increases. The IUSG's vary considerably from one event to the other, and were therefore non-dimensionalized,

$$u_*(t) = \frac{u(t)}{u_p} \tag{8}$$

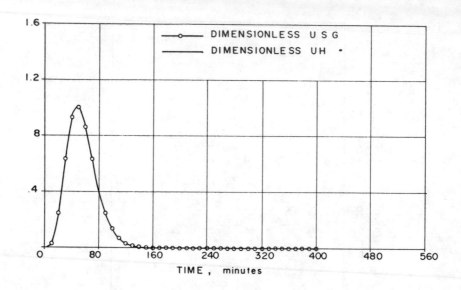

Figure 8. Dimensionless IUH and IUSG for rainfall-runoff-sediment
event of 1-10-1975 for watershed W-5, Oxford, Mississippi.

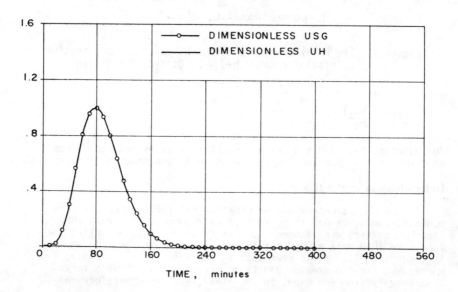

Figure 9. Dimensionless IUH and IUSG for rainfall-runoff-sediment
event of 11-27-1973 for watershed W-5, Oxford, Mississippi.

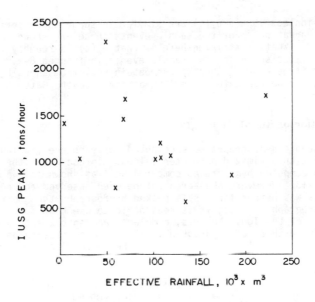

Figure 10.   The relation between the IUSG peak and the effective
runoff volume for watershed W-5, Oxford, Mississippi.

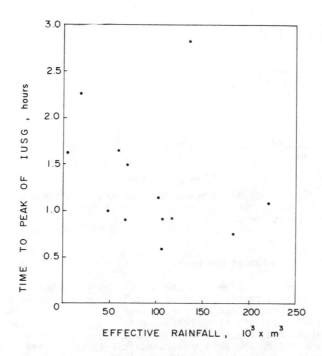

Figure 11.   The relation between the time to the IUSG peak and the
effective rainfall volume for watershed W-5, Oxford, Mississippi.

The dimensionless IUSG's $u_*(t)$ are similar in shape and experience much less variation. For two sample events these are shown in figures 8-9. What is striking here is that $u_*(t)$ is roughly the same as $h_*(t)$ as shown for two sample events in figures 8-9. This means that if we know u_p then we can obtain $u(t)$ from known $h_*(t)$. In this manner we can extend the IUSG approach to the watersheds with limited or no data.

Determination of Runoff Hydrograph

The runoff hydrographs were computed for each event using the Nash model with optimized parameter values. In each case the observed and computed hydrographs compared well as shown in two sample figures 12-13. However, the agreement between observed and computed hydrographs was better for single peaked hydrographs than multiple peaked hydrographs. Thus, it is reasonable to use Nash model for determination of the IUH. Moments, cumulants and shape factors were computed for each runoff hydrograph. It was found that shape factors remained approximately constant for all the hydrographs.

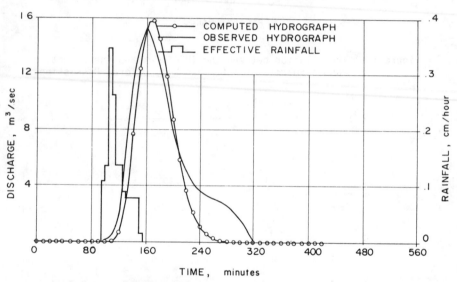

Figure 12. Comparison of observed and computed runoff hydrographs for the rainfall-runoff event of 3-20-1976 for watershed W-5, Oxford, Mississippi.

Determination of Sediment Discharge

Rainfall-runoff-sediment discharge data were plotted for each event as shown in a sample figure 14. It is seen that in each case the sediment discharge peak and runoff peak occur almost at the same time. Further, sediment discharge and runoff hydrographs possess similar shapes, and have the same duration. These characteristics must be preserved by the IUSG procedure if it is to be used as an accurate tool in prediction of sediment discharge and yield.

Employing the IUSG procedure with parameters estimated previously

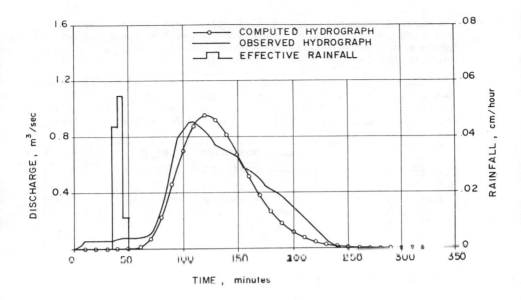

Figure 13. Comparison of observed and computed runoff hydrographs for the rainfall-runoff event of 11-27-1973 for watershed W-5, Osford, Mississippi.

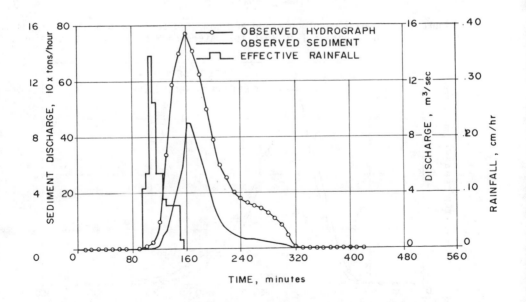

Figure 14. The effective rainfall, observed runoff hydrograph and observed sediment discharge graph for the event of 1-10-1975 for watershed W-5, Oxford, Mississippi.

the sediment discharge graph was computed for each event. The observed and computed sediment graphs compared well as shown for two sample events in figures 15-16. A comparison of observed and computed sediment peak characteristics is shown in Table 1.

Moments, cumulants and shape factors were computed for each sediment graph. The shape factors remained approximately constant for all sediment graphs. More importantly, the shape factors of sediment and discharge graph were essentially the same for each event. This

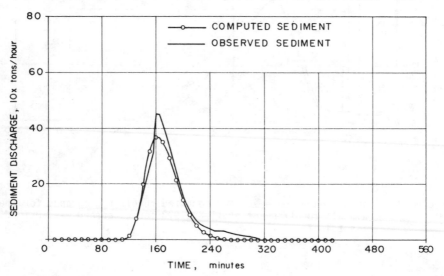

Figure 15. Comparison of observed and computed sediment discharge graphs for the event of 1-10-1975 for watershed W-5, Oxford, Mississippi.

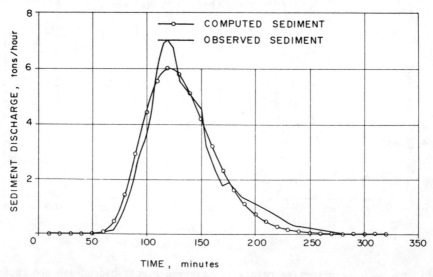

Figure 16. Comparison of observed and computed sediment discharge graphs for the event of 11-27-1973 for watershed W-5, Oxford, Mississippi.

means that the sediment discharge graph can be predicted from the runoff hydrograph and some moments of the sediment graph.

CONCLUSIONS

The following conclusions can be drawn from this study:
(1) The sediment-runoff relation is not strictly linear as is the case with rainfall-runoff relation. However, for practical purposes this assumption may not be unduly restrictive.
(2) The peak characteristics of both the IUH and IUSG are related to the volume of the effective rainfall.
(3) The parameters of the IUH are related to the volume of the effective rainfall.
(4) The SCD is only a function of time. Its parameters are related to the volume of the effective rainfall.
(5) The shape factors of the IUH remain approximately constant from one event to another. This is, however, not true with the IUSG.
(6) The shape factors of the sediment discharge graph are approximately the same as those of the runoff hydrograph.
(7) The dimensionless IUSG is approximately identical to the dimensionless IUH for a specified event.
(8) The IUSG procedure can predict sediment discharge graph reasonably accurately.

ACKNOWLEDGEMENTS

This study was supported in part by funds provided by the Office of Water Research and Technology, (Agreement No. 14-34-0001-9026-Project No. A-129-Miss), U.S. Department of the Interior, Washington, D.C., as authorized by the Water Research and Development Act of 1972.

REFERENCES

Bowie, A.J. and Bolton, G.C., 1972. Variations in runoff and sediment yields of two adjacent watersheds as influenced by hydrologic and physical characteristics. Proceedings, Mississippi Water Resources Conference, pp. 37-55, Water Resources Research Institute, Mississippi State University, Mississippi State, Mississippi.

Johnson, J.W., 1943. Distributed graphs of suspended-matter concentration. Transactions, American Society of Civil Engineering, Volume 108, pp. 941-964.

Nash, J.E., 1957. The form of the instantaneous unit hydrograph. International Association of Scientific Hydrology, Publication No. 42, pp. 114-118.

Palmer, J.R., 1969. An improved procedure for orthogonalizing the search vectors in Rosenbrock's and Swann's direct search optimization methods. Computer Journal, Volume 12, pp. 69-71.

Philip, J.R., 1957. Theory of infiltration, I. The infiltration equation and its solution. Soil Science, Volume 83, No. 5, pp. 345-357.

Rendon-Herrero, O., 1974. Estimation of wash load produced on certain small watersheds. Journal of the Hydraulics Division, Proceedings of the American Society of Civil Engineers, Volume 100, No. HY, pp. 835-848.

Rendon-Herrero, O., 1978. Unit sediment graph. Water Resources Research, Volume 14, No. 5, pp. 889-901.

Rosenbrock, H.H., 1960. An automatic method for finding the greatest or least value of a function. Computer Journal, Volume 4, pp. 175-184.

Sharma, T.C. and Dickinson, W.T., 1979a. Unit step and frequency response functions applied to the fluvial system. Journal of Hydrology, Volume 40, pp. 323-335.

Sharma, T.C. and Dickinson, W.T., 1979b. Discrete dynamic model of watershed sediment yield. Journal of the Hydraulics Division, Proceedings of the American Society of Civil Engineers, Volume 105, No. HY5, pp. 555-571.

Sharma, T.C. and Dickinson, W.T., 1980. System model of daily sediment yield. Water Resources Research, Volume 16, No. 3, pp. 501-506.

Sharma, T.C., Hines, W.G.S. and Dickinson, W.T., 1979. Input-output model for runoff-sediment yield processes. Journal of Hydrology, Volume 40, pp. 299-322.

Williams, J.R., 1978. A sediment graph model based on an instantaneous unit sediment graph. Water Resources Research, Volume 14, No. 4, pp. 659-664.

ON THE RELATION BETWEEN SEDIMENT YIELD AND RUNOFF VOLUME

Vijay P. Singh
Associate Professor of Civil Engineering
Department of Civil Engineering
Louisiana State University
Baton Rouge, LA 70803, U.S.A.

Victor J. Chen
Graduate Research Assistant
Department of Civil Engineering
Mississippi State University
Mississippi State, MS 39762, U.S.A.

ABSTRACT

The relationship between the sediment yield (wash load) produced by a rainfall storm occurring on an upland watershed and the corresponding volume of surface runoff is found to be approximately linear on a log-log paper. The two parameters, intercept and slope, of this relationship vary from one watershed to another but not to the same degree. The variability of slope suggests that the watersheds can be grouped in such a way that slope is approximately constant for watersheds within a specified group but changes from one group to another. The correlation between slope and geomorphic parameters appears to be weak.

The intercept appears to incorporate the effect of physiographic factors on sediment production. Geomorphic data from 21 watersheds, including watershed area, relief, channel length and slope, shape factor, forest area, erodibility factor, and storage percentage are analyzed to develop an empirical relation for the intercept. Watershed area, relief, channel length, channel slope and erodibility factor are found to be significantly correlated with the intercept. Utilizing the straight line relationship, with parameters estimated in the above manner, sediment yield is predicted for a number of rainfall storms in upland watersheds.

INTRODUCTION

Rendon-Herrero (1974) suggested that for small upland watersheds sediment yield (wash load) and surface runoff volume are linearly related on a logarithmic paper for a given watershed and that the slope of this straight line remains approximately constant from one watershed to another. These hypotheses were further investigated by Rendon-Herrero, Singh and Chen (1980). In these studies rainfall-runoff-sediment events were separated into winter and summer storms.

A further examination of the relation between sediment yield and surface runoff volume shows that the upland watersheds can be divided into different groups. For watersheds within a specified group the slope of the line experiences only a minor variation from one watershed to another and can, therefore, be considered approximately constant. However, the slope changes significantly from one group to another. An objective criterion for grouping of the watersheds is not yet found.

On the other hand, the intercept of the line, regardless of watershed grouping, changes appreciably from one watershed to another. It appears to incorporate the effect of physiographic factors on sediment production. A question arises: can this intercept be estimated from physically measurable watershed characteristics? The geomorphic characteristics may include watershed area, relief, channel length, channel slope, erodibility factor, forest area and area occupied by lakes and swamps. In this study we investigate this question and extend it to the slope. Furthermore, the sediment yield is predicted for a number of storms on upland watersheds to validate the logarithmically linear relationship.

SEDIMENT YIELD AND SURFACE RUNOFF VOLUME RELATION

Upland Watersheds

Twenty one watersheds were selected from ten states representing different hydrometeorologic conditions in the United States. These are small watersheds ranging in area from 45 km^2 to 2200 km^2, and are monitored by the U.S. Geological Survey (USGS). Their location and pertinent geomorphic characteristics are shown in Table 1.

Water and Sediment Discharge Data

The U.S. Geological Survey collects water and sediment discharge data on a daily basis from these watersheds and publishes them in its Surface Water Quality Records. On each watershed ten or more isolated storm events were identified for winter and summer seasons separately. By employing a standard hydrograph separation technique the surface runoff hydrograph was determined for each storm and then was determined its volume. The total sediment hydrograph was obtained for each storm from the Surface Water Quality Records. Then the direct sediment discharge hydrograph was obtained using the following relation,

$$S_{D_i} = [S_{T_i} Q_{T_i} - S_{B_i} Q_{B_i}] \times 0.0864$$

where S_{D_i} is direct sediment discharge in metric tons per day, S_{T_i} total sediment discharge in ppm, Q_{T_i} total water discharge in cubic meter per second, S_{B_i} base flow of sediment in ppm, Q_{B_i} base flow of water in cubic meter per second, and i the time period. The factor 0.0864 is used for converting the sediment discharge into metric tons per day. By integrating the sediment discharge hydrograph sediment yield of each storm was obtained.

Geomorphic Data

Eight geomorphic characteristics are used in this study which include watershed area A, main channel length L, main channel slope S_0, mean basin elevation E, erodibility factor K, watershed shape factor R, forest area A_f and storage percentage S_t. These characteristics are

Table 1. Upland watersheds and their geomorphologic characteristics.

USGS Watershed Number	State	Area A(km²)	Main Channel Length L(Km)	Main Channel Slope S0(m/Km)	Mean Basin Elevation E(m)	Erodibility Factor K	Shape Factor R	Forest Area Af(km²)	Storage Area St(%)
1-1845	N.J.	254.8558	47.4755	3.0872	120.394	0.28	0.1131	152.6000	0
1-3970	N.J.	380.7297	55.0394	5.1138	202.638	0.35	0.1257	194.1700	3.00
1-4010	N.J.	115.2549	19.3121	1.9697	93.42	0.33	0.3090	37.1121	1.20
1-4645	PA	216.5238	22.2080	0.6137	35.966	0.26	0.4390	132.0800	7.80
1-4705	PA	919.4493	65.8220	2.2159	310.900	0.20	0.2122	551.6696	0.55
1-4730	PA	722.6094	45.5604	3.2196	115.320	0.32	0.3481	209.5567	0.06
1-4810	DEL	743.3294	45.5443	3.2198	142.348	0.32	0.3584	334.5000	0.16
1-4815	VA	813.2594	68.5579	2.7462	142.350	0.32	0.1730	142.9512	0.74
1-6640	GA	1595.4388	63.8908	1.7860	240.750	0.29	0.3908	686.0387	1.00
2-3835	KY	2217.0383	119.2521	3.4090	505.160	0.28	0.1559	1806.8862	0.24
3-2170	KY	626.7795	98.4916	0.8712	289.560	0.32	0.0646	401.1389	0.02
3-2495	KY	2152.2883	214.8470	0.3977	304.94	0.35	0.0466	1614.2200	0.02
3-2915	KY	1131.8291	130.0347	0.6629	259.080	0.32	0.0669	192.4110	0.16
3-2975	KY	82.3619	23.1745	3.1249	213.567	0.37	0.1534	9.0598	0.79
3-3830	KY	660.0495	85.7778	0.3788	167.637	0.30	0.0898	515.1500	0.61
6-8880	KAN	629.3695	71.1328	1.0417	384.640	0.37	0.1244	34.3600	0.45
7-1515	KAN	2056.4584	156.1060	1.5397	466.344	0.29	0.0844	28.7904	0.20
7-2300	OKLA	665.6295	38.6242	1.0416	316.992	0.39	0.4462	246.2829	0
7-3045	OKLA	1421.9089	87.0653	1.8107	499.852	0.29	0.1876	14.2191	0
7-3135	OKLA	1458.1689	106.5383	0.9072	313.998	0.35	0.1285	58.3268	0
8-632	TEX	45.5340	12.8747	2.9357	170.999	0.36	0.2750	8.6610	0

given for each watershed in Table 1. Here the shape factor is defined as the watershed area divided by the square of the mean channel length. The geomorophic characteristics, except for the erodibility factor, were obtained for each watershed from the U.S. Geological Survey District Office located in each state. The erodibility factor was obtained for each watershed from the state Soil Scientist available in each state.

Sediment-Runoff Relationships

For the purpose of establishing a relationship between sediment yield Y and volume of surface runoff V for a given watershed, storm events were divided into summer and winter storms. The rationale for this grouping is based on Rendon-Herrero (1974), and Rendon-Herrero, Singh and Chen (1980). This relationship is shown in figures 1-3 for some sample watersheds, and can be expressed as

$$Y = aV^b \tag{1}$$

$$\text{Log } Y = \text{Log } a + b \text{ Log } V \tag{2}$$

where log a is the intercept, b the slope of the line. Y was taken in metric tons and V in cm; both are based on a unit area. Tables 2-4 provide pertinent statistical information on equation (1).

It is evident from figures 1-3 and Tables 2-4 that there does indeed exist a strong linear relationship between log-transformed values of Y and V. The correlation coefficient r varies from 0.855 to 0.985 for summer storms and from 0.779 to 0.977 for winter storms

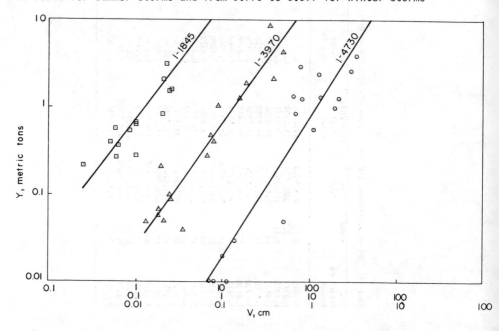

Figure 1.   The relation between sediment yield and volume of direct runoff for some sample watersheds. Here summer and winter storms are combined.

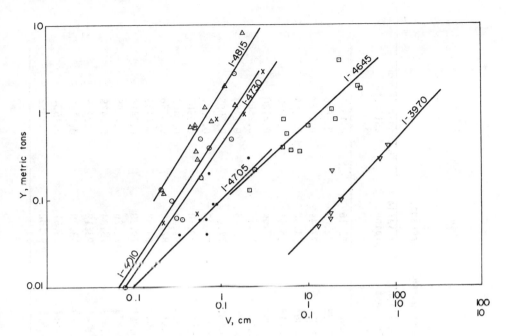

Figure 2. The relation between sediment yield and volume of direct runoff for some sample watersheds. Here only summer storms are considered.

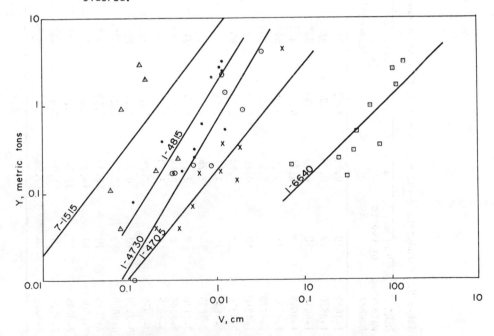

Figure 3. The relation between sediment yield and volume of direct runoff for some sample watersheds. Here only winter storms are considered.

Table 2. Parameters a and b of equation (1) for summer storms on upland watersheds.

USGS Watershed Number	Number of Events	a	b	Angle of Line (degrees)	Correlation Coefficient r	% of Variance Explained	Standard Error of Estimate S_e
1-1845	8	0.79	1.02	51°	0.960	92.16	0.1872
1-3970	7	0.48	1.02	46°	0.878	77.09	0.1606
1-4010	10	0.63	1.56	52°	0.880	77.44	0.2528
1-4645	7	0.74	0.90	42°	0.862	74.30	0.2354
1-4705	11	0.10	0.96	44°	0.818	66.91	0.2258
1-4730	7	0.41	1.47	56°	0.962	92.54	0.3120
1-4810	10	0.65	1.92	62°	0.955	91.20	0.2669
1-4815	12	1.63	1.64	59°	0.871	75.86	0.2153
1-6640	10	4.54	1.88	62°	0.881	77.62	0.3598
2-3835	8	2.25	1.46	56°	0.924	85.38	0.2771
3-2170	9	2.11	1.77	61°	0.855	73.10	0.5983
3-2495	8	2.64	1.46	56°	0.958	91.78	0.2403
3-2915	10	3.10	1.22	51°	0.911	82.99	0.3956
3-2975	10	2.88	1.42	55°	0.947	89.68	0.4145
3-3830	10	0.77	1.32	53°	0.891	79.39	0.3410
6-8880	10	43.54	1.31	53°	0.879	77.26	0.2065
7-1515	10	16.35	1.19	50°	0.868	75.34	0.3475
7-2300	10	46.29	1.41	55°	0.965	93.12	0.3258
7-3045	10	37.58	1.00	45°	0.968	93.70	0.1152
7-3135	10	13.09	0.98	44°	0.967	93.51	0.1534
8-632	8	19.58	1.16	49°	0.985	97.02	0.1352

Table 3. Parameters a and b of equation (1) for winter storms on upland watersheds.

USGS Watershed Number	Number of Events	a	b	Angle of Line (degrees)	Correlation Coefficient r	% of Variance Explained	Standard Error of Estimate Se
1-1845	7	0.54	1.75	60°	0.911	82.99	0.2158
1-3970	10	0.61	1.54	57°	0.946	89.49	0.2488
1-4010	10	0.56	1.75	60°	0.936	87.61	0.3511
1-4645	9	0.41	1.45	55°	C.934	87.24	0.2058
1-4705	10	0.18	1.25	57°	C.880	77.44	0.2652
1-4730	9	0.63	1.64	59°	0.934	87.24	0.2640
1-4810	10	0.39	1.99	63°	0.951	90.44	0.1850
1-4815	11	1.84	1.62	58°	0.890	79.21	0.2529
1-6640	10	1.27	0.98	44°	0.804	64.64	0.2662
2-3835	10	1.35	1.43	55°	0.960	92.16	0.2626
3-2170	10	0.83	1.95	63°	0.958	91.78	0.2598
3-2495	8	1.02	1.70	60°	0.928	86.12	0.3531
3-2915	8	2.57	1.66	59°	0.866	75.00	0.2101
3-2975	10	2.16	1.70	60°	0.961	92.35	0.2337
3-3830	10	0.33	1.71	60°	0.965	93.12	0.2911
6-8880	10	19.52	1.39	54°	0.381	77.62	0.3086
7-1515	8	7.78	1.30	52°	0.779	60.68	0.5497
7-2300	7	21.28	1.56	57°	0.910	82.81	0.4587
7-3045	7	31.74	1.12	48°	0.976	95.26	0.1685
7-3135	7	14.25	0.93	43°	0.977	95.45	0.0643
8-632	7	9.25	1.25	51°	C.903	81.54	0.3382

Table 4. Parameters a and b of equation (1) for summer and winter storms combined on upland watersheds.

USGS Watershed Number	Number of Events	a	b	Angle of Line (degrees)	Correlation Coefficient r	% of Variance Explained	Standard Error of Estimate S_e
1-1845	15	0.71	1.34	53°	0.932	86.86	0.2209
1-3970	17	0.66	1.36	54°	0.944	89.11	0.2380
1-4010	20	0.69	1.71	60°	0.951	90.44	0.2804
1-4645	16	0.56	1.08	47°	0.884	78.15	0.2314
1-4705	21	0.22	1.41	55°	0.912	83.17	0.2441
1-4730	16	0.76	1.59	58°	0.912	83.17	0.3684
1-4810	20	0.60	1.85	62°	0.822	67.57	0.3399
1-4815	23	1.97	1.80	61°	0.844	71.23	0.2694
1-6640	20	2.96	1.55	57°	0.822	67.57	0.3517
2-3835	18	1.62	1.22	51°	0.937	87.80	0.2848
3-2170	19	1.47	1.86	62°	0.858	73.62	0.5314
3-2495	16	1.65	1.55	57°	0.909	82.63	0.3777
3-2915	18	3.14	1.23	51°	0.923	85.19	0.3297
3-2975	20	3.21	1.41	55°	0.930	86.49	0.3751
3-3830	20	0.51	1.53	57°	0.910	82.81	0.3963
6-8880	20	24.64	1.13	48°	0.835	69.72	0.3100
7-1515	18	9.44	1.13	48°	0.780	60.84	0.4906
7-2300	17	37.28	1.52	57°	0.962	92.54	0.3624
7-3045	17	36.26	1.07	47°	0.964	92.93	0.1616
7-3135	17	16.07	1.03	46°	0.973	94.97	0.1314
8-632	15	14.14	1.28	52°	0.944	89.11	0.3020

on the 21 watersheds. The amount of variance explained by this relationship varies from 66.91% to 97.02% for summer storms and from 60.68% to 95.45% for winter storms. On the other hand, if the summer and winter storms are combined then r varies from 0.78 to 0.973 on these watersheds and the amount of variance explained from 60.94% to 94.97%.

For some watersheds r is not as high as preferable. However, an inspection of sediment discharge data indicates that for these watersheds the number of storm events is limited and that the quality of data is not very good. Nevertheless, on the average it appears that equation (1) can be a good prediction model.

Looking at figures 1-3 and Tables 2-4 it is seen that although the parameter b changes from one watershed to another, its variation is not pronounced. Furthermore, it is more or less constant for a number of watersheds. These watersheds can be grouped together and may be similar in the context of sediment production. At this point it is not clear as to how to identify such similar watersheds beforehand. On the other hand, the intercept changes greatly from one watershed to another and is related to watershed geomorphology. In this study we will attempt to determine b from geomorphic data of these watersheds.

PARAMETER ESTIMATION

The parameters, a and b, are determined in three ways. First, for each watershed summer and winter events are treated separately and the resulting parameters are related to geomorphic data. Second, summer and winter storms are grouped together and the resulting parameters are related to geomorphic data. Third, watersheds having approximately the same slope are grouped together and for each group the intercept is related to geomorphic data. We briefly discuss the three methods here.

Method 1: Separate Summer and Winter Storms

A multiple linear regression analysis was performed to correlate a and b given in Table 2 with geomorphic data of the twenty one watersheds given in Table 1. For summer storms,

$$a = -51.6986 - 0.0128A + 0.0766L - 2.5020S_0 - 0.0077A_f + 0.1174E$$

$$- 0.4491S_t + 102.9137K + 63.1112R \qquad (3)$$

with a correlation coefficient r of 0.9196 and a standard error of estimate S_e of 7.5435. Equation (3) explained 84.57% of the variance in a. By deleting 3 geomorphic characteristics in equation (3),

$$a = -50.2895 - 0.0127A - 3.3793S_0 + 0.1207E + 118.5449K$$

$$+ 49.1881R \qquad (4)$$

with r = 0.9001 and S_e = 7.4835. Equation (4) explained 81.02% of the variance in a.

Likewise for the same summer storms,

$$b = 1.1689 + 0.0002A - 0.0024L - 0.0193S_0 + 0.0002A_f - 0.0012E$$

$$- 0.0878S_t + 1.5052K + 0.2473R \tag{5}$$

with $r = 0.6819$ and $S_e = 0.2811$. This relation explained 46.50% of the variance in b. If the geomorphic characteristics were reduced

in the regression analysis,

$$b = 1.694 + 0.0001A - 0.0024L + 0.0001A_f - 0.0013E - 0.0938S_t \tag{6}$$

with $r = 0.6376$ and $S_e = 0.2648$. This relation explained 40.66% of the variance in b.

On the other hand, for winter storms,

$$a = - 21.2133 - 0.0013A - 0.0145L - 1.4924S_0 - 0.0073A_f$$
$$+ 0.604E - 0.0945S_t + 48.7213K + 20.2934R \tag{7}$$

with $r = 0.8855$ and $S_e = 5.2672$. Equation (6) explained 78.41% of the varinace in a. When the geomorphic characteristics were reduced,

$$a = - 24.0028 - 1.1956S_0 - 0.0087A_f + 0.0556E + 51.8551K$$
$$+ 22.5608R \tag{8}$$

with $r = 0.8807$ and $S_e = 4.8026$. This explained 77.57% of the variance in a.

Likewise, for the same winter storms,

$$b = 1.2957 - 0.0007A + 0.0096L + 0.0644S_0 + 0.0002A_f - 0.0002E$$
$$- 0.047S_t - 0.5083K + 1.1733R \tag{9}$$

with $r = 0.7930$, and $S_e = 0.2305$. This explained 62.89% of the variance in b. With reduced geomorphic characteristics,

$$b = 1.0861 - 0.0007A - 0.0091L + 0.0682S_0 + 0.0002A_f$$
$$+ 0.9478R \tag{10}$$

with $r = 0.7593$, and $S_e = 0.2203$. Equation (10) explained 57.66% of the variance in b. An examination of the individual plots of a and b against geomorphic characteristics revealed that better correlations can be obtained by a log transformation of a and b. Therefore, for summer storms,

$$\log a = -2.830 + 0.0007A - 0.0093L - 0.2200S_0 - 0.0005A_f$$
$$+ 0.0032E + 0.0234S_t + 10.1672K - 0.3705R \tag{11}$$

with $r = 0.9205$, and $S_e = 0.3774$. This explained 84.73% of the variance in log a. By deleting 3 geomorphologic characteristics in equation (11),

$$\log a = -2.5814 - 0.0038L - 0.2004S_0 - 0.0003A_f + 0.0042E$$
$$+ 8.6451K \tag{12}$$

with $r = 0.8956$, and $S_e = 0.3843$. This explained 80.21% of the variance in log a.

Likewise, for the same summer storms,

$$\log b = 0.0452 + 0.00006A - 0.00077L - 0.0067S_0 + 0.000076A_f$$
$$- 0.00041E - 0.0303S_t + 0.5639K + 0.0555R \qquad (13)$$

with $r = 0.7067$, and $S_e = 0.0881$. This explained 49.94% of the variance in $\log b$. If 3 geomorphologic characteristics were reduced,

$$\log b = 0.0684 - 0.00033L + 0.00009A_f - 0.0003E - 0.0287S_t$$
$$+ 0.4482K \qquad (14)$$

with $r = 0.6678$, and $S_e = 0.0829$. This relation explained 44.59% of the variance in $\log b$.

On the other hand, for winter storms,

$$\log a = -2.0093 + 0.00068A - 0.00073L - 0.13995S_0 - 0.00064A_f$$
$$+ 0.0031E + 0.0137S_t + 8.4228K - 0.2282R \qquad (15)$$

with $r = 0.9360$, and $S_e = 0.3063$. This explained 87.62% of the variance in $\log a$. When the geomorphologic characteristics were reduced,

$$\log a = -2.9280 + 0.00048A - 0.00404L - 0.00072A_f$$
$$+ 0.00302E + 8.0153K \qquad (16)$$

with $r = 0.9154$, and $S_e = 0.3135$. This explained 83.79% of the variance in $\log a$.

Likewise, for the same winter storms,

$$\log b = 0.0852 - 0.00027A + 0.0032L + 0.0226S_0 + 0.00008A_f$$
$$- 0.000013E - 0.0125S_t - 0.1828K + 0.3949R \qquad (17)$$

with $r = 0.8026$, and $S_e = 0.0700$. This explained 64.42% of the variance in $\log b$. When the geomorphologic characteristics were reduced,

$$\log b = 0.03486 - 0.00024A + 0.00289L + 0.0225S_0 + 0.00008A_f$$
$$+ 0.2993R \qquad (18)$$

with $r = 0.7790$, and $S_e = 0.0659$.

Method 2: Summer and Winter Storms Combined

When the summer and winter storms were combined for each watershed, the correlation of the parameters given in Table 4 with geomorphologic parameters given in Table 1 yield

$$a = -37.0106 - 0.0066A + 0.0301L - 1.9739S_0 - 0.0073A_f$$
$$+ 0.0852E - 0.4939S_t + 76.2341K + 46.5148R \qquad (19)$$

with $r = 0.9092$, and $S_e = 6.2508$. This explained 82.67% of the variance in a. With reduced geomorphic characteristics,

$$a = -40.8307 - 1.7846S_0 - 0.0099A_f + 0.0713E + 92.6411K$$

$$+ 39.5136R \qquad (20)$$

with $r = 0.8968$, and $S_e = 5.9705$. This explained 80.23% of the variance in a.

In a like manner,

$$b = 1.494 - 0.0002A + 0.0042L + 0.0321S_0 + 0.0001A_f$$

$$- 0.0009E - 0.0900S_t - 0.5174K + 1.0853R \qquad (21)$$

with $r = 0.7447$, and $S_e = 0.2243$. This explained 55.45% of the variance in b. When the geomorphic characteristics were reduced,

$$b = 1.7142 - 0.00005A + 0.0002A_f - 0.0012E - 0.0838S_t$$

$$+ 0.3062R \qquad (22)$$

with $r = 0.7215$, and $S_e = 0.2081$. This explained 52.06% of the variance in b.

By performing a log transformation of a and b and then correlating with geomorphic characteristics we obtain,

$$\log a = -2.5628 + 0.0006A - 0.0075L - 0.1543S_0 - 0.00055A_f$$

$$+ 0.0029E + 0.0004S_t + 8.6996K - 0.0033R \qquad (23)$$

with $r = 0.9393$, and $S_e = 0.2977$. This explained 88.23% of the variance in log a. With reduced geomorphologic characteristics,

$$\log a = -2.3370 - 0.0034L - 0.1415S_0 - 0.0004A_f + 0.004E$$

$$+ 7.576K \qquad (24)$$

with $r = 0.9107$, and $S_e = 0.3205$. This explained 82.95% of the variance in log a.

In a like manner,

$$\log b = 0.2163 - 0.00002A + 0.00053L + 0.00019S_0 + 0.00008A_f$$

$$- 0.00045E - 0.0238S_t + 0.0923K - 0.000003R \qquad (25)$$

with $r = 0.7422$, and $S_e = 0.0695$. This explained 55.08% of the variance in log b. With reduced geomorphologic characteristics,

$$\log b = 0.2625 - 0.00001 + 0.00007A_f - 0.0004E - 0.0251S_t$$

$$- 0.0000007R \qquad (26)$$

with $r = 0.7384$, and $S_e = 0.0625$. This explained 54.52% of the variance in log b.

Method 3: Similar Watersheds Lumped

The watersheds were classified into four groups according to the angle of the line: (1) 46° to 50°, (2) 51° to 55°, (3) 56° to 60°, and (4) 61° to 65°. Of 21 watersheds there are 5 in group 1, 7 in group 2, 6 in group 3 and 3 in group 4. Results of the regression analysis for determining the intercept a are as follows: For group 1,

$$a = -259.4028 + 138.0559S_0 - 0.3265E + 719.7358K \qquad (27)$$

with r = 0.999, 99.84% of explained variance and S_e = 1.1089.

For group 2,

$$a = -16.3641 - 1.5393S_0 - 0.0018E + 52.4397K + 54.3329R \qquad (28)$$

with r = 0.9426, 88.85% of explained variance and S_e = 2.828.

For group 3,

$$a = -14.7787 - 0.0457A_f + 0.196E + 0.00066R \qquad (29)$$

with r = 0.998, 99.54% of explained variance and S_e = 1.5709.

For group 4, sufficient data were not available. It is thus seen that the intercept can be estimated more accurately by watershed grouping. However, additional testing needs to be done.

PREDICTION OF SEDIMENT YIELD

The sediment yield was predicted for three sample watersheds utilizing equations (11), (13), (15), (17), (23) and (25). Figures 4-6 show a comparison of observed and predicted sediment yields for

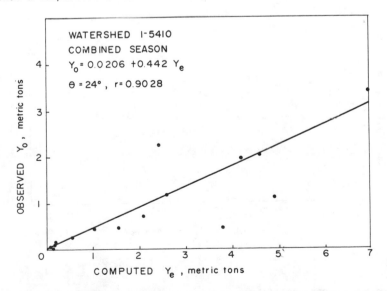

Figure 4. Comparison of observed and predicted sediment yield for a sample watershed.

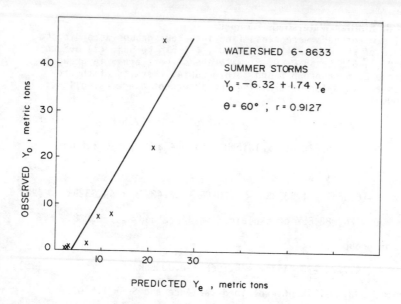

Figure 5. Comparison of observed and predicted sediment yield for a sample watershed.

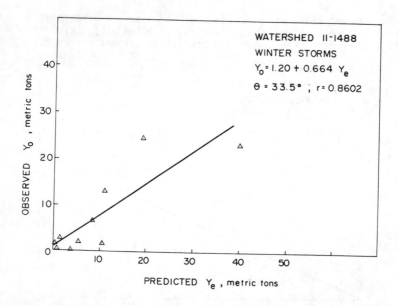

Figure 6. Comparison of observed and predicted sediment yield for a sample watershed.

these watersheds. For summer storms approximately 83.3 percent of variation in the sediment is accounted for the watershed under consideration. For winter storms, 74% of variation in sediment yield is explained. On the other hand, for combined storms, 81.5% of the variation is explained. Thus the results obtained by the simple linear analysis are very encouraging, and can be especially useful for erosion control and management. An examination of the data suggests that prediction of sediment yield can be considerably improved by augmenting the quality data base.

CONCLUDING REMARKS

The following conclusions can be drawn from this study:

(1) A strong linear relationship exists between log-transformed values of the sediment yield and the volume of the direct runoff.

(2) The variation is moderate in slope but pronounced in intercept of the linear relationship.

(3)There exists a strong correlation between the intercept and the watershed geomorphic characteristics, and a weak correlation between the slope and these characteristics.

(4) The linear relationship can account for approximately 80 percent or more of variation in the sediment yield.

(5) The linear relationship can be an effective tool in erosion control and management.

ACKNOWLEDGEMENTS

This study was supported in part by funds provided by the National Science Foundation under the project, "Free Boundary Problems in Water Resource Engineering," NSF-ENG-79-23345.

REFERENCES

Rendon-Herrero, O., 1974. Estimation of washload produced on certain small watersheds. Journal of the Hydraulics Division, Proceedings of the American Society of Civil Engineers, Volume 100, No. HY7, pp. 835-848.

Rendon-Herrero, O., Singh, V.P. and Chen, V.J., 1980. ER-ES watershed relationship. Proceedings, International Symposium on Water Resources Systems held December 20-22, 1980 at the University of Roorkee, Roorkee, India, Volume 1, pp. 11-8-41 - 11-8-47.

THE SEDIMENT TRANSFER COEFFICIENT AS DETERMINED FROM SOME SUSPENSION MEASUREMENTS IN A RIVER

Neil L. Coleman
Research Geologist, USDA Sedimentation Laboratory
Oxford, MS, USA

ABSTRACT

Measurements of suspended sediment concentration in rivers may be used to study the variation of the sediment transfer coefficient with distance from the streambed. The sediment transfer coefficient is of practical importance in formulating engineering models for simulating the operation of sediment traps and of the siltation of dredged navigation channels. A strictly descriptive study of the behavior of the sediment transfer coefficient, based on sediment measurements in the Enoree River of South Carolina, indicated that in the lower 40% of the flow depth the sediment transfer coefficient is a function of distance from the streambed. In the remaining upper part of the flow depth, the sediment transfer coefficient tends to be constant with distance from the streambed. This coefficient also is larger for larger sediment particle sizes. The results of the study are presented in absolute form, without normalization or other modification of basic data, for the convenience of workers interested in formulating their own models of the sediment transfer process.

INTRODUCTION

The suspension of sediment by turbulence in a river is generally described by an analogy with momentum diffusion. For a balanced or equilibrium condition of sediment transport by suspension, it is assumed that an upward flux of sediment occurs in the flow because of turbulence, and that this is balanced by a downward flux that occurs because of gravitational settling. Normally, the highest concentration of sediment is at the river bed, and the lowest concentration is at the water surface. The intensity of upward sediment flux is then proportional to the concentration gradient at any point along the concentration profile, and the sediment transfer coefficient is defined as the coefficient of proportionality between the flux and the gradient.

Measurements of suspended sediment concentration in rivers may be used to study the variation of the sediment transfer coefficient with distance above the streambed. The properties of the sediment transfer coefficient are of immediate practical importance in formulating engineering models for simulating the operation of sediment traps and the siltation of dredged navigation channels.

The sediment transfer coefficient distributions discussed here were calculated earlier (Coleman, 1970) by the author from some sediment concentration measurements taken in 1940 by A.G. Anderson (1941; 1942) in the Enoree River near Greenville, South Carolina. The original reason for retrieving and reanalysing these early data was to compare some sediment transfer coefficients from prototype rivers with coefficients obtained from laboratory flume studies (Coleman, 1970). Subsequently, the sediment transfer coefficient distributions obtained by the author from the Anderson data were used by Kerssens, Prins, and van Rijn (1979) in developing an engineering model for suspended sediment transport. In developing this model, these authors found, by regression, a pair of equations describing the variation of the nondimensional or normalized sediment transfer coefficient over the lower and upper halves of the flow depth, respectively. The purpose of this paper is to make the Enoree River sediment transfer coefficient distributions available in absolute form, so that they may be of further use to modelers who wish to develop more analytically or theoretically based expressions for the depthwise variation of the sediment transfer coefficient.

ANALYSING SEDIMENT SUSPENSION MEASUREMENTS

A sediment concentration profile may be plotted as a graph which is based on a series of suspended sediment measurements taken at points located in a single vertical and at varying heights above a stream bed. It is customary to plot separate concentration profiles for each particle size (as defined by sieve fractions or other methods) that is found in the suspended material.

The differential equation for sediment suspension in streams (Vanoni, Brooks, and Kennedy, 1960) is:

$$\varepsilon_s \, (d\phi/dy) + w\phi = 0 \tag{1}$$

where ε_s is the sediment transfer coefficient, ϕ is the volumetric sediment concentration, y is the height above the stream bed, and w is the particle fall velocity. Equation (1) can be solved for ε_s to obtain:

$$\varepsilon_s = - \, w\phi/(d\phi/dy) \tag{2}$$

and ε_s can be calculated using point values of ϕ and $d\phi/dy$ from a concentration profile plotted as described above. The calculation of ε_s from equation (2) requires an independent determination of the derivative $d\phi/dy$, which is the slope of the tangent to the plotted $\phi(y)$ curve at any given point y. This derivative can be found from the $\phi(y)$ graph by any one of several methods, depending on whether the plotting is done on arithmetic, semilogarithmic, or logarithmic graph paper.

In equation (2) it is sufficient to assume, at least as a first approximation, that w is constant for a given particle size in all parts of the flow. McKnown and Lin (1952) presented evidence that particle settling may actually be somewhat retarded in regions of high concentration.

Since the calculation of ε_s from equation (2) is quite sensitive to errors in determining $d\phi/dy$, a truly meaningful calculation is possible only for measurements which are reliable enough that minimal data scatter appears in the plotted profiles, so that curves can be fitted with good confidence.

SEDIMENT TRANSFER COEFFICIENTS FROM ENOREE RIVER DATA

Anderson (1941) reported a series of six suspended sediment measurements from the Enoree River. He made these measurements with specially designed equipment, which allowed the suspended material in the river to be measured at several points simultaneously. Particle size analyses were performed on the suspended material, and the concentration of each particle size fraction found was calculated. Thus, Anderson reported a total of 36 concentration profiles. Even with the improved equipment, the difficulty of measuring suspended sediment accurately in a river was so great that, of the original concentration profiles, only twelve displayed small enough scatter to justify curve fitting and analysis. Since here we are not trying to test a hypothesis, but rather are trying to find order of magnitude values of ε_s and general variation patterns of ε_s with flow depth for possible modeling application, it is permissible to analyze this obviously selected data. The major features of the twelve selected usable profiles are given in Table 1. The numerical values given in the table are those tabulated by Anderson (1942), but here they have been converted where necessary into SI unit equivalent values by the author. The particle diameters given are the geometric means of the grade size limits cited, while the fall velocities given are for spheres with diameters equal to the particle sizes cited. As indicated in Table 1, the profiles at 0.91 m depth were taken during a low water condition during July, 1940, when the water temperature was 24°C. The other profiles were taken at higher flow stages during February, 1940, and water temperatures were 7.7°C and 6.2°C for the profiles at flow depths at 1.28 m and 1.52 m, respectively.

TABLE 1. Characteristics of the sediment concentration profiles.

Flow depth (m)	Mean velocity (m/s)	Shear velocity (m/s)	Temp. (°C)	Grade-size (mm)	Particle diameter (mm)	Particle fall velocity (m/s)
0.91	0.83	0.071	24.0	0.175-0.246	0.210	.027
0.91	0.83	0.071.	24.0	0.246-0.351	0.294	.041
0.91	0.83	0.071	24.0	0.351-0.495	0.420	.064
1.28	0.88	0.096	7.7	0.175-0.246	0.210	.020
1.28	0.88	0.096	7.7	0.246-0.351	0.294	.033
1.28	0.88	0.096	7.7	0.351-0.495	0.420	.052
1.28	0.88	0.096	7.7	0.495-0.701	0.590	.078
1.52	1.15	0.112	6.2	0.124-0.175	0.150	.011
1.52	1.15	0.112	6.2	0.175-0.246	0.210	.020
1.52	1.15	0.112	6.2	0.246-0.351	0.294	.032
1.52	1.15	0.112	6.2	0.351-0.495	0.420	.052
1.52	1.15	0.112	6.2	0.495-0.701	0.590	.078

For each of the twelve profiles, the concentration ϕ was plotted against y on logarithmic graph paper, and a smooth curve was fitted through the data by eye. Tangents to this curve were then constructed geometrically at each curve point $\phi(y)$ corresponding to a data point. The slope angles θ of these tangents were then measured, and ε_s was calculated from:

$$\varepsilon_s = \frac{-w\phi}{\frac{d\phi}{dy}} = \frac{-w\phi}{\frac{\phi}{y} \frac{d(\log\phi)}{d(\log y)}} = -(wy)\cot\theta \qquad (3)$$

Figures 1, 2, and 3 show the variation of ε_s with y for the profiles taken at flow depths of 0.91, 1.28, and 1.52 m respectively. These figures show that ε_s increases with y from the riverbed up to an

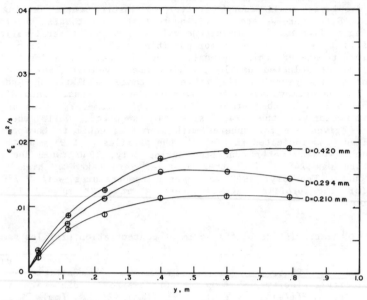

Fig. 1--Variation of the sediment transfer coefficient in a flow 0.91 m in depth.

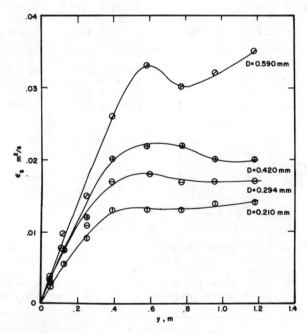

Fig. 2--Variation of the sediment transfer coefficient in a flow 1.28 m in depth.

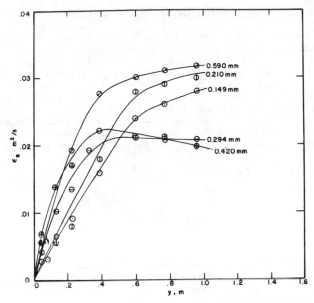

Fig. 3--Variation of the sediment transfer coefficient in a flow 1.52 m in depth.

elevation corresponding to ^{S} 30 to 50% of the total flow depth, and that above this elevation ε_s tends to be constant in the upper part of the flow. The ε_s curves in Figures 1, 2, and 3 display differences related to particle diameter, which is indicated by D in the figures. Although the curves for the particle diameters 0.149 mm and 0.210 mm are anomalous, the data otherwise show that ε_s is larger for larger particle sizes. This effect is particularly obvious in the upper flow region. Figure 4 shows the relation of ε_s to D in all flows, at an elevation equal to 60% of the total flow depth. The ε_s values in this figure were read from the smoothed curves fitted to the data in Figures 1, 2, and 3.

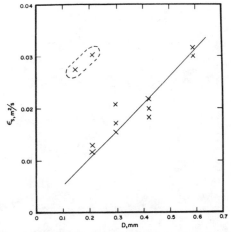

Fig. 4--Relation between the sediment transfer coefficient and particle size at 60% of the total depth, for all flows.

CONCLUSIONS

This has been a strictly descriptive study of the behavior of the sediment transfer coefficient in the Enoree River. The investigation involved the use of suspended sediment concentration profiles based on field measurements reported earlier by Anderson (1941; 1942).

The results of the study indicate that the flow in the Enoree River may be divided into a lower region, on the order of about 40% of the flow depth, in which the sediment transfer coefficient is a function of distance from the stream bed, and an upper region in which the sediment transfer coefficient tends to be constant. The sediment transfer coefficient is also shown to be larger for larger sediment particle sizes.

REFERENCES

1. Anderson, A.G., 1941, A combination suspended load sampler and velocity meter for small streams. USDA Circular No. 599, Washington: U.S. Government Printing Office, 26 pp.
2. Anderson, A.G., 1942, Distribution of suspended sediment in a natural stream. Transactions, American Geophysical Union, Part II, pp. 678-683.
3. Coleman, N.L., 1970, Flume studies of the sediment transfer coefficient. Water Resources Research, Vol. 6, No. 3, pp. 801-809.
4. Kerssens, P.J.M., A. Prins, and L.C. van Rijn, 1979, Model for suspended sediment transport. Journal of the Hydraulics Division, ASCE, Vol. 105, No. HY5, pp 461-476.
5. McKnown, J.S., and P.N. Lin, 1952, Sediment concentration and fall velocity. Proc. Second Midwestern Conference on Fluid Mechanics, Ohio State University, pp 401-411.
6. Vanoni, V.A., N.H. Brooks, and J.F. Kennedy, 1960, Lecture notes on sedimentation and channel stability. W.M. Keck Laboratory of Hydraulics, Water Resource Publication KHWR-1, California Inst. of Technology, 131 pp.

EXPERIENCE WITH A PROCESS-ORIENTED ROAD SEDIMENT MODEL

Eric Sundberg, Research Hydrologist
USDA Forest Service
Intermountain Forest and Range Experiment Station
Forestry Sciences, Laboratory
Moscow, Idaho

Robert G. Johnston, Research Hydrologist
USDA Forest Service
Intermountain Forest and Range Experiment Station
Forestry Sciences Laboratory
Logan, Utah

Edward R. Burroughs, Jr., Research Engineer
USDA Forest Service
Intermountain Forest and Range Experiment Station
Forestry Sciences Laboratory
Bozeman, Montana

ABSTRACT

Simulated rainfall events were applied to selected sections of mine haul roads to generate surface runoff and sediment for testing of ROSED (Road Sediment Model). ROSED is a process-oriented runoff and erosion prediction model developed at Colorado State University. Parameter needs for ROSED determined the format for collection of field data.

A definite improvement in sediment prediction accuracy was obtained by modifying model input to simulate wheel ruts, where applicable, in addition to the side ditch. Total runoff volume could be predicted within ±20 percent for 75 percent of the rainfall events, and runoff peak flow within ±10 percent for 67 percent of these events. Total sediment yield could be predicted within ±25 percent for 62 percent of the rainfall events.

Recommendations for improved field techniques for road erosion studies include: additional rain gages outside the plot boundary for more accurate rainfall distribution maps; and multiple event marker equipment to record beginning and ending of rainfall and runoff, rainfall intensity, and dye travel time all on a common time base.

Recommendations for future investigations with ROSED include: simulation of more complex runoff surfaces, such as wheel ruts; adjustment of model parameters for simulated snowmelt and rainfall with low kinetic energy; relationship of hydraulic roughness parameters to site characteristics; and the relationship of infiltration parameters to easily measured site characteristics.

INTRODUCTION

Impacts on water resources by road construction, operation, and maintenance are identified as concentrated surface runoff, soil disturbance and loss, disruption of surface and groundwater flow patterns, and degradation of water quality. Effects of these impacts on land use, water supplies, wildlife, and fisheries are generally considered detrimental. Environmental laws now dictate that these impacts and effects be estimated at the planning stage of road projects. Effective, workable tools are needed to quantify these impacts for forest roads and surface mine haul roads.

In 1975, as a cooperative effort between USDA Forest Service and Montana State University, evaluation began on ROSED (Road Sediment Model), a tool to estimate runoff and sediment yield from roads. We felt that ROSED had the best potential as an operable tool to meet the assessment needs. It is a process-oriented runoff and erosion prediction model developed by Simons, Li, and Shiao (1977) at Colorado State University. In the model testing, we used simulated rainfall applied to selected sections of surface mine haul roads to generate runoff and sediment yield data. Simulated rainfall was used for two reasons: (1) natural precipitation during the field season is infrequent and of short duration over much of the Northern Great Plains coal region; (2) instrumentation of a number of road sections to await natural precipitation is expensive and risks loss of data through instrument malfunction. Therefore, a modified Colorado State University rainfall simulator (rainulator) was used to generate rainfall on isolated sections of road. This report details the rainulator tests, summarizes the data, and gives an evaluation of the ROSED model. We also offer suggestions for improved field techniques for rainulator studies and recommendations for future studies and modifications to the ROSED model.

OVERVIEW OF THE RAINFALL SIMULATOR STUDY

Field procedures for rainulator operation and data collection were guided primarily by data needs for the ROSED model and by the state-of-the-art for studies using simulated rainfall (USDA-SEA, 1979). A review of ROSED documentation (Simons, Li, and Shiao, 1977; Simons, Li, and Ward, 1977; Simons et al, 1979) and our observation of road surface erosion indicated three major areas of uncertainty in the relationships between model parameters and site characteristics: (1) the interaction of loose soil stored on the road surface and the runoff detachment coefficient in the prediction model; (2) the interaction among model flow resistance parameters, the particle size gradation of road surface material, and hydraulic roughness of the surface; and (3) estimation of infiltration of water into the road surface using measurable site characteristics. Supplementary data of various types were collected to help eliminate these uncertainties in addition to the usual field procedure for rainulator studies.

The modified CSU rainulator consists of a set of sprinkler heads on 11-ft (3.35 m) risers with sprinkler pressure controlled by a regulator at the base of each riser. This rainulator calibration by Neff (1979) showed: (1) uniform areal distribution for plots up to 3,000 ft^2 (279 m^2) at wind speeds less than 7 mi/hr (11.3 km/hr) and decreasing uniformly thereafter; (2) drop size less than natural rainfall of the same intensity; (3) kinetic energy about 40 percent of natural rainfall events.

Support equipment included a 5,000-gal (18.93 m^3) water storage bag, a 250-gal/min (0.95 m^3/min) pump, and a 2,000-gal (7.57 m^3) tank truck. Measurement equipment included a triangular, supercritical Replogel flume, a FW-1 stage recorder with 120-volt electric chart drive, recording rain gage, nonrecording rain cans, nuclear density measurement instrument, a hot-wire anemometer, and water sample bottles.

FIELD PROCEDURE

Road sections for application of simulated rainfall were selected for uniformity of surface configuration, road gradient, and surface materials. Each road section was considered a unique plot and no attempt was made to replicate plot conditions. Test sections were selected with a wide variety of length, width, and slope characteristics within the normal standards for mine road design and construction.

Sections with a ditch-to-ditch width of 34 ft (10.4 m) or less could be covered entirely by the rainulator system. Maximum effective coverage of 4,420 ft^2 (0.1015 acre or 410.6 m^2) could be attained by using 24 sprinklers in 3 parallel rows 17 ft (5.2 m) apart, each with 8 sprinklers at 20 ft (6.1 m) spacings. Adjacent rows were shifted 10 ft (3.05 m) to create a staggered effect for more uniform coverage. Half widths of wider roads were used by having the mining company grade a berm along the road crown, then setting the rainulator over the berm-to-ditch width. Figure 1 shows a typical rainulator layout.

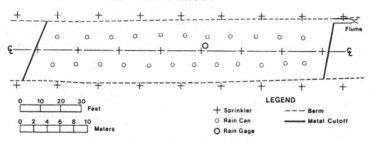

VICINITY MAP SITE 4, ROAD 2
Area = 4260.9 Ft^2 (395.9 M^2)

LEGEND
+ Sprinkler   − − − Berm
O Rain Can   ——— Metal Cutoff
O Rain Gage

Figure 1--Sample vicinity map for a rainfall simulator plot on a surface mine haul road.

Each selected road section was isolated by excavating a shallow ditch and installing metal cutoff walls across the upper and lower ends of the section. Side borders were provided by the earth berms normally constructed on the sides of mine roads in compliance with safety regulations. A chemical soil stabilizer was sprayed on the ditch-berm perimeter to prevent extraneous sediment contribution from these areas. A metal trough was placed to collect and convey water from the lowest corner of the road section to the flume.

The recording rain gage and supplementary rain cans were located in a grid upon the road section to measure rainfall distribution. Soil was scraped from the first centimeter of the road surface at random locations to accumulate a 5- to 10-lb (2.3 to 4.6 kg) composite sample. In addition, loose soil on the road surface was measured by sweeping and collecting the soil from within a 4-ft^2 (0.37 m^2) area enclosed by a metal ring 2.26 ft (0.69 m) in diameter. These samples were taken before and after rainfall application at several random locations across the section. Bulk density measurements were made (nuclear equipment) and soil moisture samples were collected before and after each rainfall application.

Reference pins 10 ft (3.05 m) apart and alined in the expected direction of overland flow were placed at various locations on the road section. The velocity of overland flow at these locations was measured periodically during rainfall application by placing a small mass of dye upslope from each 10-ft spacing and timing the passage of the center of mass between the pins. The time of each test was recorded so that these dye travel measurements could be related to the runoff hydrograph.

Water was applied at a uniform rate of approximately 2 in/hr (5.1 cm/hr) for at least 30 minutes, usually during the early morning when the air was relatively calm. Tests were suspended when wind gusts exceeded 5 mi/hr (8.05 km/hr). Field data included the times of beginning of detention storage, overland flow, entry of runoff into the flume, and times when rainfall began and rainfall and runoff stopped. At the flume outlet, 0.26-gal (1 liter) samples were collected at 1-min intervals during the rising stage and at 2-min intervals from the peak runoff through the recession limb of the hydrograph.

Rainfall was applied until runoff stabilized or until the water supply was exhausted. After rainfall stopped, the catch in the rain cans was measured and compared with the rain gage record to evaluate uniformity of coverage. If a second run with the road in a prewetted condition was desired, then the procedure was repeated the following day or during the same day if wind conditions allowed. After all runs on a road section were completed, a tape and level survey was made to map the location to the nearest 0.1 ft (3 cm) and elevation to the nearest 0.01 ft (0.3 cm) of plot boundaries, sprinklers, rain gage, and rain cans (see fig. 1).

A motion picture camera, set for a time lapse mode, photographed each plot during simulated rainfall. These films provided graphic evidence of the timing of runoff from portions of the plot and provided clues to accurate modeling of response units and the routing sequence.

REDUCTION OF FIELD DATA

All soil samples were placed in sealed containers and shipped to the Civil Engineering Department laboratories at Montana State University in Bozeman. Each sample was weighed and dried, then a particle size gradation curve was developed using American Society for Testing Materials procedures with wet and dry sieving and hydrometer analysis. Soils analysis also included determination of Plasticity Index and Liquid Limit and measurement of particle density by means of an air pycnometer.

Water samples were filtered with Millipore equipment to determine sediment concentrations (gm/cc). Sediment from each run was saved and particle size gradation curves were developed to characterize the sediment yield at intervals during the runoff hydrograph.

We used three methods to measure rainfall distribution over each road section: arithmetic mean of the rain-can catch; Thiessen polygons; and isohyets drawn with computer graphics. The arithmetic mean gave reliable results for these data with considerable savings of time. The recording rain gage record was used to distribute the total catch over 1-min intervals for the length of the simulated rainfall.

The large amount of data generated by these rainfall simulator tests required a heavy reliance on computer digitizing of rainfall data, observed hydrographs, and measured sediment yield. Several data management programs were written to expedite data analysis. Special computer graphics programs were developed (Pexton, 1979) for direct display and comparison of observed hydrographs and sedimentgraphs with predictions. These programs greatly improved our ability to rapidly evaluate the quality of field data and the accuracy of ROSED results.

Observed runoff hydrographs were digitized and each file of stage readings was converted to a file of average stage for a preselected time interval (0.5 min). This file was then linked to a program with the Replogle flume rating equation to produce a file of runoff rates (ft^3/min), runoff volume in area-inches/min, and cumulative runoff volume in ft^3. Files of sediment concentration (lb/ft^3) were multiplied by runoff rates (ft^3/min) to develop a file of sediment discharge (lb/min) for graphical display and comparison with predicted sediment-graphs.

ANALYSIS OF FIELD DATA

The ROSED time-space water and sediment routing procedure uses a conceptual road prism with five response units: cut slope, fill slope, road surface, ditch, and culvert. Three types of flow are considered: overland flow, ditch flow, and culvert flow. In this study, we considered only road surfaces and ditches that involve only overland flow and ditch flow. The ROSED model used in this study produces two distinct but interconnected outputs: (1) a hydrograph by estimating rainfall excess, overland flow routing, and ditch flow routing; and (2) a sedimentgraph by estimating soil detachment by rainfall, soil detachment by overland flow, overland flow sediment routing, soil detachment by ditch flow, and ditch flow sediment routing. All these processes of detachment, transport, and storage are modeled by incremental time units.

Our approach in this study has been to estimate ROSED model parameters from an operational standpoint using simple measurements of site characteristics whenever possible. A standardized procedure has been developed to use rainfall simulator data as a result of close communication with ROSED developers to prevent misinterpretation of model parameters. This section will only dwell on those model parameters of particular importance in this study.

The most critical step in setting up ROSED is to define the relationship of each response unit to the water and sediment routing sequence. Very simple combinations of response units and routing sequences were used in a preliminary trial with data from each road with mixed results--some good and some poor simulations. Time lapse photographs of runoff from road sections showed how water actually flowed over the surface and emphasized the importance of even minor wheel ruts in overland flow routing. Subsequent modeling runs were more successful because realistic road configurations were used taking wheel ruts into account.

The road sections fell into two groups: (1) those in which runoff moved transversely from the road crown to a lateral ditch, down the lateral to the cutoff ditch, and from there to the flume; and (2) those in which runoff moved transversely to ruts in the road surface, then longitudinally down the ruts to the cutoff ditch and from there to the flume. In case 1, there are two relatively short sections of overland flow into two relatively long side ditches; in case 2, several relatively short sections of overland flow empty into several long central ditches (wheel ruts). Figure 2 shows the configuration for case 1 for both half road widths and full widths as compared to case 2. Dimensions of overland flow length, slope gradient, ditch length and gradient, and ditch side slope were taken from contour maps made for each road section from survey data.

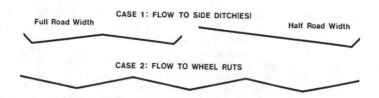

CASE 1: FLOW TO SIDE DITCH(ES)
Full Road Width   Half Road Width

CASE 2: FLOW TO WHEEL RUTS

Figure 2--Model configurations used with ROSED in
analysis of surface mine haul road data.

Soil detachment by raindrop impact can be a significant factor in surface erosion. However, the rainfall simulator used in this study produces only about 40 percent of the kinetic energy of a natural 2-in/h (5.08 cm/h) rainstorm. For this reason, in this study soil detachment from the road surface by simulated raindrops is considered to be negligible and model parameters related to this process were adjusted accordingly.

A value for overland flow resistance for use in these tests was determined by consulting with ROSED developers. Because there is no method to use site data to determine the value of this coefficient, we selected a tabular value from the documentation (Simons et al, 1979) and used this constant value on all analyses reported here.

Preliminary tests with dye travel data indicate that the overland flow resistance parameter may vary from -20 percent to +25 percent away from the constant value used in this study. We are continuing analyses designed to determine if the overland flow resistance parameter may be estimated from measurements of surface soil particle size gradation data and other site characteristics.

The runoff detachment coefficient must be determined for each overland flow and ditch segment. An empirical procedure to calculate this parameter was suggested by Simons et al (1979):

$$Df = \frac{K(D_{50})^2}{0.63} \qquad (1)$$

where: Df = soil detachment coefficient for runoff
K = soil erodibility index from the Universal
Soil Loss Equation (Wischmeier et al, 1971)
D_{50} = median soil particle size in millimeters

The ROSED procedure for calculating infiltration is based upon the Green-Ampt equation and requires estimating saturated hydraulic conductivity. Field measurements of bulk density and laboratory measurements of particle density were used to estimate void ratio and porosity. Field measurements of soil moisture were converted to percent saturation. For the highly compacted road materials used in this study, there are no readily available procedures for relating hydraulic conductivity to site characteristics in the ROSED documentation provided. In this study, the difference between the rainfall rate and the peak runoff rate is used as an estimate of the saturated hydraulic conductivity for calculations of infiltration. This procedure has the disadvantage that each rainfall simulation is "optimized" by using data from that simulation to calculate the infiltration rate. The result is that the difference between observed and predicted runoff volumes tends to be minimized.

RESULTS

Observed and predicted runoff volumes and sediment yields from selected rainfall simulations are shown in table 1. Note that the error in estimating runoff volume is usually less than the error in estimating sediment yield. One factor in more accurate estimation of runoff volume is found in the discussion in the last section.

We made 24 individual applications of rainfall. Of these, only 13 sets of data were used to evaluate ROSED. The remaining data sets were eliminated because of errors caused by wind effects, lack of rainfall intensity data, and faulty layout of cutoff ditches.

Figure 3A shows the results of a simulated rainfall test and gives the measured rainfall intensity averaged over 1-min intervals together with the observed runoff hydrograph. This shows the results of simulated rainfall on highly compacted road sections with slow infiltration rates. There is very little difference between total rainfall volume

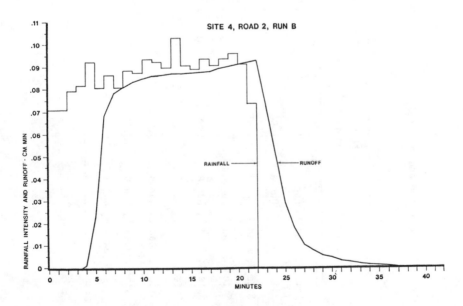

Figure 3A--Comparison of observed runoff and applied rainfall.

Table 1.--Predicted and measured values of runoff and sediment from selected sets of simulated rainfall data

Location	Predicted Runoff (m^3)	Observed Runoff (m^3)	Error (%)= $\frac{P-O}{O} \times 100$	Predicted Sediment (kg)	Measured Sediment (kg)	Error (%)= $\frac{P-O}{O} \times 100$
Site 1, Road 1, Run A	1.01	1.31	-23	13.74	16.96	-19
Site 3, Road 1, Run A	1.96	1.68	+17	41.14	52.03	-21
Site 3, Road 1, Run B	1.51	1.77	-15	31.89	46.31	-31
Site 3, Road 2, Run C	3.34	3.08	+ 8	47.17	37.51	+26
Site 4, Road 1, Run A	1.99	2.10	- 5	37.92	51.07	-26
Site 4, Road 2, Run A	2.06	2.32	-11	38.28	36.29	+ 5
Site 4, Road 2, Run B	6.57	6.43	+ 2	72.12	92.71	-22
Site 5, Road 1, Run B	10.96	11.39	- 4	122.92	89.04	+38
Site 5, Road 3, Run B	7.81	7.09	+10	88.45	80.42	+10

and total runoff volume, and between the rainfall rate and the peak runoff rate. Any error in measuring rainfall intensity can cause a significant error in estimation of saturated hydraulic conductivity according to the procedure used in this study.

Figure 3B illustrates the close estimation of hydrograph shape and peak flow rate that is possible with this model as used in this study. The rising and falling limbs of the predicted hydrographs typically lag behind their corresponding observed hydrographs. We found that the model could estimate total runoff volume with ±20 percent for 75 percent of the simulated rainfall events and estimate peak flow rates within ±10 percent for 67 percent of the events. These errors were considered to be acceptable for our preliminary evaluations of ROSED.

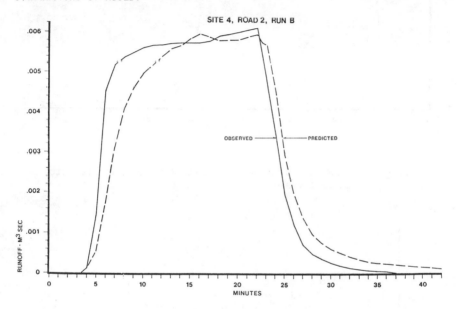

Figure 3B--Comparison of observed runoff with predicted runoff.

Figure 3C shows a typical sedimentgraph estimated by the model. In most cases the general shape of the actual sedimentgraph is matched fairly closely, although the total estimated volume of sediment is usually underestimated. Total predicted sediment yield was within ±25 percent for 62 percent of the 13 rainfall events modeled; this range of errors is also considered acceptable. The estimated sedimentgraph also shows a slower rise and recession than the observed graph. This is probably because the estimated hydrograph also lags.

Wind effects that cause uneven distribution of rainfall and fluctuations in rainfall intensity over the plot, as shown in figure 4A, can also cause erratic prediction of runoff and sediment yield. The model appears to be quite sensitive to changes in rainfall intensity as evidenced by fluctuations in the hydrograph in figure 4B as compared to the observed hydrograph. The fluctuations in the estimated hydrograph cause corresponding fluctuations in the estimated sedimentgraph (fig. 4C) that are not seen in the observed sedimentgraph.

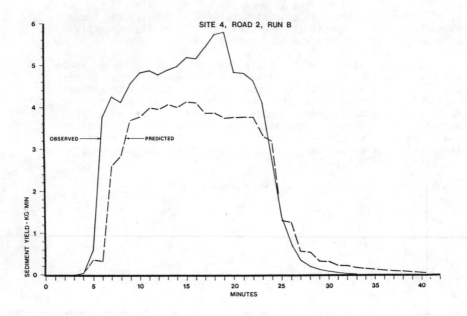

Figure 3C--Comparison of observed sediment yield with predicted
sediment yield.

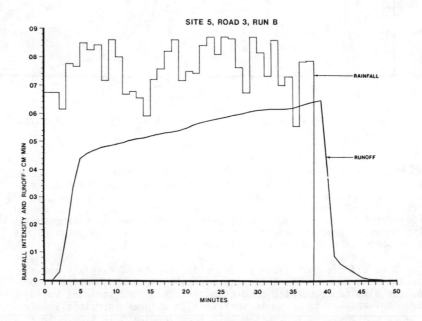

Figure 4A--Comparison of observed runoff and applied rainfall showing
wind effects.

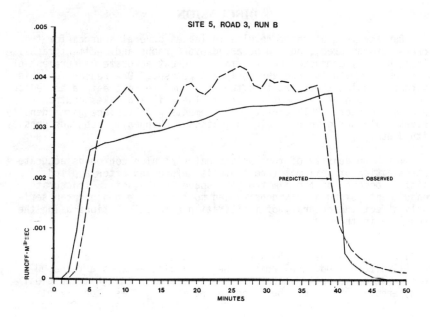

Figure 4B--Comparison of observed runoff with predicted runoff showing wind effects.

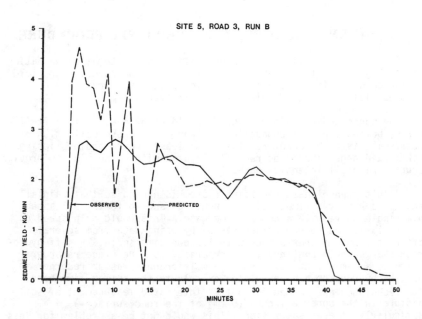

Figure 4C--Comparison of observed sediment yield with predicted sediment yield showing wind effects.

DISCUSSION

Our analysis of these results indicates several sources for the errors between predicted and observed hydrographs and sedimentgraphs. First, heavily compacted road surfaces prevent accurate measurement of bulk density with the nuclear equipment we used. One reason for this is that pounding the spike into the road surface to create a space for the gamma depth probe sometimes cracks the soil and makes neutron readings suspect. Bulk density measurements with surface gamma density meters could not give reliable estimates of changes in subgrade density with depth.

The high density of the surface material also confounds accurate determination of soil moisture contents before and after applied rainfall because of the limited pore space in the soil. Inaccurate measurements of both bulk density and soil moisture content can seriously affect the accuracy of infiltration rate calculations using the procedure in the model.

Our interpretation of the model indicates that while ROSED can be configured to represent wheel ruts, only one side slope can be used for all ditches and wheel ruts. The model also assumes a V-shaped configuration for all ditches and wheel ruts even though trapezoidal or hyperbolic shapes may be found in the field. These constraints limit the flexibility needed to simulate the actual shapes of the road surface, ditches, and wheel ruts.

The importance of the ability to simulate wheel ruts cannot be overemphasized. We observed that wheel ruts can be an extremely important feature in routing water and sediment from the forest road surface.

RECOMMENDATIONS FOR IMPROVED FIELD PROCEDURE

Simulated rainfall offers opportunities for systematic evaluation and testing of runoff and erosion models for use on roads. Our field experience with simulated rainfall indicates several areas where improved field technique would result in better data.

The nonrecording rain gage grid should be extended 3 to 7 ft (1 to 2 m) beyond the plot boundaries in order to improve estimation of the aereal distribution of rainfall. At least two recording tipping bucket rain gages should be randomly located within the plot to provide accurate rainfall intensity data.

A multichannel data recording system is needed so that rainfall records, dye travel times, time to ponding, and so forth, all appear on a single chart with a common time base. This would help speed up data analysis and reduce errors caused by using data from several different types of charts each with its own time scale. Runoff hydrograph charts would continue to be separate, but the stage recorder would use a 1-min tick mark from the multichannel master recorder so that the hydrograph can be related exactly to the common time base.

Techniques are needed for better integrated measurement of soil moisture in the upper 2 inches (5 cm) of the road surface—particularly at near saturation. This would not be a problem for less highly compacted soils. Gravitational samples taken immediately after rainfall ceased did not provide any information on soil moisture at depth. These samples were not related to a known volume so that accurate estimates of percent saturation could be made. Surface soil moisture measurements with neutron scattering equipment were tried and abandoned because they were too time-consuming, and they caused excessive disturbance of the plot surface and compromised the accuracy of succeeding simulator runs.

RECOMMENDATIONS FOR FURTHER MODEL DEVELOPMENT

Our experience with the version of the ROSED model we used indicates it cannot be used as an operational planning tool as it now stands. Our work shows several specific developmental needs that would greatly improve the utility of this model. The engineer or resource specialist concerned with estimating runoff and sediment yield from an existing or planned road has limited specific information to work with: soil particle size gradation, bulk density, road width, length, grade, storm characteristics, and so forth.

Techniques are needed to relate this elementary information to parameters used in the model. Specific model parameters requiring attention are: hydraulic roughness of the road surface, infiltration and hydraulic conductivity, and detachment by raindrops and runoff. Simons, Li, and Shiao (1976) recommended development of handbooks for field use of ROSED "to evaluate alternative routes and alternative designs of road cross sections, road gradients, and surfaces, cut slopes, embankments, and spacings of cross drains."

The addition of a snowmelt component to ROSED would greatly increase the utility of this model in the northern states, especially in mountainous regions. At this time, the mechanics of sediment movement at the soil / snowpack interface is not well understood. Modeling a snowmelt event may only require complete suppression of detachment by raindrops and overland flow in order to emphasize channel flow detachment and transport.

CONCLUSIONS

Our experience with the ROSED model is generally positive. We feel it represents a major step in the development of an operational planning tool to estimate runoff and sediment yield from low standard roads. ROSED has some flexibility to model different road prism configurations, but more is needed to model various ditch shapes and ditch side slopes.

Simulated rainfall allows testing of the ROSED model on road sections with various characteristics of slope, shape, and surfacing. These tests allow the user to determine the sensitivity of model parameters over a range of site characteristics. Further development is needed to relate model parameters to site characteristics so that ROSED may be used by land managers and resource specialists not familiar with the mechanics of this model.

REFERENCES

Neff, E.L., 1979. Simulator Activities--Sidney, Montana. in: Proceedings of the Rainfall Simulator Workshop, Tucson, Arizona, March 7-9, 1979, USDA Science and Education Administration, Agricultural Reviews and Manuals, ARM-W-10, pp. 160161, July 1979.

Pexton, M.F., 1979. Data Display System, unpublished report on file at the Dept. of Civil Engineering and Engineering Mechanics, Montana State University, Bozeman, 13 pp., September 1979.

Simons, D.B., Li, R.M., and Shiao, L.Y. 1976. "Preliminary Procedural Guide for Estimating Water and Sediment Yield from Roads in Forests," Civil Engineering Dept., Engineering Research Center, Colorado State University, Ft. Collins, CER76-77DBS-RML-LYS21, 120 pp., Nov. 1976.

Simons, D.B., Li, R.M., and Shiao, L.Y. 1977. "Formulation of Road Sediment Model," Civil Engineering Dept., Engineering Research Center, Colorado State University, Ft. Collins, CER 76-77DBS-RML-LYS 50, 107 pp., March 1977.

Simons, D.B., Li, R.M., and Ward, T.J. 1977. "Simple Procedural Method for Estimating On-Site Soil Erosion," Civil Engineering Dept., Engineering Research Center, Colorado State University, Ft. Collins, CER76-77DBS-RML-TJW38, 55 pp., Feb. 1977.

Simons, D.B., Li, R.M., Ward, T.J., and Eggert, K.G. 1979. Estimation of Input Parameters for Modeling of Water and Sediment Yields, Civil Engineering Dept., Colorado State University, Ft. Collins, CER78-79DBS-RML-TJW-KGE50, 56 pp., April 1979.

USDA-SEA. 1979. "Proceedings of the Rainfall Simulator Workshop," Tucson, Arizona, March 7-9, 1979. USDA Science and Education Administration, Agricultural Reviews and Manuals, ARM-W-10, 185 pp., July 1979.

Wischmeier, W.H., Johnson, C.G., and Cross, B.V. 1971. "A Soil Erodibility Nomograph for Farmland and Construction Sites," Jour. Soil and Water Conservation, 26:189-193.